P9-DCX-563

This book provides an up-to-date and comprehensive coverage of the properties of glasses as materials and of the vitreous state in general.

Aspects of the physics, chemistry and thermodynamics of the vitreous state are developed, covering in particular the definitions of vitreous transition, fictive temperature and various vitrification criteria, both structural and kinetic. The methods of describing disordered structures are presented in terms of different models, followed by a review of various spectroscopic and radiation scattering techniques (X-rays, neutrons) of disordered structure determination. The author also considers phase separation in glasses and the problem of middle-range order. Following a classification of the different glass types, the rheological, diffusional, electrical, optical, thermal and mechanical properties of glasses are presented. A condensed summary of glass manufacturing techniques is also given, including sol–gel processing methods.

The broad coverage of this book makes it particularly suitable as an advanced text for students of materials science studying glasses. It should also be an excellent resource for those involved in research, at both a postgraduate and an industrial level.

Glasses and the vitreous state

Cambridge Solid State Science Series

Titles in print in this series

1 Polymer Surfaces
B. W. Cherry

2 An Introduction to Composite Materials
D. Hull

3 Thermoluminescence of Solids
S. W. S McKeever

4 Modern Techniques of Surface Science
D. P. Woodruff and T. A. Delchar

5 New Directions in Solid State Chemistry
C. N. R. Rao and J. Gopalakrishnan

6 The Electrical Resistivity of Metals and Alloys
P. L. Rossiter

7 The Vibrational Spectroscopy of Polymers
D. I. Bower and W. F. Maddams

8 Fatigue of Materials
S. Suresh

9 Glasses and the Vitreous State
J. Zarzycki

J. ZARZYCKI

Professor of Materials Science, University of Montpellier, France

Glasses and the vitreous state

Translated from the French
by
William D. Scott and Claire Massart

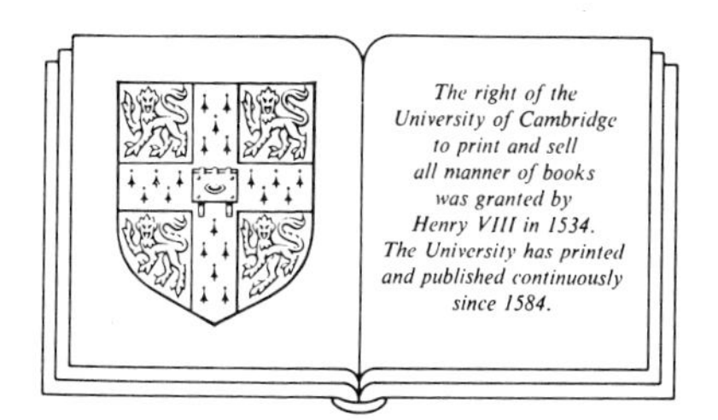

CAMBRIDGE UNIVERSITY PRESS

Cambridge
New York Port Chester
Melbourne Sydney

Published by the Press Syndicate of the University of Cambridge
The Pitt Building, Trumpington Street, Cambridge CB2 1RP
40 West 20th Street, New York, NY 10011-4211, USA
10 Stamford Road, Oakleigh, Melbourne 3166, Australia

Originally published in French as *Les Verres et l'Etat Vitreux*
by Masson, Paris, 1982

First published in English by Cambridge University Press, 1991, as
Glasses and the vitreous state

Printed in Great Britain at the University Press, Cambridge

British Library cataloguing in publication data

Zarzycki, J.
Glasses and the vitreous state.
1. Glass. Physical properties
I. Title II. Verres et l'etat vitreux. *English*
666.1

Library of Congress cataloguing in publication data

Zarzycki, J. (Jerzy)
(Verres et l'etat vitreux. English)
Glasses and the vitreous state / J. Zarzycki; translated from
French by William D. Scott and Claire Massart.
p. cm. – (Cambridge solid state science series)
Translation of: Les verres et l'état vitreux.
Includes bibliographical references and index.
ISBN 0 521 35582 6
1. Glass. I. Title. II. Series.
TP857.Z3713 1991
666.1–dc20 90-416743 CIP

ISBN 0 521 35582 6 hardback

Contents

Foreword

These remarks are to introduce the first text on the science of glasses to be published in the Cambridge Solid State Science Series since its beginnings 17 years ago. The absence of such a book hitherto is not a reflection on the editors' judgment as to the importance of the field – far from it! It reflects, instead, the fact that good general books on glasses are few and far between; they are certainly less numerous than good books on metals, ceramics, semiconductors or composites. This paucity of glass books can hardly be due to an undervaluation, by the materials community, of one of the most venerable classes of material in human civilization which has also been the focus of intensive modern research. One must conclude that good glass books are rare because they are particularly difficult to write.

I have had the good fortune to know Jerzy Zarzycki for many years past; our first meeting was due to a shared passion for materials science, at that time a novel concept. Prof. Zarzycki has been both an industrial research director and the director of a university glass research laboratory and he is a linguist and an indefatigable traveller among the purlieus of glass research. His view of the field is accordingly a comprehensive one, and this showed very clearly in his French text, *Les Verres et l'Etat Vitreux*, which appeared in Paris in 1982. As soon as I read this, I realized that here was a text that should be made available to anglophone scientists. The practicalities of arranging for this, including revision, expert translation and all the necessary permissions, necessarily stretched over years, and so it is only now, in 1991, that the English version can be presented to the public. Readers can however be assured that every care has been taken to ensure that the subject-matter remains up to date.

The field of glass science is so extensive that no single-volume text can cover every facet in equal detail. This book takes a broad view and emphasizes aspects of glass science that are general to all, or most, kinds of glasses, in preference to detailed presentation of the properties of particular categories of glass. Thus, topics like the glass transition, tendency to glass formation, rheology, optical characteristics, and the many features of glassy structure, are examples of those which are thoroughly treated here. Attention is also devoted to the modes of glass production, including especially the modern sol-gel process to the development of which Prof. Zarzycki has made major contributions.

I am particularly pleased that the Cambridge University Press can in this transparent way make its contribution to the cross-channel and transatlantic *Entente Cordiale*.

Robert W. Cahn
1990

Introduction to the English edition

Glasses constitute a fascinating group of materials both from the fundamental and applied standpoints. They are among the most ancient materials in human history and it seems paradoxical that our knowledge of their structure is far from being complete.

To the fundamentalist, glasses present the challenge of studying disordered solids which lack the spatial periodicity typical of crystals. Most of the standard interpretative methods are less effective or inapplicable and the description of the structure is limited to statistical approaches. This explains why it is only recently that the structure of glasses has begun to be better understood thanks to the simultaneous application of ever more sophisticated investigation methods.

The very existence of this type of condensed state of matter brings about difficult thermodynamic and kinetics problems which are not as yet completely resolved.

In spite of these difficulties Glass Science has made decisive progress in the last three decades and significantly in the sixties which represent the "golden age" of glass research. During this period our understanding of glass and the vitreous state experienced truly explosive growth similar to that in metallurgy a few decades earlier. Glass is no longer solely a material of primary technological value for architecture, transport, lighting or packaging. New types of glass have been discovered and these play an increasing rôle in modern optics (lasers), electronics, opto-electronics, energy conversion and in medicine (bio-materials). Glass has been promoted to the rank of a noble material, not only for "passive" but also for "active" applications. New modes of synthesis have been developed, e.g., the "sol-gel" approach in which the "wet" methods of colloidal chemistry replace the traditional high-temperature routes used in glass melting.

Glass science thus encompasses all the different aspects found in materials science and necessitates a knowledge of the various fields of physics, chemistry and mechanics.

The book *Les Verres et L'Etat Virtreux* was written in 1979, to meet teaching needs in France and to serve as a homogeneous introductory course to the various aspects of the physical chemistry of glass. It was the first textbook in French on this subject and was addressed primarily to students in universities and engineering schools confronted with the problem of glass in their materials science curricula. However, it also contained sufficient bibliography to be of interest to researchers entering this field. In response to many requests by English-speaking scientists this work has now been translated into English.

The bibliography has been thoroughly updated and additional information provided on several types of glasses. Because of their ultra-low attenuation, halide glasses are now potential candidates for optical fibers and are also particularly important for their exceptional optical characteristics which make them suitable for high power lasers in thermonuclear fusion experiments.

Metallic glasses, which are obtained by ultra-fast quench of liquid metal alloys, have shown a very rapid development because of their interesting properties which earned them the name "materials of the century" in the USA and "dream materials" in Japan. They are used in many electronic devices, in particular in audio-video techniques.

The "precursor-based" synthesis of glasses, ceramics and composites is, at the present time, one of the most rapidly progressing fields of material science and engineering. The "sol-gel" method of preparing glasses is being actively studied in leading laboratories all over the world and, in the last ten years, the number of scientific publications in this field has shown an exponential increase. This warranted an additional chapter on this subject for the present edition.

The book is divided into several parts:

Chapters 1–7 present aspects of the *vitreous state* in general. A knowledge of physics, chemistry and thermodynamics and, in particular, the theory of equilibrium diagrams is a prerequisite. The possibilities of applying various methods of studying the structure of disordered systems are concisely treated and references to appropriate monographs giving more details are indicated.

Chapter 8 presents the *classification* of various systems of practical importance.

Chapters 9–16 describe different properties of glasses considered as *materials*. Rheology (Chapter 9) and brittle fracture (Chapter 14) are treated more extensively because of their importance in glass science and technology.

Chapter 17 gives a condensed summary of *glass production techniques*. These are important even for readers who are not technologically inclined, as they determine the final characteristics of the materials obtained in practice.

Chapter 18 describes new glass-making processes based on the sol-gel approach from precursors.

Because of the vastness of the subject and the limited space available, the presentation had to be substantially condensed and, unfortunately, certain subjects had to be omitted. However, the bibliography will permit readers access to more specialized sources of information where necessary.

I wish to thank Professor William Scott of the University of Washington and Ms Claire Massart, Seattle, Washington for their excellent translation. My sincere thanks also go to the Corning Glass Co. and the Saint-Gobain Co. for the interest they have shown in this venture and for providing several original micrographs. Especially I would like to take this opportunity to acknowledge the late Dr. Ivan Pechès who first introduced me to the exciting intricacies of glass structure. Finally my greatest thanks go to my wife, Margery, for her many painstaking readings of both the French and English manuscripts and for her constant encouragement.

I hope that this work will be a useful guide to all those who are interested in the Science of Glass.

J. Zarzycki
Montpellier, 1990

Historical overview

Glass is one of the most ancient materials known to mankind. In prehistoric times, *obsidian* was used to make knives, arrow tips etc. from natural glass originating in Europe, mainly from the islands of Melos and Thera in Greece. The most ancient *glass* objects made by mankind were discovered in Egypt and are dated approximately 3000 BC. However, methods of manufacturing glass had already been discovered in Mesopotamia by approximately 4500 BC.[21]

This glass consisted essentially of a Na_2O–CaO–SiO_2 composition close to that of modern industrial glasses. It was obtained by melting together sand with alkaline flux from sea plant ashes or from minerals such as soda ash.

The use of glass as a material *per se* was preceded by the use of *enamels* on pottery, some of them dating from 12000 BC. The origins of this discovery remain unknown. Pline stated the possibility of an accidental vitrification of sand by reaction with soda ash blocks in a primitive Egyptian fireplace. It is more likely that the first glasses were simply slags from copper metallurgy.

Glass was first used as pieces cut into beads and other jewels (mummies' eyes were made out of obsidian). Glass containers were made during the reign of Touthmosis III (fifteenth century BC). They were obtained by forming a sand core coated with a potassium salt and then superficially vitrified. After cooling, the core was removed by scraping, leaving a rough internal surface. Egyptian vases were decorated in a very similar way to the vase discovered in the Ur excavations. The glass industry flourished in Egypt up to the twelfth century BC; then in Syria and Mesopotamia, until the ninth century BC. The glass products were distributed by the Phoenicians. Glass centers appeared in Cyprus, Rhodes and Greece, on the Italian peninsula (about 900 BC) and in the region around Venice (500 BC). Manufacturing techniques evolved: vases were fabricated from glass rolls (often multicolored) formed around a central core of sand then melted together. Glass pieces were also used to obtain mosaics by sintering, a technique called *"millefiori"*.

After its conquest by Alexander The Great, Mesopotamia declined and Alexandria in Egypt became predominant. New techniques were imported from Italy around 100 BC and the Syrians established new manufacturing enterprises by the beginning of our era.

Glass blowing was probably invented in Phoenicia (around 50 BC) and certainly revolutionized forming techniques by eliminating the use of a central core. This technical development stimulated art glass which spread from Persia to the Orient. In the Western world, the existence of the Roman Empire helped in establishing numerous glass centers; Syrian and Alexandrian glass blowers were working in Rome and the provinces of Saône and Rhine; the art then reached Spain, the Netherlands, Gaul and Brittany. Glass prices reduced considerably so that the use of blown glass containers became popular. Flat glass for window panes was practically unknown although excavations of Pompeii showed a few samples dated from the beginning of our era.

After the decline of the Roman Empire, refugees brought the art of glass blowing to the area around Geneva and it then spread all over Europe. Techniques hardly changed up to the eleventh century. The ashes of sea plants containing sodium were replaced by those of earth plants containing potassium.

The art of coloring glass was stimulated by the Church (stained glass).

The importance of Venice as a glass manufacturing centre grew steadily from the tenth century onwards. Murano became a large glass center where "Venice crystal" was made. Commerce with the Byzantine Empire lasted until the fall of Constantinople; interest then turned to the Western world and, until the end of the seventeeth century, Venice was predominant in the world of glass. Some groups worked first in Lorraine and Normandy, and then moved to England where coal furnaces were introduced in the seventeeth century.

The book *Arte Vetraria* by Neri published in Pisa in 1612 gives an overview of the knowledge of glass available at that time and was translated into many other languages. In Bohemia, "crystal glasses" were developed which had a particular brightness due to a high level of lead oxide.

In France, official control over the glass industry began in the sixteenth century. Henry IV conferred exclusive rights (lasting from 10 to 30 years) on Italians who were allowed to manufacture glass in several cities such as Paris, Rouen, Orleans and Nevers. Starting in 1665, Colbert made a centralised glass industry by creating a flat glass industry to produce the mirrors for the Palace of Versailles. The Saint-Gobain factory, created on the initiative of Colbert to fight the Venetian monopoly, became the Manufacture Royale des Glaces de France in 1693. France then surpassed Venice as an exporting country for mirrors. The Baccarat factories (crystal glass production) were founded in 1765.

By the end of the eighteenth century, an industrial revolution took place as a result of chemical discoveries based on the replacement of natural alkali by sodium from sea salts, obtained first by the Leblanc technique, then in 1863 by the Solvay process.

Deslandes (Saint-Gobain Co.) showed the importance of adding limestone to glass to give it chemical durability. This addition was made necessary because natural ashes which provided part of the lime had been replaced by very pure alkalis from the chemical industry.

By the end of the nineteenth century, mechanized forming processes were introduced (see Chapter 17 on technology). The invention of continuous glass drawing procedures for window glasses as well as the manufacture of plate glass were followed by the introduction of continuous casting on the surface of a bath of molten tin (the "FLOAT" process) in 1955–60, a revolution for the glass industry.

Together with technological improvements, a better understanding of the physical and chemical properties of glass emerged due originally to the use of optical glass. Glass lenses were known by the Greeks who passed their knowledge on to the Arabic world. Eye glasses were made in 1280 in Italy and the first telescope lenses were fashioned in Italy and the Netherlands around 1590. (The telescopes of Galileo and Kepler date from 1609 and 1610 respectively.) Until the end of the eighteenth century, optical glasses were treated as a sub-product of the ordinary glass industries, the glass used for this latter purpose being of poor quality. A definite improvement was obtained at the end of the eighteenth century by Guinand in Switzerland, who introduced stirring of glass to ensure good homogeneity. Working together with Fraunhofer, he widened the range of crown and flint glasses (see Chapter 12). A diversification of glass production took place by systematically introducing B_2O_3, P_2O_5 and a number of other oxides thanks to the collaboration of Abbe, Schott and Zeiss in Germany at the University of Jena around 1875. New glasses could then be produced and, within a period of ten years, progress was spectacular. Zeiss maintained a monopoly in the optical field, in particular for microscopes, up to the First World War. Starting from this period on, Parra-Mantois in France, Chance Brothers in Great Britain and Bausch and Lomb in the United States developed a competitive industry. Around 1917, Adams and Williamson resolved the problems of glass annealing and also rare earth oxides were introduced in the compositions.

Thus, the scientific approach was progressively introduced. The systematic measurement of the physical properties of various glasses was started around 1920 in Great Britain under the direction of Turner at the Department of Glass Technology at the University of Sheffield. As a result, a slow but steady generalization of the concept of glass took place and the term glass began to be applied to define a non-crystalline solid in general, not only an oxide glass. Tammann's studies in 1930 confirmed this point of view by directing research toward a general understanding of the *vitreous state* as a state of aggregation in matter. The approaches which were first of a purely phenomenological nature were then slowly directed toward more and more structural studies.

But only after the Second World War do we see strong interaction between scientific research and glass technology. In fact, the period 1950–60 can be considered as the blossoming period of true glass science. In France, glass science progressed rapidly due to the inspired guidance of Ivan Peychès, Director of the Research Laboratories, Saint-Gobain Co. All modern methods in physics and chemistry have been successively applied to the study of glass and the effort in this direction has intensified as we can see through the numerous scientific and technical works issued by different laboratories and research centers throughout the world.

Besides classical applications, in which glasses are indispensable to modern economy (construction, transportation, lighting, chemical industry etc.) we can see new glass techniques appearing (lasers, optical fiber communications, energy transformation) where glass brings original solutions.

1 Non-crystalline solids and glasses

The word *glass* has various meanings. In everyday language this term designates a fragile and transparent *material* well known since antiquity as well as certain *objects* made of this material. In scientific language its range of meaning is much larger but is difficult to define with precision, and subject to evolutionary change.

Glasses are essentially *non-crystalline solids* obtained by freezing supercooled liquids. It is recognized that numerous substances are able to solidify in this form. This led Tammann[1] to postulate the existence of a *vitreous state*, and since then glass has assumed the significance of a *physical state of matter*.

However, freezing of a liquid is not the only method of obtaining a non-crystalline solid. As we shall see, there are numerous other procedures, many of them leading to amorphous solids which cannot be obtained from a liquid. Are all these solids also to be classified as glasses? Before answering this question and giving a more precise definition of glass, we need to examine more closely the different methods of synthesis and the characteristics of the resulting non-crystalline solids. We should also bear in mind that the meaning of the word glass has recently been extended to describe a state of *disorder*; thus the term "spin glasses" is used to describe "amorphous magnetism" in certain disordered magnetic alloys.

Basically, a non-crystalline solid can be obtained by three different routes: by retaining (locking in) the structural disorder of a liquid phase, by taking advantage of the disordered character of a gaseous phase, or by disrupting the order of a crystalline phase.

1.1 Formation from a liquid phase

1.1.1 *Freezing of a supercooled liquid*

(a) *Description of vitrification*

On melting, the vast majority of elements and mineral compounds form liquids with relatively low viscosities of the order of a few centipoise. On cooling, such liquids crystallize rapidly at their freezing point even if the cooling rate is high. It is possible to supercool fine droplets of liquid, but this state cannot be maintained and crystallization inevitably occurs in the cooling process.

There are, however, substances which melt to give liquids with high viscosities of the order of 10^5–10^7 dPa s. If such a liquid is maintained slightly below its freezing point, it will have a tendency to crystallize slowly because the crystalline phase is thermodynamically more stable than the supercooled liquid. On continued cooling, starting from a temperature well above the freezing point, it may or may not crystallize, depending on the rate of cooling. For a slow cooling rate, the liquid will still be able to crystallize partially, but crystallization can be totally avoided if the cooling rate is sufficiently rapid. We then observe that the viscosity of the liquid increases progressively as the temperature is lowered and ultimately becomes so high that the final product has all the characteristics of a solid. When a liquid thus *solidifies* without crystallizing, it is currently said to form *a glass*, i.e. *to vitrify*, or to pass to a *vitreous state*.

The above description is basically a phenomenological definition of a glass. It was initially reserved for inorganic materials, but this is too restrictive: as we shall show, a large number of organic liquids also form glasses.

Since the passage from the liquid state to the vitreous state is progressive and continuous, it might be expected that the structures of the two states would be similar. This is confirmed by X-ray diffraction; the diffraction spectra of a glass show only diffuse rings without the discrete lines which characterize a polycrystalline substance (Fig. 1.1).

Like liquids, glasses thus possess a *disordered structure lacking long-range order*. That is, in a glass, there is not a regular arrangement resulting from the distribution over long distances of a repeating atomic arrangement (unit cell) which is characteristic of a crystal. There is

Fig. 1.1. X-ray diffraction spectra of (*a*) a glass (vitreous SiO_2) and (*b*) the corresponding crystalline phase (α-cristobalite).

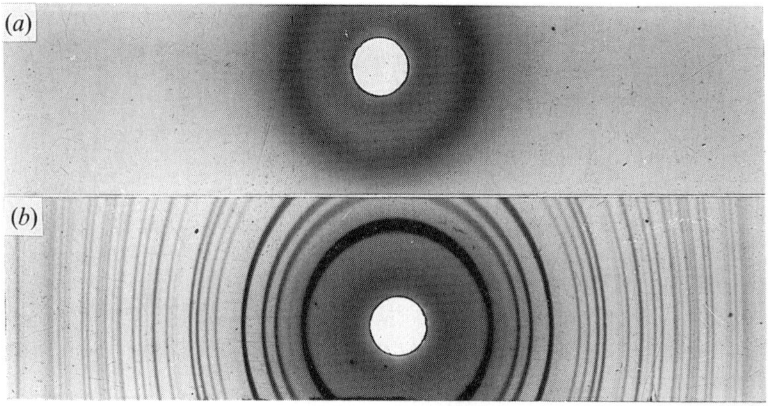

(*a*)

(*b*)

only evidence of a *short-range order* which corresponds to the mutual arrangement of nearest neighbors to a given atom and varies according to the atomic site considered.

Although the atoms or molecules of a liquid are able to undergo relatively large displacements, the displacements in a glass are restricted to thermal vibrations around an average fixed position. As a first approximation, the structure of a glass can thus be considered as a *frozen liquid* where the constituent units have lost the ability to modify their respective configurations.

(b) *Examples of vitrifiable substances*

The formation of glasses is not a very frequent phenomenon, but it occurs in very diverse classes of materials which we shall now review quickly.

1. *Elements*. The only elements able to vitrify alone are in groups V and VI of the periodic table: phosphorus, sulfur, and selenium.

2. *Oxides*. The oxides SiO_2, GeO_2, B_2O_3, P_2O_5, As_2O_3 and Sb_2O_3 (the last one rapidly quenched) transform to the vitreous state. Silica, SiO_2, is a typical example of an easily vitrified oxide, while molten boric oxide, B_2O_3, passes regularly to a glassy state and can only be crystallized with difficulty. Each of these oxides is able to form a glass either alone or in association with one or several other oxides. The mixture becomes a liquid by fusion and the liquid can pass to the vitreous state on cooling. Vitrification is generally possible only within certain limits of composition. For example, in the system SiO_2–Na_2O, glass formation is possible between 0 and 47 mole % Na_2O.

An *infinite* number of glasses can be made by mixing one or more of the above oxides in varying proportions. The oxide glasses can be considered as complex solid solutions of the constituent oxides.

The oxide glasses are particularly important from a practical point of view. Industrial glasses are almost exclusively *silicate glasses*, the approximate composition of a common glass being (in mole %) 70 SiO_2, 20 Na_2O, 10 CaO.

3. *Chalcogenides*. Glasses exist in the binary systems As–S, As–Se, P–Se, Ge–Se. More complex glasses can be formed in combination with various sulfides and selenides as well as with the elements Si, Ge, Ga, In, Sn, Te, Bi, Pb, Sb, Tl, etc.

4. *Halides*. The only halides forming glasses alone are BeF_2 and $ZnCl_2$. Various fluorides can be combined with BeF_2 to form complex fluoroberyllate glasses isomorphic with the silicate series. Other glass systems are based on ZrF_4.

5. *Molten salts.* Certain mixtures of nitrates, e.g. KNO_3–$Ca(NO_3)_2$, carbonates, e.g. K_2CO_3–$MgCO_3$, and many binary and ternary systems of phosphates, fluoroaluminates, fluoroborates, fluorophosphates, and acetates etc. present composition ranges in which glass formation is possible.

6. *Aqueous solutions of salts, acids, bases.* The formation of glasses is frequent for H_2SO_4, KOH, LiCl etc. in concentrated aqueous solutions.

7. *Organic compounds.* Among simple compounds, methanol, ethanol, glycerol, glucose, toluene, o-terphenyl, 3-methyl hexane, etc. may be cited.

8. *Organic polymers.* A large number of macromolecular compounds, for example, polyethylene, polyvinyl chloride, polystyrene, etc. form "organic glasses."

9. *Metals.* Certain metallic alloys have been vitrified. The metallic glasses can be classified into two main groups:

metal–metalloid alloys: e.g. Pd–Si; Fe–B; Fe–Ni–P–B:
metal–metal alloys: e.g. Ni–Nb; Cu–Zn.

(c) *Rôle of the quenching rate*

The preceding enumeration shows that vitrification occurs in very diverse classes of substances which do not display an obvious common relationship.

To define this *aptitude for vitrification* precisely, it is not sufficient to consider only the chemical nature of the product. Whether or not a substance forms a glass will depend not only on its chemical composition, but also on the *rate* at which the liquid is cooled, i.e. on the *quenching rate.*

Under an extremely slow cooling rate, no material will form a glass. The converse is not true: it is not certain that any given material can be vitrified with a sufficiently high cooling rate. Although in numerous cases a slow cooling rate is sufficient, in exceptional cases e.g. metal alloys, quench rates must be of the order of 10^6 °C s^{-1}. In all cases, the cooling rate depends on a number of factors (thermal diffusivity, radiation heat transfer, geometric form, condition of the surface) but the most important one is the *volume* of the sample. It is thus necessary to define precisely the conditions for vitrification and in particular the *mass* of the product in question.

(d) *Special methods of quenching*

SiO_2 based industrial glasses can be obtained easily, on a scale of hundreds of tons, by simple cooling of the liquid in air. For some substances more energetic means of cooling must be employed, e.g. quenching into a liquid bath (water, Hg, liquid N_2). By using methods of fusion such as a plasma torch or a power laser,[46] fine droplets can be quenched to a vitreous state, and glasses of TiO_2–BaO, Al_2O_3–Y_2O_3, etc. have been prepared this way.[47]

The rapid extraction of heat by contact with a high thermal conductivity solid is utilized in the "splat cooling" techniques of Duwez, Willems and Klement.[48] The liquid is projected by a shock wave (due to the expansion of a gas) onto a Cu support where it flattens and freezes into thin fragments. Quench rates of 10^5–10^9 °C s^{-1} are possible thus permitting the quenching of vitreous metallic alloys. Another arrangement due to Pietrokowsky[49] ("piston and anvil technique ") catches a freely falling molten drop between two pieces of metal which are quickly forced together producing a quenching rate of $\approx 10^5$ °C s^{-1}. These techniques permitted the quenching of $MoO_3, TeO_2, WO_3, V_2O_5$ and a number of binaries.[50] A quench between metallic rollers has also been proposed[51] and used to quench refractory oxides.[52] A technique using superficial fusion-vitrification (glazing) by a CO_2 laser has recently been proposed in the United States.[53]

1.1.2 *Other methods of formation in the liquid phase*

Quenching of a supercooled liquid is not the only way to produce a non-crystalline solid from a liquid phase.

(a) *Precipitation and coprecipitation reactions*

Examples of this are the hydrolysis and polymeric condensation of certain substances which form gels.[54,55] For example it is possible to produce a silica gel from an alkoxide solution:

$$Si(OR)_4 \xrightarrow{+H_2O} Si(OH)_4 \rightarrow SiO_2 + H_2O$$

Gels containing mixtures of several cations can be obtained, and the gel converted to a dense glass by sintering or hot pressing.[56,57] These methods of synthesis are developed in Chapter 18 (sol–gel methods).

(b) *Polymerization reactions*

Organic macromolecules are synthesized by either the addition or condensation of monomer molecules. The resulting materials can be totally or partially crystalline. These methods have great importance in the plastics industry.

(c) *Elimination of a solvent*

This is the case in the preparation of numerous organic substances or *glues* in which the evaporation of a solvent leaves a non-crystalline residue.

(d) *Electrolytic deposition*

In this case, the non-crystalline solids are thin films formed on the surface of an anode. The anodic oxidation of Ta, Al, Zr and Nb leads to the corresponding amorphous oxides.

The effect of the modes of formation on the structure of glass has recently been the object of several colloquia.[58,59]

1.2 Formation from a gas phase

1.2.1 *Non-reactive deposition*

The condensation of a vapor on a substrate maintained at a temperature sufficiently low to reduce the mobility of the deposited atoms can lead to the formation of a disordered structure, generally in the form of a *thin film.* This procedure is generally applied to metals, to Si and Ge and to alloys of metals and metalloids.

(a) *Vacuum evaporation*

The vapor here is derived from a solid or liquid heated in an enclosure in which the pressure is 10^{-4}–10^{-7} torr. The vapor condenses on a substrate which may be cooled. In practice, the heating is obtained by electrical resistance, radio frequency induction or electron beam bombardment.[60] In the case where several compounds have very different vapor pressures, "flash" evaporation can be used. By starting with alloy particles having the desired composition, a constant composition of the resulting vapor is assured.[61] The evaporation of oxides can lead to dissociation; e.g. the deposition of SiO from SiO_2.

(b) *Cathodic pulverization ("sputtering")*

The source material to be vaporized (the target) is the cathode. The substrate, which is the anode, receives particles ejected from the target by the bombardment of positive ions formed in the residual atmosphere of 10^{-2}–10^{-5} torr. A high frequency voltage can be applied to the substrate to eliminate accumulated electric charge to facilitate the deposition of dielectric materials.[62,63]

(c) *Ion implantation*

Bombardment by positive ions has been used to inject O or N into the surface of Si.[64]

1.2.2 *Reactive deposition*

By introducing sufficient activation energy (thermal or electrical) a chemical reaction can be initiated which leads to the deposition of a vitreous phase.

(a) *Chemical Vapor Deposition Process* (CVD)

This widely used technique consists of initiating a heterogeneous reaction between vapors of a halide or an organo-metallic compound and O_2 in the neighborhood of a substrate. For example a mixture of SiH_4–PH_3 and O_2 on contact with a hot wall deposits a thin glass film of P_2O_5–SiO_2.[65–7] This can be combined with the use of a microwave or radio frequency plasma.[68]

(b) *Reactive sputtering*

A variant of sputtering consists of introducing an appropriate gas into the enclosure. If the gas contains O_2 or N_2, the deposit will consist of the oxide or nitride of the vaporized metal.

(c) *Massive deposits*

The *homogeneous* oxidation of a mixture of halide vapors is the current method used to obtain large pieces of ultra–pure glass of high optical quality. For example the reaction:

$$SiCl_4 + O_2 \rightarrow SiO_2 + 2Cl_2$$

between 1300 and 1600 °C produces a particulate material ("soot") of high specific surface area which can be converted to bulk SiO_2 by viscous sintering.[69] Other halides $GeCl_4$, $TiCl_4$, BCl_3, $POCl_3$ can be added in variable proportions. This is the procedure used to fabricate glass with controlled index of refraction profiles.

The oxidation of $SiCl_4$ by O_2 in an oxy-hydrogen torch can deposit bulk SiO_2.[70]

1.3 Formation from a solid phase

In the preceding examples, the disordered solid structure was obtained by virtue of the inherent disorder of the liquid or vapor phase. In contrast, it is possible to obtain a disordered structure by processes which destroy the crystalline order, e.g. radiation damage and mechanical damage.

1.3.1 *Radiation damage*

Collisions with alpha particles or fast neutrons cause atomic displacements and the formation of defects in the lattice which, by a cumulative effect, lead to an amorphous structure. The recoil of the residual nuclei can also contribute to the development of disorder. Elevated temperatures ("thermal spikes"), which can exceed 1000 K for 10^{-10}–10^{-11} s and which affect thousands of atoms, are sufficient to cause a local fusion. The different varieties of crystalline SiO_2 are transformed into the same amorphous phase.[71]

Certain minerals in nature containing radioactive elements, e.g. thorite or gadolinite, are found in an amorphous form as a result of bombardment sustained over geological eras. Such substances are called *metamicts.*

The construction of nuclear reactors has produced numerous studies of the damage to materials subjected to radiation and has permitted systematic experiments on the formation of amorphous structures under radiation.

1.3.2 *Mechanical action*

The effects of shearing under prolonged mechanical grinding of a crystal can progressively destroy the crystalline order and lead to the formation of a non-crystalline solid. Analogous effects can be rapidly obtained by the action of a sufficiently intense shock wave. These types of studies have been stimulated by the lunar exploration programs. Much of the surface of the moon is covered by vitreous material and it is of interest to learn if this state is due to volcanic activity or meteoric bombardment.

The glasses produced by the shock effect without fusion are called *diaplectics* – they retain their original crystal shape without any trace of melting. The first diaplectic SiO_2 glasses obtained in the laboratory required pressures in excess of 350 kbar produced by an explosion.[72] Higher pressures progressively induced a fusion effect.[73]

Explosive compaction has also been used to produce large pieces of metallic glasses by instantaneous sintering of powders.[74–6]

1.4 **Definition of a glass**

To define a glass, we have a choice between:

1. An operational definition: "A glass is a solid obtained by freezing a liquid without crystallization."

This definition restricts the term "glass" to products obtained by quenching of a liquid.

2. A structural definition: "Glass is a non-crystalline solid."

Following this view, the terms "non-crystalline solid," "amorphous solid," and "glass" are synonymous. The definition encompasses all the products of the operations described previously.

Neither one of the above definitions is completely satisfactory. The apparently simple structural definition is too general. Although glass is truly a non-crystalline solid, all non-crystalline solids are not necessarily glasses – e.g. gels (which can sometimes be transformed into glass by appropriate treatment).

It is true that in a number of cases, the application of diverse processes leads to an apparently identical material. For example, vitreous SiO_2 has been prepared by different means (cf. Table 1.1) and the final product

Table 1.1. *Comparison of refractive index and density of vitreous SiO_2 formed by different methods. After Ref. 4.*

Method of preparation	Refractive index of Na D-line n_D ($\lambda = 589.6$ nm)	Density d ($g\ cm^{-3}$)
Fusion	1.458	2.203
Rendered amorphous by shock waves	1.46	2.22
Neutron bombardment ($1.8\text{–}3 \times 10^{20}\ n\,cm^{-3}$)	1.458	2.205
Deposited from organic solutions	1.40–1.45	
Reactive sputtering	1.455	2.2
Thermal decomposition	1.430	
Hydrolysis in a vapor phase	1.458	2.202

cannot be distinguished by current characterization methods. There is no reason not to consider as glasses those materials which are prepared by other processes but resemble in all respects those obtained by quenching from a liquid.

The operational definition is of little help when the origin or mode of preparation of a specimen is unknown. Moreover, certain non-crystalline materials cannot be produced by rapid quenching of liquids and require special preparation methods.

On the other hand, it is sometimes difficult to differentiate a non-crystalline structure from a microcrystalline structure with extremely fine grains. Some materials including many polymers occur in a partially crystalline form. It is also possible to have different types of structural disorder.

A solution to this dilemma is possible if a structural definition is adopted which includes a condition on the *internal stability* of the material. The non-crystalline solids obtained by the methods indicated above all contain an *excess energy* incorporated by quenching, mechanical action, radiation or chemical potential. Non-crystalline solids having an energy content greater than that of the parent crystalline phases thus correspond to *metastable* or *non-equilibrium* states of matter. The return to a stable situation by crystallization can be to a greater or lesser degree violent depending on the case. The "curing" of solids made amorphous by radiation is accompanied by a noticeable elevation in temperature (Wigner effect). Heating of amorphous deposits generally leads to their rapid crystallization or their decomposition before the melting point is reached. In contrast, "classic" glasses are characterized not only by the absence of crystallinity but above all by their ability to pass progressively and reversibly to a more and more fluid state as the temperature is increased. In the course of this change there is a modification of properties called the *glass transition*, the significance of which will be studied in detail in the following chapter. Gradual softening with increasing temperature is, moreover, extremely important in technical applications and constitutes one of the fundamental properties of a classic glass material.

Under these conditions, the following definition of a glass can be adopted: "A glass is a non-crystalline solid exhibiting the phenomenon of glass transition." The corresponding physical state is called the *vitreous state*.

This definition does not impose any restrictions relative to the manner in which the vitreous material is obtained. However, it stresses a simple and fundamental criterion to distinguish a non-crystalline material which lacks the characteristic of sufficient internal stability. Thus, for example certain thin films (Si, Ge) which always crystallize at elevated temperatures and unstable non-crystalline precipitates are not considered to be glasses. It is appropriate to reserve for them the term *amorphous*.

This partition of non-crystalline solids into *glasses* on the one hand and *amorphous materials* on the other hand constitutes a logical classification, and will be adopted in the present work.

The following presentation will mostly focus on the study of *glasses*, and more particularly, inorganic glasses. Sometimes, the classification limits of glass and *amorphous materials* will be very close according to their theoretical and practical importance. The methods of investigation are equally applicable to "amorphous materials" and to liquids which constitute the natural extension of glass to high temperatures. It is necessary, moreover, to recognize that the borderline between the two classes of non-crystalline solids will sometimes be indistinct.

2 The vitreous transition

2.1 Phenomenological aspects

2.1.1 *Definition of the transition temperature T_g*

The classical way to produce glass consists of cooling a liquid so quickly that crystallization does not have time to occur. As the temperature decreases, the continuous increase of viscosity results in a *progressive freezing* of the liquid to its final solidification.

To study this process with more precision, it is convenient to follow the evolution of a thermodynamic variable, e.g. the specific volume, V as a function of temperature (Fig. 2.1(*a*)).

Starting with a liquid at an elevated temperature, the lowering of the temperature first causes a contraction. When the point of solidification (or freezing) T_f is reached two phenomena may occur – either the liquid crystallizes and a discontinuity ΔV_f (generally a contraction) is introduced in the curve, or crystallization is avoided and the liquid passes to a supercooled state. In the latter case, the representative point follows an extension of the liquid curve which passes the temperature T_f without discontinuity. It is as if the system "ignored" the melting point.

In the first case, on completion of crystallization, as the temperature decreases the crystalline solid contracts again, the slope of the curve now being less than that of the initial liquid (about 1/3). In the second case, the decrease in the temperature at first causes a contraction of the supercooled liquid with a coefficient identical to that of the original liquid.

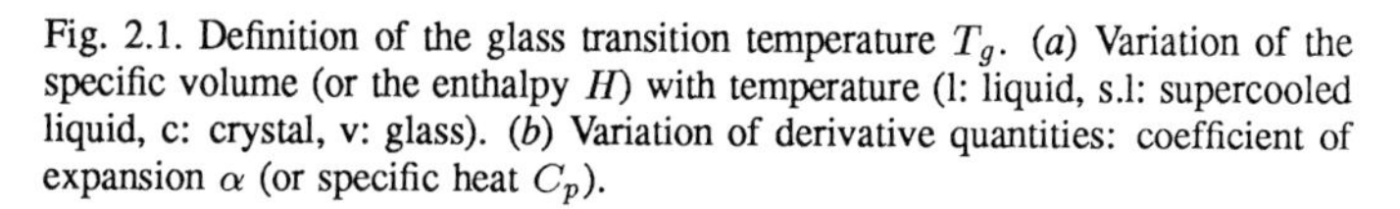

Fig. 2.1. Definition of the glass transition temperature T_g. (*a*) Variation of the specific volume (or the enthalpy H) with temperature (l: liquid, s.l: supercooled liquid, c: crystal, v: glass). (*b*) Variation of derivative quantities: coefficient of expansion α (or specific heat C_p).

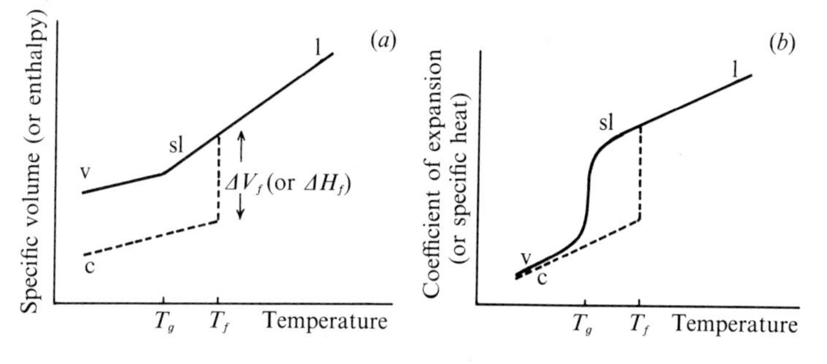

Then, starting at a certain temperature T_g, the slope of the curve decreases to become close to that of the crystalline solid. This break in the cooling curve marks the passage from a supercooled liquid to a glass. The temperature T_g is called the *transition temperature* or the *glass transformation temperature*. The viscosity of the liquid increases continuously as the temperature decreases, and the passage through T_g corresponds to a viscosity in the neighborhood of 10^{13} dPa s. Note that, in contrast to V, the coefficient of expansion α shows a rapid change on passing through T_g (Fig. 2.1(*b*)).

2.1.2 *The transition interval*

At constant pressure, the position of the transition point T_g is not fixed as in the case of T_f but varies slightly with the *rate* at which the liquid is cooled. Rapid cooling has the effect of shifting the break defining T_g towards higher temperatures while slow cooling displaces T_g towards lower temperatures (Fig. 2.2) For this reason it is preferable to substitute for T_g the concept of *a transition interval* or *a transformation interval* $[T_g]$ where the upper and lower limits are defined respectively by the highest and lowest cooling rates used to determine T_g.

If, in place of V, other physical variables characterizing the system are considered, there is continuity between the liquid and the supercooled liquid on passing through T_f, and, in contrast, a singularity when T_g is crossed. Figure 2.3 shows the variations of the specific mass, refractive index, static permittivity, and thermal conductivity of glycerol which occur

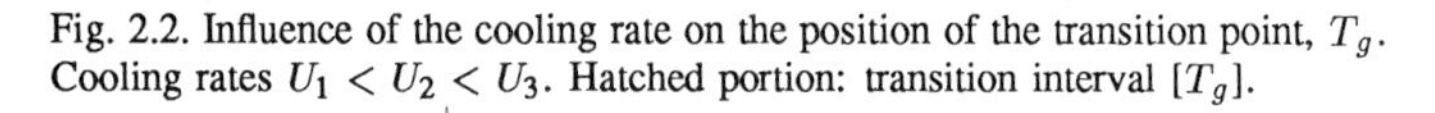

Fig. 2.2. Influence of the cooling rate on the position of the transition point, T_g. Cooling rates $U_1 < U_2 < U_3$. Hatched portion: transition interval $[T_g]$.

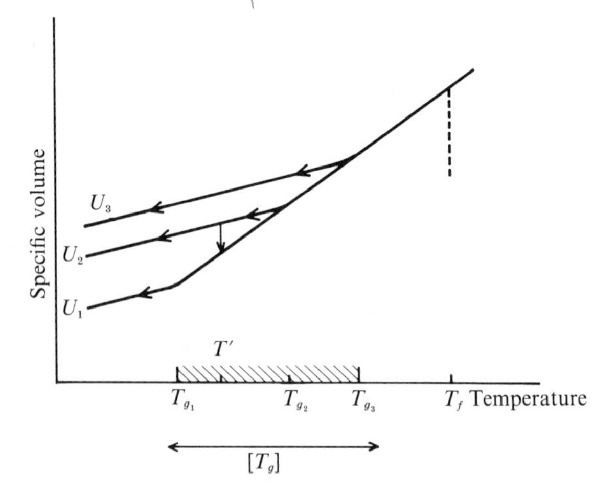

Table 2.1. *Average values of the glass transition temperature T_g (K).*

SiO_2	1500–2000
Na_2O–CaO–SiO_2 glass (window glass)	800–820
B_2O_3	470–530
S	244
Se	302–308
Glucose	280–300
Glycerol	180–190
C_2H_5OH	90–96

either on crystallization or during passage to the vitreous state.

The values of T_g resulting from different experiments can be slightly different. Therefore, to be completely rigorous in speaking of T_g, it is necessary to define the method used for its determination. In practice, standard methods are used (cf. § 2.5).

Table 2.1 gives examples of average values of T_g for a number of substances forming glass by cooling (quenching) of a liquid. It is seen that

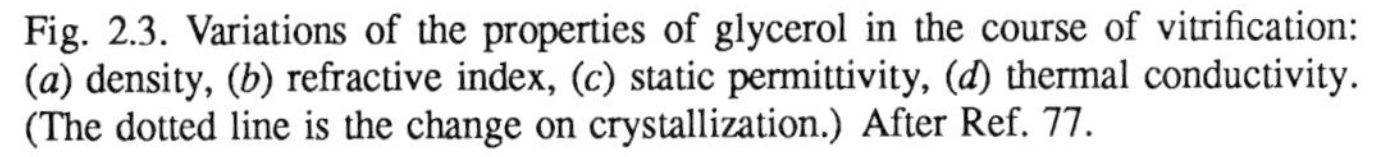

Fig. 2.3. Variations of the properties of glycerol in the course of vitrification: (*a*) density, (*b*) refractive index, (*c*) static permittivity, (*d*) thermal conductivity. (The dotted line is the change on crystallization.) After Ref. 77.

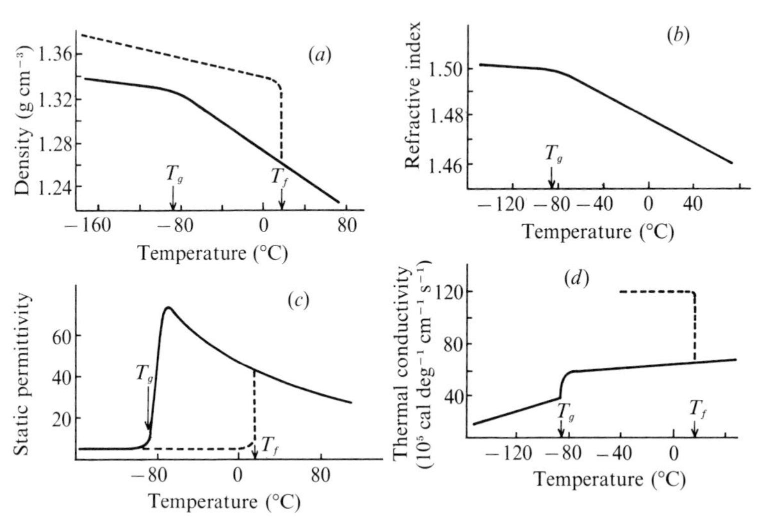

the values of T_g cover a temperature range from 1500 K for SiO_2 to 90 K for C_2H_5OH. In all cases, however, T_g corresponds to the temperature at which the viscosity attains 10^{13} dPa s.

Given the fact that the temperature of T_g varies with the cooling rate, it can only have an operational significance. T_g seems to depend logarithmically on the rate of cooling: q.

$$q = q_0 \exp\left(-E_a/RT_g\right)$$

Figure 2.4 shows the variations of T_g with q for three very dissimilar substances.

2.1.3 *The phenomenon of stabilization*

Consider a system cooled at a rate U_2 (Fig. 2.2) into the transformation interval and subjected to an *isothermal* treatment at a temperature T'. It is found that the glass contracts to a point on the curve represented by a linear extrapolation of the supercooled liquid. This phenomenon is given the name *stabilization.*

Consider two samples of the same glass stabilized in the transformation interval at respective temperatures T_1 and T_2, $T_1 < T_2$, and then taken suddenly to an intermediate temperature T_3 (Fig. 2.5). The glass stabilized at T_2 *contracts* while the one stabilized at T_1 *expands*, both approaching

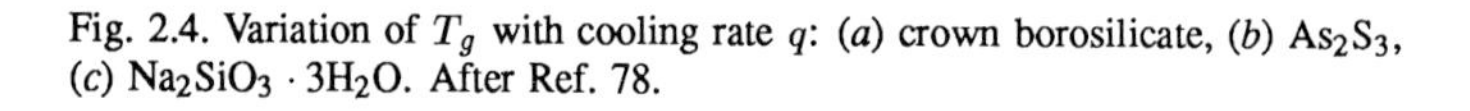

Fig. 2.4. Variation of T_g with cooling rate q: (*a*) crown borosilicate, (*b*) As_2S_3, (*c*) $Na_2SiO_3 \cdot 3H_2O$. After Ref. 78.

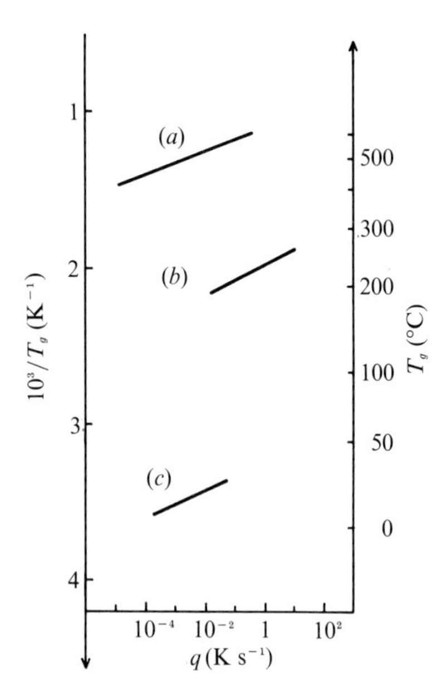

Fig. 2.5. Stabilization at a new temperature T_3 of glasses previously stabilized at T_1 and T_2.

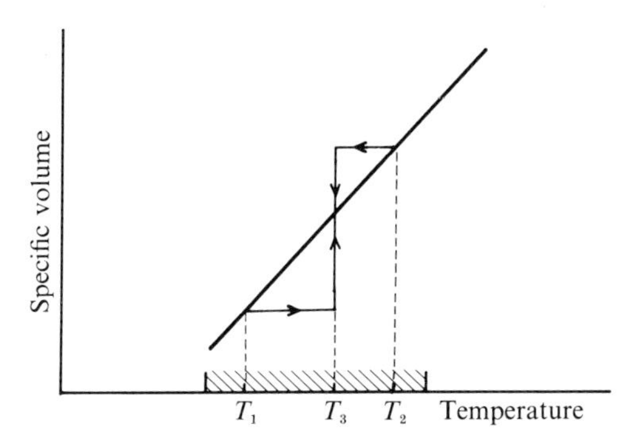

a common limit corresponding to T_3 which may be attained either from higher or lower values.

The phenomenon may be easily followed by measuring the refractive index of the glass in the course of stabilization (Fig. 2.6). This shows that a stabilized glass in the transition interval is in a state of *metastable* equilibrium at the stabilization temperature while, at a lower temperature, the glass is in a non-equilibrium situation.

2.2 Thermodynamic aspects

2.2.1 *The entropy of glass – excess entropy*

Consider the preceding system. It can either crystallize at the temperature T_f or pass to a vitreous state at T_g depending on the rate of cooling. The variations in the specific heat (Fig. 2.1(*a*)) are in all respects analogous to those variations in volume already studied with the following exception: the discontinuity $\Delta H_f = L_f$ corresponding to the latent heat of fusion L_f is always positive while $\Delta V_f = V_l - V_c$ may, depending on the particular case, be either positive or, more rarely, negative (subscripts l and c refer to liquid and crystal respectively).

The specific heat at constant pressure, $C_p = \left(\partial H / \partial T\right)_p$ is obtained by derivation from the preceding curves (Fig. 2.1(*b*)). As the system vitrifies, the passage through T_f does not show a discontinuity. When C_p diminishes rapidly in the neighborhood of T_g, the specific heat of the glass tends toward that of the crystal.

The variation of entropy is:

$$dS = \frac{C_p dT}{T} = C_p d \ln T$$

Fig. 2.6. Change in the refractive index of a glass in the course of stabilization at three different temperatures. After Ref. 79.

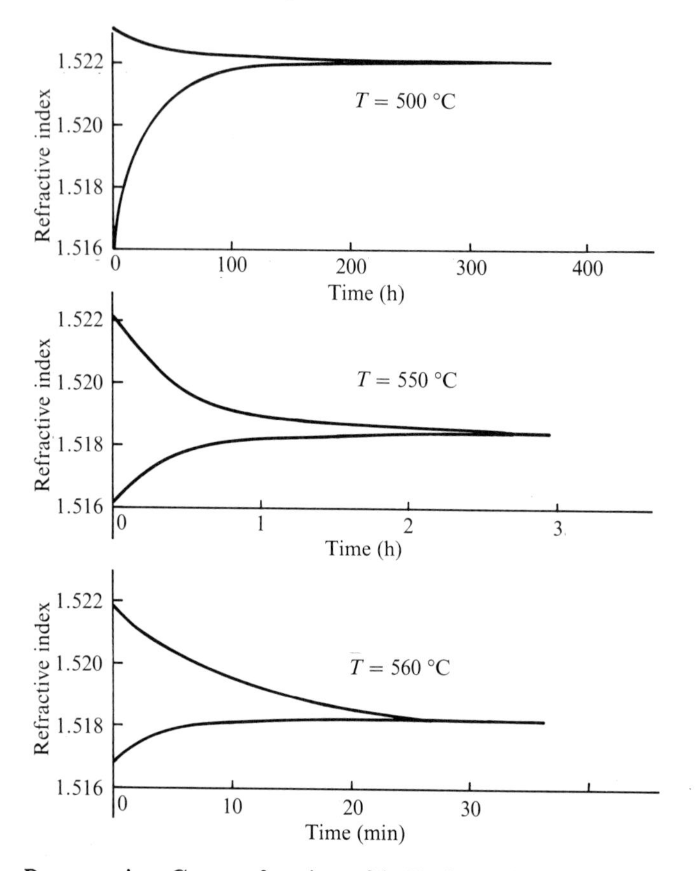

Representing C_p as a function of $\ln T$, the variations in the entropy can be deduced from the areas under the curves.

The specific heat of the liquid, supercooled liquid or glass is designated by C_{pl}, and the specific heat of the crystalline solid is designated by C_{ps} (Fig. 2.7).

The entropy of the liquid at a temperature $T > T_f$ is evaluated by integration from absolute zero following first the path crystal → liquid:

$$S_l = S_0 + \int_0^{T_f} C_{ps} \frac{\mathrm{d}T}{T} + \frac{L_f}{T_f} + \int_{T_f}^{T} C_{pl} \frac{\mathrm{d}T}{T}$$

For the crystal, the entropy S_0 at absolute zero is zero following the third law of thermodynamics.

The same value for the entropy of the liquid, S_l must be obtained following this time the path glass → liquid:

$$S_l = S'_0 + \int_0^{T_g} C_{pl}\frac{\mathrm{d}T}{T} + \int_T^{T_f} C_{pl}\frac{\mathrm{d}T}{T} + \int_{T_f}^{T} C_{pl}\frac{\mathrm{d}T}{T}$$

By setting the two values of S_l equal, it is possible to deduce the unknown value of the entropy S'_0 of the glass at absolute zero:

$$S'_0 = \frac{L_f}{T_f} - \int_0^{T_f} (C_{pl} - C_{ps}) \frac{\mathrm{d}T}{T}$$

which corresponds to the difference between the entropy of fusion of the crystal and that represented by the shaded area between the two paths shown in Fig. 2.7. Experiments show that S'_0 is not zero, and thus the third law of thermodynamics does not apply to glasses. Experiments on glycerol[80] give the value

$$S'_0 = 5.6 \text{ cal} \,^\circ\text{C}^{-1} \text{ mole}^{-1}$$

It is then easy to evaluate the entropy difference

$$\Delta S = S_{liquid\ or\ glass} - S_{crystal}$$

for temperatures $T < T_f$

$$\Delta S = \frac{L_f}{T_f} - \int_T^{T_f} (C_{pl} - C_{ps}) \frac{\mathrm{d}T}{T}$$

Fig. 2.7. Determination of the excess entropy of a glass. The variation of C_p for a glass cooled more slowly is shown by the dashed line and the dotted line is the extrapolation limit for a cooling rate tending toward zero. The rectangle represents the entropy of fusion of the crystal.

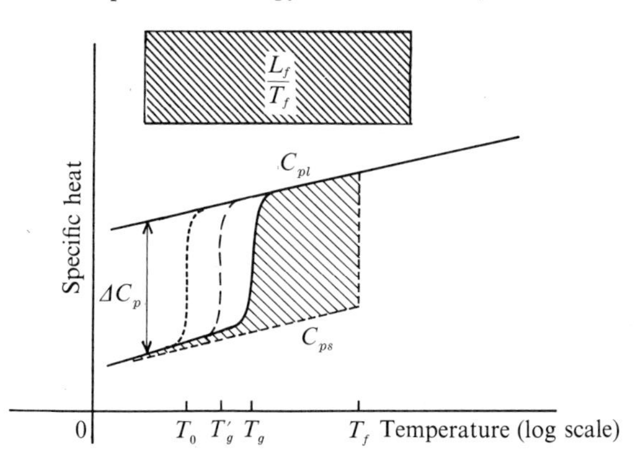

Figure 2.8 shows the variations of ΔS as a function of the temperature. In the course of the transition, the system seeks unsuccessfully to reduce ΔS and some entropy of fusion is retained at T_g. This corresponds to the *configurational disorder* of the liquid which is frozen in at the moment of glass formation.

Fig. 2.8. Variation of the excess entropy ΔS with temperature. Dashed line is for a slower cooling rate.

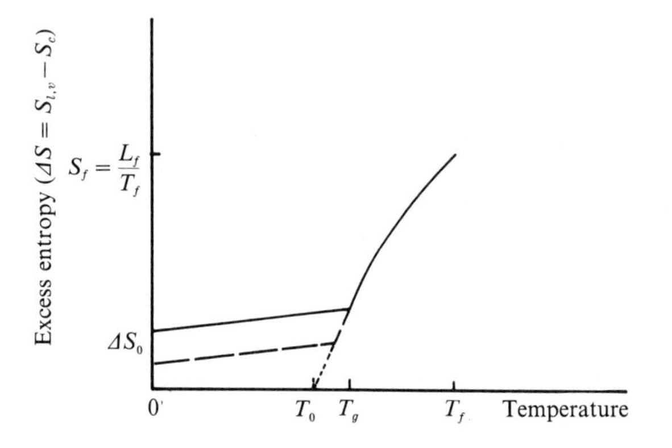

2.2.2 *The relaxational character of the vitreous transition*

When the liquid is in a temperature domain where changes in the relative positions of the atoms or molecules can occur rapidly, the system readjusts to imposed variations in temperature by modifying the relative positions to minimize the free energy at each temperature. As long as the time for readjustment (the relaxation time) to attain a new configurational position is not too great relative to the speed of the external changes, equilibrium is assured.

For a liquid, the variations of the volume V or the enthalpy H may be separated into two parts: vibrational and configurational:

$$\alpha_l = \frac{\mathrm{d}V_l}{\mathrm{d}T} = \left(\frac{\partial V}{\partial T}\right)_{vib} + \left(\frac{\partial V}{\partial T}\right)_{con}$$

$$C_{pl} = \frac{\mathrm{d}H_l}{\mathrm{d}T} = \left(\frac{\partial H}{\partial T}\right)_{vib} + \left(\frac{\partial H}{\partial T}\right)_{con}$$

Since the short-range order in the liquid is similar to that of the crystal, it may be assumed that the *vibrational* terms of the liquid and the crystal are very nearly the same. However, depending on the structure (the space group of the crystal), the *configurational* term of the liquid may be much higher than that of the crystal.

When a liquid is able to vitrify on cooling, its viscosity increases very rapidly, and at a certain temperature, probably due to the increase of the mutual blockage of the structural units, the relaxation time necessary to attain an equilibrium configuration increases suddenly. This temperature corresponds to T_g.

For temperatures below T_g, the relaxation time becomes so large relative to the normal time of observation that the configurational positions can be considered as frozen into arrangements which depend on the thermal history of the system.

According to the preceding model, the glass may be considered as a liquid in which the configurational arrangement is no longer able to attain a free energy minimum. The configurational entropy remains at a constant value which has been fixed near the temperature T_g.

As we have seen, the vitreous transition is produced when the viscosity attains a value near to 10^{13} dPa s. For reasonable values of the shear modulus G, this value of the viscosity corresponds to a Maxwell relaxation time $\tau = \eta/G$ of about 10 min (see Chapter 9).

This demonstrates the *relaxational character* of the vitreous transition. The "accident" which defines T_g is simply the result of an equality between the time constants of the relaxing structure and those of the experiment (i.e. the rate of the extraction of heat during cooling). Glasses thus do not represent states of thermodynamic equilibrium with regard to configurational changes in their molecular structure – they are truly *non-equilibrium* systems. On the other hand, the vibrational energy is equilibrated in the glass. The temperature is uniform, and the thermal conduction ensures the equilibrium of the local distribution of energy.

For substances which can be extensively supercooled, the properties of the supercooled liquid are a continuous extension of those of the stable liquid above T_f. In the transition zone, the *stabilized* glass, which constitutes a natural extension of the supercooled liquid, is thus in a state of *metastable* equilibrium. Below the transition zone, the system becomes incapable of reaching an equilibrium configuration and is thus in a state of *non-equilibrium.*

2.2.3 *The Kauzmann paradox*

A slower cooling rate moves T_g towards a lower temperature and reduces the excess entropy ΔS (Fig. 2.8). What would happen for a system cooled infinitely slowly? Note that if the curve ΔS is extrapolated beyond a temperature T_0, ΔS would be negative, i.e. the liquid (glass) would have an entropy smaller than that of the crystal.

This constitutes the paradox of Kauzmann[81] who was the first to call attention to this difficulty. It is necessary therefore to admit the existence of a temperature T_0 near which ΔS disappears. Since the entropy of the liquid at T_0 remains above that of the crystal, the free energy $H - T_0 S$ is also greater. This excludes the existence of an equilibrium between the crystal and the liquid undercooled to T_0. A transformation leading to the glass must have occurred between T_f and T_0. The temperature T_0 may be considered as the thermodynamic limit of the glass transition, impossible to attain experimentally and always hidden in practice by a relaxation process.

However, there might be an alternative, as yet unknown experimentally, where the heat capacity of the liquid decreases gradually towards that of the solid without sudden transitions.

While T_g is related to kinetics, T_0 is a thermodynamic parameter.

2.3 Theories of vitreous transition

Whereas the relaxational character of the "apparent" glass transition at T_g is understood, the nature of the transition towards T_0, which has characteristics similar to those of a second-order transition (in the sense of Ehrenfest), still remains to be fully clarified.

Two theories have been proposed which are essentially related to theories of the viscosity. These are:

1. The free volume theory of Turnbull and Cohen.
2. The cooperative relaxation theory of Adam and Gibbs.

2.3.1 *The free volume theory*

The Kauzmann paradox suggests that there is a uniquely defined disordered state at low temperatures characterized by a minimum in the specific volume–temperature or energy–temperature relations lying above the corresponding relations for the crystal. This state would be realized if the liquid were cooled very slowly without crystallization to a temperature well above absolute zero.

Turnbull and Cohen[82] identify the vitreous state with this unique amorphous state. Since no structural change seems to accompany the vitreous transition and since this transition occurs in a fixed interval of temperature, it is deduced that the liquid and the glass belong to a unique thermodynamic phase. If the glass transition is not a thermodynamic phase change, what is its nature?

Fox and Flory[83] advanced the hypothesis that this transition results from the diminution of the free volume of the amorphous phase below a certain characteristic value.

On the basis of this work, Williams, Landel and Ferry[84] showed that in the region of the glass transition, the fluidity Φ (inverse of viscosity) obeys the empirical equation of Doolittle:[85]

$$\Phi = A \exp\left[-(qv_0/v_f)\right]$$

where q is a constant near unity, v_0 the volume of a molecule, and v_f the "free volume" defined by $v_f = \overline{v} - v_0$, $\overline{v}$ being the specific volume.

Cohen and Turnbull[86] developed a relation for the diffusion coefficient similar to the Doolittle equation. They essentially considered a system of molecules moving with a velocity u' from the kinetic theory of gases but remaining most of the time confined inside the cages formed by their immediate neighbors.

Occasionally, by virtue of a fluctuation, a "hole" is formed in the cage permitting a significant displacement of the molecule. This displacement only gives rise to a diffusion movement if another molecule jumps into the hole before the first molecule returns to its original position. The movement by diffusion proceeds principally by molecular jumps in the free space (holes) created at the expense of *excess volume* present in the liquid. The jump can only happen if the "hole" has a volume greater than a critical value v^* formed by the redistribution of the *free volume, defined as that part of the excess volume which can be redistributed without expending energy*. In the liquid, there would be a random distribution of cavities of different sizes, and only those with a volume greater than v^* contribute to the transport. The authors cited have developed an expression giving the free volume distribution probability for cavities of different sizes:

$$p(v) = \frac{\gamma}{v_f} \exp\left(-\frac{\gamma v}{v_f}\right) \tag{2.1}$$

In this expression, $p(v)$ is the probability of the existence of a cavity with a volume between v and $v + \mathrm{d}v$, v_f is the average free volume per molecule and γ is an "overlap factor" between 1/2 and 1. The diffusion coefficient D can be written in a general form:

$$D = gu \int_{v^*}^{\infty} a(v)p(v)\mathrm{d}v \tag{2.2}$$

where g is a geometric factor, u the kinetic velocity of the gas and $a(v)$ the characteristic jump distance which is about the diameter of the cage. Combining equations (2.1) and (2.2)

$$D = ga^* u \exp\left(-\frac{\gamma v^*}{v}\right)$$

where a^* is approximately the diameter of a molecule. The fluidity Φ and the self-diffusion coefficient D are related by the Stokes-Einstein equation (cf. Chapter 10):

$$D = \frac{kT}{3\pi a_0}\Phi$$

where a_0 is the molecular diameter, hence the expression for Φ and the viscosity; $\eta = 1/\Phi$. This has the form of the Doolittle equation, and the problem is reduced to that of the dependence of the free volume on temperature. This requires a more precise definition of the concept of free volume.

Assuming that the interaction of a molecular pair is described by a Lennard–Jones (L–J) potential, the work $V(R)$ necessary at 0 K to extract a molecule from the center of its cage will depend on the radius R of this cage as shown in Fig. 2.9. The value of $V(R)$ is minimum at R_0 and increases quickly for $R < R_0$. For $R > R_0$, the increase is more gradual and a linear region exists around the point of inflection. The

Fig. 2.9. Theory of Turnbull and Cohen. (*a*) The work $V(R)$ of extraction of a molecule from the center of the cage. (*b*) Internal potential $U(r)$ of a molecule in the interior of a cage. (*c*) Representation of a cage $\overline{\omega}$ for increasing radii R.

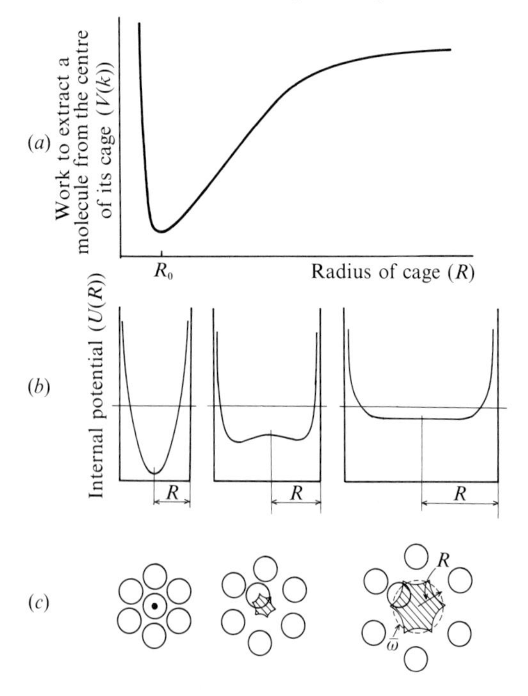

internal potential $U(r)$, where r is the displacement of the molecule from the center of the cage, has been evaluated for different pair interaction laws.[87] For the (L–J) interaction and different values of the radii of the cage R, the results shown schematically in Fig. 2.9 are obtained.

For $R \sim R_0$, $U(r)$ is parabolic for small values of r with a minimum at $r = 0$. For larger values of R, $U(r)$ is less deep and wider. The level of the potential minimum corresponds to $V(R)$ and thus increases with R. A central volume $\overline{\omega}$ is accessible to the center of the molecule. In the central part of $\overline{\omega}$, the difference $U(r) - U(0)$ is small compared to kT and the molecule moves with a velocity comparable to that of the kinetic theory of gases.

In the peripheral part of $\overline{\omega}$, $U(r)$ presents an abrupt increase and the movement will be vibratory. The average value $\overline{R}$ will increase with temperature. For $\overline{R}$ not too far removed from R_0, the total relative expansion $\Delta\overline{v}/v_0 = (\overline{v} - v_0)/v_0$ is proportional to $\overline{R} - R_0$. How is this excess volume $\Delta\overline{v}$ distributed among the different cages? The form of the function $V(R)$ indicates that when $\overline{R} \sim R_0$ the potential energy of the system will be minimal if $\Delta\overline{v}$ is distributed uniformly over all the cages. Any deviation of R involves a significant expense of energy. In contrast, if $\overline{R}$ is close to the zone of inflection of $V(R)$, a non-uniform redistribution can be established without a corresponding increase in energy. The "free volume" v_f of the system is the part of the excess volume which can be redistributed without an energy increase.

Following Turnbull and Cohen,[82] two regions can be distinguished in the thermal expansion of an amorphous phase. In the first, at low temperatures, all the thermal expansion arises from the anharmonicity of the vibratory movement of the molecules. The redistribution of the energy is significant and the added volume tends to be uniformly distributed among the cages:

$$\Delta\overline{v} \sim \Delta v_c \quad \text{and} \quad v_f \sim 0$$

where Δv_c is the the expansion of the crystalline solid.

As the temperature increases, the expansion reaches a value $\Delta\overline{v}_g$ which corresponds to the linear portion of $V(R)$. For $\Delta\overline{v} > \Delta\overline{v}_g$ the second regime appears and the major part of the added volume will be "free" for redistribution. A non-uniform distribution which is random reduces the free energy of the amorphous phase by increasing the configurational entropy of the system. Such a random distribution of free volume is possible in an amorphous phase but not in a crystalline phase. This results in an amorphous phase which can be more stable than a crystalline phase of the same volume when $\Delta\overline{v}$ is slightly greater than $\Delta\overline{v}_g$. Passing beyond a critical value of $\overline{R}$ affects the transport properties and causes an increase in the fluidity and the diffusion coefficient.

Thus the glass transition temperature can be considered as the approximate temperature at which the excess volume attains a value $\Delta\overline{v}_g$ and the free volume v_f begins to appear.

In the relation:

$$\Delta\overline{v} = v_f + \Delta v_c$$

v_f can be replaced by:

$$\Delta\alpha\overline{v}_m (T - T_0)$$

where $\overline{v}_m$ is the average molecular volume and $\Delta\alpha = \alpha_l - \alpha_g$ is the difference in the expansion coefficients of the liquid and the glass.

2.3.2 *The theory of cooperative relaxations*

The concept of "free volume" is difficult to define and, especially in the case of polymers, could be interpreted as a function of the inter and intramolecular interactions as well as of the internal topology. Using statistical mechanical arguments, first Gibbs and Di Marzio[88,89] and then Adams and Gibbs[90] sought to relate the relaxation properties of glass-forming liquids to their quasi-static properties such as the glass transition temperature T_g and the specific heat of the equilibrium glass and liquid.

T_g represents the temperature below which the molecular relaxation times are too long for equilibrium to be established even in the course of the slowest experiments. This increase in the relaxation times is related to the significant reduction in the number of configurations accessible to the system. This is apparent if one considers the small value of the excess entropy near T_g. Gibbs and Di Marzio attempted to resolve the Kauzmann paradox by showing the existence of a second-order transition at a temperature T_2 where the configurational entropy disappears. Below T_2 this entropy remains zero instead of taking negative values which have no significance. According to this theory, the transport properties in the liquids, diffusion and viscosity, are manifestations of *cooperative rearrangements*.

For any given temperature T, there is a minimum sized region which can undergo a cooperative rearrangement and, independently of its environment, transform into a different configuration under the effect of an energy (enthalpy) fluctuation.

For a region containing z molecules, which can be rearranged, the probability of a transition $W(T)$ can be calculated from statistical mechanical considerations. The theory, initially developed for the case of linear polymers and then generalized, shows that

$$W(T) = A \exp\left(-z\Delta\mu/kT\right)$$

where $\Delta\mu$ represents the potential energy per molecule (or monomer unit) opposing the rearrangement. To produce at least one configurational change, the existence of a minimum size, z^*, is necessary, and only clusters containing at least z^* molecules create a non-zero transition probability.

Adam and Gibbs showed that the average transition probability is of the form:

$$\overline{W}(T) = \overline{A} \exp\left(-z^*\Delta\mu/kT\right)$$

where $\overline{A}$ is a new frequency factor approximately independent of temperature. This result shows that the majority of transitions occur in regions which are nearly the same size as the smallest size z^* allowing the transition.

The critical size z^* of the cooperative region can be related to the molar configurational entropy, S_c of the macroscopic specimen by the equation:

$$z^* = Ns_c^*/S_c$$

where N is Avogadro's number and s_c^* is the critical configurational entropy which cannot decrease below $k \ln 2$ (because at least two configurations must be possible for z^*); here k is Boltzmann's constant.

The resulting transition probability is of the form :

$$\overline{W}(T) = \overline{A} \exp\left(-C/TS_c\right)$$

where C is a constant. The equation explicitly shows the dependence of $W(T)$ on the configurational entropy of the glass-forming liquid. It is clear, moreover, that as long as C has a non-zero value, the system cannot attain the situation where $S_c = 0$ in a cooling process of finite time.

For the same reason, the experimental T_g is not equal to the theoretical T_2 but is always higher. This difference increases with C, i.e. with the barrier $\Delta\mu$ opposing the rearrangement. Thus real glasses have a residual configurational entropy that increases along with the difficulty of rearrangement in the structure. The residual entropy can be estimated by calorimetric measurements.

By taking measurements near 0 K and comparing the results for a vitreous specimen structurally frozen at T_g to those for a crystallized specimen, the configurational entropy S_c frozen-in at T_g is determined in the absence of contributions from thermal vibrations:

$$\Delta S_0 = (S_{glass} - S_{crystal})_{T=0} = S_c(T_g)$$

On the other hand, if ΔC_p is the difference between the specific heat of the equilibrium liquid and the glass, the relation is:

$$S_c(T_g) = \int_{T_2}^{T_g} \frac{\Delta C_p}{T} dT$$

Assuming ΔC_p is constant and using $S_c(T_2) = 0$:

$$S_c(T_g) = \Delta C_p \ln \left(T_g / T_2 \right)$$

the ratio T_g/T_2 is calculated from ΔS_0 and ΔC_p. Bestul and Chang[91] made such measurements for twelve different materials and found an average value:

$$T_g/T_2 = 1.29 \pm 10.9\%$$

Such a determination is unambiguous for single-component glasses. In the case of multicomponent glasses, difficulties arise in estimating the residual entropy of mixing which does not vanish even at 0 K in the absence of the total separation of phases.

Without such information, an estimate of the *deviation from ideal behavior* can be obtained from measurements of the transport properties which depend essentially on $\overline{W}(T)$. Because the probability of transition $\overline{W}(T)$ is inversely proportional to the relaxation time, τ, the Adam–Gibbs relation contains the important part of the temperature dependence of all normal mass transport processes. The diffusion coefficient D, the fluidity Φ , the equivalent conductance Λ are all proportional to $\overline{W}(T)$.

By relating the configurational entropy $S_c(T)$ as a function of temperature by the equation:

$$S_c(T) = \Delta C_p \ln \left(T / T_2 \right)$$

where ΔC_p is assumed constant, it can be shown that, for temperatures $\left[(T - T_2)/T_2 \right]^2 << 1$:

$$\overline{W}(T) = A \exp \left(-\frac{B}{T - T_2} \right)$$

where A and B are constants.

Experimentally it is found that $D, \Phi = 1/\eta$ and Λ follow relations of this type, the relation for η being known as the Vogel–Fulcher–Tammann relation (V–F–T). For *ionic liquids*, the value of T_2 from these measurements is close to those obtained from calorimetric measurements.[92]

In the case of oxide glasses, difficulties set in because in the neighborhood of T_g the V–F–T formula applies less well and the behavior follows the Arrhenius type: $A \exp(-B/T)$, (i.e. $T_2 \to 0$). Moreover, the behavior of ΔC_p which plays an important rôle is not well known near T_g.

As noted by Angell,[92] in the classical theories of transport in ionic liquids, the transition probability $W(T)$ is generally a product of two terms:

$$W(T) = A' \exp\left(-\frac{E_1}{kT}\right) \exp\left(-\frac{E_2}{kT}\right)$$

with E_1 being the activation energy necessary for a jump and E_2 that necessary for the formation of a hole, i.e. a favorable configuration. In the Adam–Gibbs theory, the two terms are combined into one which does not permit the separation of E_1 and E_2.

A more complete treatment of the vitreous transition has been proposed by Cohen and Grest using a free volume approach.[93]

2.4 Relaxational behavior in the transformation interval

2.4.1 *The notion of "fictive temperature" – Tool's equations*

The fact that T_g (vitrification point) depends on the cooling rate results in glasses produced by different cooling rates having *different properties*. Figure 2.2 clearly shows that the specific volume of glass increases with the cooling rate. Glasses cooled rapidly are less dense corresponding to the more "open" structure of the liquid at a higher freezing temperature.

The same observation can be made for other properties: rapidly solidified glasses will have a lower index of refraction, a lower viscosity, a higher conductivity, etc.

The properties of solid glass, which is a *non-equilibrium* system, cannot be described by means of the usual thermodynamic variables. It is necessary to add supplementary parameters which can characterize *the state of configuration* (degree of order) of the liquid at the moment of freezing-in. The simplest representation is the use of a *single* supplementary parameter z.

Consider, following Jones,[9] the variation of the volume of an idealized system as a function of temperature at constant pressure (Fig. 2.10). AB

represents the equilibrium curve of a stabilized glass, i.e. one obtained in experiments of unlimited time. Consider three points on this curve a, b and c for which the degree of order is respectively z_1, z_2 and z_3. For experiments carried out in a finite time interval, the liquid follows the path Bbb', the break at point b representing, in an idealized manner, the vitreous transition at temperature T_2. For experiments in a shorter time interval, the path will follow Bcc' with the transition occurring at c corresponding to T_3. The points a, m and n represent the vitreous states at temperature T_1 having respectively the degrees of order z_1, z_2 and z_3. In the field of the vitreous states (to the left of AB), rapid changes in temperature cause a point representing the system to trace lines $z = Constant$ parallel to aa' or bb'. On the line AB or to the right, the state can be modified reversibly in the normal manner.

The parameter z can be taken to be equal to the equilibrium temperature of the liquid at the moment of solidification corresponding to the idealized point T_g on each curve of Fig. 2.10. This configurational parameter will be called the "fictive temperature" of the system. Thus, the glasses in Fig. 2.10 will have the respective fictive temperatures $\overline{T}_1, \overline{T}_2, \overline{T}_3$.

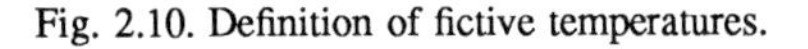
Fig. 2.10. Definition of fictive temperatures.

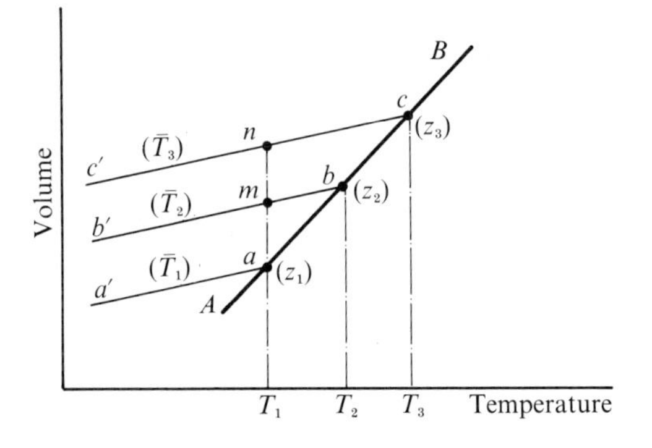

The above formulation permits an explanation of the "anomalies" of the quantities V (or H) observed in glasses subjected to reheating. Consider three glasses formed with respective rates $U_1 < U_2 < U_3$ and reheated at the intermediate rate U_2 (Fig. 2.11(a)). Glass 2, having a fictive temperature $\overline{T}_2$ and reheated at the same rate as its formation rate, is constantly in equilibrium and will show a transition point T_g at T_2. In contrast, glass 1 has the lowest fictive temperature $\overline{T}_1$ and, in the course of softening, will seek to increase its configurational temperature before joining the equilibrium line leading to a trace with a higher slope. Before joining the equilibrium line, glass 3 first attempts to lower its configurational temperature $\overline{T}_3$ which is too high, with the effect of producing the minimum seen on Fig. 2.11(*a*).

The same phenomenon occurs in the derivative quantities α and C_p (Fig. 2.11(*b*)). Glass 2 shows a nearly identical reheat curve while glass 1 gives rise to the formation of a maximum and glass 3 to a minimum.

This phenomenon of "pursuit" of the configurational temperature during reheating is clearly shown by dilatometry experiments (Fig. 2.12). By cutting samples of glasses 2 and 3 to the same length, one is in fact comparing non-equivalent lengths because glass 3 is less dense than glass 2.

If the rate of heating during the dilatometry experiment is the same as the rate of freezing of glass 2, the experiment leads to an unambiguously defined point $T_g(= \overline{T}_2)$. Glass 3, however, leads to a minimum. Each of the two curves terminates in a hook which is an *artifact* of the experimental conditions and corresponds to the flow of the sample under the stress exerted by the spring of the measuring system.

Fig. 2.11. The behavior of glasses having different fictive temperatures: (*a*) volume or enthalpy, (*b*) expansion or specific heat. The hatched portion corresponds to the transformation interval.

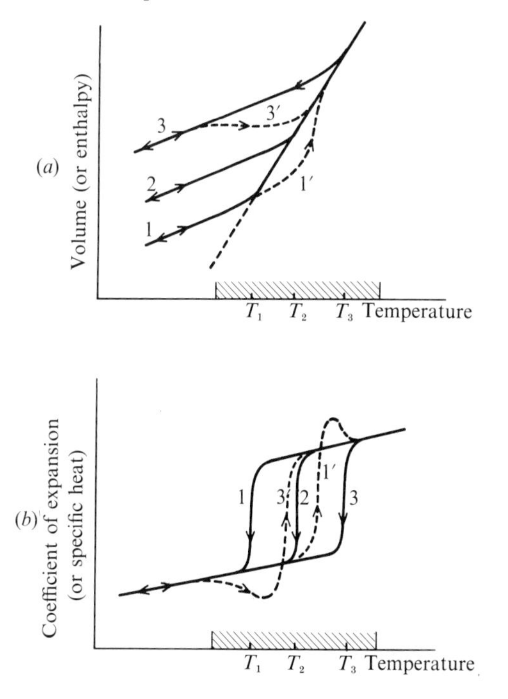

The hook corresponds to a certain standard viscosity characteristic of the softening of the specimen (dilatometric viscosity, cf. Chapter 9) and thus indicates the same state in both glasses. By translating the curves to make the hooks coincide, the actual arrangement of the curves becomes apparent, as indicated schematically in Fig. 2.12.

Tool,[94] who introduced the concept of "fictive temperature" to interpret experiments on glass stabilization, proposed the following empirical expressions:

$$\frac{d\overline{T}}{dt} = K(T - \overline{T})\exp\left(\frac{T}{k}\right)$$

or

$$\frac{d\overline{T}}{dt} = K(T - \overline{T})\exp\left(\frac{T}{g} + \frac{\overline{T}}{h}\right)$$

where K, k, g, and h are constants.

Tool's second formula implies that for a given temperature and for a given deviation from equilibrium, the rate of approach to equilibrium will be greater when $\overline{T} > T$ than when $\overline{T} < T$. It conforms with the experimental results in Fig. 2.6 where the approach from high temperature is faster.

The recent book by Scherer[95] presents a detailed discussion of relaxation phenomena in glasses.

2.4.2 *Thermodynamic treatment*

Davies and Jones[96] developed a thermodynamic treatment using De Donder's formalism (see e.g. Ref. 97) which is applicable to irreversible phenomena.

Fig. 2.12. Thermal expansion curves for specimens corresponding to cases 2 and 3 in Fig. 2.11.

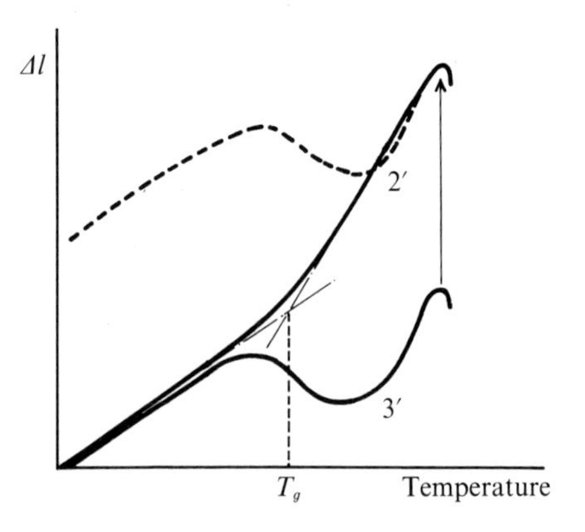

(a) *Fictive temperature*

Starting with the fundamental thermodynamic equation:

$$\mathrm{d}Q = \mathrm{d}U + p\mathrm{d}V = T\mathrm{d}S - A\mathrm{d}z$$

where A is the affinity and z is the order variable, the fictive temperature $\overline{T}$ is defined by the condition:

$$A(p, \overline{T}, z) = 0$$

Rewriting De Donder's theory in terms of $\overline{T}$, and introducing the simplifying linearization that assumes the system is sufficiently close to equilibrium:

$$A(p, \overline{T}, z) \simeq (z - \overline{z}) \left(\frac{\partial A}{\partial z} \right)_{p,T} \simeq (T - \overline{T}) \left(\frac{\partial A}{\partial T} \right)_{p,z}$$

the following general formulae are obtained:

$$T\mathrm{d}S_{irr} \equiv A\mathrm{d}z = -V\Delta\alpha(T - \overline{T})\mathrm{d}p + (\Delta C_p/\overline{T})(T - \overline{T})\mathrm{d}\overline{T}$$

$$\mathrm{d}Q = C_{pg}\mathrm{d}T - VT\alpha\mathrm{d}p + \Delta C_p\mathrm{d}\overline{T}$$

$$\mathrm{d}V/V = \alpha_g\mathrm{d}T - \beta\mathrm{d}p + \Delta\alpha\mathrm{d}\overline{T}$$

in which:

$$\Delta\alpha = \alpha_l - \alpha_g$$

$$\Delta C_p = C_{pl} - C_{pg}$$

designate the differences between the properties of equilibrium liquid (l) and glass (g) and where S_{irr} refers to the irreversible entropy.

The total increase in entropy is:

$$\mathrm{d}S = \mathrm{d}Q/T + A\mathrm{d}z/T = \mathrm{d}S_m + \mathrm{d}S_{irr}$$

For changes at constant pressure:

$$\mathrm{d}S_{irr} = \Delta C_p \left(\frac{1}{\overline{T}} - \frac{1}{T} \right) \mathrm{d}\overline{T}$$

(b) *Fictive pressure*

We defined the fictive temperature by $A(p, \overline{T}, z) = 0$. In the same way we can define the "fictive pressure" $\overline{p}$ by the condition that a glass with the value $\overline{p}$ is in equilibrium when instantaneously brought to $\overline{p}$ at constant temperature.

Analytically:

$$A(\overline{p}, T, z) = 0$$

With the approximation:

$$A(p, T, z) \simeq (p - \overline{p}) \left(\frac{\partial A}{\partial p} \right)_{T,z}$$

the following set of equations is obtained:

$$T\mathrm{d}S_{irr} \equiv A\mathrm{d}z = V\Delta\beta(p - \overline{p})\mathrm{d}\overline{p} - V\Delta\alpha(p - \overline{p})\mathrm{d}T$$

$$\mathrm{d}Q = C_p\mathrm{d}T - VT\alpha_g\mathrm{d}p - VT\Delta\alpha\mathrm{d}\overline{p}$$

$$\mathrm{d}V/V = \alpha\mathrm{d}T - \beta_g\mathrm{d}p - \Delta\beta\mathrm{d}\overline{p}$$

where $\Delta\beta = \beta_l - \beta_g$ designates the difference in the compressibility of liquid and glass.

It can be shown[96] that the fictive temperature $\overline{T}$ and the fictive pressure $\overline{p}$ are related by the equivalent formulae:

$$(T - \overline{T})\Delta\alpha = -(p - \overline{p})\Delta\beta$$

$$(T - \overline{T})\Delta C_p = -(p - \overline{p})TV\Delta\alpha$$

A rapid isobaric change of temperature δT which leaves $\overline{T}$ and $\overline{p}$ unchanged is equivalent to a variation of the pressure δp such that:

$$\delta p = -(\Delta C_p/TV\Delta\alpha)\delta T$$

2.4.3 *The rate of stabilization – volume viscosity*

Using the preceding formulae, the rate of entropy production can be written:

$$\left(\frac{\mathrm{d}S}{\mathrm{d}T} \right)_{irr} = \frac{V(p - \overline{p})}{T} \left[\alpha_g \frac{\mathrm{d}T}{\mathrm{d}t} - \beta_g \frac{\mathrm{d}p}{\mathrm{d}t} - \frac{1}{V} \frac{\mathrm{d}V}{\mathrm{d}t} \right]$$

The expression in the square brackets can be considered as a "displacement" and its coefficient as a "force." As with other irreversible phenomena it is assumed that close to equilibrium the displacement is proportional to the force, which permits the kinetic equation to be written:

$$\alpha_g \frac{\mathrm{d}T}{\mathrm{d}t} - \beta_g \frac{\mathrm{d}p}{\mathrm{d}t} - \frac{1}{V}\frac{\mathrm{d}V}{\mathrm{d}t} = \frac{p-\overline{p}}{\eta_v} = \frac{p'}{\eta_v}$$

where p' is the *effective pressure* and η_v the *volume viscosity* which, close to equilibrium will determine the kinetic properties of glasses.

At constant p and T there is a pure volume relaxation; the defining equation is written taking into account that $\mathrm{d}V/V = -\Delta\beta \mathrm{d}\overline{p}$.

$$\frac{\mathrm{d}\overline{p}}{\mathrm{d}t} = -\frac{p-\overline{p}}{\Delta\beta\eta_v}$$

where

$$\frac{\mathrm{d}p'}{\mathrm{d}t} = -\frac{p'}{\tau}$$

with a relaxation time $\tau = \Delta\beta\eta_v$.

The molecular processes which accompany the changes tending toward equilibrium must be similar to those of normal flow under the action of a shear stress which defines the classical viscosity coefficient (cf. Chapter 9).

It can be shown that the ratio of the viscosity coefficients η_v and η which respectively control volume "flow" and flow in shear is independent of the temperature for a given substance.

In particular, the activation energies of η_v for glycerol (25 kcal mole^{-1}) and for glucose (130 kcal mole^{-1}) are equal to those of η. The ratio η_v/η is of the order of 10 for glycerol and 200 for glucose.[96]

2.4.4 *Prigogine's relations*

The two theories (free volume and configurational entropy) have been applied with some success to the study of glass-forming liquids, but they both have their limitations.

They have been criticized by Goldstein.[98] The free volume theory predicts vitreous transition when the free volume falls to a minimum while the theory of cooperative rearrangements associates T_g with a minimum in the excess entropy.

If the excess free volume and the excess entropy are designated V_e and S_e respectively:

$$V_e = V_l - V_g \quad \text{and} \quad S_e = S_l - S_g$$

(the indices l and g correspond to liquid and glass), the variations of the excess quantities can be expressed by the relations:

$$\mathrm{d}V_e = \left(\frac{\partial V_e}{\partial T}\right)_p \mathrm{d}T + \left(\frac{\partial V_e}{\partial p}\right)_T \mathrm{d}p$$

$$\mathrm{d}S_e = \left(\frac{\partial S_e}{\partial T}\right)_p \mathrm{d}T + \left(\frac{\partial S_e}{\partial p}\right)_T \mathrm{d}p$$

At temperature T_g, the quantities $\mathrm{d}V_e$ and $\mathrm{d}S_e$ are independently zero in the two models, thus:

$$0 = V_g \Delta\alpha \mathrm{d}T_g - V_g \Delta\beta \mathrm{d}$$

$$0 = \frac{\Delta C_p}{T_g} \mathrm{d}T_g - V_g \Delta\alpha \mathrm{d}p$$

If the excess free volume leads to a true description of T_g, it follows that:

$$\frac{\mathrm{d}T_g}{\mathrm{d}p} = \frac{\Delta\beta}{\Delta\alpha} \tag{2.3}$$

while if the excess entropy gives a correct description, then:

$$\frac{\mathrm{d}T_g}{\mathrm{d}p} = \frac{T_g V_g \Delta\alpha}{\Delta C_p} \tag{2.4}$$

In the ideal situation, where S_e and V_e are equivalent descriptions, it then follows:

$$\frac{\Delta\beta}{\Delta\alpha} = \frac{T_g V_g \Delta\alpha}{\Delta C_p}$$

The preceding equations have the same form as Ehrenfest's equations relative to a second-order phase transition. They are presented here assuming that the transition, although kinetic or relaxational in nature, is produced when a parameter which specifies the thermodynamic state attains a critical value.[96]

These equations can be shown[98] to be valid only if the liquid can be described by a *single* order parameter. In this case, the ratio of Prigogine and Defay,

$$R = \frac{\Delta C_p \Delta\beta}{T_g V_g (\Delta\alpha)^2} = 1$$

It is found experimentally that in a large number of cases equation (2.4) is satisfied while for relation (2.3) $\mathrm{d}T_g/\mathrm{d}p < \Delta\beta/\Delta\alpha$, the values being about two times too high. Thus:

$$\frac{\Delta\beta}{\Delta\alpha} \geq \frac{T_g V_g \Delta\alpha}{\Delta C_p}$$

whence $R \geq 1$. This shows that the representation with only one parameter is insufficient and implies the need to introduce a *spectrum* of configurational parameters.[98] (For a more complete discussion, see Ref. 99.)

2.5 Practical laboratory determination of T_g

In practice, dilatometry, Differential Thermal Analysis (DTA) and measurement of electrical conductivity are used to determine the vitreous transition T_g of a glass.

2.5.1 *Dilatometry*

The procedure consists of measuring the expansion of a sample of glass as a function of temperature. It is necessary to use a well-annealed glass to avoid the effects of quenching shown in Fig. 2.12. The intersection of the straight lines corresponding to the expansion of the glass before and after the transition determines T_g. For industrial glasses, the procedure is the subject of French standard (B 30-010) and the German standard (DIN 52-238).

2.5.2 *DTA*

The temperature T is varied linearly as a function of time and the difference, ΔT, between the glass specimen and an inert reference is measured.

On reheating, because of the increase ΔC_p of the heat capacity, an endothermic effect is seen. The temperature T_g corresponds to the inflection point on the curve $\Delta T = f(t)$ (Fig. 2.13(*a*)). (For details and different applications, consult the work of Robredo.[100])

2.5.3 *Electrical conductivity*

The method consists of measuring the resistance of a specimen as a function of the temperature T. The curve $\log R = f\left(1/T\right)$ shows a break which permits the determination of T_g (Fig. 2.13(*b*)). The procedure is the subject of a German standard (DIN 324).

Fig. 2.13. Determination of T_g by: (*a*) DTA, (*b*) electrical conductivity.

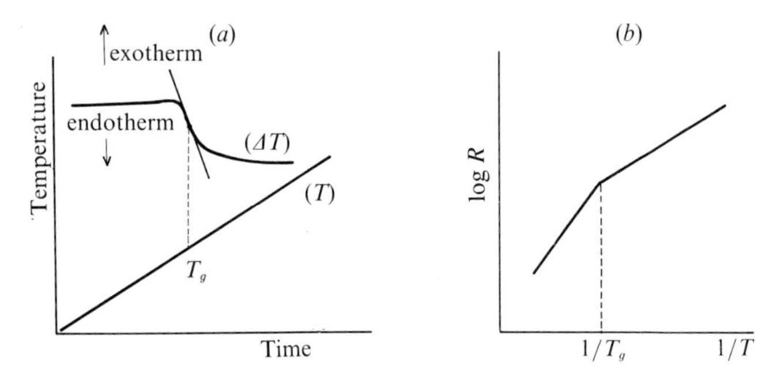

3 Conditions for vitrification

The numerous attempts to explain the formation or the non-formation of glasses can be classified into two approaches: one is based on *structural* considerations, i.e. geometry of the constituent entities of the glass, the bond forces, etc., the other on general *kinetic* considerations which at first neglect the structure.

Historically, the structural approaches were developed first and were the origins of various "criteria for vitrification."

3.1 Structural theories

3.1.1 *Theories based on crystal chemistry concepts*

Given the large diversity of substances which form glasses, it is difficult to find criteria which can be applied equally in every case. Because of their practical importance, oxide glasses have been given particular attention.

(a) *Goldschmidt's criterion (1926)*

In searching for the vitrification conditions for simple oxides with the stoichiometric formula A_mO_n, Goldschmidt[101] thought that the criterion could be the ratio of the ionic radii r_A/r_O of the cation and oxygen. For the glass-forming oxides, this ratio should be between 0.2 and 0.4. Following classical considerations of crystal chemistry for ionic structures, the ratio r_A/r_O is directly related to the coordination number of the central cation (Fig. 3.1). The ratio proposed above thus implies tetrahedral coordination.

(b) *Zachariasen's rules (1932)*

A more complete examination of different cases shows that Goldschmidt's criterion is inadequate: the oxide BeO, for example, satisfies the criterion, but cannot be vitrified. Zachariasen[102] reconsidered the problem and, by empirical reasoning, established a set of rules which have had substantial impact on glass research.

His analysis was based on the following considerations:

1. The interatomic bonding forces in glasses and crystals must be similar given the similar mechanical properties of the two types of solids.
2. Like crystals, glasses consist of an extended three-dimensional "network" but the diffuse character of the X-ray diffraction spectra shows that

the network is not symmetric and periodic as in crystals (i.e. there is no long-range order).

The glass network may be compared to a unique molecule or a system with a giant unit cell. The disorder of the network introduces a distribution of bond forces; their progressive rupture on heating explains the gradual decrease of viscosity. The disorder would also explain the fact that the energy content is higher than that of a crystal.

The structure can be analyzed in terms of coordination polyhedra of cations surrounded by a variable number of oxygen ions (Fig. 3.1). In crystalline oxides, the polyhedra can have common corners, edges or faces.

Zachariasen sought the manner in which the polyhedra could be joined to build a disordered network related to that of a crystal. In the case of the different *crystalline* forms of SiO_2 (quartz, cristobalite, tridymite, etc), the network is built with SiO_4 tetrahedra joined at their corners. In the case of vitreous SiO_2, the network is built with the same SiO_4 units joined by their corners but the mutual orientation of the consecutive tetrahedra is variable.

Figure 3.2 is a schematic illustration of a hypothetical oxide structure A_2O_3 built from triangular AO_3 units. Such a two-dimensional "lace" structure permits the introduction of some disorder without the cations approaching each other too closely which would cause a large increase in internal energy in the disordered glass network. For example, if a crystalline structure were built from the same AO_3 units bonded by their edges (this would be the case of a stoichiometric AO oxide) it would be impossible to introduce disorder without excessive internal energy increase.

Fig. 3.1. Regions of stability of coordination polyhedra with coordination number CN according to the ratio r_c/r_a, of the radii of the cation and anion. After Ref. 29, p. 531.

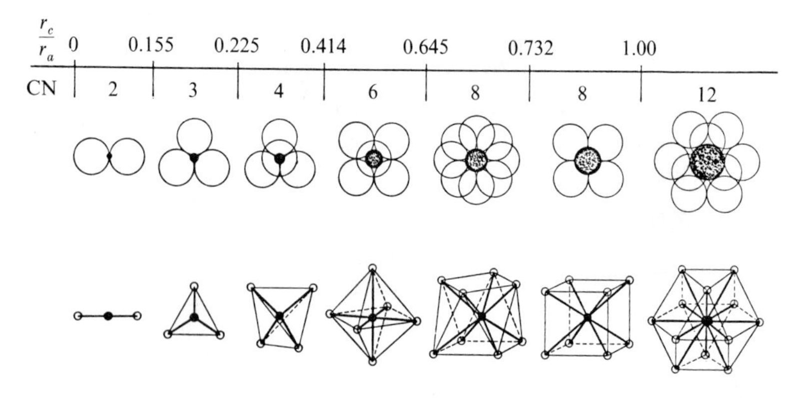

After systematic examination of the structures formed by different coordination polyhedra, Zachariasen showed that a glass-forming oxide must satisfy the following group of rules:

1. The number of oxygens around an A atom must be small.
2. None of the oxygens may be shared by more than two A cations.
3. The polyhedra must bond at corners but not on edges or faces.
4. At least three corners of each polyhedron must be shared with other polyhedra.

Zachariasen next reviewed the possibilities of vitrification according to the stoichiometry of the oxide. Oxides of the formulae A_2O and AO are not able to satisfy the preceding rules and thus are not able to form glasses. In fact, none of the oxides of group I and II elements form glass (except perhaps H_2O).

Rules 2, 3, and 4 are satisfied:

(*a*) in the oxides A_2O_3 when the oxygens form triangles around the A atoms.

(*b*) for the oxides AO_2 and A_2O_5 when the oxygens form a tetrahedra and,

(*c*) for the oxides AO_3 and A_2O_7 when they form octahedra.

Because there were no known examples of glass formation in this group,

Fig. 3.2. Two-dimensional schematic representation of the structure of: (*a*) a hypothetical crystalline compound A_2O_3, (*b*) the vitreous form of the same compound. After Ref. 102.

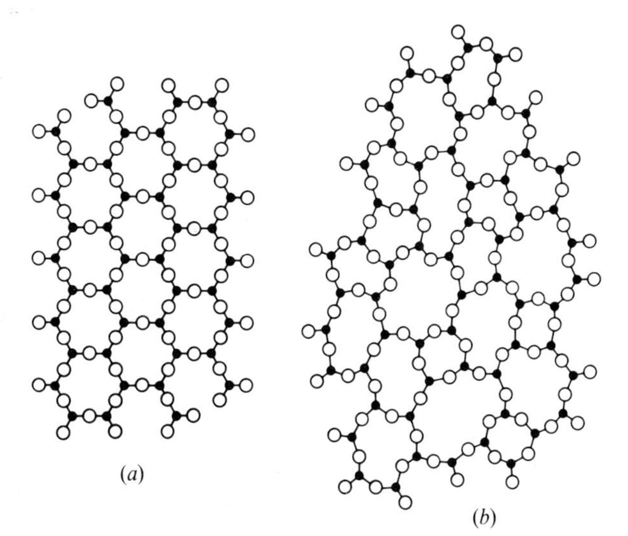

Zachariasen concluded that only triangular and tetrahedral arrangements satisfy the conditions of rule 1 which he made more specific:

1. The number of oxygens around A must be 3 or 4.

Zachariasen systematically examined the coordination properties of cations in different *crystalline* oxides and concluded that only B_2O_3, SiO_2, GeO_2, P_2O_5, As_2O_5, Sb_2O_3, V_2O_5, Sb_2O_5, Nb_2O_5, and Ta_2O_5 were capable of forming glass. At this time, only B_2O_3, SiO_2, GeO_2, P_2O_5, As_2O_5 and As_2O_3 had actually been vitrified. All of these satisfy the rules, the structure of SiO_2 and GeO_2 being based on tetrahedra (AO_4) and B_2O_3 and As_2O_3 on triangles (AO_3).

Applying the same principle to fluorides it is found that only BeF_2 is susceptible to glass formation, with a structure based on tetrahedra.

Zachariasen next addressed the problem of the formation of a more complex oxide glass obtained by the addition of several other oxides, alkali metal, alkaline earth, etc., to oxides of the preceding type. He slightly modified the rules for simple oxides. To form a complex oxide glass it is necessary that:

1. the sample contains a sufficient percentage of cations surrounded by tetrahedra or oxygen triangles.
2. the tetrahedra or triangles have only corners in common.
3. some oxygen atoms are only bonded to two of these cations and do not form new bonds with other cations.

This means that oxide glasses must contain an appreciable proportion of cations able to form vitrified oxides by themselves with the other cations being able to replace them in an isomorphic manner.

The cation Al^{3+}, which replaces Si^{4+} isomorphically can be added to the cation list B^{3+}, Si^{4+}, Ge^{4+}, P^{5+}, As^{5+}, As^{3+}, P^{3+}, Sb^{3+}, V^{5+}, Sb^{5+}, Nb^{5+}, Ta^{5+}, but Al_2O_3 is not able to form a glass by itself.

Zachariasen has given the name *network-forming cations* to the preceding ions which, in association with oxygen, form the "vitreous network" of the glass.

The term *network former* has been adopted for an oxide which belongs to the vitreous network, and *network modifier* for an oxide which does not participate directly in the network. The latter term is justified by the way the oxides behave in the structure as will be examined next.

When a non-glass-forming oxide, such as Na_2O, is added to SiO_2, the additional oxygens participate in the network and cause the rupture of a specific number of bonds. The process represented in Fig. 3.3 is produced by each molecule of Na_2O introduced. An Si–O–Si bond is broken and the added oxygen saturates the unsatisfied bond of one Si and two $Si{-}O^-$ are formed. The two negative charges on the oxygens are compensated by the nearby presence of a pair of Na^+ cations ensuring the electrostatic neutrality of the ensemble. In the course of fusion which leads to glass formation, the primitive network of SiO_2 is thus progressively broken – depolymerized – and the alkali metal cations lodge in the neighborhood of the broken bonds.

The Si–O–Si *bridge rupture* mechanism leads to a loosened network structure with two types of oxygens: an oxygen bonded to two Si is called a *bridging oxygen* and an oxygen bonded to one Si is called *non-bridging*.

The same mechanism applies to the introduction of an oxide of a divalent cation, for example CaO. In this case, a single cation Ca^{2+} is sufficient to compensate the two negative charges of the non-bridging oxygens. Several types of cations can coexist in the structure. The oxides which take this rôle are called network *modifying oxides*. These are essentially alkali metal and alkaline earth oxides.

Certain oxides can function either as glass-formers or as modifiers depending upon the glass compositions involved. They are called *intermediate oxides*. Table 3.1 classifies the principal oxides of practical importance.

The "chemical formula" of an oxide glass is thus A_mB_nO. In this

Fig. 3.3. Rupture of a Si–O–Si bridge by the modifier oxide Na_2O, (*a*) intact SiO_2 network, (*b*) formation of a pair of non-bridging oxygens.

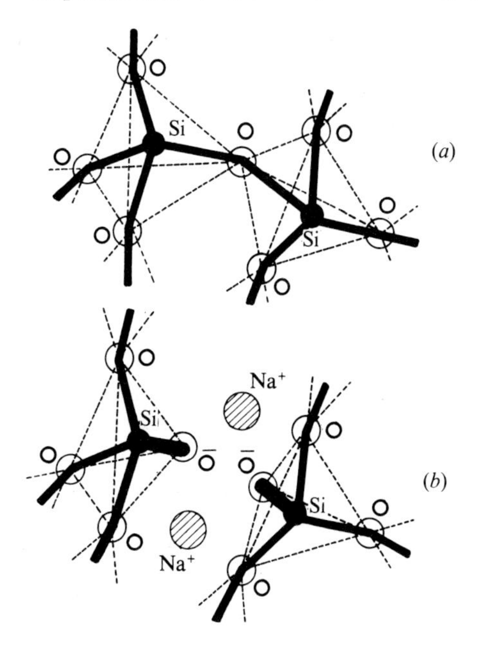

Table 3.1. *Classification of the oxides after Zachariasen.*

Glass-formers	Modifiers	Intermediates
SiO_2	Li_2O	Al_2O_3
GeO_2	Na_2O	PbO
B_2O_3	K_2O	ZnO
P_2O_5	CaO	CdO
As_2O_3	BaO	TiO_2
As_2O_5		
V_2O_5		

formula, m and n are usually not integers and represent simply the number of atoms A (modifiers) and B (glass-formers) per oxygen atom. If at least three corners are shared in a network formed by tetrahedra, n will be between 0.33 and 0.50 in the formula.

Zachariasen further suggested that modifying cations occupy the "holes" which form together with the glass network and that the cations are distributed at random. He proposed his model before work had been done on the structure of glass using X-ray diffraction methods. Shortly afterwards Warren and his collaborators undertook such studies and strongly supported Zachariasen's ideas.[140]

The Zachariasen–Warren model of a continuous disordered network, shown schematically in Fig. 3.4, has dominated glass science for several decades and is still found to be useful. As will be shown later, diffraction methods cannot actually provide definite proof for such a model; they can only confirm that the results do not contradict this hypothesis. Other models assuming a certain kind of aggregation, in particular the *crystallite* model and the *paracrystal* model, are also compatible with the experimental results (cf. Chapter 7).

Another point to keep in mind is that the Zachariasen model was developed specifically for *oxide glasses*, and is not applicable to other types of glasses, e.g. chalcogenides or glasses formed with molecules (e.g. aqueous solutions).

It has since been found that there are numerous exceptions to the Zachariasen model; for instance, oxide glasses based on octahedral coordination (titanate or telluride glasses) are possible. Nevertheless, Zachariasen's hypothesis has prevailed for a long time and to a certain degree had a restricting effect on research by codifying the structural conditions

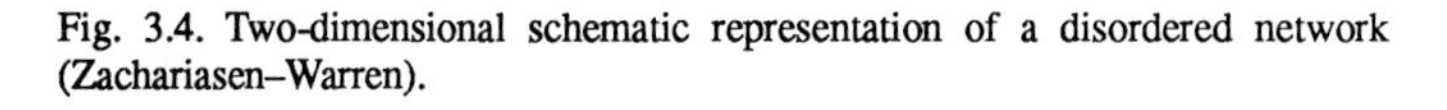
Fig. 3.4. Two-dimensional schematic representation of a disordered network (Zachariasen–Warren).

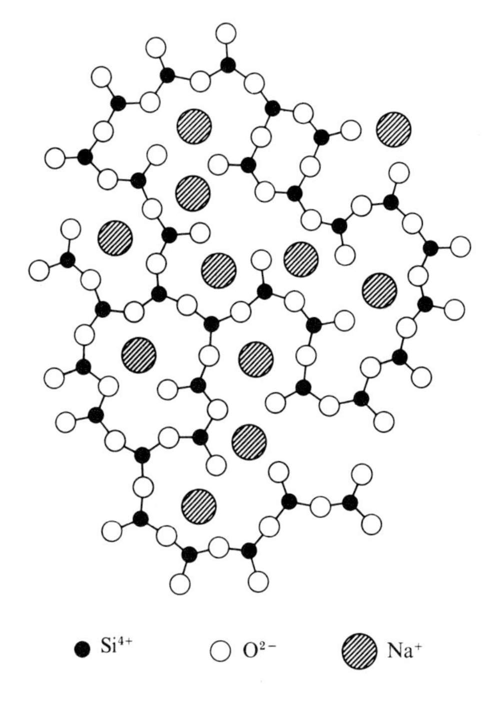

for vitrification in an apparently definitive manner.

Because the preceding model lacks generality, attempts have been made to relate vitrification to the interatomic bonding in the structure.

3.1.2 *Correlations between the ability to vitrify and the type of bond*

A number of semi-empirical rules have been proposed to correlate the ability to vitrify and the nature of the interatomic bonds. A few examples are cited below:

(a) *The necessity of mixed bonds*

Smekal[103] held that the presence of mixed bonds is essential to allow a disordered arrangement. Purely covalent bonds (e.g. the C–C bonds in diamond) with fixed and rigid directions, as well purely ionic or metallic bonds which do not have directional character, cannot lead to disordered structures.

Smekal separated substances which vitrify into three classes:

– inorganic compounds (e.g. SiO_2, B_2O_3) where the bonds A–O are par-

tially ionic and partially covalent.
– elements (e.g. S, Se) forming chain structures with covalent bonds within the chains and van der Waals interactions between the chains.
– organic compounds with intramolecular covalent bonds and intermolecular van der Waals bonds.

According to him, it is also important that no regular arrangements be made through continuous distortion of the chains without having to break valence bonds. This agrees with Hägg's ideas.[104]

(b) *Electronegativity criterion*

For oxides, Stanworth[105] showed that there is a quantitative correlation between the degree of covalency of the A–O bond and the ability of the oxide to vitrify. According to Pauling,[106] the difference in electronegativity values of the two elements allows the ionic percentage of the bond to be estimated (Fig. 3.5). The smaller the difference, the more covalent the character of the bond. For example, the bond Si–O is estimated to be 50% ionic. Taking $x_O = 3.5$ for the electronegativity of oxygen allows the elements commonly found in glass to be classified into three groups as shown in Table 3.2.

The oxides belonging to group 1 are capable of forming glass by themselves. Those of group 3 can form glasses together with those of group 1, and group 2 is intermediate. This agrees with Zachariasen's classification of group 1 being "glass-formers", group 3 "modifiers" and group 2 "intermediates." Using this approach, Stanworth predicted the possibility of glass formation based on TeO_2 as Te has the same electronegativity as phosphorus. Experiments have confirmed this; TeO_2 is not a glass-former,

Fig. 3.5. Estimation of the degree of ionicity of a bond from differences in electronegatvity

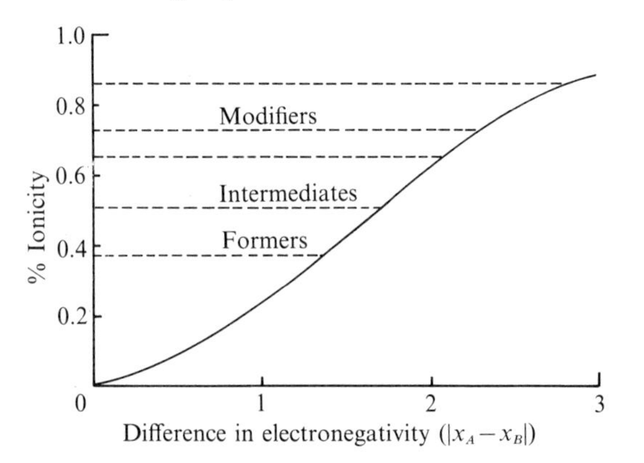

Table 3.2. *Classification of the oxides according to their electronegativity relative to oxygen. After Stanworth, Ref. 105.*

Group 1		Group 2		Group 3	
B	2.0	Be	1.5	Mg	1.2
Si	1.8	Al	1.5	Ca	1.0
P	2.1	Ti	1.6	Sr	1.0
Ge	1.8	Zr	1.6	Ba	0.9
As	2.0	Sn	1.7	Li	1.0
Sb	1.8			Na	0.9
				K	0.8
				Rb	0.8
				Cs	0.7

but glasses have been made in the binary systems TeO_2–Al_2O_3, TeO_2–BaO, etc.

However, this criterion is not without exceptions even in the limited field of oxide glasses. Sb has the same electronegativity as Si whereas the glass Sb_2O_3 is more difficult to obtain than vitreous SiO_2.

Outside the oxides, the correlation is even less satisfactory. For example. the Se–Se bond is purely covalent but Se easily forms a glass. On the other hand Be–F is 80% ionic and BeF_2 vitrifies easily.

3.1.3 *Correlation between the ability to vitrify and bond strength*

(a) *Sun's criterion*

Because the processes of atomic rearrangement in the course of crystallization involve the rupture of bonds, Sun[107] attempted to establish a correlation between the strength of these bonds and the ability of the oxide to vitrify.

The "strength" of the bond M–O in an oxide MO_x was calculated by dividing the energy, E_d, to dissociate the crystalline oxide into its elements in the gaseous state by the number of oxygen atoms surrounding the atom M in the crystal (or in the glass), i.e. by its coordination number, Z. Values for E_d have been obtained from spectroscopic measurements.

Calculated "bond strengths" are given in Table 3.3. Note that the "glass-forming" oxides have "strengths" greater than 90 kcal mole^{-1} and the "modifiers" strengths less than 60 kcal mole^{-1}. The "intermediates" have strengths of 60–73 kcal mole^{-1}.

Table 3.3. *Bond strengths of oxides. After Sun, Ref. 107.*

M in MO_x	Valence	Energy E_d of dissociation for a MO_x unit (kcal mole^{-1})	Coordination number Z	Bond strength $B_{M-O} = E_d/Z$ (kcal mole^{-1})
Glass-formers				
B	3	356	3	119
Si	4	424	4	106
Ge	4	431 (?)	4	108
Al	3	402–317	4	101–79
B	3	356	4	89
P	5	442	4	88–111
V	5	449	4	90–112
As	5	349	4	70–87
Sb	5	339	4	68–85
Zr	4	485	6	81
Intermediates				
Ti	4	435	6	73
Zn	2	144	"2"	72
Pb	2	145	"2"	73
Al	3	317–402	6	53–67
Th	4	516	8	64
Be	2	250	4	63
Zr	4	485	8	61
Cd	2	119	"2"	60

Continued on next page

When making use of this criterion, it is necessary that all the bonds be of the same type. If it is applied to CO_2 for example, although the C–O bond strength is 120 kcal mole^{-1}, molecules of CO_2 do not form a glass. The strength of the C–O bond does not compensate for the weak intermolecular van der Waals forces.

(b) *Rawson's criterion*

Rawson[108] improved Sun's criterion by noting that relating the ease of vitrification to the breaking of bonds at the melting temperature

Table 3.3. Continued: *Bond strengths of oxides. After Sun, Ref. 107.*

M in MO_x	Valence	Energy E_d of dissociation for a MO_x unit (kcal mole^{-1})	Coordination number Z	Bond strength $B_{M-O} = E_d/Z$ (kcal mole^{-1})
Modifiers				
Sc	3	362	6	60
La	3	406	7	58
Y	3	399	8	50
Sn	4	278	6	46
Ga	3	267	6	45
In	3	259	6	43
Th	4	516	12	43
Pb	4	232	6	39
Mg	2	222	6	37
Li	1	144	4	36
Pb	2	145	4	36
Zn	2	144	4	36
Ba	2	260	8	33
Ca	2	257	8	32
Sr	2	256	8	32
Cd	2	119	4	30
Na	1	120	6	20
K	1	115	9	13
Rb	1	115	10	12
Hg	2	68	6	11
Cs	1	114	12	10

must include not only consideration of the strength of the bond, but also the thermal energy available for bond breaking. The melting temperature (or the liquidus temperature) indicates the quantity of available energy. Rawson suggested the parameter obtained by dividing the energy of the bond by the melting temperature, expressed in degrees Kelvin, as the criterion for glass formation.

Table 3.4 contains the values of the Rawson parameter. This new classification gives a better separation for a glass-former such as SiO_2 for

Table 3.4. *Classification of oxides. After Rawson, Ref. 108.*

Oxide	Bond strength B_{M-O}(kcal mole^{-1})	Melting point T_m(K)	Rawson parameter B_{M-O}/T_m
B_2O_3	119 or 89	723	0.164 or 0.122
SiO_2	106	1993	0.053
GeO_2	108	1388	0.078
P_2O_5	88–111	843	0.104–0.131
V_2O_5	90–112	943	0.095–0.119
TiO_2	73	2123	0.034
ZrO_2	81	2923	0.023
MoO_3	92	1068	0.086
WoO_3	103	1748	0.059
TeO_2	68	1006	0.067
MgO	37	2913	0.013
CaO	32	2773	0.011
BaO	33	2193	0.015

which the bond strength according to Sun is 106 kcal mole^{-1} and ZrO_2 with 81 kcal mole^{-1} but which , due to its high melting point, has a lower Rawson parameter and does not form a glass. Note in particular the high value of this parameter for B_2O_3 which is very difficult to crystallize.

As with Sun's criterion, the Rawson criterion has numerous exceptions.

3.2 Kinetic theories

We have seen in the preceding sections that a supercooled liquid can produce a glass if crystallization is avoided. The quest for the conditions of vitrification thus becomes that for non-crystallization, and kinetic theories of phase transformation can be applied.

3.2.1 *The mechanism of crystallization*

The crystallization of a homogeneous phase, liquid or glass, is not a transformation which is produced throughout the whole volume at once: it begins and extends progressively from *discrete centers* distributed throughout the mass.

The following stages can be distinguished:

(a) *Nucleation*

In this stage the agglomerations which can serve as the starting points for the development of ordered regions are formed. These agglomerations or *embryos*, which form and disappear according to structural fluctuations produced by thermal agitation, have different and constantly fluctuating sizes. We will show that such an embryo must attain a certain critical size to be able to serve as a starting point for the development of the new (crystalline) phase; i.e. to constitute a *nucleus*.

Nucleation which occurs in a totally random manner throughout the entire system is said to be *homogeneous*. The necessary condition for this to happen is that all volume elements of the initial phase must be structurally, chemically and energetically identical.

This is, of course, only possible if the entire volume of the phase is chemically homogeneous and devoid of structural imperfections. In practice this is difficult to achieve: the surface itself already constitutes an inevitable imperfection, and foreign particles (impurities, etc.) may also be present. In this case, the energy necessary for the formation of a nucleus is lowered for these sites, and the nucleation will preferentially occur at the interface. This process is called *heterogeneous nucleation*.

In practice, heterogeneous nucleation is difficult to avoid and it is questionable if totally homogeneous nucleation is attainable. Nevertheless, it is an ideal case which will be considered first.

(b) *Crystal growth*

An embryo that has become a nucleus proceeds to grow by the successive addition of atoms from the liquid phase. This leads to the formation of a crystalline particle which grows at a certain rate at the expense of the surrounding phase.

3.2.2 *General kinetic conditions for vitrification*

The number of nuclei, I, produced in a unit volume per unit time is called the *nucleation rate*, and the speed u with which these particles grow is called *growth rate*. Both depend on temperature as shown schematically in Fig. 3.6. The form of the dependence will be calculated more exactly in later sections.

For a liquid to form a glass, it must be cooled fairly rapidly to avoid crystallization. Above the temperature of fusion, T_f, the liquid is the stable phase. When the liquid is supercooled below T_f, crystal growth is theoretically possible between T_f and T_3. However, the initial nuclei formation, necessary before growth is possible, occurs between T_2 and T_4.

The critical region is thus between T_2 and T_3, and the possibility for crystallization will depend on the manner in which the curves overlap and also on the absolute values of the rates in the region of overlap. If, in the common interval T_2–T_3, I or u, or both, are too small, no perceptible crystallization will occur and the system will pass to a vitreous state. If both rates I and u are high (and there is substantial overlap of the curves), total crystallization cannot be avoided. If, in the interval T_2–T_3, I is low and u is high, the crystallization will lead to a small number of crystals distributed in a vitreous phase. However, if I is high and u is low, the result will be a partially crystalline material with very fine grains.

3.2.3 *The classical theory of homogeneous nucleation*

(a) *Thermodynamic barrier to nucleation*

At temperatures where there is appreciable atomic mobility, there is a continual rearrangement of the atoms through thermal agitation. If the phase is thermodynamically unstable, these domains of rearrangement have a temporary existence then they are destroyed and replaced by others. When the phase is metastable, such fluctuations are potential sources of a stable phase and can become permanent. The fluctuations may differ in size, form, structure or composition.

In the simplest classical model, proposed by Volmer and Weber,[109] and Becker and Döring,[110] it is assumed that the embryos have uniform structure, composition and properties identical to those of the future phase and differ only in *form* and *size*.

The form taken is that which will result in the minimum formation energy which is intimately connected with the nature of the interface. If we

Fig. 3.6. Variations of the rate of nucleation I and the rate of growth u as a function of temperature.

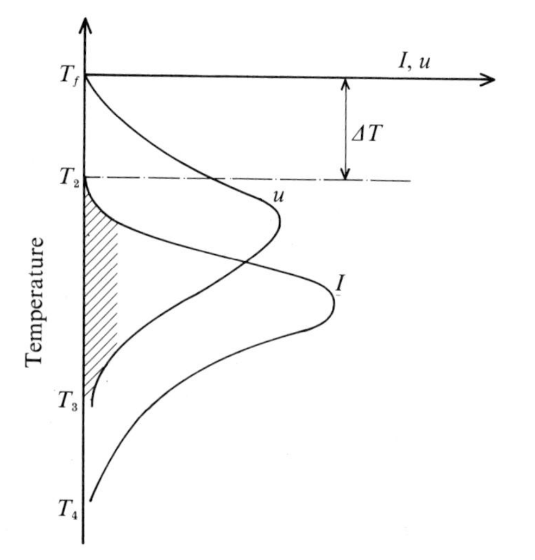

assume as a first approximation that the interfacial energy is independent of the crystallographic orientation and that the energy due to elastic deformation is negligible, the embryos will have a spherical form. The size of the embryos is a function of the conditions of thermodynamic stability.

Figure 3.7 shows the molar free energy G of the liquid and crystalline phases. The curves intersect at the temperature of fusion T_f which represents the equilibrium between the phases. At this point, the difference in free energy G between the two phases is zero. The portions of the curves corresponding to the stable liquid and crystalline phases are indicated by solid lines. The metastable phases, supercooled liquid and superheated crystal are indicated by dashed lines.

Fig. 3.7. Molar free energy of the crystal (c) and the liquid (l) in the neighborhood of the melting point T_f. The portion l' of the curve corresponds to a supercooled liquid and c' to a superheated crystal.

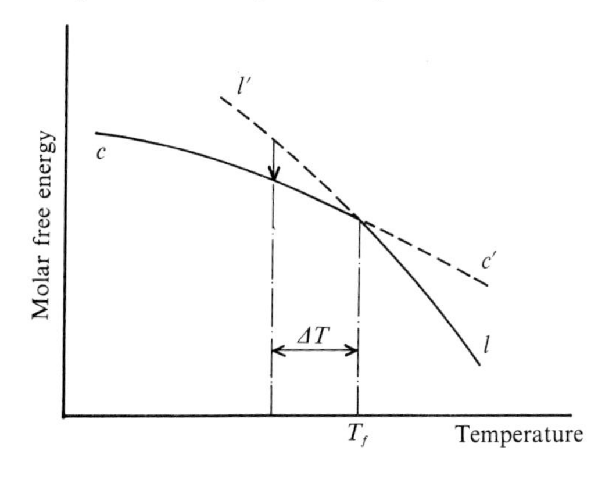

For a temperature $T < T_f$ or a *degree of supercooling* $\Delta T = T_f - T$, the supercooled liquid is in metastable equilibrium with respect to the crystal. The transformation from the supercooled liquid to the crystal is accompanied by a negative free energy change Δg_v per unit volume. For a spherical embryo of radius r, this corresponds to the *liberation* of a quantity of energy equal to $\frac{4}{3}\pi r^3 \Delta g_v$.

However, the formation of an embryo is accompanied by the creation of an interface which has an energy Δg_s per unit area (which one may take, as a first approximation, equal to the macroscopic surface energy γ). For an embryo of radius r, an expenditure of energy to create the interface is equal to $4\pi r^2 \Delta g_s$.

The total energy Δg_r required for the formation of a nucleus of radius

Fig. 3.8. Variation of the free energy Δg_r for the formation of a nucleus as a function of the radius r of the nucleus. The dashed curves correspond to different values of the reduced supercooling ΔT_r.

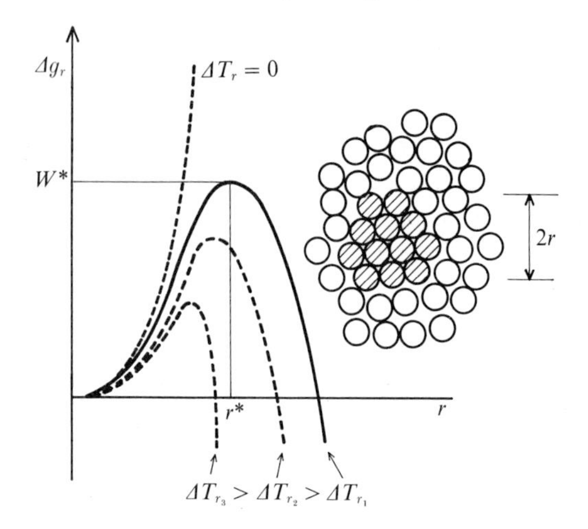

r is given by the formula:

$$\Delta g_r = \frac{4}{3}\pi r^3 \Delta g_v + 4\pi r^2 \Delta g_s$$

Figure 3.8 shows the form of the variations of Δg_r with r. For small nuclei, the surface term (in r^2) predominates, and Δg_r increases, whereas for large nuclei the negative volume term (in r^3) predominates. Between these two situations, there is a maximum corresponding to a critical nucleus size r^* : any size increase beyond r^* is accompanied by a decrease in Δg_r, and any fluctuation having passed this state would have a strong possibility of growth. Therefore, r^* indicates the size of the embryo beyond which the embryo may become a nucleus where r^* is defined by the condition:

$$\frac{\partial \Delta g_r}{\partial r} = 0$$

which gives:

$$r^* = -\frac{2\Delta g_s}{\Delta g_v}$$

Corresponding to the size r^* is a critical value W^* of Δg_r equal to:

$$W^* = \frac{16\pi}{3}\frac{\Delta g_s^3}{\Delta g_v^2}$$

An embryo contains $\frac{4}{3}\pi r^3 N/V_m$ molecules, V_m being the molar volume of the crystal and N being Avogadro's number, which gives in the case of a critical nucleus the number of molecules, n_c^*.

$$n_c^* = \frac{32\pi}{3}\left(\frac{\Delta g_s}{\Delta g_v}\right)^3 \frac{N}{V_m}$$

To see how r^* and W^* depend on temperature, one makes the approximation that the free enthalpies of the solid and liquid vary linearly in the neighborhood of the melting temperature T_f.

At any temperature, by definition:

$$\Delta G = \Delta H - T\Delta S$$

ΔH and ΔS being the differences in the molar enthalpy and entropy between the solid and the liquid. By treating ΔH and ΔS as independent of temperature in the neighborhood of T_f, ΔG may be evaluated, since at T_f :

$$0 = \Delta H_f - T_f \Delta S_f$$

Therefore

$$\Delta G = \frac{\Delta H_f}{T_f}(T_f - T) = \Delta S_f \Delta T$$

where ΔH_f and ΔS_f are respectively the enthalpy and entropy of fusion and ΔT is the degree of supercooling, $\Delta T = T_f - T$. Under these conditions:

$$\Delta g_v = \frac{\Delta S_f \Delta T}{V_m} = \frac{\Delta H_f}{T_f}\frac{\Delta T}{V_m}$$

For a larger degree of supercooling ΔT it may be shown[111] that a better approximation is obtained by taking:

$$\Delta G = \frac{T}{T_f}\Delta S_f \Delta T$$

In these conditions:

$$\Delta g_v = \frac{\Delta H_f}{T_f}\frac{\Delta T}{V_m}\left(\frac{T}{T_f}\right)$$

However, the evaluation of the surface energy term poses problems.

Turnbull[112] showed that the energy Δg_s relating to the liquid–crystal interface can be connected with ΔH_f by the relation:

$$\Delta g_s = \alpha \Delta H_f N^{-1/3} V_m^{-2/3}$$

where α is a dimensionless parameter for a particular type of fluid. For metals, $\alpha \sim 1/2$; for more complex materials $\alpha \sim 1/3$. The value of α generally lies within an approximate interval of $1/4 < \alpha < 1/2$. Physically, α represents the number of monomolecular layers per unit of crystal surface which would be melted at T_f by free energy equal to Δg_s.

One then introduces a second dimensionless parameter β by the relation:

$$\beta = \frac{\Delta S_f}{R} = \frac{\Delta H_f}{RT_f}$$

where R is the gas constant; β lies between 1 and 10 with the lower values corresponding to monatomic liquids. More complex structures have larger entropies of fusion with β approaching 10.

To evaluate the influence of temperature, it is convenient to introduce the *reduced temperature* $T_r = T/T_f$ and the *reduced supercooling* $\Delta T_r = \Delta T/T_f = 1 - T_r$. This permits r^* and W^* to be expressed by the following formulae (valid for small supercooling):

$$r^* = \frac{2\alpha V_m^{1/3}}{N^{1/3}\Delta T_r}$$

and

$$W^* = \frac{16\pi R}{3N} \frac{\alpha^3 \beta T_f}{(\Delta T_r)^2}$$

Figure 3.8 shows the form of the function Δg_r for different degrees of supercooling. At equilibrium, for $\Delta T_r = 0$, the curve does not have a maximum; r^* is infinite, as is W^*. As ΔT_r increases, the critical radius diminishes and nucleation becomes possible.

(b) *Nucleation rate in steady-state conditions*

The minimum energy W^* required to form a stable nucleus is called the thermodynamic barrier to nucleation.

To evaluate the rate of homogeneous nucleation, it is assumed that the critical nuclei are formed by random molecular fluctuations. Assuming that the size distribution and the formation of the nuclei are governed by statistical fluctuations, the probability that a fluctuation would be critical is proportional to $\exp(-W^*/kT)$, and, at equilibrium, the number of nuclei per unit volume would be approximately equal to $n\exp(-W^*/kT)$ with $n = N/V_m$.

The growth of a nucleus results from the addition of one or several units through a diffusion mechanism. If the jump across the nucleus–matrix interface is controlled by a free energy of activation, $\Delta G'$, the rate of nucleus formation I (the number of nuclei formed per second per unit volume) is given by an expression of the form:

$$I = K\exp\left(-\frac{W^*N}{RT}\right)\exp\left(-\frac{\Delta G'}{RT}\right)$$

where $\Delta G'$ represents the *kinetic barrier* to nucleation and K is a constant.

Absolute reaction rate theory permits the calculation of K. Thus Turnbull and Fisher[113] found:

$$K = \left[\left(\frac{4r^{*2}\Delta g_s}{9kT}\right)^2 n^*\right] n\nu$$

in which n^* is the number of atoms on the surface of a critical nucleus, n is the number of atoms or molecules per cubic centimeter of the precipitating phase, and ν represents the fundamental vibration frequency. In most of the situations encountered in practice, the term between the square brackets is about unity, so that:

$$I \simeq n\nu\exp\left(-\frac{\Delta G' + NW^*}{RT}\right)$$

The process of transport across the interface can be described by a diffusion coefficient D_n (cf. Chapter 10):

$$D_n = a_0^2\nu\exp\left(-\frac{\Delta G'}{RT}\right)$$

where a_0 is the molecular jump distance.

Combining the last two expressions gives:

$$I \simeq \frac{nD_n}{a_0^2}\exp\left(-\frac{NW^*}{RT}\right)$$

For the evaluation of the diffusion constant D_n, two cases may be distinguished:

1. Non-reconstructive transformations. This is the case where liquids crystallize without change of composition, e.g. systems in which there is only one component. All the necessary structural units are always present in the neighborhood of the interface, and the process of long-range diffusion is not necessary. The activation energy, $\Delta G'$, is then of the same order as that for viscous flow where similar processes of displacement and molecular orientation take place. D_n may be taken as equal to the coefficient of self-diffusion D for a supercooled liquid, which is related to the viscosity coefficient by the Stokes–Einstein equation (cf. Chapter 9):

$$D = \frac{kT}{3\pi a_0 \eta}$$

where a_0 is the diameter of the diffusing species which is of the same order of size as the preceding jump distance.

2. Reconstructive transformations. In systems with several components, crystallization often requires rupture of bonds and long-range diffusion processes. This is called a *reconstructive* transformation. In the case of SiO_2, for example, the Si–O bonds must be broken before the structural rearrangement. Since this must also happen in the processes of viscous flow and self-diffusion, the activation energy of these processes can be substituted for $\Delta G'$ and the Stokes–Einstein equation may be applied. However, in the case of large changes of composition, the long-range diffusion processes are necessary to carry the elements to the interface in the required proportions. In this case, it is the diffusion of the least mobile species which limits the speed, and $\Delta G'$ is related to the activation energy for diffusion of this species.

(c) *Temperature dependence*

In the case of small values of supercooling the nucleation rate may be written as

$$I = \frac{K_n}{\eta} \exp\left[-\frac{16\pi\alpha^3\beta}{3T_r(\Delta T_r)^2}\right]$$

where η can be expressed in the Vogel–Fulcher–Tammann (VFT) form already considered, and K_n is a constant.

Figure 3.6 shows how the nucleation rate I depends on temperature. For small supercoolings, ΔT, the barrier W^* is high and I is very small. I then increases rapidly to the temperature at which $W^* \sim \Delta G'$, which corresponds to a maximum nucleation rate. For lower temperatures where $\Delta G' \gg W^*$, the diffusion phenomena dominate and I is progressively reduced. Note in Fig. 3.6 the metastable zone immediately below T_f where I is very small.

(d) *Dependence on the nature of the liquid*

Figure 3.9 after Turnbull,[114] shows the form of the curves I as a function of reduced temperature T_r calculated for different values of the product $\alpha\beta^{1/3}$. To give an upper limit to I, η was taken equal to 10^{-2} P and independent of the temperature. The pre-exponential coefficient, K_n, was taken equal to 10^{30} dyn cm (10^{23} N m).

For small undercoolings, ($T_r \sim 1$) where the theory is the most valid, I is negligible which makes verification of the theory impossible close to equilibrium. I then increases rapidly towards a flat maximum which extends to $T_r \sim 0.3$ and then falls to zero for 0 K.

For liquids of normal density in which $n \sim 10^{21}$ cm^{-3} and the fundamental frequency is about 10^{11} s^{-1}, the normal values of $\alpha\beta^{1/3} \sim$ 0.5–0.25 lead to a high nucleation rate of the order of 10^{25} cm^{-3} s^{-1} over a large temperature interval so that nucleation cannot be avoided even for high rates of cooling.

In contrast, substances with high surface energy or large entropy of fusion, $\alpha\beta^{1/3}$ can reach values as high as 0.9 ($\log I < 0$) and homogeneous nucleation is never observed in practice. To avoid the formation of glasses, nuclei have to be introduced into such fluids.

Fig. 3.9. Rate of nucleation I as a function of the reduced temperature T_r. The vertical line aa' is at the limit of detection. After Ref. 114.

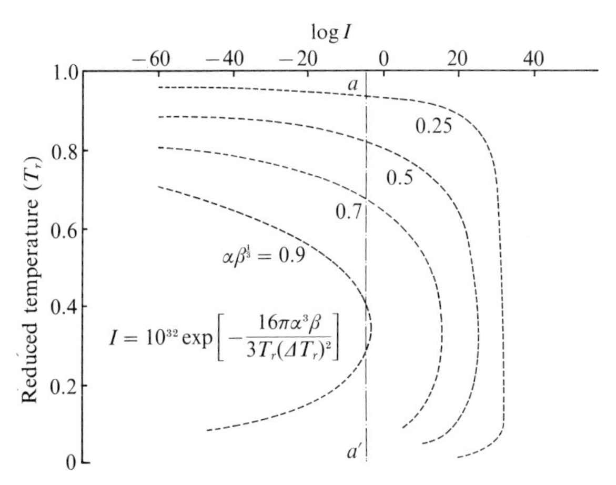

Figure 3.10 shows the form of I when the variation of viscosity is taken into account. $\alpha\beta^{1/3}$ is taken equal to 0.5 and the viscosity is assumed to follow the VFT form where T_g is close to T_0 (cf. Chapter 2).

The curves drawn for different hypothetical values of the reduced transition temperature $T_{rg} = T_g/T_f = T_0/T_f$ show that the tendency for nucleation decreases as T_{rg} increases. Liquids with T_g higher than $\frac{2}{3}T_f$ would have only a limited ability to nucleate and only in a limited temperature range. They would thus easily form a glass.

On the other hand, if for instance $T_g \sim T_f/2$, the nucleation will only be avoided within very small volumes. The calculation gives a maximum diameter of droplets of $\sim$ 60 μm for cooling rates of the order of 10^6 K s^{-1} ("splat cooling").

3.2.4 *Nucleation in the transient state*

The possibility of a supercooled liquid forming a glass depends upon the speed with which the system can pass through the zone in which both nucleation and growth occur (cf. Fig. 3.6).

For high cooling rates, the time at each temperature may be insufficient for the establishment of the Boltzman distribution of embryos, i.e. the actual rate of nucleation may not be the same as that calculated earlier for equilibrium conditions.

The problem of nucleation in a transient state has been treated by Hillig[115] and Hammel[116] who obtained an equation of the type:

$$I_t = I \exp\left(-\tau/t\right)$$

Fig. 3.10. Rate of nucleation I as a function of the reduced temperature T_r for different values of the reduced transition temperature T_{rg}. The line aa' is the limit of detection. After Ref. 114.

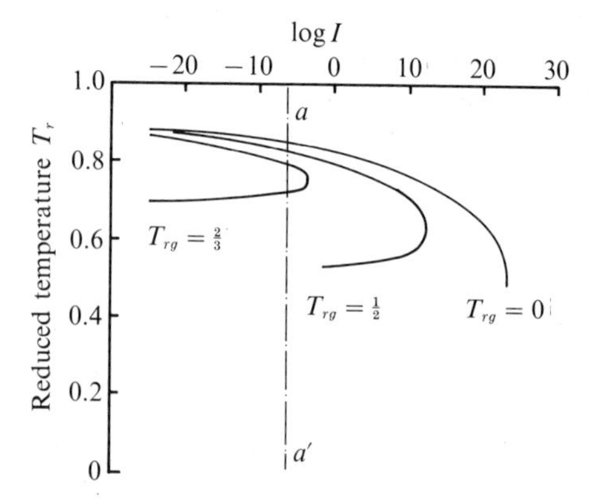

where I_t is the rate of transient nucleation at time t and I the rate of nucleation in the steady state.

The time constant τ given by Collins[117] for the ideal case of instantaneous cooling from T_f to T is:

$$\tau = \frac{144kTT_f V_m^{1/3}\alpha}{a_0^2 \Delta S_f (\Delta T)^2 N^{1/3} D}$$

where the diffusion coefficient D is given by the Stokes–Einstein equation.

Hillig[115] used random walk theory to calculate the mean time $\bar{t}$ for the formation of a nucleus. According to this approach:

$$\tau = \bar{t} = \frac{\pi}{4D}\left(\frac{V_L}{V_m}\right)^2\left(\frac{r^*}{X}\right)^2$$

where V_L and V_m are the molar volumes of the liquid and the precipitating phase, D is the appropriate diffusion coefficient and X is the mole fraction of the precipitated phase. (For crystallization without change in composition, $X = 1$.) The time constant from Hillig's equation does not take into account the instability of the embryos and $\bar{t}$ calculated in this manner is less than τ given by Hammel.

Figure 3.11 shows in a qualitative way the effect of these transient phenomena on the nucleation. The rate of nucleation in the temperature zone between T_f and T_g corresponds to specimens cooled instantly from above T_f to the given temperature and maintained for the indicated time t_i. The different curves show the sharp reduction of the nucleation rate at low temperatures. Their envelope corresponds to the steady state (where $t \to \infty$). We see that longer and longer times are required to reach steady-state conditions as the temperature decreases.

This figure shows that cooling from T_f to T_1 does not produce any nucleation, regardless of the cooling rate. Some nucleation is possible between T_1 and T_2. To avoid nucleation, the region T_1–T_2 must be traversed in a time less than t_2, etc. If the material could be quenched to T_6 without crystallization, it would be necessary to maintain it for a time t_6 to produce detectable nucleation. There is no risk of detectable nucleation beyond the temperature T_∞.

3.2.5 *Heterogeneous nucleation*

Other surfaces present in the liquid, e.g. the container walls or foreign particles, reduce the energy barrier for nucleation W^* by reducing the surface energy between the liquid and the solid.

The complete calculation is complicated because of variation in interfacial energies and the surface geometry of the embryo. Following the

Fig. 3.11. Influence of transient effects on the rate of nucleation (schematic). The line aa' is the limit of detection.

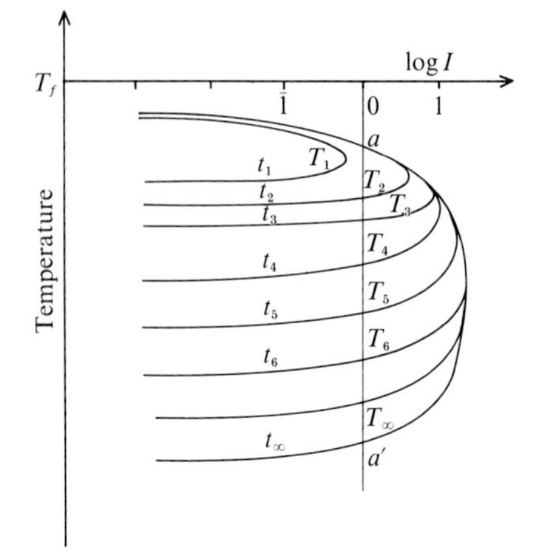

usual practice we will take a simplified treatment. Assume that a crystalline nucleus in the shape of a spherical segment C (Fig. 3.12) forms from the liquid L in contact with a solid impurity surface P suspended in L. In addition to considering the creation of the crystal–liquid interface for which the interfacial energy is $\gamma_{CL} = \Delta g_s$, it is necessary to consider the interfacial energies of the solid–impurity interface γ_{CP} and the liquid–impurity interface γ_{LP}.

If θ is the contact angle (wetting angle), then, at equilibrium:

$$\gamma_{LP} = \gamma_{CP} + \gamma_{CL} \cos\theta$$

Using the calculation for the formation of a critical spherical nucleus by homogeneous nucleation and applying it to the present case, leads to the following result:

$$W^*_{heter} = W^* f(\theta)$$

where $f(\theta)$ is a function which depends on the contact angle θ (Fig. 3.12):

$$f(\theta) = \frac{(2 + \cos\theta)(1 - \cos\theta)^2}{4}$$

The thermodynamic barrier, W^*_{heter} is thus $< W^*$. The barrier reduction depends on the value of θ, i.e. the degree of *wetting* of surface of the impurity by the solid which precipitates in the presence of the liquid. For complete wetting ($\theta \sim 0$), $W^*_{heter} \rightarrow 0$. This results in a reduction in the size of a critical nucleus and a significant reduction in the region of supercooling.

The treatment is further complicated by the fact that the nucleation rate must also be affected by the concentration of pre-existing heterogeneous sites.

These considerations have great practical importance in catalyzed nucleation which leads to glass-ceramics (cf. Chapter 16)

3.2.6 *Kinetics of crystal growth*

Stable nuclei formed either by homogeneous or heterogeneous nucleation grow at a rate which depends both on the speed with which the required atoms are able to diffuse towards the crystal surface and the manner in which they cross the interface. The growth may be limited either by the interface or by diffusion. The first case is normally found for single-component systems in which only short-range rearrangement is necessary. The second case applies when the crystallization is accompanied by significant change in composition, as, for example, in systems containing several components.

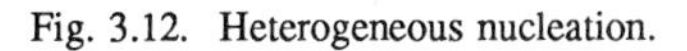

Fig. 3.12. Heterogeneous nucleation.

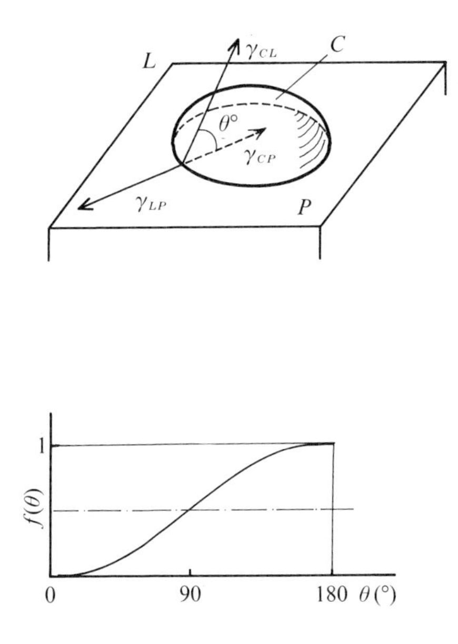

(a) *Growth rate limited by the interface*

The following treatment is after Turnbull.[118]

The crystal–liquid interface may be represented by two potential wells separated by the jump distance a_0 (Fig. 3.13). The change in molar free energy corresponding to the crystallization is:

$$\Delta G_v = G_c - G_l \qquad (\Delta G_v < 0)$$

Consider the thermally activated process for the transfer of atoms in the direction liquid → crystal and the inverse process, crystal → liquid. In the first case the free energy barrier to be overcome is $\Delta G''$ and in the second case $|\Delta G_v| + \Delta G''$.

The net growth rate u will be proportional to the difference between the frequencies of the transitions in the two directions, ν_{ls} and ν_{sl}. According to the theory of thermally activated processes,

$$\nu_{ls} = \nu \exp(-\Delta G''/RT)$$

$$\nu_{sl} = \nu \exp[-(|\Delta G_v| + \Delta G'')/RT]$$

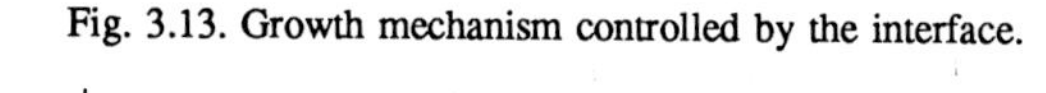

Fig. 3.13. Growth mechanism controlled by the interface.

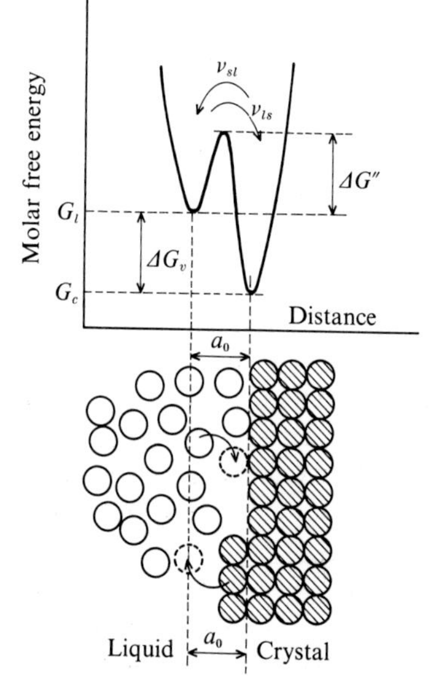

Introducing f, the fraction of sites on the surface which are available for growth ($(0 < f < 1)$), the growth rate is given by:

$$u = f a_0(\nu_{ls} - \nu_{sl})$$

that is,

$$u = f a_0 \nu \exp(-\Delta G''/RT)[1 - \exp(\Delta G_v/RT)]$$

with $\Delta G_v = R\beta\Delta T$.

Figure 3.6 shows the form of the curves for u as a function of T. For $T = T_f, u = 0$, and for small supercooling, u varies linearly with ΔT. As the supercooling increases, u increases up to a maximum due to the fact that the kinetic factor $\exp(-\Delta G''/RT)$ becomes predominant and decreases to zero at lower temperatures.

In most cases the preceding equation cannot be used directly because the value of $\Delta G''$, the activation energy for the movement of an atom across the interface, is not known. $\Delta G''$ is not necessarily equal to, or even of the same order of magnitude as, the activation energy $\Delta G'$ for nucleation. The growth may be controlled by long-range diffusion, while for nucleation only processes occurring over relatively short distances from the nucleus are important.

However, Turnbull and Cohen[119] suggest that at least for simple substances it is reasonable to assume that the activation energy $\Delta G''$ is close to that for viscous flow. As in the case of nucleation, this leads to consideration of the self-diffusion coefficient D, related to the viscosity η by the Stokes–Einstein equation. This gives the following equation for the growth rate:

$$u = f\frac{RT}{3\pi a_0^2 \eta N}\left[1 - \exp\left(\frac{\Delta G_v}{RT}\right)\right]$$

(b) *General case. Application to complex glasses*

Experimental results show that for simple substances, e.g. glycerol, this expression gives the correct order of magnitude, but for large supercooling, u varies more rapidly with T than predicted by the formula. For SiO_2 in particular, the calculated results are about 100 times less than the experimental observations.

This equation would thus not be directly applicable to complex systems. On the other hand the differences observed for SiO_2 may be explained by the presence of impurities which exert a catalytic or accelerator effect.

Different models for the parameter f have been proposed by Jackson[120] and Uhlmann.[121] Uhlmann also pointed out the influence of the entropy of fusion on the resulting glass–crystal interface morphology. The growth rate is critically dependent on the molecular structure of the crystal–liquid interface which itself depends on ΔS_f.

For materials characterized by small ΔS_f ($\beta \sim 2$), the interface is rough and the growth kinetics are normal ($f \sim 1$). Observed microscopically, the morphology is without facets. When $\beta > 4$, the normal model does not apply and the morphology shows facets.

3.2.7 *Kinetic criteria*

(a) *Determination of the minimum activation energy for glass formation*

Having established analytic expressions for I and u, it is now possible to give a more quantitative formulation of the conditions for glass formation presented earlier in this chapter. The problem is to determine the minimum values of $\Delta G'$ and $\Delta G''$ necessary for glass formation.

Maximum values must be assumed for the nucleation rate I_{max} and growth rate, u_{max}, which must not be exceeded in the course of cooling. These values are to some extent arbitrary since they depend on the anticipated cooling rate.

Turnbull and Cohen[119] take:

$I_{max} = 1$ nucleus cm^{-3} s^{-1}
$u_{max} = 10^{-5}$ interatomic distances s^{-1}.

The conditions for non-crystallization may then be formulated as follows:

1. Either there is no temperature $T'(< T_f)$ such that $I > I_{max}$.
2. Or, if such a temperature exists, the growth rate u always remains less than u_{max}.

An algebraic analysis of the system indicates that for these conditions to be satisfied, it is necessary that $\Delta G'$ and$\Delta G''$ shall both be greater than $30RT_f$. These conditions are relatively insensitive to the choice of u_{max}; e.g. taking u_{max} equal to 0.1μm s^{-1} (this is 10^8–10^9 higher than the preceding value) the minimum values for $\Delta G'$ and $\Delta G''$ are only lowered to $20RT_f$.

For *non-reconstructive* crystallization processes, $\Delta G'$ and $\Delta G''$ are comparable to the activation energy for viscous flow, ΔG_η, or the activation energy for self-diffusion, ΔG_D. For simple liquids, these free energies are much less than $20RT_f$ and glass formation is less probable. However, if the parameter $\alpha\beta^{1/3} \gg 0.5$, as in the case of complex or asymmetric molecules, glasses can form.

In the case of *reconstructive* crystallization, in which strong interatomic bonds are broken, the bond energies are generally greater than $20RT_f$, and the formation of glasses is probable.

The difficulty in the application of the preceding formulae to complex systems seriously limits their practical value. The preceding approximations and simplifications are no longer adequate and the thermodynamic barrier to nucleation cannot be calculated with precision. The chemical potentials of the components and their variation with temperature and composition as well as the surface energies for the different crystal–liquid pairs must be known.

The estimates of the kinetic barriers $\Delta G'$ and $\Delta G''$ are also more problematic.

(b) *Calculation of the critical cooling rate for glass formation*

The importance of the cooling rate in glass formation is well known – e.g. systems thought to be non-vitrifiable have finally been vitrified by extremely rapid quenching ("splat cooling") – the latest examples being V_2O_5 [52] and several refractory systems containing La_2O_3 which have been vitrified by fusion with a powerful laser followed by free fall cooling.[46] Therefore, as stated by Uhlmann,[121] the question is not whether a system will form a glass when it is cooled, but rather *at what speed* it must be cooled to avoid detectable crystallization. Uhlmann has estimated that the minimum crystalline fraction detectable by the usual methods (X-ray diffraction) is of the order of 10^{-6}.

The volume fraction y precipitated as a function of the nucleation rate I and the growth rate u for different modes of growth can be calculated according to Johnson–Mehl and Avrami's classic theory familiar to metallurgists (cf. e.g. Ref 122). Assuming I is constant, it can be shown that for interface controlled growth, the volume fraction transformed y or *degree of transformation* is:

$$y = \frac{\pi}{3} u^3 I t^4$$

While for a process controlled by diffusion:

$$y = \frac{8\pi}{15} S^3 D^{3/2} I t^{5/2}$$

where S is a coefficient depending on the degree of supersaturation and D is the diffusion coefficient.

The TTT curves (Time-Temperature-Transformation) are constructed using the preceding equations to calculate t_y, the time necessary to transform (crystallize) a given volume fraction y at a temperature T. Figure 3.14 shows the curves calculated for salol for $y = 10^{-8}$ and $y = 10^{-6}$. Figure 3.15 is the TTT diagram for vitreous SiO_2 for $y = 10^{-6}$. The critical cooling rate for the formation of a glass, i.e. the quench necessary to avoid a given crystalline fraction, can then be estimated.

Taking the fraction $y = 10^{-6}$ as the limit of detection, the critical quench speed is given by:

$$\left(\frac{\mathrm{d}T}{\mathrm{d}t}\right)_c = \frac{T_f - T_N}{t_N}$$

where t_N and T_N correspond to the "nose" of the TTT curve for $y = 10^{-6}$.

These calculations tend to overestimate the critical quenching speed because there is an implicit assumption that the kinetics of crystallization in the entire interval between the melting temperature and the nose of the curve is as rapid as that at the temperature of the nose. A more realistic estimate may be obtained by constructing a network of curves corresponding to continuous cooling. These curves, developed from the TTT curves using a procedure described by Grange and Kiefer,[123] are widely used in metallurgy in the heat treatment of steels. The curves can be constructed for constant cooling rates or for other rates, e.g. logarithmic rates where $\mathrm{d}T/\mathrm{d}\ln t = Constant$. The latter is applicable for the cooling of large volumes.

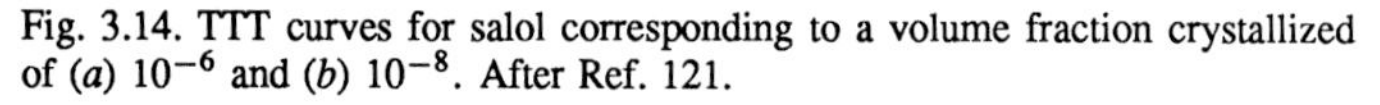
Fig. 3.14. TTT curves for salol corresponding to a volume fraction crystallized of (*a*) 10^{-6} and (*b*) 10^{-8}. After Ref. 121.

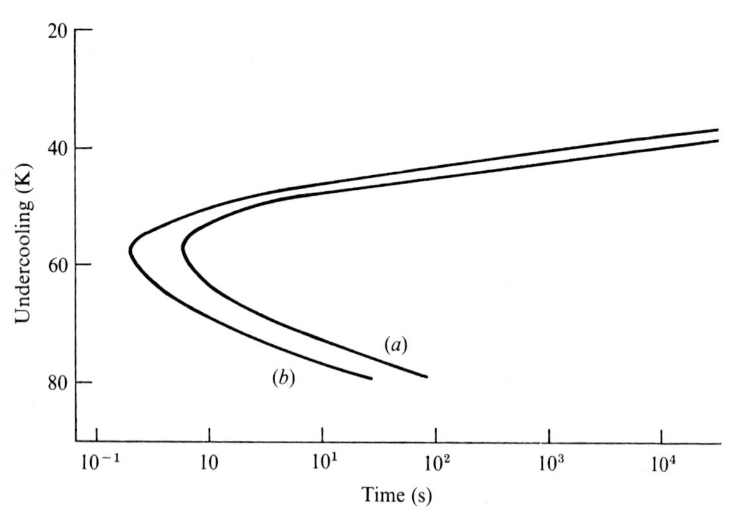

Fig. 3.15. TTT curves for SiO_2 corresponding to a crystallized fraction of 10^{-6}. After Ref. 121.

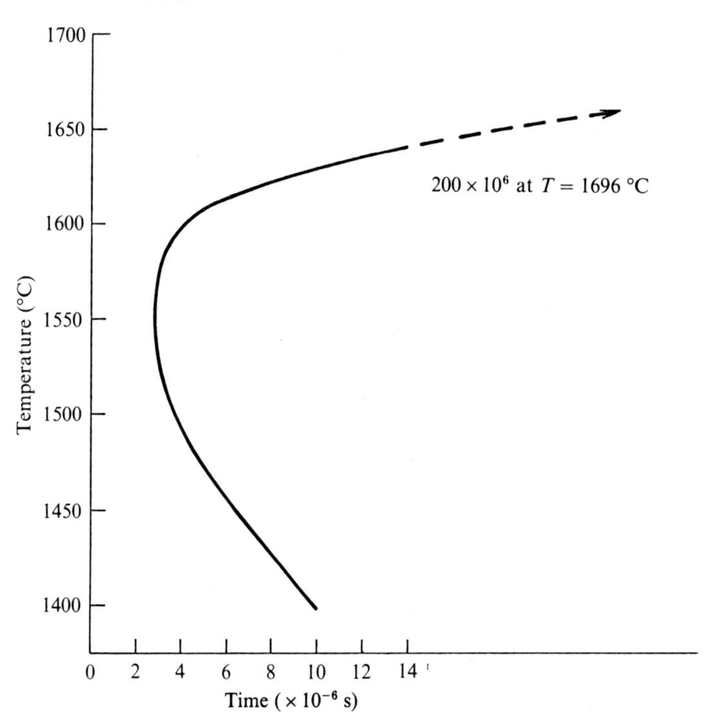

Carrying the analysis further, it is possible to define a crystal size distribution function $\psi(r, t, R)$ which gives the number of crystals with radii between R and $R + \mathrm{d}R$ present in a volume $\mathrm{d}V$ at position r. This function[124] contains all the basic statistical information on system crystallinity.

Uhlmann has also estimated the thickness of specimens which may be vitrified by quenching. Neglecting heat transfer at the surface, he has shown[121] that the thickness Y_c is of the order of:

$$Y_c \simeq (D_{TH} t_N)^{1/2}$$

where D_{TH} is the thermal diffusivity.

Table 3.5 gives the critical quenching speeds and the thicknesses Y_C calculated for different substances. This allows a numeric estimate of their tendency to vitrify.

These kinetic treatments are more likely to be valid when, in the course of crystallization, the atomic movements are closely related to viscous

Table 3.5. *Critical quenching rates. After Ref. 121.*

Substance	Critical quenching rate ($K\,s^{-1}$)	Thickness Y_c which can be cooled in the form of a glass
SiO_2	2×10^{-4}	400 cm
GeO_2	7×10^{-2}	$\sim$ 7 cm
Salol	50	0.07 cm
H_2O	10^7	$\sim$ 1 μm
Ag	10^{10}	100 nm

flow processes or interdiffusion. This is the case for more complex molecular liquids while for simple liquids (e.g. metals), the interfacial processes may differ from those operating in flow.

The formation of glass is easier in more complex systems, e.g. oxide glasses containing a variety of cations; in addition to lowering the melting point, the complexity necessitates diffusional redistribution of diverse constituents before crystallization can begin.

3.2.8 *Correlations between certain material properties and their ability to vitrify*

The kinetic treatments described previously permit an advanced quantitative analysis but they require a knowledge of numerous physical characteristics of the materials. It is interesting to review different attempts at simple correlations proposed in the past to evaluate the probability of vitrification of a given substance.

It appears, in general, that a high viscosity at the melting point and a rapid increase in viscosity as the temperature falls below the melting point are essential.

(a) *Relation between T_g and T_f*

For classes of materials which have similar viscosity curves, a low melting point (or liquidus) favors vitrification. This corresponds to a high T_g/T_f ratio. Figure 3.16 shows this effect for materials of the type $2SiO_2 \cdot Na_2O$ with different melting points. For a large number of glass-forming liquids, it has been recognized[126] that $T_g/T_f \sim 2/3$, and this is a useful approximation for a number of organic liquids and molten oxides (Fig. 3.17).

In general, however, such a simple relation should not be expected: the melting point depends on the properties of the liquid and the crystal, while

Fig. 3.16. TTT diagram for materials of the type $2SiO_2 \cdot Na_2O$, having different melting points: T_E is the equlibrium temperature. After Ref. 125.

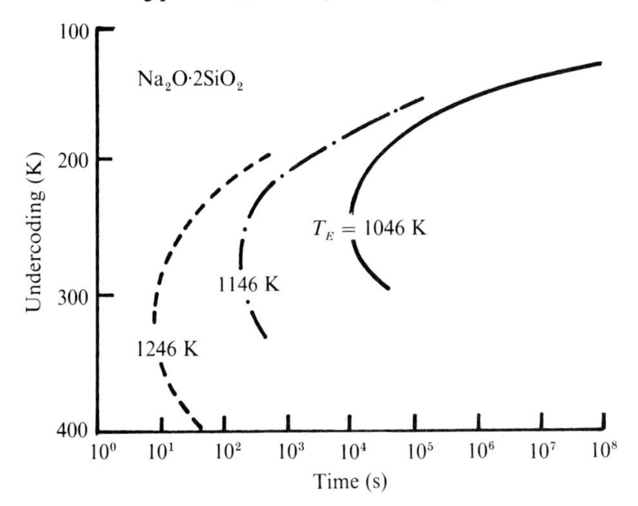

T_g depends primarily on the properties of the liquid. Moreover, as noted by Turnbull,[114] $T_g/T_f \ll 2/3$ for liquids which do not form glasses at normal cooling rates. Indeed, the value of T_g/T_l (T_l = the liquidus temperature) is closer to 0.46–0.56 for glass-forming metallic alloys. But in the absence of a simple correlation, the ability to vitrify is expected to increase with T_g/T_f.

(b) *Relation between T_B and T_f*

Cohen and Turnbull[127] pointed out that T_g is directly related to the *cohesion energy* of the liquid and thus with the boiling point T_B for a given class of materials.

Data exist for a number of organic liquids and it is found that for $T_B/T_f > 2$ vitrification normally occurs. When $T_B/T_f \sim 1.8$, it is still possible to form glasses from fine droplets. For $T_B/T_f < 1.6$ glasses are generally not obtained. However, for $ZnCl_2$, which easily forms a glass, $T_B/T_f = 1.7$. This approach must be refined for other classes of materials.

A better correlation may be obtained by introducing the concept of "corresponding temperatures."[128] Since the cohesion energy of the liquid is directly related to the heat of vaporization $\Delta H_{vap} = N\Delta h_{vap}$, a reduced temperature scale $\tau = kT/\Delta h_{vap} = RT/\Delta H_{vap}$ directly related to this concept can be adopted.

Each class of liquids is then characterized by a unique viscosity relation

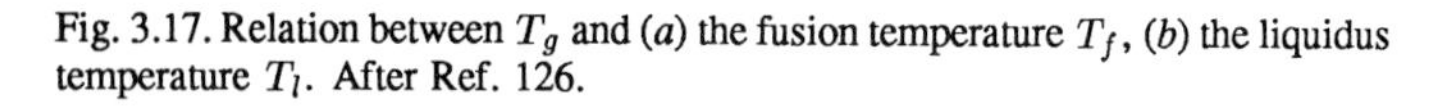

Fig. 3.17. Relation between T_g and (*a*) the fusion temperature T_f, (*b*) the liquidus temperature T_l. After Ref. 126.

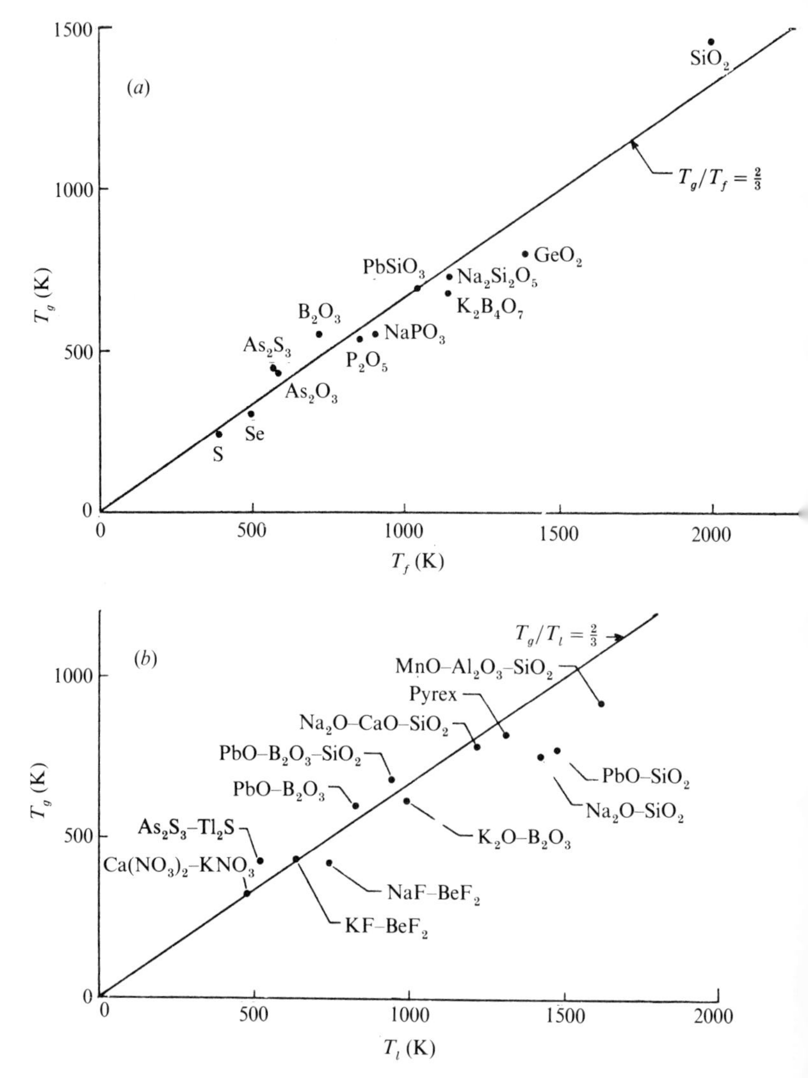

$\eta = f(\tau)$ which leads to a reduced temperature for vitrification defined by $\eta = 10^{13}$ P. On the other hand, Trouton's law equates $\tau_B = Constant$ for the same class in contrast to the reduced melting temperatures τ_f which may be quite different. For a *given class*, vitrification is more probable if τ_f is situated in the lower part of the interval $\tau_B - \tau_g$; i.e. when the melting point, T_f, is lower for a given *constant cohesion energy*. Given that $T_f = \Delta H_f / \Delta S_f$, this means that the tendency for vitrification is enhanced if less energy ΔH_f is required to produce a given quantity of disorder, ΔS_f. This is consistent with Rawson's criterion (cf. Section 3.1.3(b)).

This rule is well demonstrated in a series of organic liquids.[128] Greater asymmetry of the molecules increases the tendency for vitrification because the entropy of the liquid phase is increased so that a larger amount of disorder may be produced with little energy. The classic case is that of xylene; m-xylene forms a glass while the more symmetric p-xylene does not. The cohesion energies of the two liquids are similar (T_B points are similar) but the T_f for m-xylene is much lower, corresponding to its much higher entropy of fusion.

The influence of lowering the fusion temperature is demonstrated in the case of eutectic compositions which vitrify easily. Sometimes glass formation is only possible in a narrow composition band around the eutectic while the separate constituents are impossible to vitrify. The systems $CaO–Al_2O_3$ and $PbO–V_2O_5$ are classic examples.

3.2.9 *Experimental methods*

(a) *Nucleation rate*

Because of the small size of crystal nuclei, it is not possible to observe them directly. In practice, one seeks to "reveal" the latent nuclei by growing them to observable dimensions where they can be seen and counted. The specimens of glass to be examined are therefore submitted, after quenching, to a double heat treatment: the first at a variable nucleation temperature T_n during the variable time t_n, followed by a second at a conveniently chosen "development" temperature $T_d > T_n$. The count of the number of crystals $n(T_n)$ rendered visible after the second treatment permits the tracing of the curve $n(T_n) = f(t_n)$ and thus the determination of I. Such measurements are rare: Figure 3.18 shows the results obtained for a glass in the system $SiO_2–Na_2O–CaO$[129] and $SiO_2–Li_2O–TiO_2$.[130a] Figure 3.19, for the system $2SiO_2–Li_2O$, also shows the time constant τ for transient nucleation.[130b]

It is difficult to separate the effects due to heterogeneous nucleation. Impurities which induce preferential nucleation may be removed from the liquid by repeated partial crystallization and removal of the precipitated phases. This is impractical in the case of solid glasses. Moreover, oxide glasses are excellent solvents for various nucleating agents including water which is always present and which clearly plays a rôle in promoting nucleation.

Fig. 3.18. Rates of nucleation and growth. (*a*) SiO_2–Na_2O–CaO system. After Ref. 129. (*b*) SiO_2–Li_2O–TiO_2 system. After Ref. 130a.

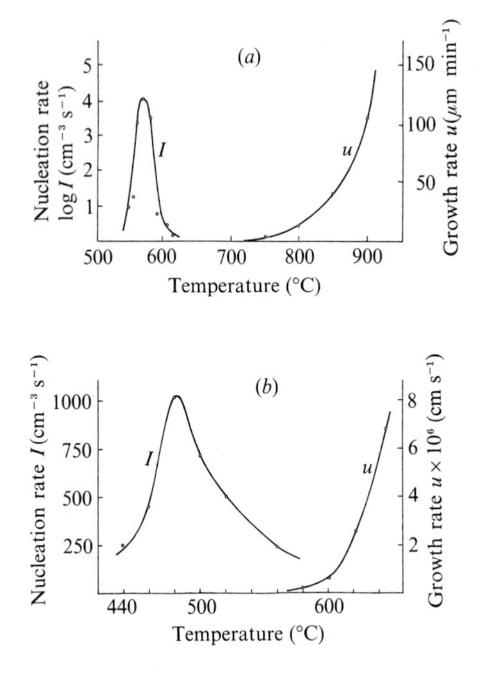

Fig. 3.19. Rate of nucleation in the $2SiO_2$–Li_2O system. The time constant τ is also shown. After Ref. 130b.

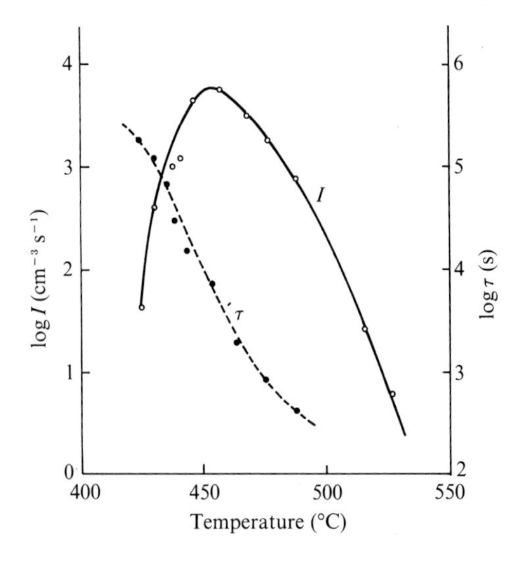

Turnbull used an elegant procedure in his studies on nucleation in liquids. It consisted of sub-dividing the material being studied into very small volume elements so that the heterogeneous sites (impurities) would be isolated in certain particles which would behave differently. By studying the nucleation properties of a liquid emulsion in a non-miscible matrix, the crystallization of individual droplets may be studied either by an optical method or, globally, by high sensitivity dilatometry which will show steps corresponding to the crystallization of a series of particles containing different types of impurities.

(b) *Crystal growth rate*

The linear crystal growth rate may be measured by direct and continuous microscopic observation of a specimen heated to a temperature T_d or by comparison of a series of specimens heated isothermally at various temperatures T_d (the parameters for nucleation, T_n and t_n being maintained constant). To save time, a series of specimens may be heat-treated simultaneously in a thermal gradient furnace. This permits the determination of the crystal growth rate $u = f(T_d)$. Figure 3.20 shows an example for the crystallization of devitrite in an industrial glass.

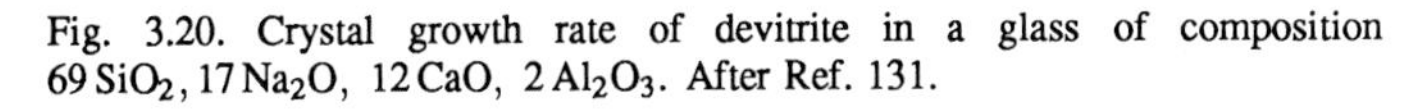

Fig. 3.20. Crystal growth rate of devitrite in a glass of composition 69 SiO_2, 17 Na_2O, 12 CaO, 2 Al_2O_3. After Ref. 131.

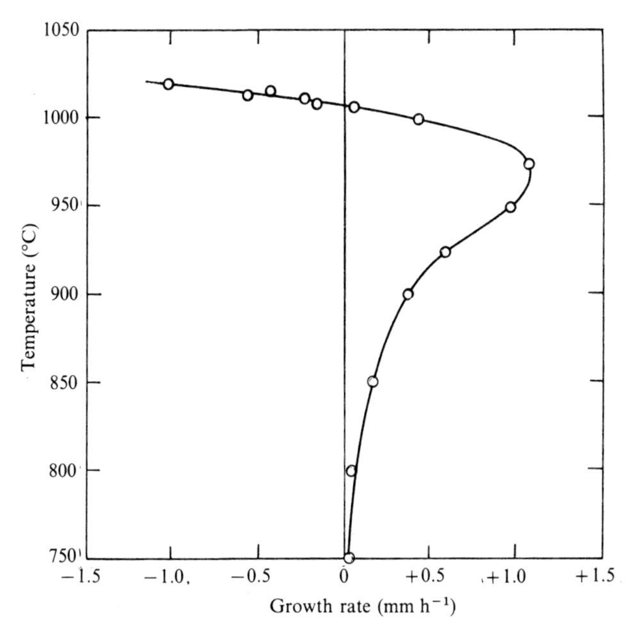

(c) *Global evaluations*

The conditions for crystallization (devitrification) have great importance in the glass industry where it is desirable to avoid this accident in the furnaces or during forming. It is generally sufficient to measure a global devitrification rate without separating I and u. The experiments are made on quenched samples which have been held at a given temperature for various times and then sectioned for microscopic examination. The different crystalline phases are characterized by X-ray diffraction and petrographic methods.

Figure 3.21 is the portion of the SiO_2–Na_2O–CaO ternary system of practical importance, showing the boundaries of the primary phase fields for the crystallization of quartz, tridymite, cristobalite (SiO_2), wollastonite (CaO · SiO_2) and devitrite (Na_2O · 3CaO · $6SiO_2$) etc. Figure 3.22 shows micrographs of several crystalline phases resulting from devitrification.

Fig. 3.21. Part of the equlibrium phase diagram for the system SiO_2–Na_2O–CaO. After Ref. 132.

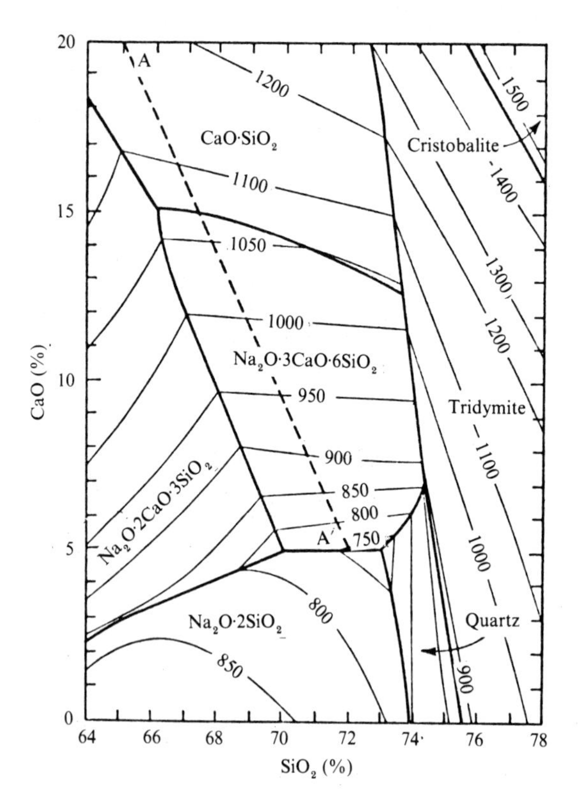

Fig. 3.22. Photomicrographs of the crystalline phases resulting from the devitrification of sheet glass (by courtesy of *Saint-Gobain Co.*): (*a*) cristobalite, initial formation of dendrites; (*b*) cristobalite, fully crystallized; (*c*) devitrite needles; (*d*) devitrite needles associated with cristobalite hexagons; (*e*) β-wollastonite; (*f*) α-wollastonite.

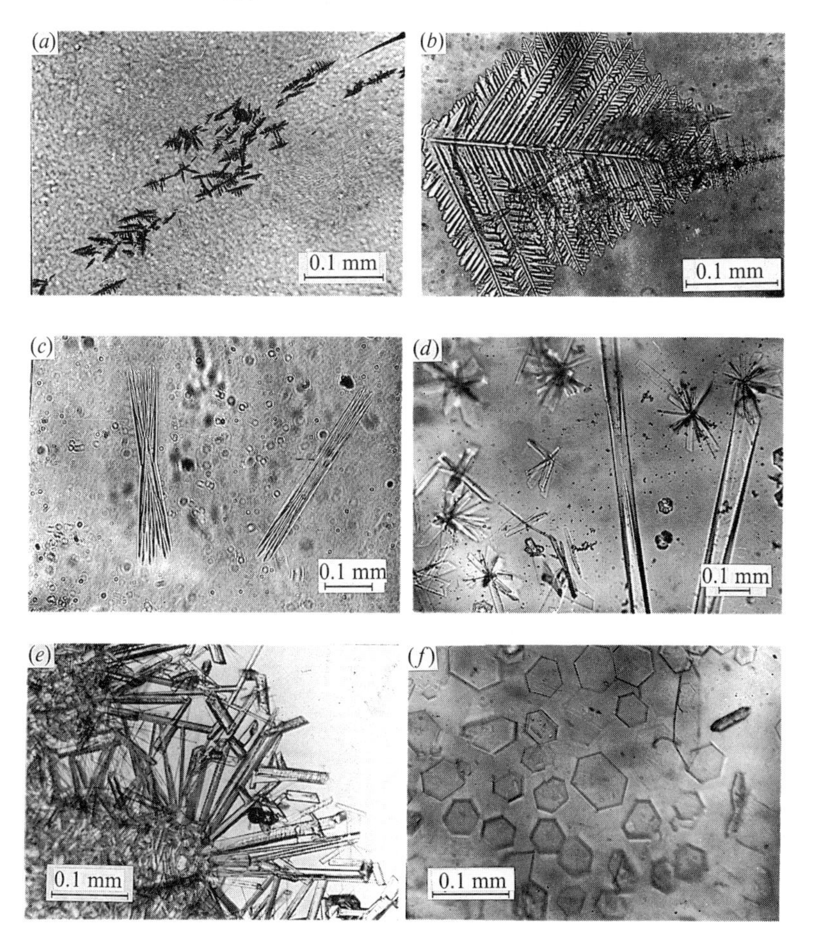

4 Structure of glass: Methods of study using radiation scattering

4.1 Problem of the description of a disordered structure

4.1.1 *Various modes of description*

In the case of a crystal, the atomic arrangement is periodic in three dimensions. The detailed description of such a structure is thus complete by giving the dimensions and the content of the unit cell – the disposition of all the atoms being determined by translation of this cell following the axes $0x, 0y, 0z$ (Fig. 4.1(*a*)).

The description of a perfect crystal is the object of X-ray crystallographic determinations in combination using point and translation group theories.

The *real* crystal differs from the ideal structure thus defined. Above absolute zero, the atoms are subject to thermal vibrations and the system presents point and extended defects in the network. For example, there are defects in the occupation of certain atomic sites (vacancies), local substitutions (substitutional defects) and the additions of extra atoms (interstitial defects). The array of atoms can undergo displacements as in the case of dislocations.

Fig. 4.1. Description of a crystalline structure: (*a*) perfect crystal, (*b*) paracrystal. Top: two-dimensional schematic representations, bottom: definition of the unit cell. The P represents probable location of the extremities of the vectors defining the cell.

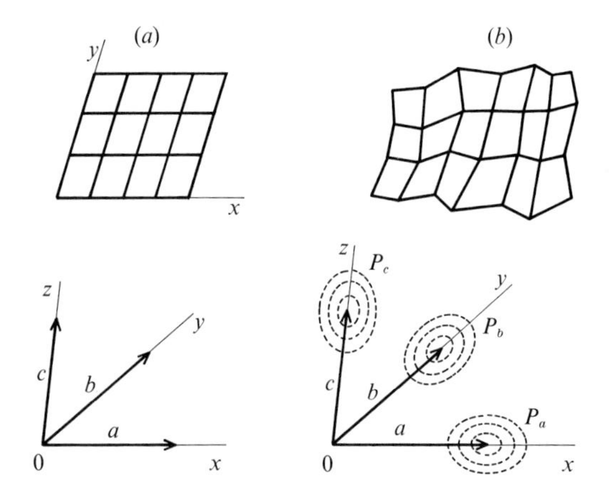

All these deviations can be described with reference to the ideal network of a perfect crystal. In the case of extended defects, the concept of the Burgers vector can be introduced. In certain crystals, the unit cell presents fluctuations in its angles or dimensions. It is then necessary to describe the distribution probability of the extremities of the vectors defining the unit cell. The network constructed of these unit cells then becomes distorted. Such a structure is called *paracrystalline* (Fig. 4.1(*b*)).[133] This is found in the case of certain complex organic substances, namely natural polymers.

At the other extreme, the case of complete disorder is given by a gas where the molecules are sufficiently separated from each other to have an identical probability of occupying all the space available to them.

The description becomes more complex when the elements of the disordered structure are close enough to each other to influence each others' positions. This is exactly the case for *liquids* and for the *glasses* which come from them. In this case, it is possible to describe the structure from two points of view:

1. A portion of the disordered structure is considered as a giant molecule and the position and nature of all its elements are defined.

2. The structure is characterized by a *statistical* description through the distribution of certain distances.

The first approach consists of building a *detailed model* by providing the coordinates $M_i(x_i y_i z_i)$ for the structure. In the case of a liquid where the coordinates depend on time, such a model corresponds to an instantaneous configuration. The validity of the model is tested by calculating various properties and comparing the results obtained with experimental values. Currently, this effort is greatly facilitated by the use of computers which permit calculations on populations of several thousand atoms, as will be discussed in detail in Chapter 7.

The second approach consists of searching for a *global* representation of the structure using statistical methods to deduce certain results accessible to experiment.

It is also possible to use the second approach to characterize and compare various models.

4.1.2 *Radial analysis – short-range order*

One possible description of the structure consists of the probability of finding a pair of atoms M_i M_j defined by the vector $\vec{r}_{ij} = \vec{r}_j - \vec{r}_i$, where $\vec{r}_i$ and $\vec{r}_j$ are the position vectors of the atoms M_i and M_j (Fig. 4.2(*a*)).

In an isotropic structure, as is the case for a liquid or a glass, if all orientations of $\vec{r}_{ij}$ are assumed equivalent, a description based on the distance $r_{ij} = |\vec{r}_{ij}|$ and a scalar distribution function is sufficient. This is the principle of radial analysis.

(a) *Structures containing only one kind of atom (homoatomic)*
A function $\varrho(r)$ is defined by the expression:

$$4\pi r^2 \varrho(r)\,\mathrm{d}r$$

which represents the average number of atomic centers situated in a spherical shell of radius r and of thickness $\mathrm{d}r$ centered on any atom of the structure (Fig. 4.2(*b*)).

The function $\varrho(r)$ is obtained by successively taking each atomic center of the structure as the origin and calculating the resulting average value. The term *radial density* is applied to $\varrho(r)$, which represents the average number of atoms per unit volume situated at some distance from any atom in the structure taken as the origin. The form of the function is given in Fig. 4.3(*a*). $\varrho(r) \to 0$ for the values of $r < D$ where D is the minimum distance of approach (contact distance).

For $r \simeq D$, the function presents a maximum corresponding to the presence of the nearest neighbors. The successive maxima correspond to the most frequent average distances in the structure, the minima to the distances least frequently encountered. The regular arrangement of the neighbors close to the atom at the origin indicates the existence of a *short-range order* in the structure.

For $r \to \infty$, $\varrho(r)$ tends toward the value $\varrho_0 = N/V$ which is the average density of N centers distributed in volume V.

The function:

$$\varrho(r)/\varrho_0 = g(r)$$

$g(r)$ is called the *pair distribution function* and represents the probability of the presence of a pair of atoms at distance r.

Fig. 4.2. Description of a disordered structure.

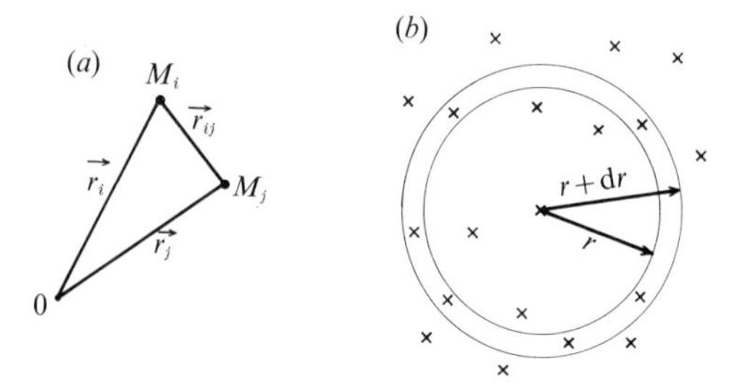

The quantity $G(r) = g(r) - 1$ is called the *pair correlation function* (Fig. 4.3(*b*)). For $r \to \infty$, $g(r) \to 1$ and $G(r) \to 0$ which shows that for larger distances , the structural effect disappears and all positions for the atoms are equally probable.

The function $4\pi r^2 \varrho(r)$ is called the *Radial Distribution Function* (RDF) (Fig. 4.3(*c*)). It shows maxima and minima oscillating around the parabola $4\pi r^2 \varrho_0$ corresponding to the average distribution of centers in the structure and tending asymptotically toward it at $r \to \infty$. The area under the curve representing the RDF permits the calculation of the average number of centers situated between the two given distances, $r_1 r_2$.

It is instructive to apply the preceding method of description to the case of an *ideal crystal* structure. Figure 4.4(*a*) shows the calculated RDF for cubic, face-centered cubic, body-centered cubic and hexagonal close packed crystal structures.

The atoms are grouped on a series of spherical surfaces called *coordination spheres* centered on the atom chosen as the origin and following a regular numerical progression resulting from the geometry of the crystal lattice.

The numbers N_r of neighbors situated at these well-defined distances are called *coordination numbers*, and are indicated by the lines on the RDF diagrams. The number of nearest neighbors (in the first coordination sphere) is the principal coordination number of the crystalline structure considered. Note that this number is the highest (12) for the face-centered cubic and hexagonal close packed lattices.

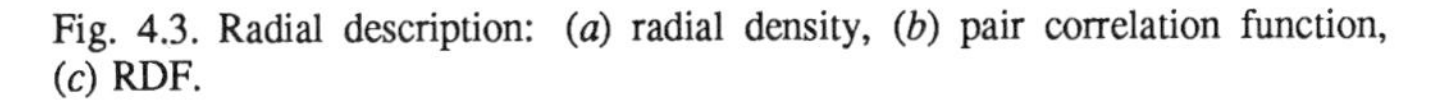
Fig. 4.3. Radial description: (*a*) radial density, (*b*) pair correlation function, (*c*) RDF.

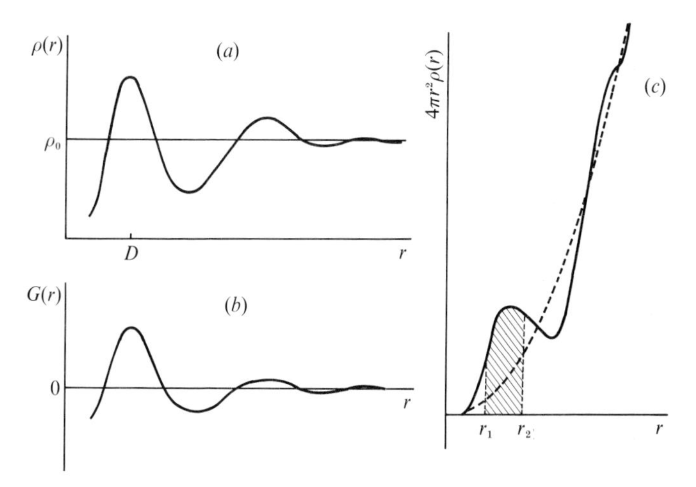

By virtue of the periodicity of the lattice, the regular arrangement continues indefinitely for larger and larger values of r, corresponding to the existence of *long-range order* in the crystalline arrangement.

It is found, however, that the different coordination spheres come closer together as r increases – the differences between the radii of the successive coordination spheres become less and less – which is an effect inherent to the radial calculation method utilized.

Assume now that the atoms undergo displacements relative to their positions in the ideal crystal (e.g. the effect of thermal vibrations around the rest position at absolute zero). The atoms are no longer situated on the ideal spheres considered previously, but are distributed in the spherical shell situated between r and $r + \mathrm{d}r$ from the ideal rest position.

Let the number of atoms contained in such a shell be denoted by $h(r)\,\mathrm{d}r$. Figures 4.4(*b*) and (*c*) show the form of the function $h(r)$. The ideal vertical lines have been replaced here by peaked curves, the width of

Fig. 4.4. (*a*) Radial distribution functions for different lattices: (1) cubic, (2) face-centered cubic, (3) body-centered cubic, (4) hexagonal close packed. (*b*) The effect of variations in atomic positions. (*c*) Disordered structure.

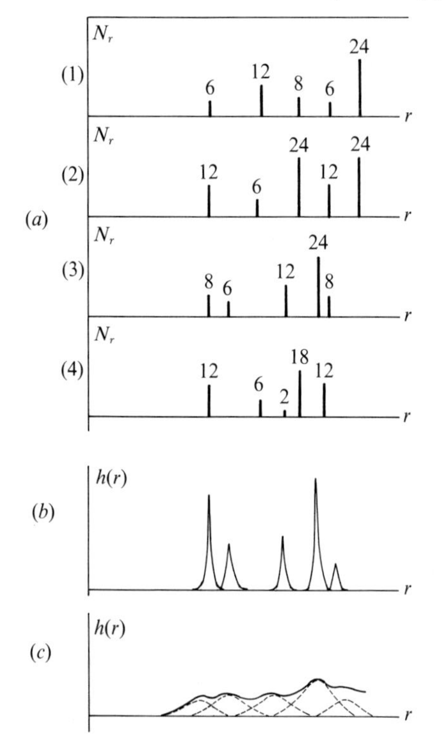

which increases with the amplitude of the displacements. The area under each peak corresponds to the number of atoms contained by the ideal sphere for the unperturbed structure. The structure still possesses both short-range and long-range order, and it is possible to distinguish the different coordination spheres.

If the temperature is increased, the vibrational amplitude increases and, at a well-defined temperature, the regular network breaks down. This is the phenomenon of fusion.

The radial distribution function thus corresponds to the case of a disordered state already studied. The discrete distribution gives way to a continuous distribution where it is no longer possible to distinguish the successive coordination spheres with precision. It is not possible to define the exact coordination numbers. It is at the most possible to speak of *average coordination shells* to which can be associated *average coordination numbers*. The area under a RDF can be sub-divided into portions assigned to different maxima, which is the same as attributing approximate average numbers to coordination shells. This deconvolution is somewhat arbitrary, because the areas present some successive overlap. If the deconvolution has a meaning for the nearest neighbors, it loses all significance for the higher-order coordination shells. This is the consequence of the disappearance of long-range order as the structure blends statistically into a continuum.

(b) *Structures containing several kinds of atoms (heteroatomic)*

When the disordered structure consists of several different kinds of atoms, as is generally the case, it becomes necessary to introduce RDFs for different atomic pairs, i.e. to define the partial pair functions.

In the case of a structure containing two species, A and B, there would be three independent functions g_{A-A}, g_{A-B}, g_{B-B} for the pairs A–A, A–B, and B–B respectively. For three species, A, B, C, there would be six pair functions:

$$g_{A-A},\ g_{B-B},\ g_{C-C},\ g_{A-B},\ g_{A-C},\ g_{B-C}$$

It is apparent that their number increases rapidly. In the general case of n species, it is necessary to have $n(n+1)/2$ pair functions to define the structure. This imposes a practical limit for the determination of polyatomic disordered structures, because, as we shall see, scattering methods only permit the determination of the weighted sum of the different partial pair functions.

This, along with the fact that the information is *one-dimensional* constitutes a serious limitation to this description method. The disordered structure is insufficiently defined, and for a complete determination it would be necessary to know the higher-order correlation functions (triplets etc.) locating precisely the simultaneous position of three atoms and more. Unfortunately, these functions are not accessible through currently available experimental methods.

The preceding correlation functions have been defined from a static model, the structure being assumed rigid. In reality, the atoms oscillate around the average equilibrium positions while in a liquid the configuration changes with time with the atoms undergoing displacements. The preceding functions correspond to the average with respect to time for the period of observation.

In order to study the movements in a liquid in detail, it is necessary to introduce the space-time correlation functions $G(\vec{r}, t)$ of Van Hove[134] describing the relation between the structure at time t and that at $t = 0$. For glasses, apart from exceptional cases, the structure can be taken as fixed, and it suffices to consider the instantaneous functions $G(\vec{r}, 0)$ which become identical to the previously defined correlation functions.

4.2 Methods using radiation scattering

Radiation scattering methods are used to determine the pair correlation functions.

4.2.1 *Radiation–matter interaction*

When monochromatic radiation interacts with a solid, part is absorbed and the rest emerges deviated to a greater or lesser extent relative to its initial trajectory, which constitutes the phenomenon of scattering. The scattered radiation is distributed in space in a manner which is related to the structure of the medium. A fraction can be concentrated in well-defined directions depending on the crystalline structure. This is the phenomenon of *diffraction* with the maximum intensities corresponding to the Laue condition. The remainder of the radiation, scattered in a more isotropic manner, constitutes *diffuse* scattering.

When the scattered radiation has the same energy (or the same wavelengths which is equivalent) as the incident radiation, the scattering is then called *elastic*. If there is a change in the wavelength (or energy) of the radiation the scattering is said to be *inelastic*.

In the case of elastic scattering, the angular distribution of the intensity of the scattered radiation depends on the spatial distribution of the scattering centers. Inelastic scattering provides information on the *dynamics* of the scattering system.

If $\vec{k}_0$ and $\vec{k}$ are respectively the wave vectors of the incident and scattered radiation, the scattering vector is defined as:

$$\vec{Q} = \vec{k} - \vec{k}_0$$

for which the modulus is:

$$|\vec{Q}| = \frac{4\pi \sin\theta}{\lambda}$$

where λ is the wavelength of the radiation and θ the half-angle of the scattering.

For X-rays, the scattering centers are the electrons and thus a scattering experiment leads to an *electronic* correlation function.

The amplitude of the coherent scattering, or atomic scattering factor $f_i = Z_i\Phi(Q)$ for each atom in the structure is approximately proportional to the atomic number Z_i, and is a function of the modulus of the vector $\vec{Q}$, i.e. of $(\sin\theta)/\lambda$. This angular variation results from the size of the atomic electron orbits being of the same order as the wavelength of the incident λ.

For *neutrons*, in the case where there are no magnetic effects, the scattering centers are the atomic nuclei, and the correlation function thus deduced will relate to the atomic centers. The corresponding scattering factor, or "scattering length" b_i, is independent of $\vec{Q}$, i.e. independent of θ which simplifies interpretations. Furthermore, unlike X-ray scattering, the scattering factor for neutrons does not increase systematically with Z_i. Different atoms and even different iotopes can have markedly different scattering factors, so the information obtained can be varied by the systematic substitution of appropriate isotopes. The methods of X-ray scattering and neutron scattering are thus complementary.

The *inelastic* scattering of neutrons has been utilized in a limited manner to determine the relations for the dispersion of phonons in glass.

The scattering of *electrons* has been used less because of their strong absorption by the medium which limits their use to thin films. Moreover, the interpretations are difficult because of multiple scattering and insufficient knowledge of the scattering factors. The electronic scattering factor is approximately:

$$f_i = Z_i[1 - \Phi(Q)]/Q^2$$

4.2.2 *Scattered intensity – Debye's relation*

Consider a disordered assemblage of atoms in which the instantaneous position of each atom m is represented by the position vector $\vec{r}_m$. The coherent intensity I_c scattered by this group, expressed in electronic units, (i.e. compared to the intensity scattered by one electron) is obtained by summing the amplitude scattered by each atom multiplied by the conjugate quantity:

$$I_c = \sum_m f_m \exp\left(\mathrm{i}\vec{Q}\cdot\vec{r}_m\right) \sum_n f_n \exp\left(-\mathrm{i}\vec{Q}\cdot\vec{r}_n\right)$$

In the case of a crystal, the summation can be evaluated because of the periodicity of the arrangement. For a disordered substance, there is no simple relation between the position vectors, $\vec{r}_m$.

Introducing $\vec{r}_{mn} = \vec{r}_m - \vec{r}_n$ relative to the pairs m, n:

$$I_c = \sum_m \sum_n f_m f_n \exp\left(\mathrm{i}\vec{Q} \cdot \vec{r}_{mn}\right)$$

Let us calculate the scattered intensity when $\vec{r}_{mn}$ takes all orientations in space with the same probability.

The locus of the extremities of $\vec{r}_{mn}$ is replaced by a sphere (Fig. 4.5) and the average is:

$$\exp\left(\mathrm{i}\vec{Q}\vec{r}_{mn}\right) = \frac{1}{4\pi r_{mn}^2} \int_0^\pi \exp(\mathrm{i}Qr_{mn}\cos\phi) 2\pi r_{mn} \sin\phi \, \mathrm{d}\phi = \frac{\sin Qr_{mn}}{Qr_{mn}}$$

The intensity of coherent scattering by the assemblage of atoms is:

$$I_c = \sum_m \sum_n f_m f_n \frac{\sin Qr_{mn}}{Qr_{mn}}$$

the double summation being extended to all the atomic pairs. This formula due to Debye[135] is the basis for all structural determinations of disordered substances.

In practice, there are two possible ways to use this relation.

1. The Debye relation is applied directly to a *model* of the structure considered as a giant molecule, and the calculated intensity I is compared to the experimentally observed intensity after various corrections.

Fig. 4.5. Geometric conditions for scattering.

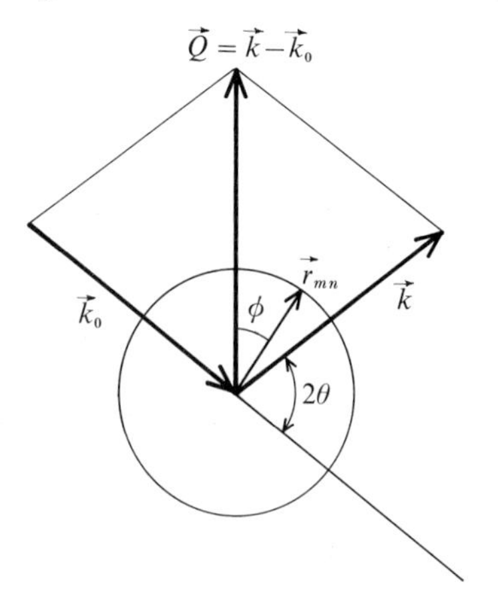

This is always possible, and the calculations are facilitated by the use of computers that allow a number of rapid trials to adjust the model. However, a unique solution cannot be assured as different models can lead to calculated values of I which are very close.

2. The second way consists of extracting from the function I the information concerning the pair distribution functions by means of a Fourier transform without previous hypotheses about the structure. References 136 and 137 offer a "unified" treatment applying to both scattering of X-rays and neutrons. We shall limit ourselves in the following to a more traditional approach and treat in greater detail the case of X-ray scattering.

4.2.3 *Determination of the correlation functions from the scattered X-ray intensity*

(a) *Case of homoatomic structures: exact solution*

To simplify the treatment let us consider first the case where all the scattering centers are identical. In this case, for N atoms with scattering factor f, Debye's formula becomes:

$$I_c = Nf^2 \left(1 + \sum_{m \neq n} \sum \frac{\sin Qr_{mn}}{Qr_{mn}}\right)$$

By introducing the radial distribution function of the centers, $4\pi r^2 \varrho(r)$, this formula becomes:

$$I_c = Nf^2 \left[1 + 4\pi \int_0^\infty r^2 \varrho(r) \frac{\sin Qr}{Qr} \mathrm{d}r\right]$$

Introducing the ratio:

$$i(Q) = \frac{I_c - Nf^2}{Nf^2}$$

which expresses the relative difference between the total scattered coherent intensity and the independent coherent intensity, $I'_c = Nf^2$ which would be scattered by the same N scattering centers assuming sufficient separation between them so that interference is negligible (as in an atomic gas).

Introducing the difference between $\varrho(r)$ and the average density ϱ_0 one thus obtains:

$$Qi(Q) = 4\pi \int_0^\infty r[\varrho(r) - \varrho_0] \sin Qr \, \mathrm{d}r + 4\pi \int_0^\infty r \varrho_0 \sin Qr \, \mathrm{d}r$$

It can be shown that the second integral is zero except for very small, experimentally inaccessible, values of Q. By a Fourier transform, the radial distribution function is obtained directly:

$$4\pi r^2 \varrho(r) = 4\pi r^2 \varrho_0 + \frac{2r}{\pi} \int_0^\infty Qi(Q) \sin Qr \, dQ$$

This equation obtained by Zernike and Prins[138] is applicable to liquids and glasses containing only one kind of atom and thus allows the calculation of the radial distribution function from the coherent scattered intensity $I_c(Q)$ without other assumptions. The experimental problem consists of measuring $I_c(Q)$ and expressing it in electronic units (e.u.) to form the function $i(Q)$.

The theory assumes that the wavelength λ of the scattered radiation is constant and it is necessary in cases of disordered substances to operate with strictly *monochromatic* radiation. When the scattered radiation is recorded photographically, a crystal monochromator is indispensable. Balanced filters have sometimes been employed using a proportional counter and a pulse-height analyzer. When a counter is employed, the monochromator can be placed in the incident beam or in the scattered beam.

The experimentally measured intensity $I_{exp}(Q)$ consists of the sum of the coherent intensity $I_c(Q)$ and the incoherent intensity $I_i(Q)$ scattered inelastically with wavelength modification (Compton radiation). The value of $I_{exp}(Q)$ is obtained in arbitrary units. For high values of Q, the interferences are reduced, and I_c tends toward $I_c' = f^2(Q)$, and this is used to express $I_{exp}(Q)$ in electronic units (e.u.). The sum $I_{cal} = f^2 + I_i(Q)$ can be calculated with the aid of tables providing a way to express I_{exp} in electronic units by making the tail of the curve I_{exp} coincide with I_{cal} for a unit scattering.

The function $i(Q)$ is then obtained from the relation $(\overline{ab} - \overline{ac})/\overline{ac}$ (Fig. 4.6). This procedure called *normalization* is very important because small errors in adjustment can lead to very noticeable modifications of $i(Q)$.

Moreover, because of the multiplication by Q, the weak terminal oscillations of $i(Q)$ are strongly amplified in $Qi(Q)$ and greatly influence the Fourier transform. The importance of a careful normalization is apparent. It requires a determination as precise as possible of the tail of $I_{exp}(Q)$ – the most difficult part to measure in practice because of the weak scattered intensities. Calculation procedures have been developed to make the normalization more objective.[139]

The difference $I_c - I_c'$ tends toward zero when Q increases, and is not measurable beyond a value Q_m where the intensity curves are said to

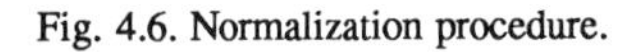
Fig. 4.6. Normalization procedure.

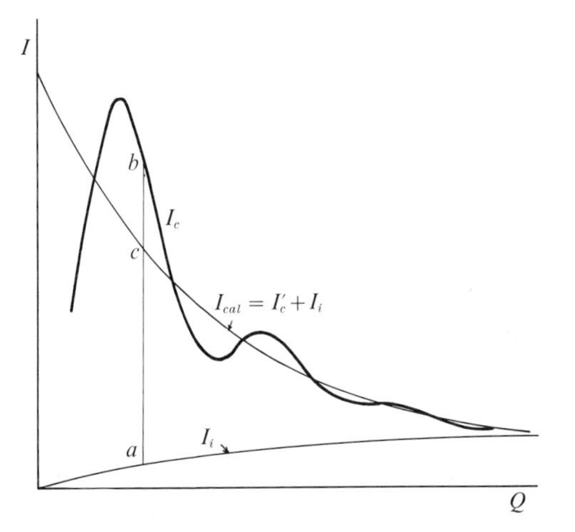

converge. However $Qi(Q)$, obtained by multiplying this difference by Q and dividing by f^2 can present significant variations and uncertainties near Q_m which might introduce artificial peaks in the RDF. To eliminate this, $Qi(Q)$ is multiplied by a *convergence factor* $\exp(-\alpha^2 Q^2)$, where α is an adjustable constant, to progressively attenuate the detrimental variation in the intensity function.

In practice, the Fourier transform is not made by integrating between 0 and ∞, but between 0 and a certain finite value Q_0 of Q. This is equivalent to multiplying the function to be transformed by a function $U(Q)$ such that $U(Q) = 1$ for $Q < Q_0$ and $U(Q) = 0$ for $Q > Q_0$. It has the effect of introducing a "termination error" in the form of parasitic oscillations in the RDF which can be confused with the real maxima relating to the structure.

The resolution of an RDF increases with Q_0; but Q is at most equal to $4\pi/\lambda$. Thus, it is necessary to operate with several wavelengths: Cu $K_\alpha = 1.54$ Å for the region of average Q, then with Mo $K_\alpha = 0.71$ Å, Ag $K_\alpha = 0.56$ Å or Rh $K_\alpha = 0.61$ Å to obtain the details of $Qi(Q)$ up to large values of Q. While in most cases Q_0 is of the order of 10–20 $Å^{-1}$ for X-rays, in recent measurements Q_0 values of 40–50 $Å^{-1}$ have been obtained using pulsed neutron sources.

(b) *Case of heteroatomic structures: approximate solution*

The formula of Zernike and Prins has been extended by Warren, Krutter and Morningstar[140] to the general case of disordered substances containing several atom species. They have shown an approximate method which is very often used in practice.

First of all a "composition unit" (c.u.), e.g. a stoichiometric unit, is chosen. The specimen is assumed to contain N groups. Debye's formula gives the average coherent scattering intensity:

$$I_c(Q) = N \sum_m f_m^2 + \sum_{p \neq q} \sum f_p f_q \frac{\sin Q r_{pq}}{Q r_{pq}}$$

where the first summation is extended to all the atoms of a group and the second summation relates to all the atom pairs without taking into account the group to which they belong.

The *weighted density functions*, $\varrho_m(r)$ (where ϱ_m relates to an atom of type m), are then defined as follows: spheres of radii r and $r + \mathrm{d}r$ are traced around an atom m (taken as the center), let $a_m, a_n, \ldots$ be the average number of atoms $m, n, \ldots$ contained in this layer with scattering factors $f_m, f_n, \ldots$.

By definition, one takes:

$$4\pi r^2 \varrho_m(r)\mathrm{d}r = \sum_{c.u.} a_m f_m$$

the summation being extended to all atoms on a composition unit (c.u.). Since there are N groups:

$$I_c(Q) = N \left[\sum_{c.u.} f_m^2 + \sum_{c.u.} f_m \int 4\pi r^2 \varrho_m(r) \frac{\sin Qr}{Qr} \mathrm{d}r \right]$$

Because f_m and ϱ_m are functions of Q, a direct solution is not possible unlike in the monoatomic case.

The approximate method consists of defining an *effective electron number* K_m for each kind of atom by the relation:

$$K_m(Q) = f_m(Q)/f_e(Q)$$

where $f_m(Q)$ is the scattering factor of atom m of atomic number Z_m and f_e an average electron scattering factor defined for the group considered by the relation:

$$f_e(Q) = \frac{\sum_{c.u.} f_m(Q)}{\sum_{c.u.} Z_m}$$

Because K_m varies with Q, in order to obtain the Fourier transform, calculated average values $\overline{K}_m$ for the experimental $0 - Q_0$ interval of the spectra are introduced:

$$\overline{K}_m = \frac{1}{Q_0} \int_0^{Q_0} K_m \mathrm{d}Q$$

This approximation, called the "constant effective K," actually consists in assuming that the scattering factors are all proportional to a single electronic factor f_e(Q):

$$f_m = \overline{K}_m f_e$$

Then distribution functions $g_m(r)$ for the species m are introduced by the relation:

$$\varrho_m(r) = f_e g_m(r)$$

Under these conditions, because $\overline{K}_m$ and g_m only depend on Q, the solution becomes possible. Proceeding as in the monoatomic case, the classic expression is obtained:

$$\sum_{c.u.} \overline{K}_m 4\pi r^2 g_m(r) = \sum_{c.u.} \overline{K}_m 4\pi r^2 g_0 + \frac{2r}{\pi} \int_0^\infty Qi(Q) \sin rQ \, \mathrm{d}Q$$

The radial distribution function appears as the summation of the electronic radial distribution functions weighted by the coefficients $\overline{K}_m$ calculated *a priori* for a given unit of composition.

The intensity function $i(Q)$ is:

$$i(Q) = \left[I_c/N - \sum_{c.u.} f_m^2 \right] \Big/ f_e^2$$

and the average electronic density g_0:

$$g_0 = \frac{dN}{M} \sum_{c.u.} Z_m$$

where d is the specific mass (macroscopic), M the molar mass, and N Avogadro's number.

Experimental intensity is normalized as described for the monoatomic case, $I_c(Q)$ and $I_i(Q)$ being evaluated for a unit of composition.

The weighted RDF oscillates around the parabola of the average electronic density. The different maxima correspond to the sequence of the first most frequently encountered distances in the disordered structure.

The areas under the different peaks are proportional to the successive coordination numbers; the proportionality coefficients being simple functions of $\overline{K}_m$.

If x_n^m is designated as the coordination number relative to the distribution of atoms n around a central atom m, it can be shown that the area, expressed in (electrons)^2 is equal to:

$$\sum_{c.u.} \overline{K}_m \overline{K}_n x_n^m$$

4.2.4 *Examples of applications*

Warren and his collaborators were the first to apply these methods to the study of simple glasses: SiO_2, B_2O_3, SiO_2–Na_2O, etc. Since that time, many measurements have been made on diverse systems and the experimental techniques as well as the calculation methods have been considerably improved.

In the monograph of Wright[137] a complete compilation of all the studies made up to 1973 can be found. In the following paragraphs we will only give some typical examples to provide a better understanding of the problems.

To interpret the information furnished by a RDF, one generally compares the short-range order defined by the RDF with the short-range order of an assumed parent crystalline arrangement. Peaks of the RDF are sequentially assigned to interatomic distances in the arrangement. An additional verification is furnished by the set of average coordination numbers calculated from the areas under the corresponding peaks.

Several examples of studies of the vitreous network as a function of temperature will be given later showing the influence of temperature on short-range order in the substance.

Figure 4.7 shows the X-ray scattering spectra for solid (vitreous) B_2O_3 at 20 °C and liquid B_2O_3 at 1200 and 1600 °C. The corresponding radial distribution functions are shown in Fig. 4.8.[141]

The composition unit here is B_2O_3. The first maximum at 1.30 Å at 20 °C corresponds to the the B–O distance, the second, at 2.40 Å can be interpreted as being due to O–O which leads to a coordination of 3 for the boron with the structure being formed from (BO_3) units.

The area A_1 under the first maximum is

$$A_1 = 2\overline{K}_B\overline{K}_O n_O^B + 3\overline{K}_O\overline{K}_B n_B^O$$

and since by virtue of the stoichiometry $3n_B^O = 2n_O^B$:

$$n_O^B = A_1/4\overline{K}_B\overline{K}_O$$

Similarly for the second maximum:

$$n_O^O = A_2/3\overline{K}_O^2$$

the contribution from the B–B pair being negligible.

Fig. 4.7. X-ray scattering spectra from B_2O_3 solid at 20 °C and liquid at 1200 and 1600 °C. After Ref. 141.

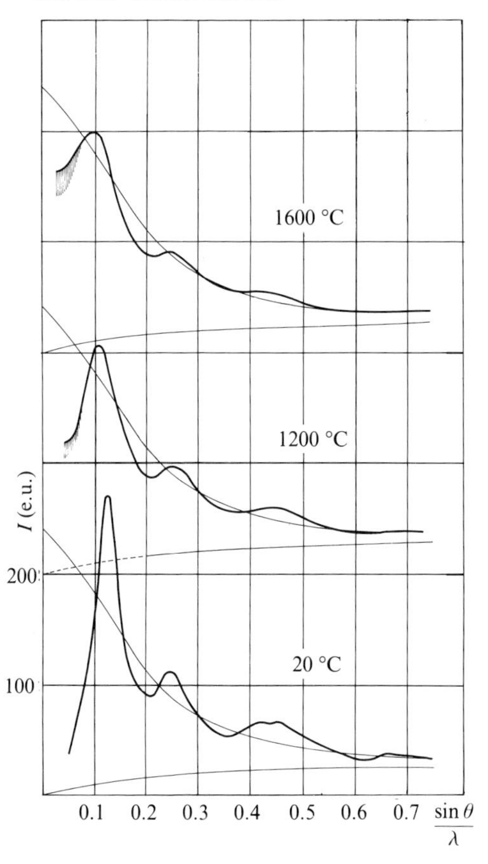

Table 4.1 shows the evolution of the coordination numbers associated with distances r_{B-O} and r_{O-O} as a function of temperature. It is apparent that the distances increase and the coordination decreases which corresponds to a breakup of the network.

A similar study[141] of GeO_2, vitreous at 20 °C and liquid at 1200 °C, resulted in the spectra of Fig. 4.9. The corresponding RDF (Fig. 4.10) show a tetrahedral coordination; the network thus consists of (GeO_4) units and the bond ruptures at 1200 °C produce a lowering of n_O^{Ge} which changes from 4.4 to 3.4.

Finbak[142] reported that the introduction of the factor $1/f_e^2$ in the interference function $i(Q)$ is equivalent to concentrating the electronic density into the centers of the atoms. This factor plays the rôle of a "sharpening

Fig. 4.8. Radial distribution functions for B_2O_3 corresponding to the spectra of Fig. 4.7.After Ref. 141.

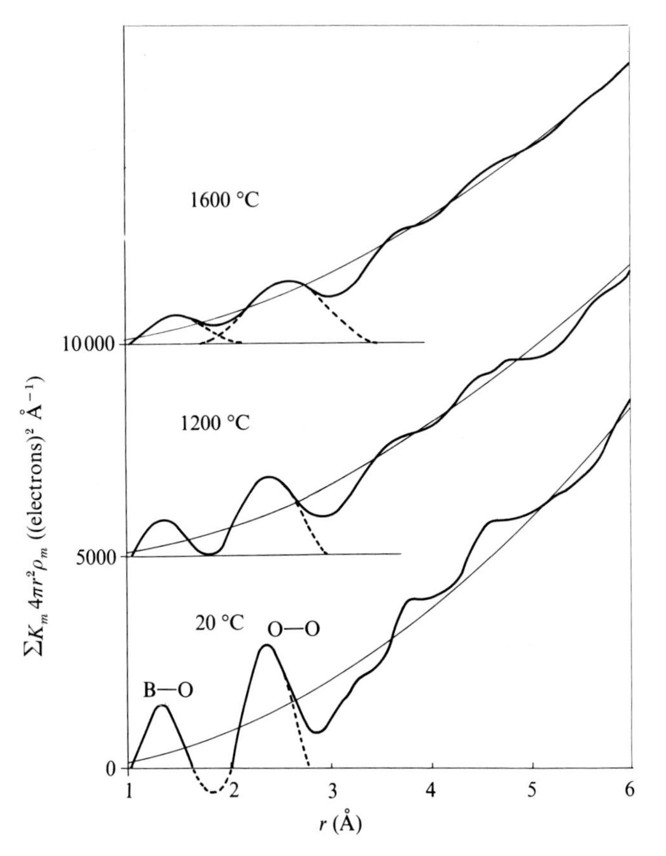

function" and the inversion of $Qi(Q)$ actually leads indirectly to a distribution function for the atomic centers. By simply inverting the difference $I_c - \sum_{c.u.} f_m^2$ an *electronic density distribution* is obtained within a factor k:

$$\sigma(r) = kr \int_0^\infty Q \left(I_c - \sum f_m^2 \right) \sin rQ \, \mathrm{d}Q$$

where the parasitic effects are less apparent so that the evaluation of the interatomic distances is more precise. On the other hand, it does not provide information on the coordination numbers, and this advantage of the RDF is thus lost.

Table 4.1. *Evolution of the structure of* B_2O_3 *with increasing temperature. After Ref. 141.*

Bond		Temperature (°C) 20	1200	1600
B–O	A_1 (el^2)	520	365	350
	n_B^O	3.3	2.3	2.2
O–O	A_1 (el^2)	1350	1190	1090
	n_O^O	6.7	6.0	5.4

Fig. 4.9. X-ray scattering spectra from GeO_2 solid at 20 °C and liquid at 1200 °C. After Ref. 141.

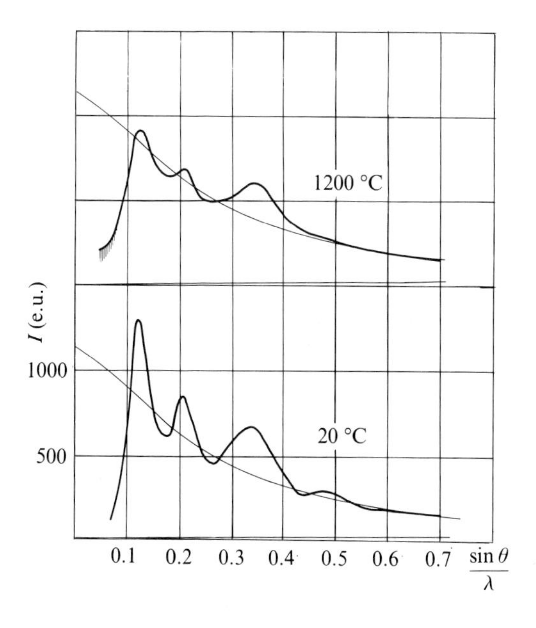

We have used this treatment for vitreous SiO_2 at 20 °C and 1600 °C and for GeO_2 which is vitreous at 20 °C and liquid at 1200 °C (Figs. 4.11 and 4.12) to calculate, from the principal interatomic distances, the bond angles Si–O–Si and Ge–O–Ge between two consecutive tetrahedra in the network.[143]

In vitreous SiO_2, the average Si–O–Si angle is between 127° and 160° i.e. 143° ±17° and is apparently unaffected by temperature. For GeO_2, vitreous at 20 °C, Ge–O–Ge is between 135° and 152°, and, for GeO_2 liquid at 1200 °C, between 137° and 180°. The progressive opening of this angle is thus easier to see for GeO_2, which is isostructural with silica but "weaker."

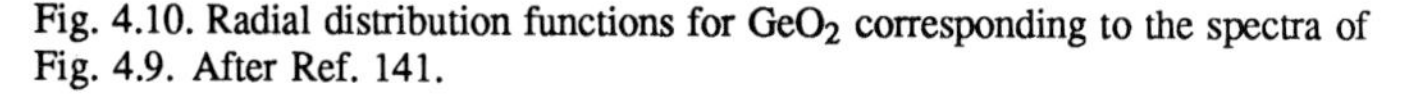

Fig. 4.10. Radial distribution functions for GeO_2 corresponding to the spectra of Fig. 4.9. After Ref. 141.

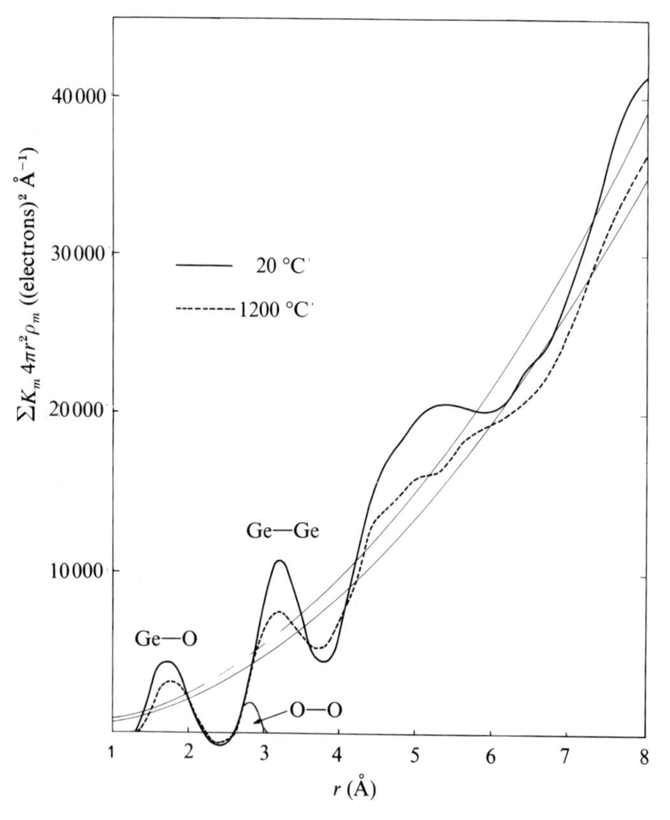

Figure 4.13 shows the RDF for vitreous K and Cs titanates which have been proven to have coordination $n_O^{Ti} \sim 6$ for the Ti^{4+} ion.[144] This result contradicts the Zachariasen–Warren theory which does not include the possibility of an octahedral coordination for a "network-former."

The preceding examples relate to a network structure where the coordination of the principal atom: B, Si, Ge appeared obvious. Consider now the case of an ionic liquid having a "dense" structure where this coordination is unknown in the liquid state, for example, molten NaF. Even though NaF does not form a glass, the study of the structure of liquid NaF is instructive because it completes the information provided by RDF studies. This can be compared with results obtained by dynamic molecular methods (cf. Chapter 7).

Figure 4.14 shows the RDF for molten NaF at 1000 °C.[145] This compound undergoes a large expansion on melting, ($\Delta V/V \sim 0.21$). Examination of the RDF discloses a series of peaks corresponding to the successive interatomic distances which vary as $\sqrt{1}, \sqrt{2}, \sqrt{3}, \sqrt{4}, \ldots$ thus following the progression of interatomic distances of the crystalline arrangement (NaCl type structure). Moreover, the first distance (Na–F) $\sim$ 2.3 Å is about equal to that in crystalline NaF. The large volume in-

Fig. 4.11. "Electronic" radial distribution centers. curves for vitreous SiO_2 at 20 °C and 1600 °C. After Ref. 143.

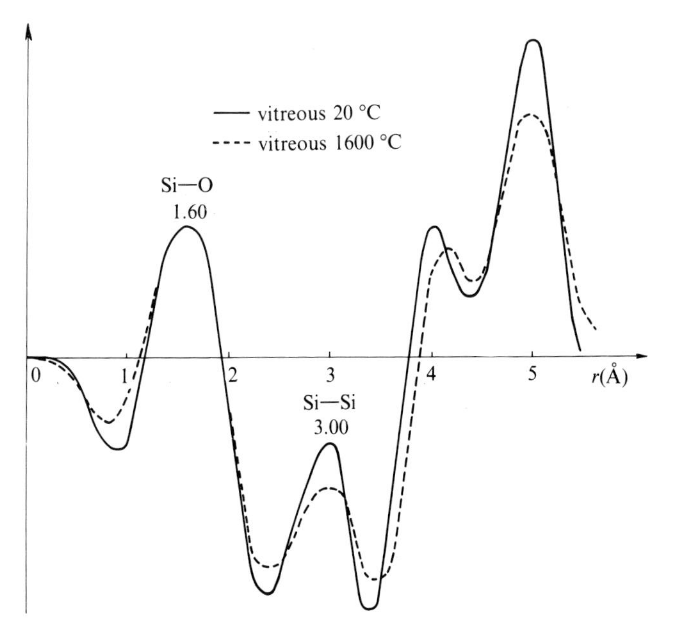

Fig. 4.12. "Electronic" radial distribution curves for vitreous GeO_2 at 20 °C and liquid at 1600 °C. After Ref. 143.

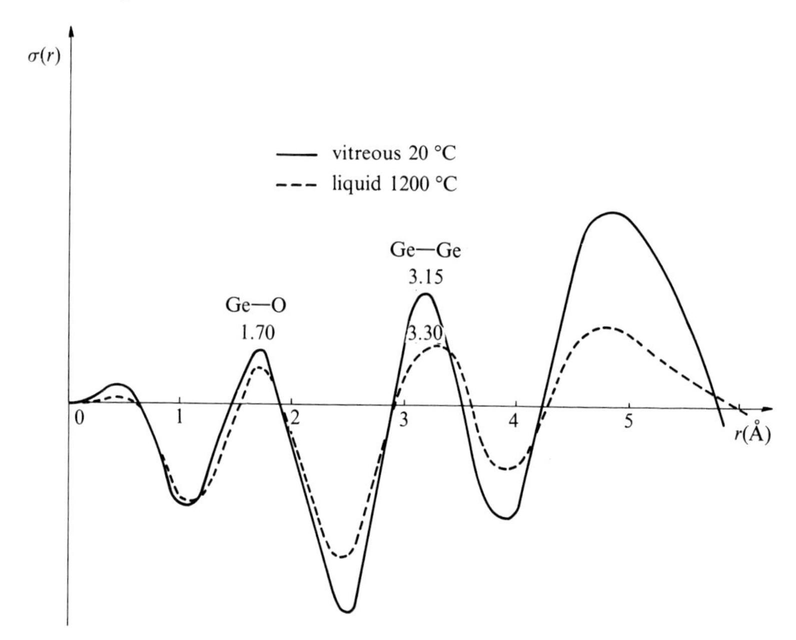

crease on melting thus cannot be interpreted as a uniform expansion of the interatomic distances but corresponds to the formation of voids (holes) in the liquid structure.

This is confirmed by calculation of the average first coordination number n_F^{Na} which is nearly equal to 4 (while in the crystal it has a value 6). It undergoes an *average* decrease because Na^+ ions in the neighborhood of the holes are no longer surrounded by 6 F^- ions but by a smaller number of ions. Further considerations permit the selection of a final model where the liquid is formed by clusters with a crystalline arrangement separated by "fluctuating fissures."[145]

4.2.5 *Studies by neutron scattering*

In the case of neutron scattering, the scattering factors $f_i(Q)$ are replaced by the "scattering lengths" b_i independent of Q which simplifies calculations. In particular, the normalization of the scattering spectra is facilitated because the scattered intensity at large angles tends toward a constant value.

The b_i do not vary systematically with the atomic number Z_i, certain of them can even be negative. This has been used for the determination

Fig. 4.13. Radial distribution functions for (*a*) vitreous $2\,TiO_2 \cdot K_2O$ and (*b*) vitreous $2\,TiO_2 \cdot Cs_2O$. After Ref. 144.

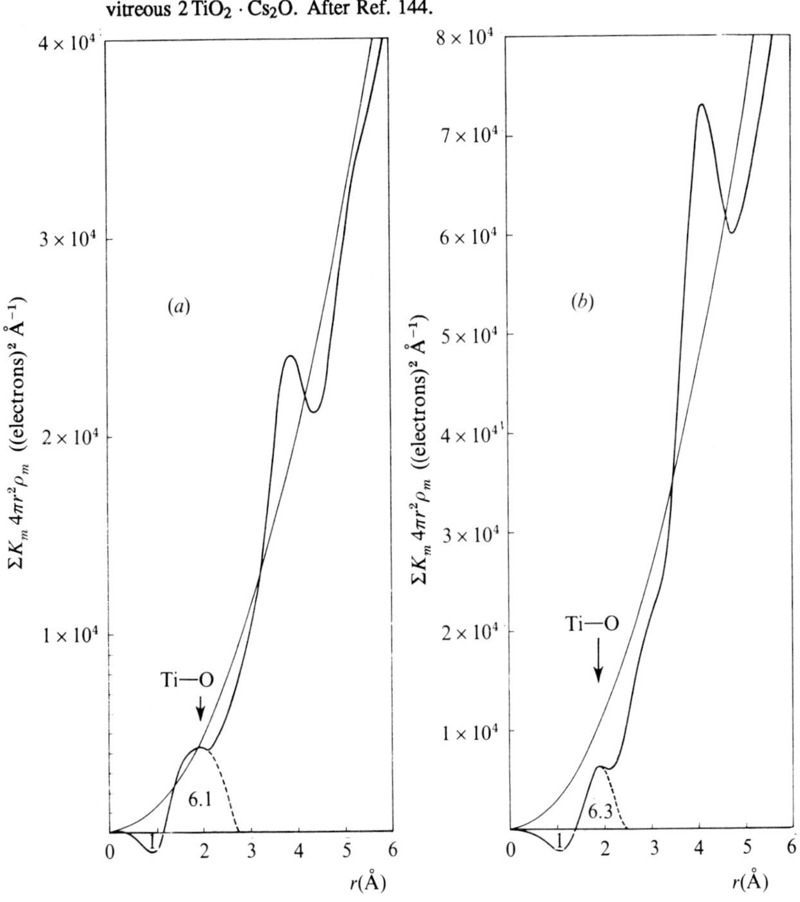

of the sites of Li (Ref. 146) and Ti (Ref. 147) by employing isotopes ^7Li and ^{48}Ti.

Figure 4.15 shows the RDF obtained from X-ray and neutron scattering for vitreous SiO_2. Note the better resolution of the O–O distance for the neutron case.[148]

4.2.6 *Various methods for the study of heteroatomic structures*

(a) *Additivity methods*

In the general case of glasses containing more than one kind of atom, the RDFs are superpositions of the partial functions from the different pairs. This rapidly leads to a lack of resolution and prevents the

Fig. 4.14. Radial distribution function for molten NaF at 1000 °C. After Ref. 145.

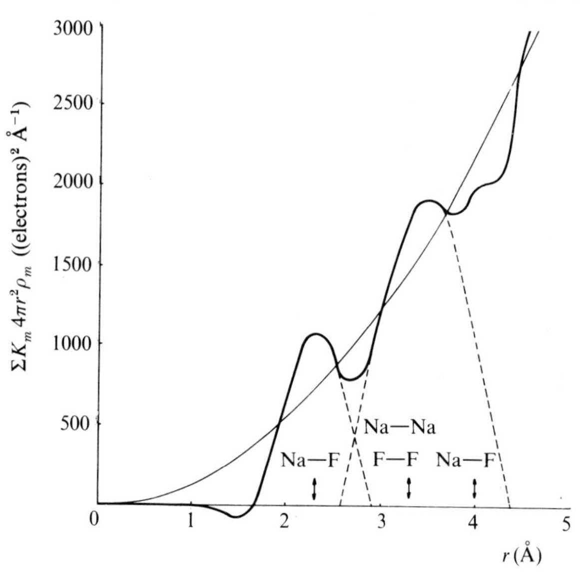

Fig. 4.15. Radial distribution functions for vitreous SiO_2 obtained by (*a*) X-ray scattering and (*b*) neutron scattering. After Ref. 148.

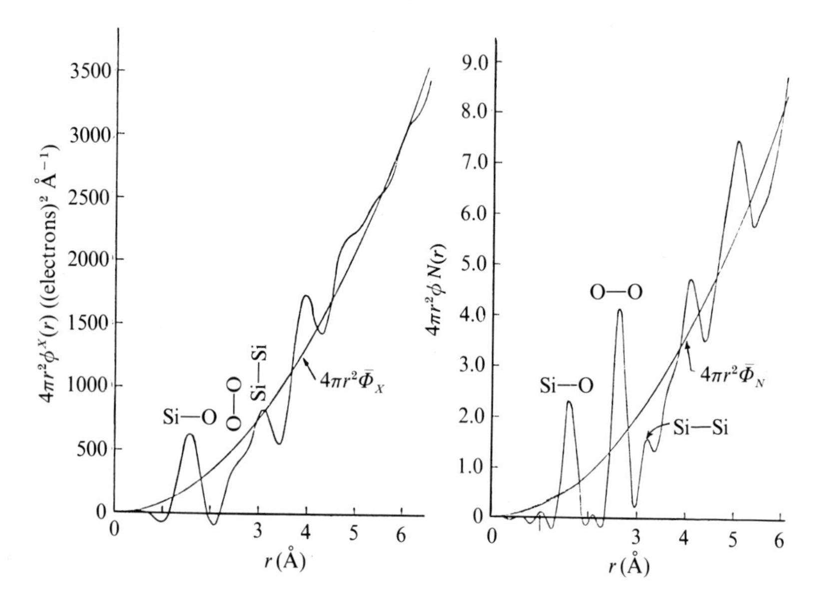

determination of the principal interatomic distances. Additivity (or difference) methods have been recommended[149] to extract this information. For example, for a glass $2\,SiO_2 \cdot Na_2O$, the distances Na–O and Na–Si could be obtained by subtracting the RDF of this glass from that of $2\,SiO_2$.

This approach is certainly questionable when the attempt is made to represent a complex disordered structure as the sum of two partial structures. It can, however, be justified in the case where the substance truly consists of a juxtaposition of two distinct structures. This is the important case of polyphasic glasses resulting from phase separation (cf. Chapter 6).

The study of a series of vitreous alkali metal fluoroberyllates[150] presenting large immiscibility gaps has been carried out with this method allowing the alkali metal cations to be located (Fig. 4.16).

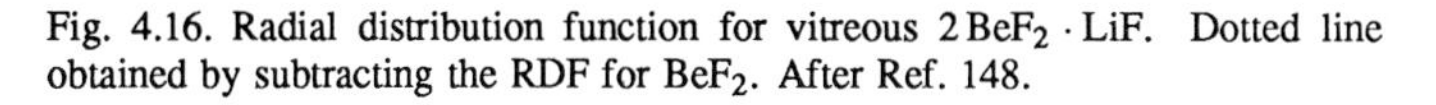

Fig. 4.16. Radial distribution function for vitreous $2\,BeF_2 \cdot LiF$. Dotted line obtained by subtracting the RDF for BeF_2. After Ref. 148.

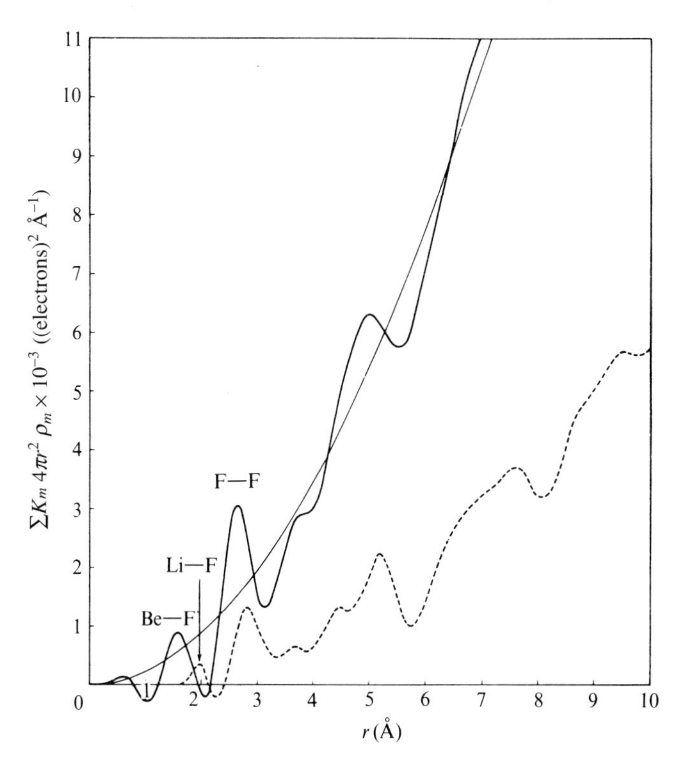

(b) *Substitution methods*

Assigning the peaks of a RDF can be facilitated by the study of a series of glasses where an atom, whose position is sought, is progressively replaced in the composition by a "heavy" atom, i.e. an atom having a higher scattering factor which will show up clearly in the scattering spectra and in the resulting RDF. In such a comparative series, an obvious risk is that the structure also changes with the substitution. To avoid this, ion exchange experiments have sometimes been used (cf. Chapter 10), e.g. by substituting Ag for Na in a SiO_2–Na_2O–CaO glass at rather low temperatures ($< T_g$) so that relaxation of the structure does not take place.

The RDF of the substituted glass seems to indicate that the Ag atoms (thus also the original Na sites) are not distributed uniformly but grouped at least as pairs.

But in this type of analysis, it is necessary to be careful since larger peaks generally come from the superposition of several different distances where the heavy atom plays a rôle as shown in a histogram of the interatomic pairs calculated from the crystalline arrangement. Figure 4.17 shows, for example, in the case of Ba silicate, that the Ba–Ba distance interferes with the distances Si–Ba and Ba–O. This brings into question the determination of the Ba–Ba separation distance obtained from the RDF.[151]

(c) *Joint use of X-rays and neutrons*

In the case of a biatomic substance, A–B, it is possible in principle to separate the three partial correlation curves relating to the pairs A–A, A–B, and B–B if the results of *three* independent scattering experiments are available. This is theoretically possible by making one X-ray and two neutron measurements with a different isotopic substitution in one of the cases to modify a scattering length b_i. Such a complete experiment is very difficult in practice and has not yet been attempted for oxide glasses. The only experiments carried out for SiO_2 consisted of one X-ray and one neutron measurement.[148]

Sadoc and Dixmier[152] obtained three partial RDFs for an amorphous alloy $Co_{81}P_{19}$ with one experiment using X-rays and two experiments with polarized neutrons (Fig. 4.18).

4.2.7 *Exact method in the heteroatomic case*

The preceding methods, which all contain the fundamental approximation of average effective $\overline{K}_i$, are the only ones which have been systematically employed up to recent years. However, an exact method exists as shown by Finbak[142] as well as by Waser and Schomaker[153]. Credit goes again to Warren for calling attention to this method and applying it to the case of glasses. We will only outline the principle of the

Fig. 4.17. (*a*) Histogram of the interatomic distances in crystalline $2B_2O_3 \cdot BaO$. (*b*) The same, spread 0.1 Å. (*c*) RDF for the crystalline compound. (*d*) RDF for the corresponding glass. After Ref. 151.

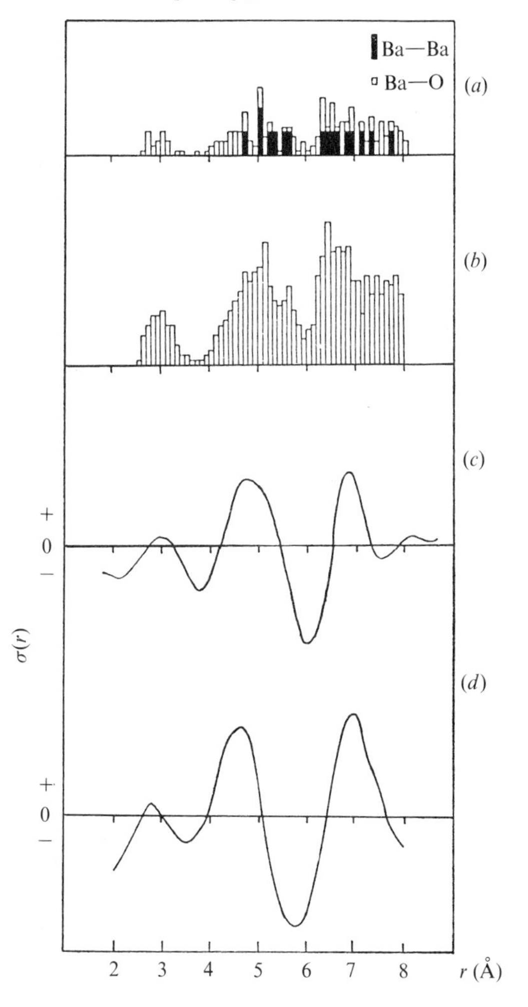

method. The reader is referred to the work of Warren[154] for the details and calculations.

This method consists of synthesizing a distribution curve from the "pair functions" P_{ij} which take into account the variations of f_i with Q, the correction factor $g^2(Q)$, and the chosen convergence factor $\exp(-\alpha^2 Q^2)$. The truncation effects, etc. are thus all incorporated into the P_{ij} and the method automatically takes into account the occurrence of secondary

Fig. 4.18. Partial RDFs for $Co_{81}P_{19}$. After Ref. 152.

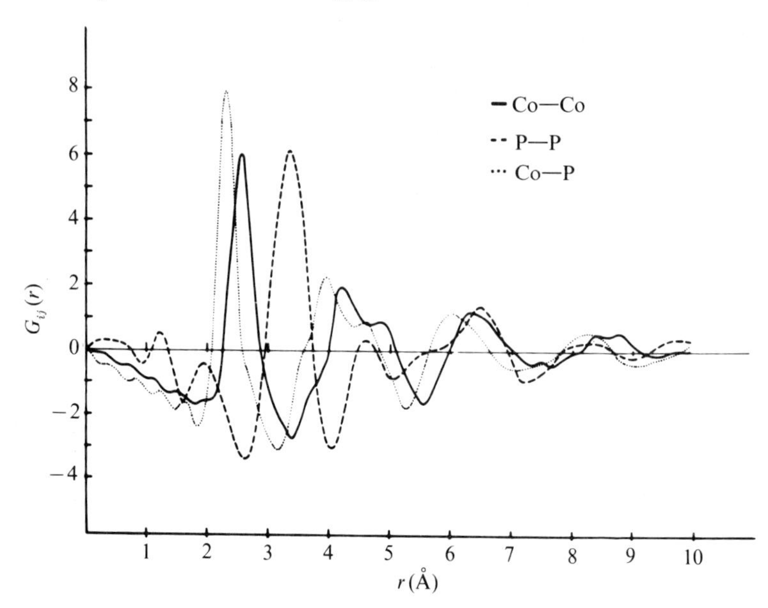

maxima (satellites) which are so troublesome in the preceding approximate methods. More precisely, the pair function P_{ij} is defined by the expression:

$$P_{ij} = \int_0^{Q_0} \frac{f_i f_j}{g^2(Q)} \exp(-\alpha^2 Q^2) \sin Q r_{ij} \sin Q r \, \mathrm{d}Q$$

which can be solved by means of an auxiliary function $Q_{ij}(x)$:

$$Q_{ij}(x) = \frac{1}{2} \int_0^{Q_0} \frac{f_i f_j}{g^2(Q)} \exp(-\alpha^2 Q^2) \cos xQ \, \mathrm{d}Q$$

taking into account the relation:

$$P_{ij}(r) = Q_{ij}(r - r_{ij}) - Q_{ij}(r + r_{ij})$$

where the second term can generally be neglected.

The distribution function is written as a function of the P_{ij} terms:

$$\sum_{\mathrm{c.u.}} \frac{N_{ij}}{r_{ij}} P_{ij}(r) = 2\pi r \varrho_0 \sum_{\mathrm{c.u.}} Z_j + \int_0^{Q_0} Qi(Q) \exp(-\alpha^2 Q^2) \sin rQ \, \mathrm{d}Q$$

In this expression, N_{ij} represents the average number of atoms i in the shell situated at a distance r_{ij} from an atom of type j; ϱ_0 is the average electron density, Z_j the atomic number of atom j, the function $g^2(Q)$ is the correction factor and $\exp(-\alpha^2 Q^2)$ is the convergence factor used. Q_0 represents the largest value of Q (truncation). The method has been successfully applied to vitreous SiO_2[155] and to vitreous B_2O_3.[156]

Figure 4.19 shows the auxiliary functions $Q_{ij}(x)$ calculated for the three interatomic pairs and in Fig. 4.20 the global distribution curve for the pair functions and the separate contributions from the principal pairs are indicated. In the derivation of this function, the distribution function of the Si–O angles from Fig. 4.21 has been introduced which gives satisfactory agreement and which is maximum for $\sim 145°$. In the case of B_2O_3, a similar calculation seems to indicate the presence of (B_3O_6) groups in the structure.

It is certain that the exact method presents a considerable advantage relative to the traditional methods and should be more widely employed.

Fig. 4.19. Auxiliary correlation functions $Q_{ij}(x)$ for SiO_2. After Ref. 155.

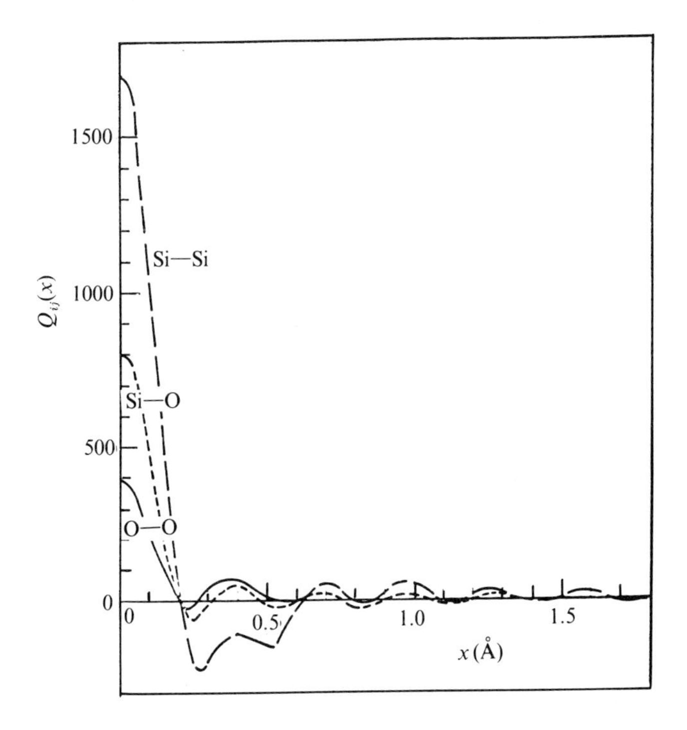

Fig. 4.20. Pair distribution function for vitreous SiO_2. The contributions of the principal pairs have been separated below. After Ref. 155.

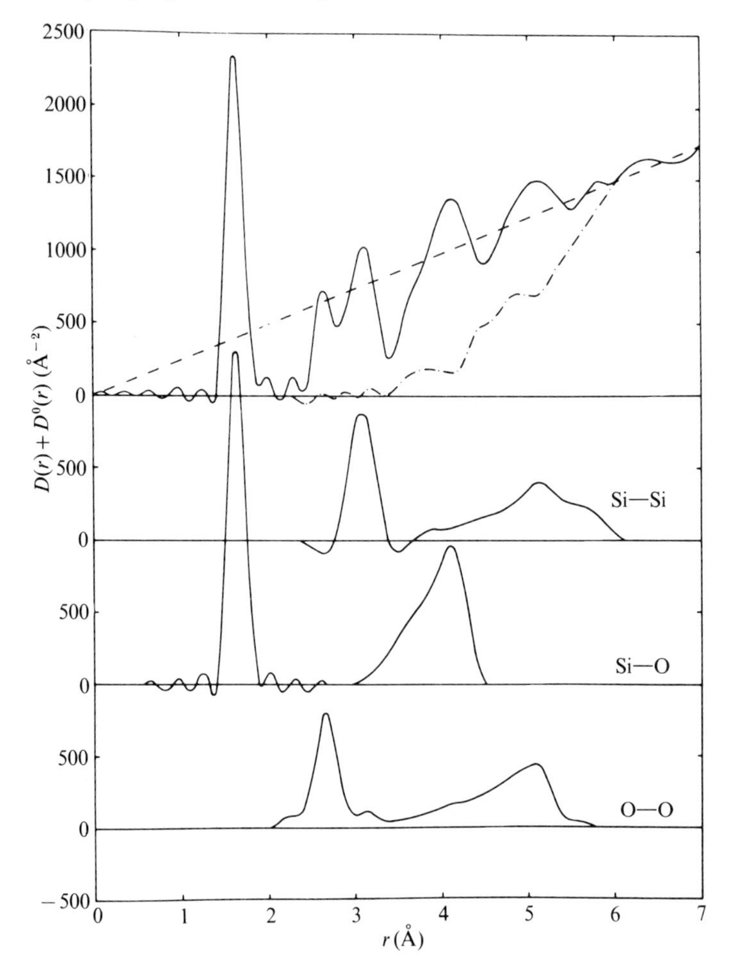

It ends the criticism relative to the average effective K approximation and above all permits an increase in the resolution *in extending the limit* Q_0 *without risk of introducing artificial peaks into the RDF, which have no structural significance* (because they are automatically taken into account in the Q_{ij} calculated in advance). Other sources of experimental errors must be minimized to obtain the precise measurement of I_c within as large a range of Q as possible.

The elimination of the incoherent intensity I_i (Compton scattering) can

Fig. 4.21. Distribution function of the angles Si–O–Si in vitreous SiO_2. After Ref. 155. The arrow indicates the average angle deduced from steric considerations. After Ref. 143.

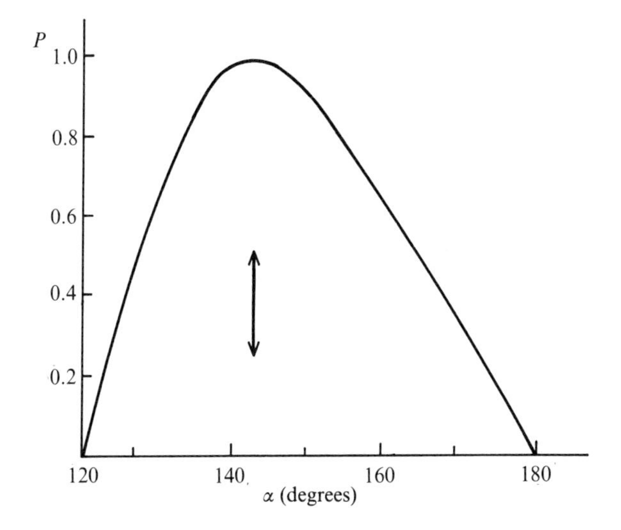

be approximated by the calculation as indicated previously. However, in the case of light elements, like B, I_i can greatly exceed I_c for high Q, causing significant errors. A method indicated by Warren consists of experimentally separating the Compton radiation:[157] using Rh K_α radiation, the flux scattered by the sample falls on a Mo foil exciting X-ray fluorescence which is measured. The Compton radiation (the longer wavelength) cannot excite the fluorescence and is thus eliminated from the recorded intensity.

The present tendency is to maximize the recorded limit of the scattered spectrum Q_0 to increase the resolution. Originally Q_0 was limited to values $\sim$ 16 Å^{-1} for Mo K_α and 20 Å^{-1} for Rh K_α. Modern sources of pulsed neutrons combined with time-of-flight measurements allow measurements up to 60 Å^{-1}.[158] Figure 4.22 shows the function $Qi(Q)$ for SiO_2 measured up to $Q_0 = 40$ Å^{-1}: the oscillations of this function are still appreciable even near to Q_0.

Needless to say these improvements expand our knowledge of short-range order in glasses but detailed interpretation requires a serious modeling effort. (cf. Chapter 7). More recent developments can be found in Refs. 159,160 and Ref. 161 discusses the newer possibilities offered by pulsed neutron sources.

Fig. 4.22. Interference function $Qi(Q)$ for vitreous SiO_2 obtained from neutron scattering. After Ref. 158.

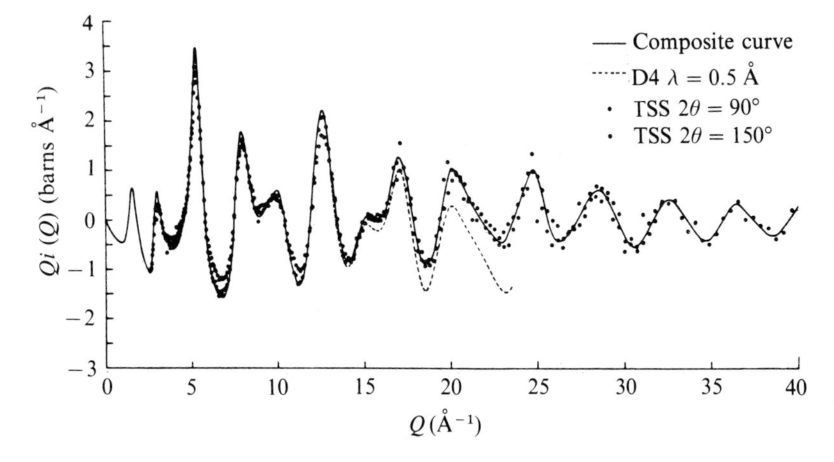

5 Structure of glass: Use of spectroscopic methods

5.1 Analysis of the symmetry of sites

Methods using radiation scattering lack resolution. Therefore other methods are used which can provide information on the disposition of nearest neighbors of a given ion, which then functions as a *probe*. This is the case for certain *spectroscopic* methods which enable the analysis of atomic *site symmetry*. Among the most frequently employed are infrared (IR) and Raman spectroscopy, Nuclear Magnetic Resonance (NMR), Electron Paramagnetic Resonance (EPR), Mössbauer spectroscopy, visible light spectroscopy using ligand-field theory, fluorescence spectroscopy, Extended X-ray Absorption Fine Structure (EXAFS) etc.

This imposing arsenal of investigative methods, currently used in the study of crystals, suffers from theoretical limitations in the case of disordered substances like glasses, because the interpretation is considerably complicated by the lack of periodicity in the material. Moreover, the methods are usually *specific* and only provide information on certain kinds of ions. It is generally necessary to use a whole set of methods to collect data on the different kinds of atoms present in a given glass.

A detailed description of all the above-mentioned methods is outside the scope of this work. Only typical examples directly connected with the most frequent applications and the particular interpretation methods applied to glass systems will be presented here.

The limits of these methods will also be discussed to allow the reader to choose between different techniques for the solution of problems of interest.

Readers wishing a deeper understanding of a particular topic should consult classical sources on various spectroscopies. The excellent work of Wong and Angell[15] includes an extensive bibliography on spectroscopic studies of glass up to 1974. More recent results are to be found in the proceedings of various conferences.[35–45]

5.2 Vibrational spectroscopies: IR and Raman

Molecular groups or glass networks have a certain number of characteristic vibrational modes determined by the masses of the constituent atoms, the interatomic forces and geometric arrangements, i.e. the structure. The vibrations modify the electric dipole moment leading to the absorption of electromagnetic waves in the range of 10000–100 cm^{-1}: i.e. 1–100 μm corresponding to the domain of *IR spectroscopy*. Studying the IR spectrum thus provides, in theory, certain information on structure.

Similarly, when light passes through a transparent material, the associated electric field causes the local electronic charges to vibrate. These oscillating electrons functioning as dipole emitters produce scattered light of the same frequency as the incident radiation (Rayleigh scattering). If, however, a coupling occurs between the tensor of the electric polarizability and vibratory modes of the material, the energy of the scattered photons can be increased (or decreased) by absorption (or creation) of a quantum of vibrational energy. This is the *Raman effect* and the supplementary lines (Stokes and anti-Stokes) are shifted relative to the Rayleigh line by a quantity which corresponds to the energies of the different vibrational modes.

The shifts cover the same range of frequencies as the IR spectrum and thus both spectroscopies reach the fundamental vibrations of the structure. Because the *rules of selection* are not the same in all cases, certain vibrations may not appear in both methods, making them complementary.

5.2.1 *General methods of analysis*

Even in the case of crystals, the direct determination of the structure from vibrational spectra is not possible. The structure must be known in advance using X-ray crystallographic methods before attributing different spectral bands to different vibrational modes. The spectral analysis is based on the use of group theory, applicable to isolated molecules or to three-dimensional crystalline networks. In the case of disordered networks, this approach is impossible and increasingly complex approximations are used.

(a) *Isolated groups*

SiO_4 tetrahedra, for example, are assumed to vibrate independently from the rest of the network. Group theory (for group T_d appearing in this system) can then predict that there will be only four bands in the vibration spectrum corresponding to the modes shown in Fig. 5.1: a symmetric stretching mode of frequency ν_1, a doubly degenerate bending mode ν_2, a triply degenerate stretching mode ν_3, and a triply degenerate bending mode ν_4. While all modes are active in the Raman effect, only ν_3 and ν_4 are active in the IR.

(b) *Coupled groups*

The groups in a crystal interact with their neighbors. If it is assumed that the coupling is weak and that the nearest neighbors of a unit

Fig. 5.1. Vibrational modes on an isolated tetrahedron. Type A: symmetric; E: doubly degenerate; F: triply degenerate.

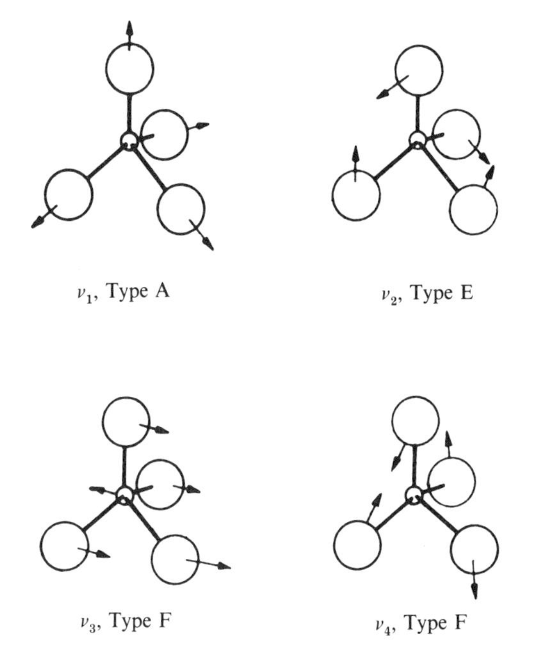

form a group of fixed symmetry, a *site group* can be defined if the crystal space group is known, as well as the point group of the molecule and the number of molecules in the unit cell. The symmetry of the site may be less than that of a molecule which can lead to relaxation of the selection rules – forbidden bands can appear and certain bands can split.

(c) *Limited network models*

Vibrational modes of structural units forming more and more complex links, e.g. chains or sheets of SiO_4 tetrahedra can be analyzed. Different predicted modes can be compared to observed experimental results (Fig. 5.2).

This enables evaluation of the distortion in the SiO_4 tetrahedra in a network containing non-bridging oxygens[162] and comparison of the force constants of the bonds Si–O, Ge–O and Be–F.

(d) *Statistically disordered networks*

The statistical model of Bell and Dean (cf. Chapter 7) served as a basis to calculate vibrational spectra, leading to a mode density distribution which can be compared to IR or Raman spectra (Fig. 5.3).

Fig. 5.2. IR absorption spectra of glasses: (*a*) $2\,SiO \cdot Li_2O$, (*b*) $2\,SiO_2 \cdot Na_2O$, (*c*) $2\,SiO_2 \cdot K_2O$, (*d*) $2SiO_2 \cdot Rb_2O$. After Ref. 162.

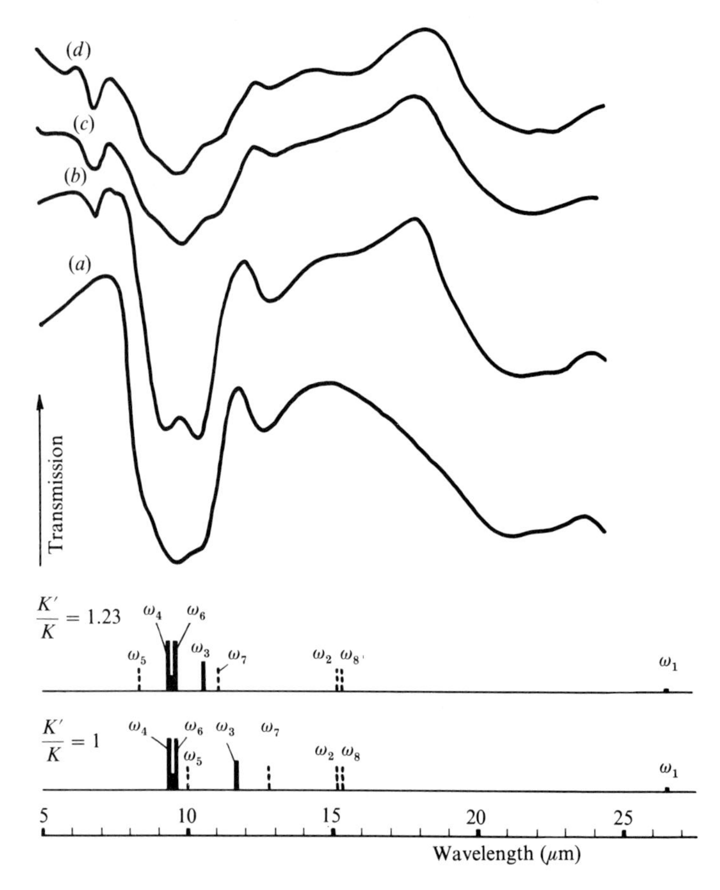

An important result of this study is that the observed bands may not be attributed to particular vibrational modes; the maxima always appear to be determined by mixing of different modes in variable proportions.[163]

In trying to match the experimentally determined IR and Raman spectra with calculated spectra by different methods, the problem of the validity of a chosen model is encountered. The final result of the calculation consists of an ensemble of force constants governing the relative movement of atoms in a chosen unit. The model is more likely to be correct if it satisfactorily predicts the effects of an isotopic substitution and if the group of calculated force constants is "self-consistent." It means in particular that the force constants must be "reasonable" and vary inversely with the length of the bonds, etc. . . .

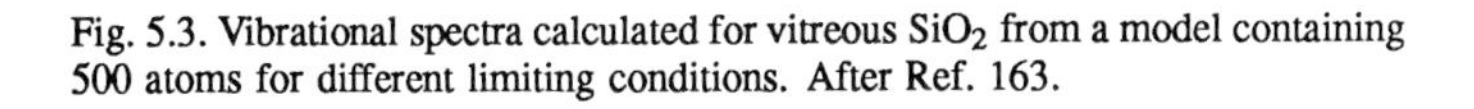

Fig. 5.3. Vibrational spectra calculated for vitreous SiO_2 from a model containing 500 atoms for different limiting conditions. After Ref. 163.

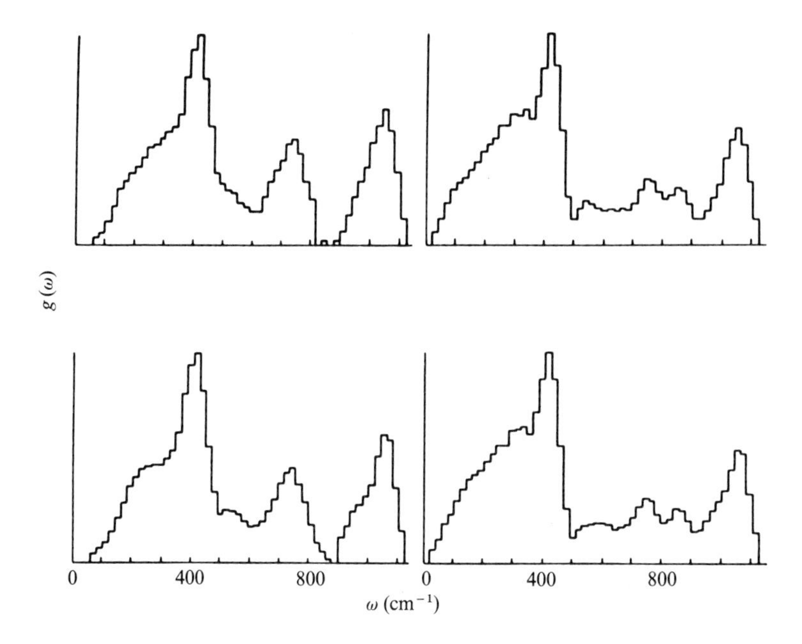

(e) *Empirical comparisons*

More frequently, the interpretation of Raman or IR spectra from glasses is done by comparing results obtained from corresponding *crystalline* materials. It comes down to locating bands in the glass spectra which can be identified with vibration bands in the crystal spectra, calculated from the structure known in advance.

5.2.2 *Examples of studies by IR spectroscopy*

(a) *Determination of coordination numbers*

If we take the examples of the IR spectra of vitreous SiO_2 and different crystalline allotropic varieties: quartz, cristobalite, etc. (Fig. 5.4), all the spectra appear to be similar and contain a principal band near 9.1 μm (1098 cm^{-1}). Since the structure of all the varieties is based on 4-coordinated silicon, this band is associated with a fundamental vibration of the SiO_4.

In contrast, stishovite, the only allotropic variety of crystalline SiO_2 where Si is in coordination 6, has a clearly different spectrum, thus confirming that in vitreous SiO_2 the coordination of Si is 4 and not 6.

Similar results are obtained for vitreous GeO_2 (Fig. 5.5), where the

Fig. 5.4. IR spectra of vitreous SiO_2 (top) and crystalline allotropic varieties. After Ref. 164.

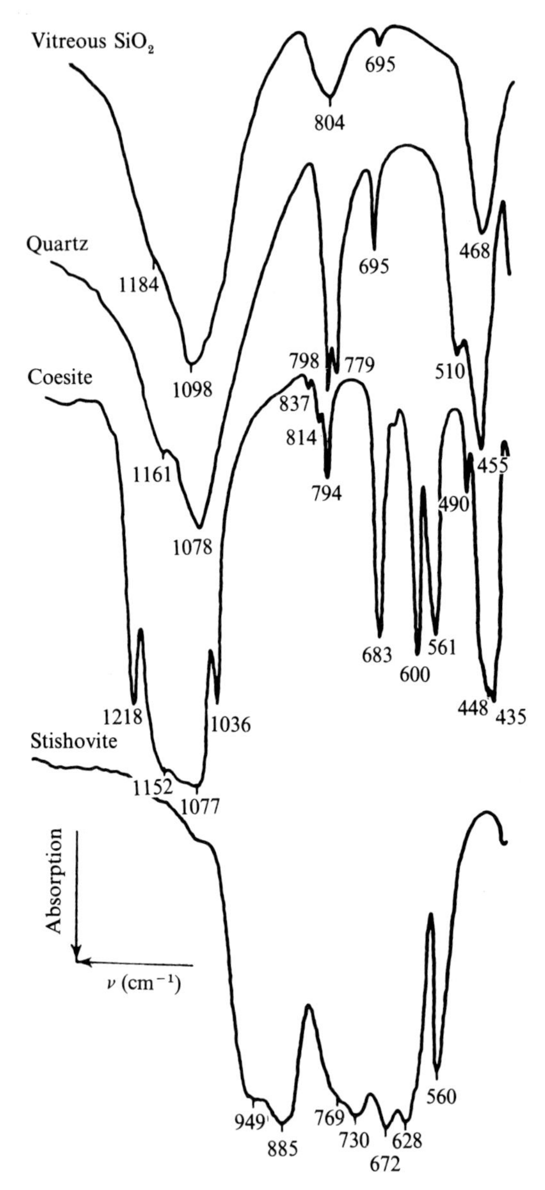

comparison is made with crystalline varieties of GeO_2: hexagonal (Ge in coordination 4) or insoluble tetragonal (Ge in coordination 6). Here again, the results clearly indicate a coordination 4 for vitreous GeO_2.

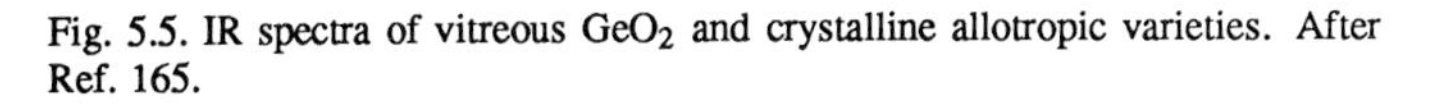

Fig. 5.5. IR spectra of vitreous GeO_2 and crystalline allotropic varieties. After Ref. 165.

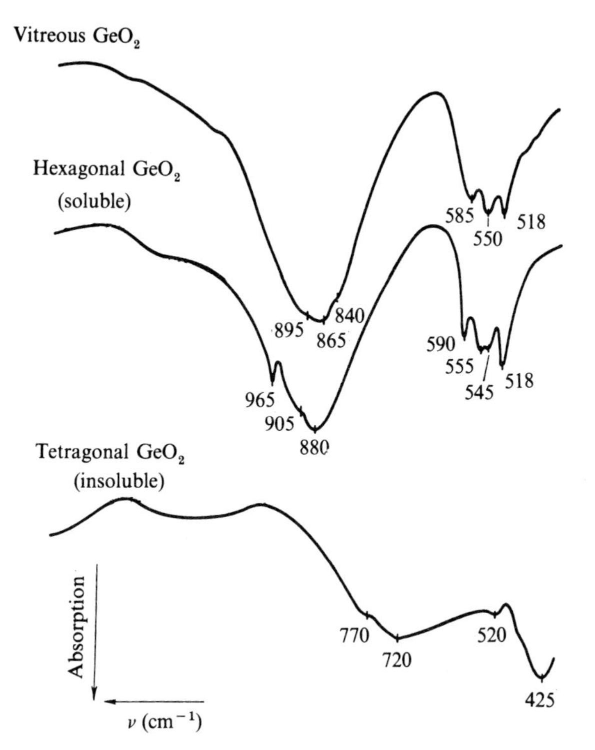

Generally speaking, the frequency ranges of coordination groups determined from a series of crystalline compounds can be used to determine the coordination number of certain cations in glasses. Tarte[166] used the method to identify the coordination of Ti in particular. But this type of analysis is to be used with caution. The couplings between different coordination groups and the diffused character of the absorption bands, especially at low frequencies, reduces the sensitivity of the method.

Table 5.1 shows the influence of the coordination number n on the characteristic frequencies of XO_n groups deduced from empirical studies.

A frequently used method consists of progressively modifying the composition of a series of glasses. The evolution of bands, the appearance

Table 5.1. *Characteristic frequency ranges for coordination groups in oxide glasses. After Ref. 166.*

	Characteristic ranges (cm^{-1})			
	Isolated groups		"Condensed" groups	
Cation X	XO_4	XO_6	XO_4	XO_6
Si	1050–800	?	1200–1000	950–900
Ge	850–680	< 500 ?	900 and less	700 and less
Ti	800–690	< 500 ?	?	600–500
Al	800–650	500–400	870–700	650 and less
Fe^{III}	650–550	400–300	700–550	550–400
Cr^{III}		450–300		650 and less
Ga	700–570	~ 400	750–600	600–500
Zn	500–450		600–400	
Mg	600–500			480 and less
Fe^{II}	~ 450			~ 320
Mn^{II}	~ 450			~ 320
Li	500–400	300?	600–400	< 300?

of new bands, etc. are compared with hypothetical modifications of the network based, for instance, on a model of a partially depolymerized network. For example, in the system SiO_2–Na_2O (Fig. 5.6), the band near 9 μm is attributed to a stretching mode of the Si–O while the band near 10.5 μm is attributed to non-bridging oxygen bonds $Si–O^-$. This latter band appears independent of the associated cation,[167] even in structures containing hydroxyl groups. Vibrational methods provide information on the coordination number of those cations which are bonded to their neighbors by mainly covalent bonds (as is the case of "glass-forming" elements for oxide glasses). On the other hand they do not provide information on the coordination of ionically bonded cations ("modifiers") which require the use of alternative methods.

Vibrational studies of chalcogenides have been limited to several simple glasses:

$$Se, \quad As_2S_3, \quad Tl_2Se–As_2Te_3, \quad CdGeAs_2$$

and glasses of the binary system Ge–S, which are often studied for their transmission properties.[168]

Fig. 5.6. Comparison of IR absorption spectra of the glass $2\,SiO_2 \cdot Na_2O$ (top) and SiO_2 gel (below). (*a*) Spectrum calculated for a "sheet" of (SiO_4) tetrahedra; (*b*) The preceding spectrum after substitution of OH (or OD) for the singly bonded oxygen; (*c*) Spectrum calculated for a "chain" of SiO_4 tetrahedra after substitution of OH (or OD) for the singly bonded oxygen. After Ref. 167.

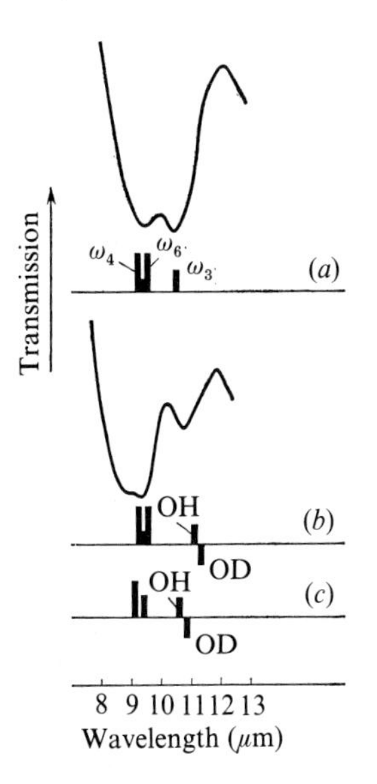

Interpretations are made on the basis of *molecular* models with the units, which are weakly coupled to each other, vibrating in an independent manner. In As_2S_3 the "molecule" consists of the AsS_3 group with S forming a bridge.

(b) *Water in glasses*

Glasses containing traces of water strongly absorb IR radiation in the spectral region 2.7–4.5 μm (3700–2250 cm^{-1})[169] (Fig. 5.7). These bands are currently used for quantitative analysis of the water content of glasses. Water is generally present in the form of hydroxyl groups, and in silicate glass the various bands are attributed to OH groups either singly bonded to a Si (2.7 μm) or hydrogen bonded to a non-bridging oxygen ($\sim$ 3.5 μm) or, as in the case of glasses with high alkali metal content, to more mobile SiO_4 groups ($\sim$ 4.5 μm).

The study of glass surfaces and the processes of absorption is carried out by IR spectroscopy. Studies have been made in particular on H_2O and NH_3 using porous glass with a large specific surface area.

Fig. 5.7. IR absorption bands due to the presence of water in a glass $70\,SiO_2 \cdot 30\,Na_2O$ After Ref. 169.

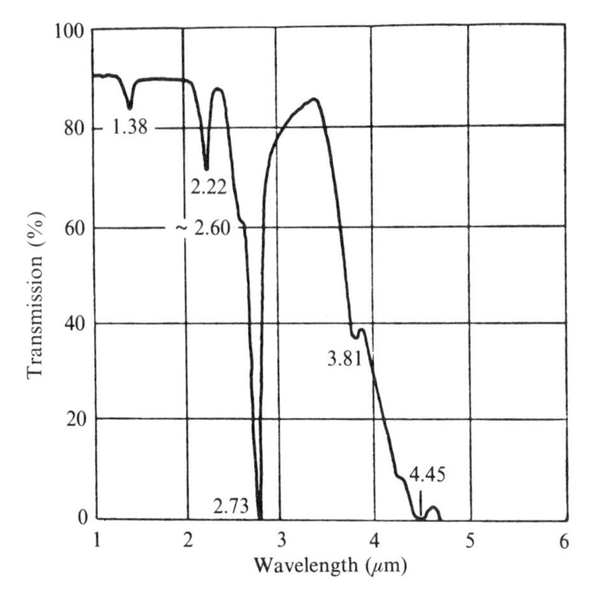

(c) *Temperature dependence*

IR spectra play an important rôle in glass technology by permitting the evaluation of heat transfer during glass fusion or forming processes. Only a few studies have been made of the evolution of IR spectra as a function of temperature; Fig. 5.8 shows an example for vitreous SiO_2:[170] raising the temperature to 1500 K produces a depolymerization effect equivalent to that resulting from the addition of 13 % Na_2O (indicated by the broken line).

5.2.3 *Raman spectroscopy*

Although Raman spectroscopy is complementary to IR spectroscopy, its satisfactory application to glass problems is more recent and dates from the advent of intense monochromatic sources (lasers).

Compared with IR spectra, Raman spectra have the following advantageous characteristics:

– The observed peaks are generally well defined, limited in number and often polarized.

– The spectra are simpler than IR spectra and depend strongly on the composition.

Fig. 5.8. Evolution of the IR spectra (index of absorption) of vitreous SiO_2 as a function of the temperature. After Ref. 170.

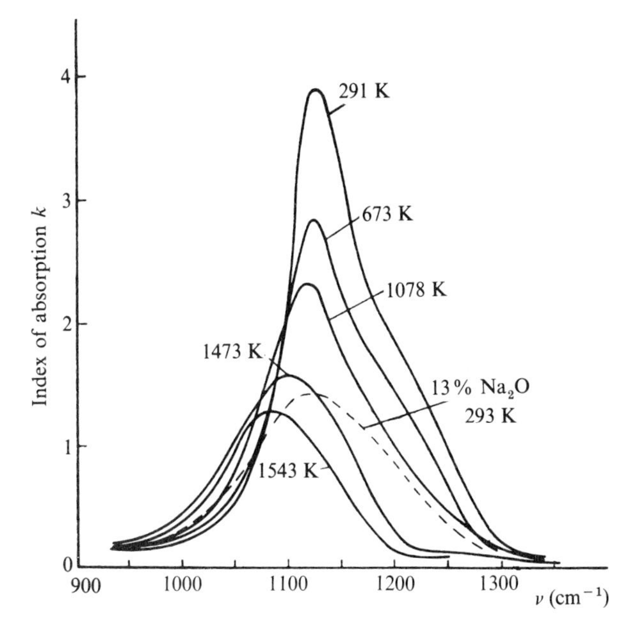

– The low frequencies are easier to investigate.

– The spectra have a low sensitivity to surface contamination and water content in the glasses.

– Because samples are larger, volume effects rather than thin film effects are measured.

– Measurements at elevated temperatures are easier.

Fig. 5.9 is an example of Raman spectra for the system B_2O_3–Na_2O.

As for the IR spectra the interpretation is made by relation to groups existing in the parent crystalline compounds of known crystallographic structure. Raman spectroscopy is helpful to probe different glasses but does not generally provide unambiguous results as to the different structural elements present. For B_2O_3 for instance, it cannot prove by itself the presence of boroxol units and its conclusions must be substantiated by those obtained from other techniques.

For monographs on IR and Raman techniques, consult Refs. 15, 172 and 173. For a detailed study of the vitreous network structure of SiO_2, see Ref. 174.

Fig. 5.9. Raman spectra of glasses $(1-x)B_2O_3 \cdot xNa_2O$. After Ref. 171.

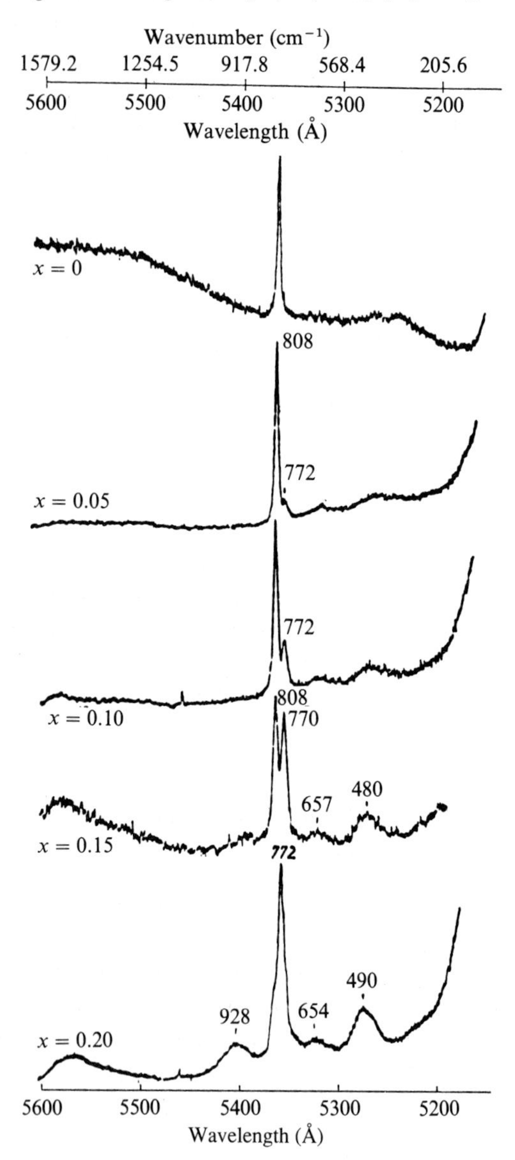

5.3 Nuclear Magnetic Resonance (NMR)

NMR has been used to determine the coordination numbers of certain cations in glasses. This method studies transitions between energy levels of the nuclear magnetic moment in the presence of an external magnetic field.

Thus, the nucleus serves as a *probe* and provides information on its immediate surroundings. Table 5.2 lists the nuclei which have been studied in glasses by NMR. Most of the atomic nuclei possess a dipole magnetic moment:

$$\vec{\mu} = g\mu_0 \vec{I}$$

where $\vec{I}$ is a vector of modulus $\sqrt{I(I+1)}$, I is the nuclear spin, μ_0 Bohr's magneton and g is the "g factor" of the nucleus. The interaction between the dipole moment and an applied magnetic field $\vec{H}$ leads to a set of Zeeman levels:

$$E = -\vec{\mu} \cdot \vec{H} = -g\mu_0 \, |\vec{H}| \, m_I$$

where m_I is the magnetic quantum number, which can take $2I+1$ values: $I, I-1, \ldots, -I$. In the absence of other interactions, these levels are equally spaced. The transition between adjacent levels ($\Delta m_I = \pm 1$) can be induced by subjecting the system to electromagnetic radiation of frequency ν_0 which satisfies the resonance condition:

$$h\nu_0 = g\mu_0 H$$

The levels being equally spaced, there is only one single resonance frequency in this case. The transition frequency is called the Larmor frequency.

By varying the frequency or the intensity of the field, an absorption line is obtained at resonance which is broadened by various interactions.

By applying the electromagnetic energy in the form of pulses and measuring the magnetization as a function of time, the *network-spin relaxation time* T_1 is obtained. This characterizes the influence of the relative movements of the nuclei and of the lattice.

5.3.1 *Dipole–dipole interaction*

The local magnetic field on the nucleus due to neighboring nuclei produces a dipole–dipole interaction which varies from site to site for a given type of nucleus. This distribution causes a *broadening* of the resonance line around ν_0; the exact calculation of the line form is difficult.

Table 5.2. *Properties of the nuclei which have been studied by NMR in glasses. After Ref. 175.*

Nucleus	Natural abundance (%)	Larmor frequency (MHz) (for 10^4 G)	Magnetic moment (nuclear magnetons)	Spin (units of h)	Electric quadrupole moment (10^{-14} cm^2)
^{1}H	99.9844	42.577	2.7927	1/2	
^{7}Li	92.57	16.547	3.2560	3/2	-4.2×10^{-2}
^{9}Be	100	5.983	−1.7740	3/2	2.0×10^{-2}
^{11}B	81.17	13.660	2.6880	3/2	3.55×10^{-2}
^{19}F	100	40.055	2.6273	1/2	
^{23}Na	100	11.262	2.2161	3/2	0.1
^{27}Al	100	11.094	3.6385	5/2	0.149
^{29}Si	4.70	8.460	−0.55477	1/2	
^{31}P	100	17.235	1.1305	1/2	
^{45}Sc	100	10.343	4.7491	1/2	
^{51}V	~100	11.193	5.1392	7/2	0.3
^{73}Ge	7.61	1.485	−0.8768	9/2	−0.2
^{75}As	100	7.292	1.4349	3/2	0.3
^{111}Cd	12.86	9.028	−0.5922	1/2	
^{113}Cd	12.34	9.444	−0.6195	1/2	
^{125}Te	7.03	13.45	−0.8824	1/2	
^{133}Cs	100	5.585	2.5642	7/2	< 0.3
^{205}Tl	70.48	24.57	1.614	1/2	
^{207}Pb	21.11	8.899	0.5837	1/2	

In practice, the broadening is measured as well as the moments of the function of the form $f(\nu - \nu_0)$ of the line. The "second moment" M_2 is the most frequently employed. It is defined by the expression:

$$M_2 = \int_{-\infty}^{+\infty} (\nu - \nu_0)^2 f(\nu - \nu_0)\,\mathrm{d}\nu \Big/ \int_{-\infty}^{+\infty} f(\nu - \nu_0)\,\mathrm{d}\nu$$

For a glass, considering all orientations of the position vector $\vec{r}_{jk}$ of the atomic nuclei j, k as equally probable, the calculation leads to:

$$M_2 = K_1 \sum_j r_{ij}^{-6} + K_2 \sum_k r_{ik}^{-6}$$

where the first term includes the interactions between identical nuclei, and the second the interactions between different kinds of nuclei. K_1 and K_2, which depend on the properties of the nuclei, can be evaluated. The method consists of comparing M_2 from a model with M_2 obtained from the form of the experimental curve.

When the nuclei surrounding the chosen nucleus are in motion, the resonance lines are narrowed. The passage from glass → liquid can thus be studied through the obvious accompanying decrease in the width of the line and the second moment (Fig. 5.10).[176,177]

Fig. 5.10. Variations of the second moments as a function of temperature for a glass $HNO_3 \cdot H_2O$: T_g, glass transition temperature; T_f, fusion temperature; T_c, crystallization temperature on reheating the glass. The various symbols correspond to several experiments. After Ref. 177.

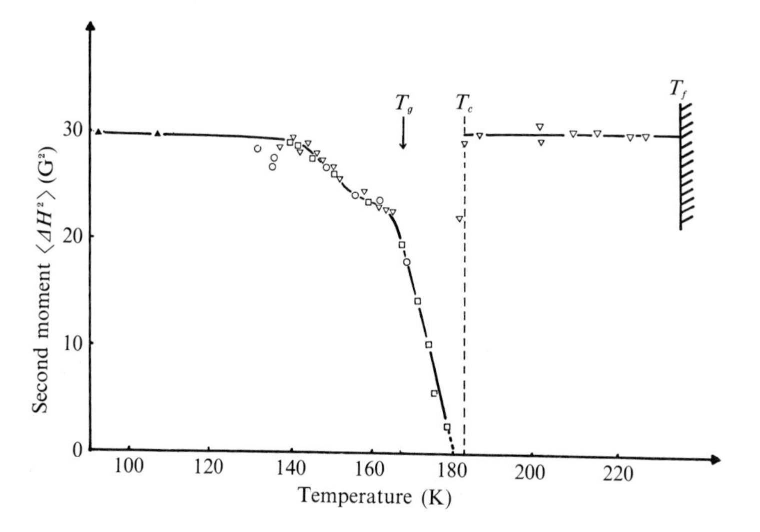

A study in the glass system SiO_2–Na_2O of the width of dipole lines in the NMR spectra from ^{29}Si and ^{23}Na nuclei which have magnetic moments has shown that the Na^+ ions are not uniformly distributed but grouped in pairs or larger groups.[178]

5.3.2 *The chemical shift effect*

In an atomic system, the effective field $\vec{H}_{\text{eff}}$ acting on the nucleus is generally less than the applied field $\vec{H}_0$ because of the screening effect of the electrons around the nucleus, such that:

$$\vec{H}_{\text{eff}} = \vec{H}_0(1 - \sigma_{ij})$$

where σ_{ij} is the *chemical shift* tensor.

The magnetic resonance of a nucleus in a given exterior field will occur at different frequencies in different chemical compounds.

The study of the shift of ^{29}Si in the glasses and compounds of the system SiO_2–K_2O confirmed that structural groups characteristic of crystalline compounds are present in the glass.[179]

The chemical shift, which is small in the case of ^{29}Si, is higher in the case of ^{205}Tl and ^{207}Pb and has been used to study atomic bonds in glass.[180,181]

5.3.3 *Quadrupole interaction*

All nuclei with spin $I \geq 1$ have an electric quadrupole which can interact with the electrostatic field at the nucleus. Such gradients originate from either the distribution of electrons in the bonds around the nucleus or the charges present on other atoms or ions. The quadrupole interaction thus constitutes a sensitive means of detection of bonding modes and the atomic environment.

The general theory of quadrupole effects in NMR enables the calculation of the perturbation introduced in the resonance. The interaction energy depends upon the electric field gradient tensor ∇E, the quadrupole moment eQ and the orientation of the magnetic field $\vec{H}$ relative to the principal axes of ∇E. When the latter has an axial symmetry, first-order perturbation theory shows that the unique resonance is split into a spectrum of $2I$ lines. For an isotropic disordered substance like glass, the envelope of all responses for all mutual orientations of ∇E (considered as equally probable) must be taken into account and the relative spectra of each transition must be superimposed.

The method consists of establishing structural models to allow the calculation of the coupling constant and of the asymmetry parameter of ∇E. One tries to obtain the best possible agreement between the experimental and calculated spectra by varying adjustable parameters.

Most of the studies have been on glasses containing B because of the naturally high abundance (81.17 %) of the isotope ^{11}B which has a high quadrupole moment (cf. Table 5.2).

When the atom ^{11}B is 3-coordinated with O, the coupling constant is 2.41 to 2.81 MHz whereas in tetrahedral coordination, this constant is about 50 kHz. This large difference between coupling constants enables the separation of effects due to these resonances and even their quantitative measurement in materials where both structural types coexist. The most remarkable application has been the measurement of the fraction N_4 of B atoms in 4-coordination in alkali metal borate glasses with the "boron anomaly" (cf. Chapter 8). Contrary to the results of more or less empirical calculations, the fraction N_4 increases regularly up to 40 mol. % of modifying oxide, then decreases (Fig. 5.11).[182]

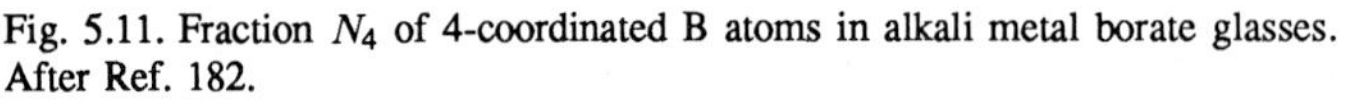

Fig. 5.11. Fraction N_4 of 4-coordinated B atoms in alkali metal borate glasses. After Ref. 182.

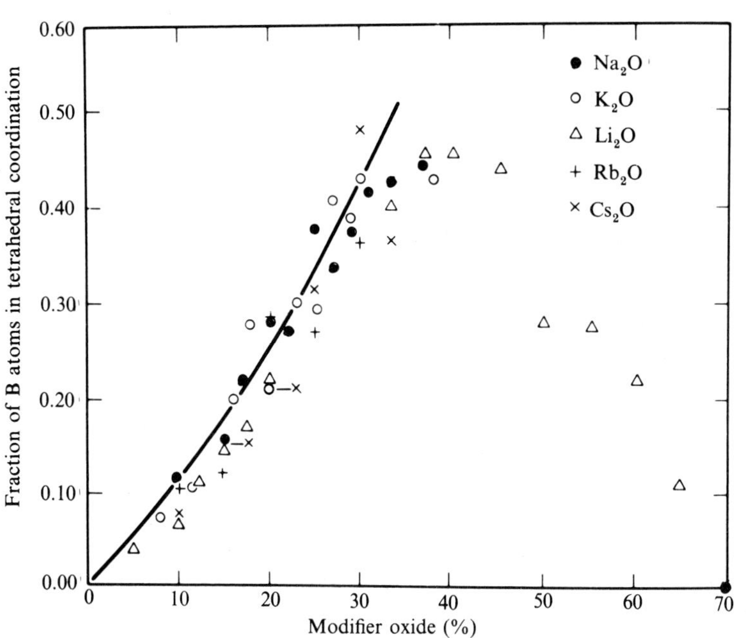

The study of B_2O_3 glasses enriched to 37 % in isotope ^{17}O (normal abundance: 0.037 %) has provided spectra in accord with a model based on boroxol units.

The preceding examples relate to the transitions $m = \frac{1}{2} \leftrightarrow -\frac{1}{2}$ which are affected to a second order by quadrupole interactions. Calculations on spectra of nuclei with integer spins have recently been made. The transitions $0 \leftrightarrow -1$ and $1 \leftrightarrow 0$ of the isotope ^{10}B $(I = 3)$ give spectra which are particularly sensitive to the structure. The study of B_2O_3–Na_2O glasses seems to confirm the existence in the glasses of groups present in the parent crystalline compounds.[179,183,184]

Application of NMR techniques to silicate glasses is less favorable because the isotope ^{29}Si does not have a quadrupole moment $(I = 1/2)$ and its natural abundance is low (4.7 %). In contrast, ^{51}V (abundance 100 %) possesses a quadrupole moment close to that of ^{11}B and appears to be an ideal probe; it has been used in studies of semiconductor glasses based on vanadium phosphates. Applications to chalcogenide glasses are less numerous and of a provisional nature. Reviews may be found in References 185–8.

5.4 Optical absorption spectra of transition metal ions

Transition metal ions in which the 3d electron levels are incomplete and contain 1–9 of the 10 possible electrons (Ti^{3+}–Cu^{2+} ions) can be used as sensitive "probes" for the symmetry of the site which they occupy in the structure. These ions absorb light at well-defined wave lengths and the polarization of the ions in the structure of a crystal or a glass modifies the energy levels of the outer electrons in a characteristic manner. The coordination changes are thus related to color changes due to absorption effects.

A quantitative description of these effects is possible using crystal field or ligand-field theories. In the crystal field approach, the bonds are assumed ionic and the separations of the levels are calculated from the symmetry of the electric fields acting on the ion. Another approach starts with considering the molecular orbitals and covalent bond effects. The two treatments form the bases for the "ligand-field theory." The immediate neighbors of an ion are called ligands. In the case of oxide glasses, the ligands are the oxide ions.

To understand the basic ideas of this theory, consider first the simplest case of the ion Ti^{3+} which has a single electron in the d level. This electron can occupy one of five d orbitals corresponding to the five possible values of the magnetic quantum number. Figure 5.12 shows forms of these orbitals: the first three d_{xy}, d_{xz}, and d_{yz} have lobes directed at 45° off the main axes; the last two $d_{x^2-y^2}$ and d_{z^2} follow the axes. These figures give an idea of the angular distribution of the wave function.

For a free Ti^{3+} these five orbitals belong to the same degenerate level. However, when the ion is placed in an electric field the orbitals no longer have the same energy but separate into distinct groups. Thus, for Ti^{3+} placed in an *octahedral* environment (Fig. 5.13(*a*)), the six oxygen ligands placed along the axes direct negative charges toward the central ion. The electron of the Ti^{3+} will tend to avoid the $d_{x^2-y^2}$ orbital the lobes of which point directly along the axes and would preferably occupy the orbitals d_{xy}, d_{xz} d_{yz}. Calculation shows that d_{z^2} is degenerate with $d_{x^2-y^2}$.

The *octahedral* crystal field cancels the degeneracy and separates the five levels into two groups. The energy difference between levels is designated Δ (or 10 Dq) and constitutes an empirical parameter characteristic of the ligand strength.

In *tetrahedral* symmetry (Fig. 5.13(*b*)), the situation is reversed. The lobes of the orbitals d_{xy}, etc. pointing toward the ligands are avoided;

Fig. 5.12. The d orbitals.

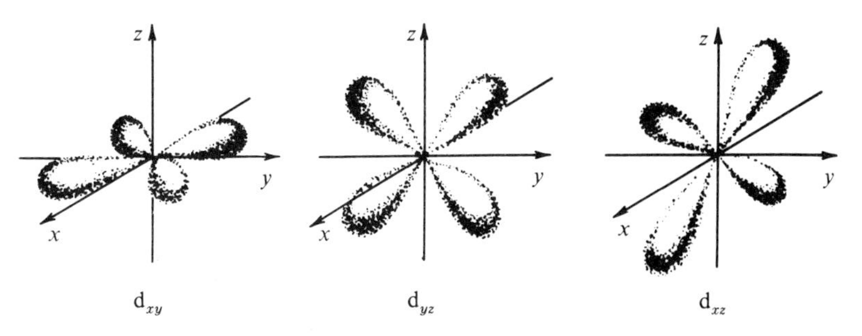

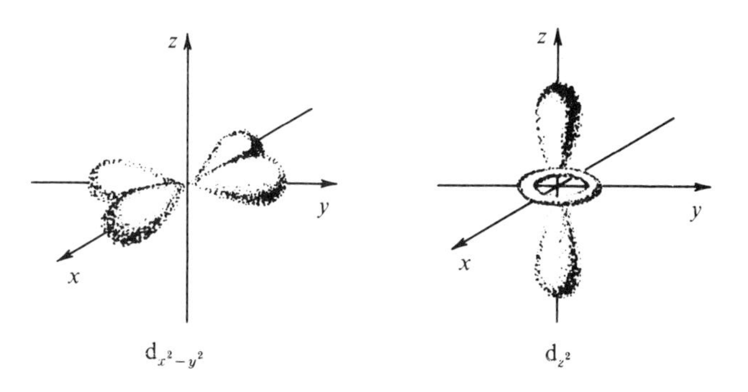

Fig. 5.13. An ion placed in an environment: (*a*) octahedral, (*b*) tetrahedral.

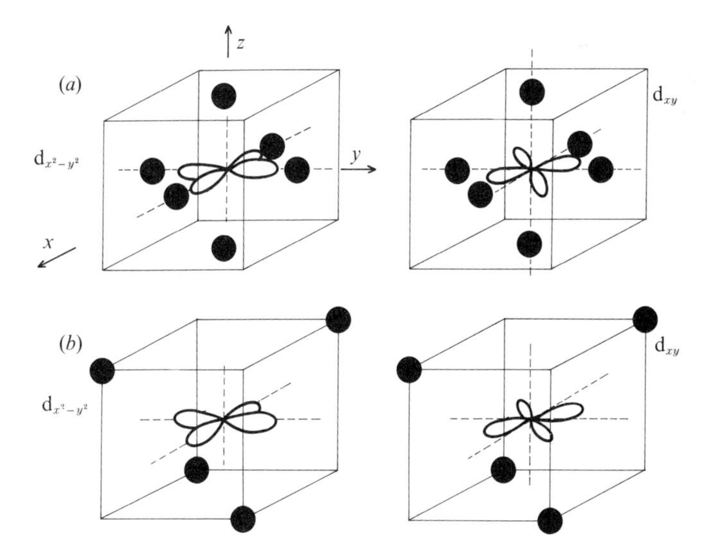

the separation of the levels is indicated in Fig. 5.14 and the calculation shows that Δ_{tet} is weaker than in the octahedral case: $\Delta_{tet} = -\frac{4}{9}\Delta_{oct}$. The electronic transitions between these two energy levels correspond to a band situated in the visible or near-IR region of the spectrum and measurement of the band position allows the determination of 10 Dq.

Fig. 5.14. Separation of the energy levels of the Ti^{3+} ion in an octahedral or tetrahedral environment.

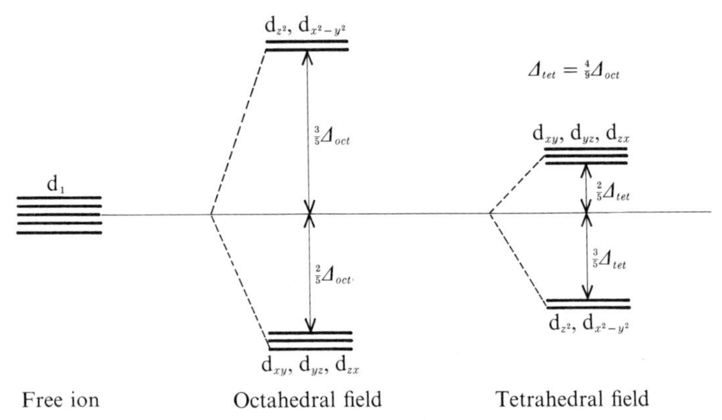

This case is not significant because the presence of a single transition does not inform us about the disposition of the ligands. When an ion has more than one d electron, the situation is complicated since the spins tend to be parallel and the electrons tend to occupy the lower energy orbitals. The use of group theory allows the determination of the manner in which the degeneracy is raised. This leads to the construction of diagrams giving the energy of different orbitals as a function of Δ for given environmental symmetries. Figure 5.15 gives, for instance, such a diagram (Orgel's diagram) for a Cr^{3+} ion in an octahedral environment.

The method consists of finding the value of 10 Dq for which the different theoretical transitions correspond quantitatively to the bands observed in the spectrum. In this test, 10 Dq is systematically varied and if agreement cannot be found, the diagram of another configuration is tried (e.g. tetrahedral). A detailed description of this method may be found in the article by Bates.[189]

Table 5.3 collects the main results obtained in this manner for the coordination numbers (CN) for transition metal ions in different oxide glasses. Except for the ions Co^{2+} and Ni^{2+} which can exist simultaneously in coordination 4 and 6, octahedral coordination (CN=6) is the general rule for all glasses studied up to the present time.

Fig. 5.15. Orgel's diagram for the Cr^{3+} ion in an octahedral environment. After Ref. 189.

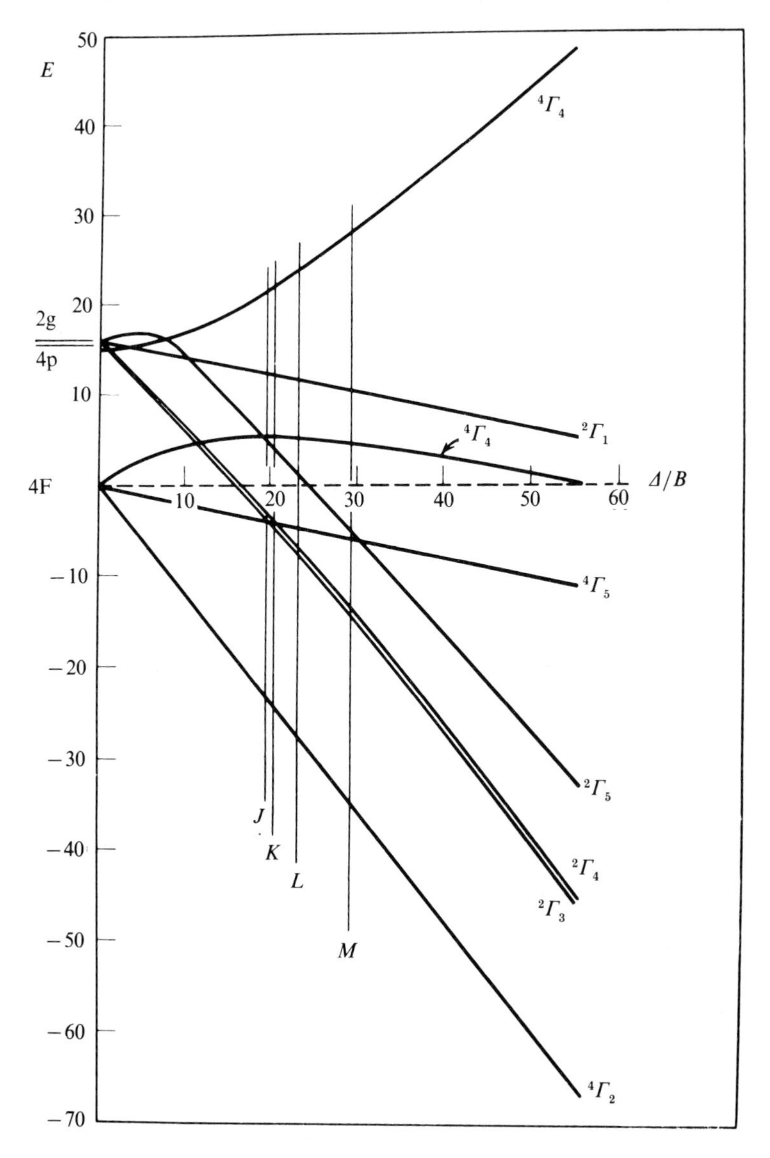

Table 5.3. *Coordination of transition metal ions determined by ligand-field theory. After a compilation in Ref. 165.*

Electronic configuration	Ion	Type of glass	Coord. number CN	Glass color
$3d^1$	Ti^{3+}	phosphates, borosilicates	6	violet-brown
	V^{4+}	silicates	6	blue
$3d^2$	V^{3+}	silicates, borates, phosphates, borophosphates	6	green
$3d^3$	Cr^{3+}	silicates, borates, aluminophosphates	6	light green
	V^{2+}	silicates, aluminophosphates	6	green
$3d^4$	Mn^{3+}	silicates, borates	6	violet
$3d^5$	Mn^{2+}	silicates	4 or 6	pale yellow
	Fe^{3+}	silicates, borates	4 or 6	yellow-brown
$3d^6$	Fe^{2+}	silicates, borates, aluminophosphates	4 or 6	blue-green
$3d^7$	Co^{2+}	borates rich in alkali,	4	blue
		borates poor in alkali,	6	pink
		silicates	4	blue
$3d^8$	Ni^{2+}	borates rich in alkali,	4	blue
		borates poor in alkali,	6	yellow-brown
		silicates, aluminophosphates	6	violet
$3d^9$	Cu^{2+}	silicates, borates, aluminophosphates	6	blue

An interesting application is in the study of high-pressure effects on the local symmetry of cation sites in various vitreous silicates and phosphates. It can be shown in particular that tetrahedral sites are less compressible than octahedral sites and, as the pressure increases, the ions Co^{2+} and Ni^{2+} pass from tetrahedral to octahedral sites.

The method is limited because large quantities of transition metal ions (coloring elements) cannot be introduced into the structure without an excessive increase in the optical density. Because of their very low concentrations, the cations studied in this manner must often be considered as impurities rather than elements intrinsic to the structure; the symmetry of the cation sites may be reduced by this fact. Distortions of the site, in particular by the Jahn–Teller effect, make the interpretation of the spectra even more complex.

In a similar manner, *rare earth ions* give colors due to the separation in their 4f levels. The situation is more complex because there are seven orbitals (instead of five 3d) and the levels are more deeply situated in the interior of the ions which significantly reduces the effect of the ligands.

5.5 Electron Paramagnetic Resonance (EPR)

The EPR method is extremely sensitive for the detection of certain atomic sites in the structure and the study of their symmetry. It is frequently combined with transmission optical spectra previously discussed. The method consists of studying the separation of electron levels of the atoms in the presence of an external magnetic field.

When a free atom or ion has a resultant kinetic moment $\vec{G}$, its dipole magnetic moment will be:

$$\vec{\mu} = \gamma \vec{G}$$

where γ is the gyromagnetic ratio given by the formula:

$$\gamma = -\frac{ge}{2mc}$$

where e and m are the charge and the mass of the electron and c is the velocity of light. g is a pure number the value of which depends on the relative contributions of the orbital and the spin. For a free electron $g = g_0 = 2$, or, taking into account a quantum electrodynamic correction:

$$g_s = 2.002322$$

If only the orbital moment $\vec{L}$ is present in an atom:

$$\vec{G} = \hbar \vec{L} \qquad g = g_L = 1$$

If only the electronic spin $\vec{S}$ is present:

$$\vec{G} = \hbar \vec{S} \qquad g = g_S$$

When both $\vec{L}$ and $\vec{S}$ are present, the value of $g = g_J$ is a function of coupling and is, for example, (LS coupling):

$$g_J = \frac{J(J+1)(g_L + g_S) + \{L(L+1) - S(S+1)\}(g_L - g_S)}{2J(J+1)}$$

with $\vec{J} = \vec{L} + \vec{S}$.

The resulting electronic magnetic dipole moment is:

$$\vec{\mu}_J = -\left(g_J \frac{e}{2mc}\right)\hbar\vec{J} = g_J\beta\vec{J}$$

where the quantity $\beta = e\hbar/2mc$ is the electronic Bohr magneton.

In the presence of a magnetic field $\vec{H}$ the Zeeman interaction is:

$$E = -\vec{\mu}_J \cdot \vec{H} = g_J\beta H M_J$$

where M_J is the component of the total angular moment $\vec{J}$ in the direction of the field. Magnetic dipole transitions occur according to the selection rule $\Delta M_J = \pm 1$ and for an alternating field of frequency ν applied perpendicularly to $\vec{H}$, a quantum of energy is absorbed:

$$h\nu = g_J\beta H$$

when the electron spin passes from a position parallel to the magnetic field to an anti-parallel direction. An emission is produced in the inverse process.

At thermal equilibrium, a larger number of spins is in parallel position corresponding to a lower energy and there is a net absorption of energy.

The separation $g_J\beta H$ between the Zeeman levels increases linearly with H. For $g = 2$ and $H = 3300$ G $\nu \sim 10$ GHz which corresponds to the "X band" of wave guides. Experimentally, ν is held constant and the field H is varied to achieve resonance.

The observed transitions depend, in a complex manner, on the ion or the given paramagnetic center, the symmetry of the crystal field, the spin–orbital coupling effects and the hyperfine interactions between the electron and the nucleus. In practice, the concept of the "spin-Hamiltonian" is utilized to describe the observed resonance effects (cf. Ref. 190). It can be uniquely determined if the atomic structure around the paramagnetic center is known – the converse is not true, even though various possible models can be tested.

The g factor of a solid differs from the free spin value g_s as a result of spin–orbital coupling. Perturbation theory permits the definition of a tensor quantity $\Delta g_{ij} = (g_{ij} - g_s)$ called the *g shift* which is analogous to the chemical shift in NMR.

Experimentally, the factor g, also called the *spectroscopic separation factor*, is deduced from the formula $h\nu = g\beta H$ through measurement of ν and H. It depends on the orientation of the external magnetic field relative to the crystalline electric field; therefore, it is a tensor quantity.

The spin-Hamiltonian will for instance take the form:

$$\mathcal{H} = \vec{H} \cdot g \cdot \vec{S} + \vec{I} \cdot A \cdot \vec{S}$$

where the spin–orbit and Zeeman–orbit interactions are included in the first term and the hyperfine interactions in the second term.

The principal components of the tensor g and those of the hyperfine tensor A are measured experimentally.

Synthesis of the spectrum is sought by assuming a spin-Hamiltonian, and adjusting the values of the components of g and A by introducing distribution functions, to take into account the effect of disorder. However, in practice, very different distribution functions lead to nearly identical calculated spectra. The choice of the model depends finally on many other physical arguments.

In the case of glasses, EPR has been applied to study two types of paramagnetic centers:

1. substitutional impurities such as transition metals or rare earths;
2. paramagnetic centers produced by irradiation.

An excellent summary of studies up to 1974 is in the book by Wong and Angell[15] which covers the subject in depth. Only a general outline of this subject will be given here.

5.5.1 *Study of substitutional impurities*

In the case of transition metal ions, the interaction of the outer 3d electrons with the ligands is so strong that only the electron spin contributes to the magnetism. (In the case of rare earths, the interaction of the 4f layers is weaker and the paramagnetism can be considered similar to that of free ions with contributions from both orbital movement and spin.)

The studies can be classified according to the electronic configuration of the external d^n layers. Table 5.4 gathers some of the most characteristic results along with g values and indications of site symmetry. Moreover, Ti^{3+} and Mn^{2+} have also been used as site symmetry indicators to study the glass crystallization process. The symmetrization and ordering of paramagnetic centers becomes apparent. This is indicated by the spectra long before the effect can be detected through X-rays.

Table 5.4. *Transition metal ions studied in glass by EPR. After data from Ref. 15.*

Configuration	Ion	Glass	*g*	Site
	Ti^{3+}	silicates	1.922–1.930; 1.94	distorted-octahedral
	Zr^{3+}	silicates	1.89; 1.960	
d^1	V^{4+}	borates, phosphates	1.936; 1.976	distorted-octahedral, tetragonal
	Cr^{5+}	silicates, borates	1.98	
	Mo^{5+}	borates, phosphates	1.91	tetragonal
	W^{5+}	silicates, phosphates	1.64–1.72	distorted-tetragonal
d^3	Cr^{3+}	phosphates	2–6	rhombic
	Mo^{3+}	phosphates		rhombic
	Mn^{2+}	silicates, borates, phosphates	4.3; 3.3; 2.0	octahedral, distorted-cubic
d^5		chalcogenides	4; 2	
	Fe^{3+}	silicates	6; 4.2	predominently tetrahedral
		phosphates	4.2; 2	predominently octahedral
d^7	Co^{2+}	fluorberyllates	4; 2	tetrahedral
d^9	Cu^{2+}	silicates, borates, phosphates	~ 2	octahedral

5.5.2 *Defects produced in solids by irradiation*

High-energy electromagnetic radiation, charged particles and neutrons can bring about permanent structural effects when interacting with matter. The resulting "defects" depend in the first place on the kind of radiation. Ultraviolet (UV), X-rays and γ-rays penetrate significant thicknesses (order of 1 cm) and have mainly electronic effects; the same is true of fast electrons even though their depth of penetration is less ($\sim$ 1 mm).

Charged particles (D^+, α) penetrate less (order of μm), but because of their large collision cross-sections they cause considerable damage, mainly electronic, and induce an artificial radioactivity. Fast neutrons cause nuclear and electronic displacements in significant depths ($\sim$ 1 cm) and may introduce artificial radioactivity for higher doses.

The interactions can thus be very complex, and according to the particular case, can result in the atom's displacement from its normal position or can produce ionizing effects. New impurities can also be created *in situ* by nuclear transmutations (e.g. $^{10}B \rightarrow {}^{7}Li$ where a network modifier locally substitutes for a network former). When projectile particles remain embedded in the solid, the process is called *ion implantation.* An originally crystalline material can be progressively converted into an amorphous material by high-dose irradiation (metamict state). All the preceding effects can be extremely complex. Modern nuclear technology has stimulated this research branch and numerous methods have been developed to help determine structures.

5.5.3 *The concept of a network defect in glass*

Since a disordered material does not possess long-range order, the idea of an extended network defect, e.g. a dislocation, does not have any meaning for a glass. Only the concept of a point defect may be used for which consideration of short-range order is sufficient.

Taking as an example a network of SiO_2 or a more complex glass resulting from its modification by the introduction of alkali metal cations, several possibilities can be distinguished *a priori*:

(*a*) The absence of certain atoms from their normal positions, i.e. *vacancies*: this is less probable in the case of Si, but common for bridging and, especially, weakly bonded non-bridging oxygens.

(*b*) Additional atoms in positions different from those normally expected i.e. *interstitials* – because of the "open" network structure, this category is difficult to define rigorously. Thus non-bridging oxygens accompanied by alkali metal cations could be considered as interstitials relative to the pure SiO_2 network. Oxygens not belonging to the network can be considered as interstitials.

(*c*) Atoms of a nature different from those normally present in the network are *substitutional.* This frequently occurs for SiO_2 where the impurities Ge^{4+}, Al^{3+}, P^{5+}, etc. occupy Si^{4+} sites in the network. In the case of more complex glass containing several kinds of glass-formers, this concept loses its meaning except for impurities present at low concentration. The imperfections can combine and even trap the *electrons* or *holes* (positive charges) to ensure electrostatic neutrality.

Figure 5.16 shows schematically a few examples taking the perfect SiO_2 structure as a starting point. A detailed nomenclature for defects in oxide glasses was proposed by Stevels.[191]

Fig. 5.16. Examples of point defects in glasses; (*a*) reference network; (*b*) non-bridging oxygen; (*c*) oxygen vacancy; (*d*) substitutional impurity.

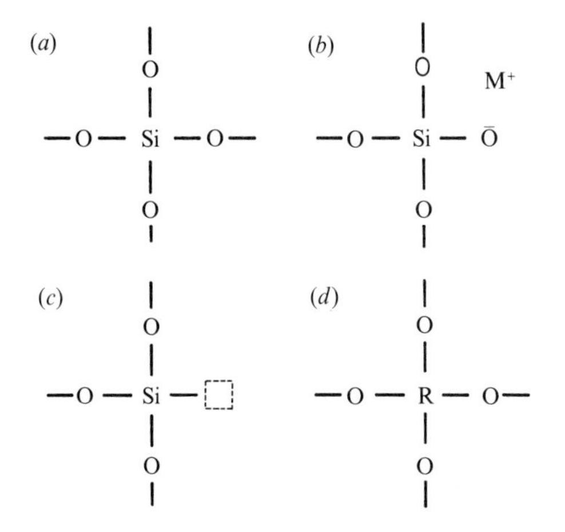

Electrons displaced from their normal orbitals can interact with network defects or impurity atoms which act as *traps* or *recombination centers* for mobile electrons. This gives rise to optical effects — an absorption band in the UV or visible region or a *color center*, and also to paramagnetic resonance phenomena due to unpaired electrons.

In the case of oxide glasses, the most common defects are non-bridging oxygens associated with alkali metal cations, oxygen *vacancies* or substitutional impurities in the network (Fig. 5.16). Such sites are modified by irradiation as shown in Fig. 5.17.

Oxygen is susceptible to the loss of an electron by ionization which corresponds to the capture of a hole. This ionization will be more probable for a more weakly bonded *non-bridging oxygen* (Fig. 5.17(*a*)).

Fig. 5.17. Modification of a network defect by irradiation: (*a*) hole trapping at non-bridging oxygen; (*b*) electron trapping in an oxygen vacancy near Ge^{4+} impurity; (*c*) hole trapping near Al^{3+} substitutional impurity.

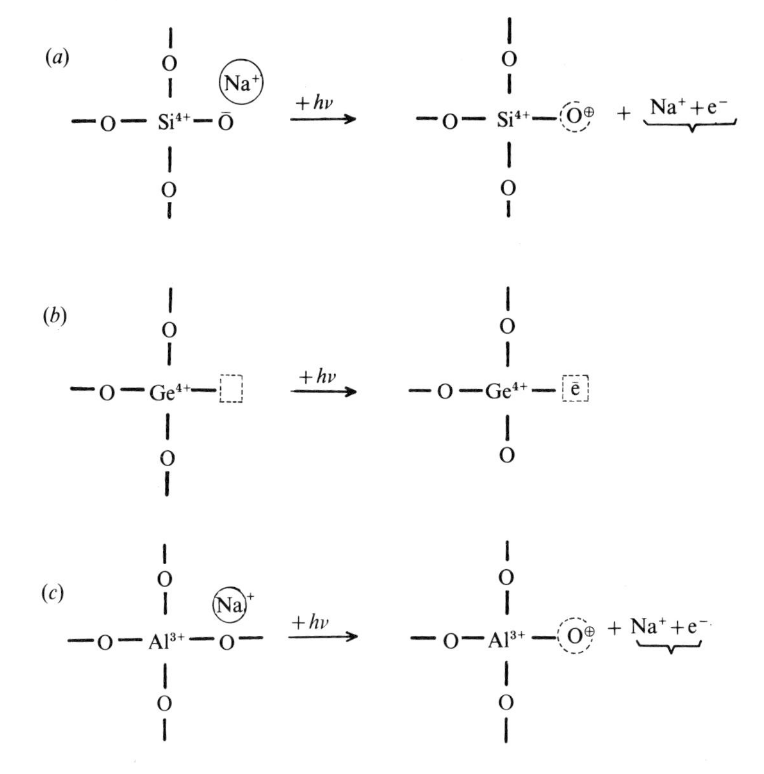

An *oxygen vacancy* in the network constitutes a site for the trapping of an electron ejected from another site by ionization. This results in the formation of a paramagnetic center (called an E_1' *center* by Weeks) in SiO_2 irradiated by neutrons.[192] The resonance spectrum has two anisotropic lines: a large one with a $g = 2.0090 \pm 0.007$ and $\Delta H = 40$ G, and a fine line with $g = 2.0013 \pm 0.006$ and $\Delta H = 1.76$ G. The latter results from a resonance similar to one found in crystalline quartz, averaged for all spatial orientations of the site. The E_1' center is related to an optical absorption band near 2120 Å. The presence of OH^- groups in silica inhibits the formation of E_1' by the formation of Si–OH units which hinders electron trapping. In general, an electron trapped at the site of an oxygen vacancy attached to a cation B^{3+}, Si^{4+}, P^{5+}, Ti^{4+}, Al^{3+} is a very common EPR center.

The effects resulting from *substitutions* can be more complex. Consider a site consisting of a Ge atom replacing an Si atom in the network (Fig. 5.17(*b*)). Ge and Si carry the same charge (4+) and are therefore electrostatically equivalent. Because the Ge–O bond is weaker than the Si–O, an oxygen vacancy can be produced more easily near a Ge; by trapping an electron, this vacancy gives rise to a UV absorption band near 2960 and 2600 Å.

When the substitutional impurity has a charge different from Si, e.g. Al^{3+}, a local charge compensation is necessary. This is provided by the presence of an alkali metal ion (generally Li^+ or Na^+) or a proton. In this case, irradiation leads to the capture of a hole near the Al, and the electron is trapped by an alkali metal ion which moves away from the site (Fig. 5.17(*c*)). The hole is characterized by a $g \sim 1.96$ and absorbs in the visible near 5500 Å independently of the compensating alkali metal (Fig. 5.18). It is the same defect which gives the two bands resolved at 4600 and 6200 Å present in natural quartz called "smokey quartz."

Among silicate systems, the EPR spectra of irradiated alkali metal silicate glasses have been the most intensively studied. Four paramagnetic centers were observed in very pure glasses: all were attributed to hole trapping. These centers are also related to optical absorption bands near 4600 and 6200 Å.

A recent discussion of various effects influencing optical absorption can be found in Ref. 196. (For a general review, see Refs. 194,195.)

5.5.4 *General methods of studying color centers*

Optical absorption spectra and EPR are used jointly to study color centers. To separate bands due to different centers which overlap in the spectra, optical bleaching methods can be used. The methods consist of producing a recombination by selective radiation absorption. The bands are bleached by illuminating with light of the appropriate wavelength, enabling separation.

The *thermoluminescence* methods are based on the principle that electronic traps filled by irradiation can be emptied by a progressive temperature increase of the sample. The return of electrons to lower energy levels can be accompanied by radiative transitions. By imposing a linear increase of temperature with time, the intensity of emitted light is measured for specific wavelengths. From this kind of thermoluminescence spectrum the depth of traps responsible for absorption can be deduced.

The method of *thermal bleaching* is used to determine the thermal energy necessary for electrons or holes responsible for absorption at a given frequency to escape their traps and be rapidly eliminated.

Since the different traps have different thermal stabilities, the spectrum resulting from irradiation with a certain dose will depend on the

Fig. 5.18. Absorption spectra of SiO_2–X_2O after irradiation by X-rays (200 KV, 20 mA, 10^6 R). After Ref. 193.

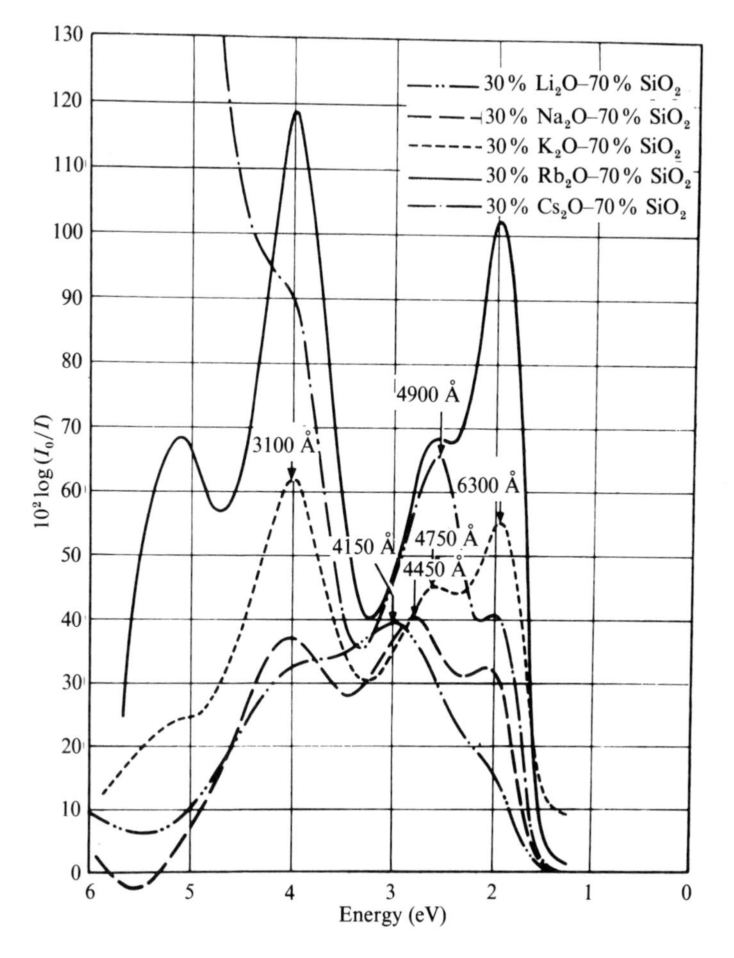

temperature at which the irradiation occurred. Spectra obtained by high-temperature irradiations will be *a priori* simpler because only the deepest traps will be occupied and contribute to the absorption.

For a given glass, the spectra depend on a number of factors among which the most important are the trace impurities and *redox conditions* (partial pressure of oxygen above the melt) present during glass preparation.

In order to determine whether a given color center is due to electron or positive hole trapping, *doping* techniques using ionic species which selectively trap either holes or electrons are employed. Any easily oxidized species can serve as a trap and will compete with intrinsic hole traps present in a given glass. Generally speaking, for an ion:

$$M^{n+} + \square \rightarrow M^{(n+1)+}$$

where $\square$ designates a "free" hole in the structure.

In practice, the Ce^{3+}, the most effective hole trap, is frequently used. The disappearance of an absorption band in the presence of Ce^{3+} or an equivalent species (e.g. Mn^{3+}) is a strong indication that this band is due to a hole trap.

Similarly, doping with easily reduced ions, for example Eu^{3+}, has been used to establish that a given center is due to electron trapping:

$$Eu^{3+} + e^{-} \rightarrow Eu^{2+}$$

Other elements: Ag^{+}, Cd^{2+} and Pb^{2+} have also been used for this purpose.

The disturbances introduced by irradiations represent a departure from initial quasi-equilibrium states toward which the system tends to relax. (In all studies of this type, both the *duration* of the irradiation and *temperature* must be defined.) Since the relaxation time depends on the temperature, the time scale and the temperature of the observation must be considered as equivalent.

In order to distinguish natural (intrinsic) traps in the initial structure from the traps produced by irradiation, a process of repetitive irradiation followed by complete optical bleaching is used. Increase of the absorption after each cycle indicates that each new irradiation produces supplementary defects which become available as traps for charges liberated in the subsequent irradiations.

The study of color centers in glasses has been stimulated by recent optical applications. In glasses used for lasers and optical fibers (cf. Chapter 12) color centers must neither be present initially nor develop during use.

Glasses of extremely pure synthetic SiO_2 are very insensitive to radiation. Absorptions equivalent to those produced in a common SiO_2 require doses 1000 times higher, resulting mainly in E' centers which absorb in the ultraviolet and are less troublesome for the above-mentioned applications.

5.6 The Mössbauer effect

The Mössbauer effect is a nuclear resonance phenomenon occurring in the γ-ray region. In this technique there is an emission without recoil effect on the nucleus and a resonant absorption of γ-rays by nuclei bonded in the solid. The γ-rays produced by fluorescent emission between the nuclear levels in the "source" or "reference solid" are absorbed between the same levels in "the absorbing solid." The technique is based on Mössbauer's discovery that in a solid, the recoil energy is essentially zero. This is in contrast to the case of free nuclei where the recoil energies accompanying the emission and absorption are sufficient to hinder the observation of resonance.

The experiment is performed in transmission. The fluorescent rays from a material (source) traverse a second material (absorber) and the number of transmitted γ photons is measured as a function of the wavelength (or energy) of the emitted γ-rays. The emitter, or "source" is a solid containing a radioactive isotope (e.g. ^{57}Co) while the absorber is a solid containing a corresponding stable isotope (i.e. ^{57}Fe if the source contains ^{57}Co).

In practice the source consists of a solid solution of ^{57}Co in stainless steel, Cu or Pd. The sample being studied then contains ^{57}Fe (its abundance in natural iron is 2.14%). To increase sensitivity samples enriched with ^{57}Fe can be used.

By imposing a relative velocity of the order of 1 cm s^{-1} on one of the two solids, a change of energy is produced by the Doppler effect which is sufficient to scan the spectral region of interest. The movement can be a uniform linear translation (sliding on rails) or an alternating sinusoidal movement (loud-speaker membrane driven by a generator providing the desired wave form). A multichannel analyzer is used to count the impulses detected by the scintillation counter. For details of principles and instrumentation, see Refs. 197 and 198.

Among the possible nuclei, ^{57}Fe, ^{119}Sn, ^{121}Sb, ^{125}Te, ^{129}I, ^{151}Eu, ^{169}Tm have been studied in vitreous matrices: borates, silicates, phosphates, chalcogenides, vitrified aqueous solutions, etc. Information concerning the application of this method to glasses can be found in Ref. 199.

Although exterior disturbances have a very small influence on nuclei levels, the extreme resolution ($1 : 10^{12}$) of this technique permits the study of changes in valence, chemical bonding and coordination.

The principal parameters provided by Mössbauer spectroscopy are: chemical shift, quadrupole coupling, hyperfine magnetic splitting and linewidth.

The *chemical* or *isomeric shift* δ results from the interaction between the nuclear and the electronic charge producing a shift of the nuclear energy levels without raising the degeneracy of the nuclear spin.

The shift δ expresses the energy difference between the same nuclear transition in the two atomic nuclei for which the electron wave functions of the nuclei are different:

$$\delta = \frac{2}{5} Ze \left[R_{ex}^2 - R_g^2\right] e \left[|\Psi_a|^2 - |\Psi_e|^2\right]$$

where Ze is the nuclear charge, Ψ_a and Ψ_e the electron wave functions originating in the absorber and emitter, and R_{ex} and R_g are the nuclear radii of the excited and ground state. The difference between the electronic densities $-e|\Psi|^2$ is an atomic or chemical parameter because it is affected by the valence state of the atom.

It has been shown that δ is a monotonic function of the s-electron density on the nucleus and decreases linearly as the latter increases (i.e. an increase in the density causes a shift of the resonance line toward negative velocities).

In the case of ^{57}Fe, δ is mainly determined by the degree of occupation of the 3s and 3d levels: for example, the shift for Fe^{2+} ions is more positive than that for Fe^{3+} because in the divalent state 3s electrons are more strongly shielded by the supplementary 3d electrons.

The *quadrupole coupling* is essentially similar to that observed in NMR. The nuclear quadrupole moment Q reflects the deviation from spherical symmetry. The quadrupole coupling results from the interaction of nuclear quadrupole moments with the electric field gradient on the nucleus. Such gradients can originate from the electrons in the atomic bonds associated with the nucleus under consideration and from the charges on other atoms or ions. The nuclei with spins 0 and $+\frac{1}{2}$ have a spherical symmetry and do not exhibit quadrupole coupling. Nuclei with spin greater than $\frac{1}{2}$ can show this effect (e.g. ^{57}Fe in which the first excited state has spin of $+\frac{3}{2}$).

In the case of Fe^{3+} in the state $^6S_{5/2}\,3d^5$ the electrons have a spherical distribution and will not exert an electric field gradient on the nucleus. If the normal tetrahedral or octahedral site is not distorted, quadrupole coupling would not be observed. The presence of a quadrupole doublet would then be an indication of *deviation from spherical symmetry*.

In the case of Fe^{2+} in the state $^5D_4\,3d^6$ with an additional electron in the half-filled spherical shell, the resulting gradient is from contributions of the bonds and network. The significant quadrupole couplings generally observed for Fe^{2+} ions are due to the raising of degeneracy by axial distortions.

Hyperfine magnetic interaction occurs in the presence of an internal or external magnetic field which raises the degeneracy of the ground and excited states. For the ^{57}Fe absorber and ^{57}Co emitter, six sub-levels appear which produce by the Zeeman effect a spectrum of lines with six maxima. The effect can be used to calibrate the observed field and the velocity scales.

In addition to the preceding parameters, the *linewidth* is important in the study of vitreous solids. Since the environment of individual atoms as well as the crystal field is variable, the linewidth allows an *estimation of the similarity of sites.*

The most frequently studied nucleus is ^{57}Fe. For Fe^{2+} chemical shifts of the order of 0.9–1.2 mm s^{-1} were obtained for tetrahedral and octahedral coordination respectively. Similarly Fe^{3+} gives respective displacements of $\sim$ 0.3 and 0.5 mm s^{-1} for tetrahedral and octahedral coordination (shifts measured relative to ^{57}Co in Fe). Generally, for inorganic systems containing Fe^{2+} and Fe^{3+}: $\delta_{(divalent)} > \delta_{(trivalent)}$ and for a given valence: $\delta_{(octahedral)} > \delta_{(tetrahedral)}$.

Experiment shows that Fe^{3+} has a preference for tetrahedral coordination in alkali metal silicate glasses and octahedral coordination in phosphate glasses. Figure 5.19 shows Mössbauer spectra for alkali silicate glasses containing ^{57}Fe measured with a ^{57}Co source in Cu. The number of counts per channel depends on the relative absorber/source velocity. The quadrupole separation is given by the velocity difference of the peaks, and the chemical shift through the velocity change of the center of gravity of the doublet relative to the velocity $v = 0$. The two upper spectra have displacements δ of 0.008 and 0.006 cm s^{-1}, and quadrupole separations of 0.008 and 0.078 cm s^{-1}, typical for Fe^{3+}. For the lower spectrum of a glass melted under reducing conditions, $\delta = 0.083$ cm s^{-1} and a quadrupole separation of 0.206 cm s^{-1} indicates the presence of Fe^{2+}. For low Fe concentrations a hyperfine magnetic interaction can be observed.

Figure 5.20 shows that when the temperature is lowered to 4 K, a six component hyperfine spectrum is superimposed on the spectral lines near $v = 0$ (a non-resolved doublet due to Fe^{3+} and two weaker lines due to a Fe^{2+} doublet). Measurement of the quadrupole coupling and the linewidth indicates the presence of a series of different sites in the glass.

The study of Sn by Mössbauer spectroscopy indicates that it is found in octahedral coordination[200] in borosilicate glasses.

The Mössbauer effect implies the existence of a rigid network and cannot be produced in low viscosity liquids where the movements of the particles are not restrained. The study of aqueous solutions of $SnCl_4$ containing the isotope ^{119}Sn and frozen in vitreous form shows that the Mössbauer effect decreases rapidly when $T > T_g$ then reappears at higher temperatures when the superheated metastable liquid crystallizes.[201] A similar effect is observed in CH_3OH glasses in which ^{57}Fe in $FeCl_2$ is introduced as an indicator.[202]

Mössbauer studies on ^{119}Sn, ^{121}Sb, ^{125}Te, and ^{129}I were conducted on chalcogenide glasses: Ge–Se–Sn and mixtures of As_2Se_3–As_2Te_3; results

Fig. 5.19. Mössbauer spectra (^{57}Fe) for alkali metal silicate glasses containing Fe. After Ref. 199.

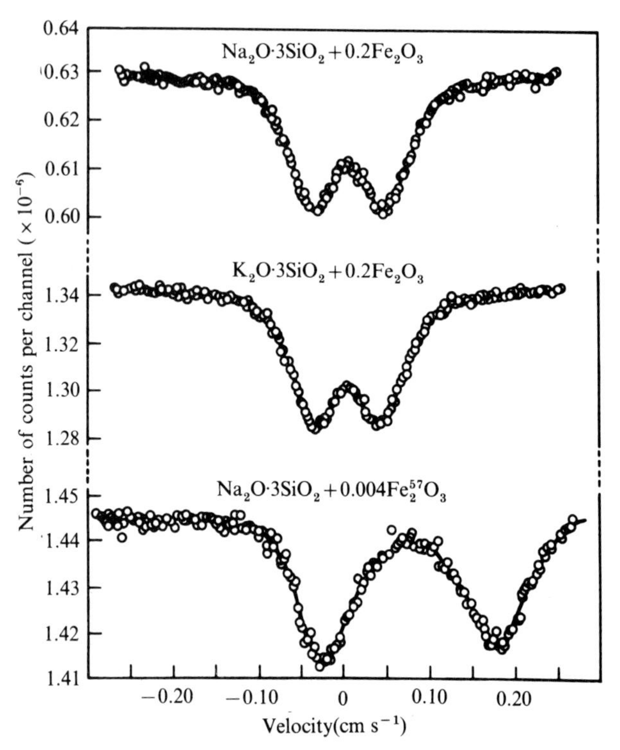

are provisional. See Ref. 15, p. 121.

The Mössbauer effect is still a recent technique in the study of glass structure and, given its potential, systematic studies should be conducted.

5.7 X-ray emission spectroscopy

Besides its usefulness as a tool for chemical analysis (fluorescence spectra), X-ray K emission can give information on coordination, oxidation state, bond length and bond forces. The K emission spectra of Mg, Al and Si are widely used in mineralogy. The general method consists of establishing a correlation between the position of an emission peak and the structure in known crystalline compounds which then serve as standards for the interpretation of the results obtained from glasses.

Figure 5.21 is an example of the results obtained for a series of glasses in the system Al_2O_3–B_2O_3–CaO.[203] The $\Delta(2\theta)$ shift of the Al K_α peak relative to the standard peak is plotted as a function of B_2O_3 concentra-

Fig. 5.20. Mössbauer spectra of $3\,SiO_2 \cdot Na_2O$ ($0.1\,Fe_2O_3$) as a function of temperature. After Ref. 199.

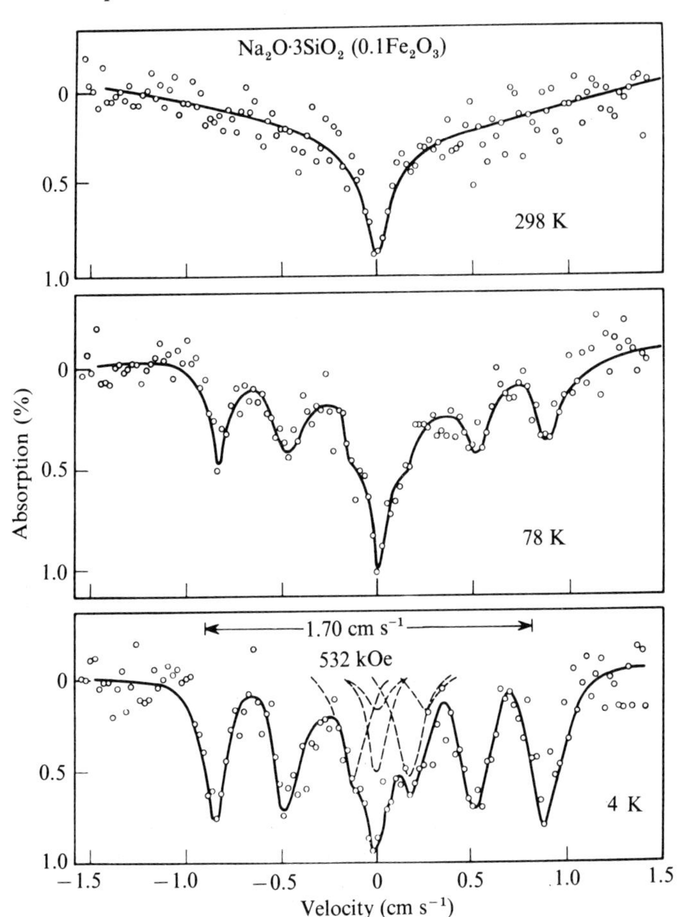

tion. The results are bracketed by potassium feldspar (where Al has a coordination 4) and α-Al_2O_3 (coordination 6). The results clearly show an increase in the average coordination number of Al from 4 to 6 as the concentration of B_2O_3 increases.

In alumino-phosphate glasses the coordination of Al is the same as that in crystalline compounds of similar composition. The most acidic oxides B_2O_3 and P_2O_5 increase the coordination of Al more effectively than SiO_2.

By standardizing the Si K_β emission using crystalline silicates, a corre-

Fig. 5.21. Chemical shift of the Al K_{α} line in $CaO–B_2O_3–Al_2O_3$ glasses. After Ref. 203.

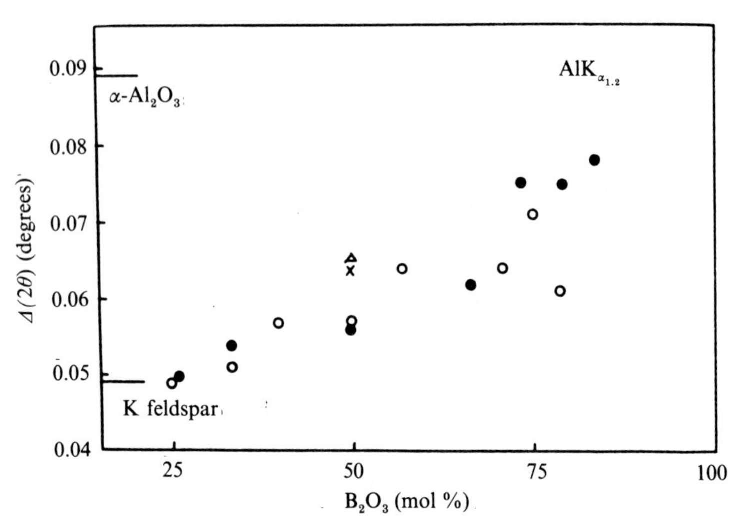

lation between the average Si–O bond length and the $\Delta(2\theta)$ displacement can be obtained. The application to glasses of the $SiO_2–Al_2O_3–Li_2O$ system has shown that the Si–O bond weakens when the SiO_2 concentration decreases.[204]

5.8 X-ray absorption spectroscopy – EXAFS and XANES

X-ray absorption techniques have been increasingly used in recent times in the study of non-crystalline materials due both to theoretical advances and new experimental facilities which made the application of these well-known spectroscopies practical.

In the photoelectric effect, an incoming photon can eject an electron from an atom, provided its energy $h\nu$ is greater than the binding energy E_b of the electron. For X-ray photons of energy higher than 200–300 eV, a series of discontinuities is observed in the absorption coefficient $\mu(E)$ as a function of energy; the edges K, L, corresponding to the excitation of 1s, 2s, and 2p electrons toward continuum states. The absorption spectra considered near the edge may be further sub-divided into two zones. The immediate vicinity of the edge ($h\nu$ < 50 eV above the edge) corresponds to XANES (X-ray Absorption Near Edge Structure). For higher spectral ranges (40 eV $< h\nu <$ 50 eV above the edge) the absorption coefficient shows an oscillatory behavior. This is the region of EXAFS.

In XANES, the transitions of the photoelectron are mixed with oscillations resulting from multiple diffusion involving essentially the atoms of the first coordination sphere. XANES can thus, in principle, furnish information on the valence and the site symmetry of the absorbing atom as well as average distances to the surrounding ligands. However, the calculation of $\mu(E)$ is difficult and, in spite of theoretical efforts, the XANES region is much less studied than the EXAFS region. The EXAFS zone corresponds to simple retrodiffusion effects involving first neighbors. In this zone the photoejected electrons are excited far above the Fermi level and a single scattering (i.e. weak scattering) approximation can be used to calculate the final state wave functions.

Figure 5.22 shows the EXAFS spectra of two allotropic crystalline forms of GeO_2 and of vitreous GeO_2 in the vicinity of the Ge K edge. Note the similarity between the spectra of vitreous GeO_2 and hexagonal GeO_2 in contrast to the spectrum of tetragonal GeO_2. (See similar effects observed by IR spectroscopy, Fig. 5.5.)

Fig. 5.22. X-ray absorption spectra for Ge (EXAFS) in vitreous GeO_2 and crystalline varieties. After Ref. 205.

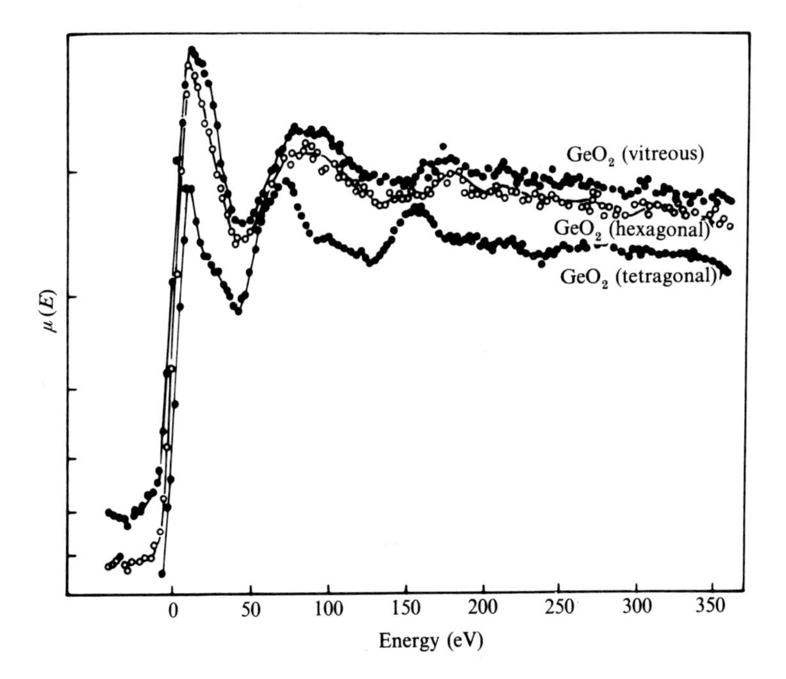

The oscillatory behavior of $\mu(E)$ in the EXAFS spectra is evaluated considering the expression:

$$\chi(k) = (\mu - \mu_0)/\mu_0$$

where μ_0 is a smoothly varying background and

$$k = (2\pi/h)\sqrt{2m(E - E_0)}$$

where m is the electron mass, h Planck's constant and E_0 the threshold of energy of a particular electronic shell of the atom. It has been shown[205,206] that $\chi(k)$ is related to the atomic parameters by an expression of the form:

$$\chi(k) = \sum_j \left(\frac{N_i}{kr_j^2}\right) f_j(k) \sin[2kr_j + \phi_j(k)] \exp(-2k^2\sigma_j^2) \exp(-2r_j/\lambda)$$

where r_j is the distance from the central (absorber) atom to the jth coordination shell, N_j the coordination number of the jth shell, σ_j a Debye–Waller factor and λ a mean free path of the order of 3.5–5 Å. The amplitude function $f_j(k)$ and the phase function $\phi_j(k)$ can be calculated or determined empirically from compounds of known structure.

In principle, the parameters N_j, r_j, σ_j and λ can be determined for the first few shells. In practice, while distances r_j are obtained with a precision of ± 0.01 Å, the number of the surrounding atoms N_j is measured to an accuracy of about 10–20% and different atomic species can only be distinguished if their atomic numbers differ by 10 units. The σ_j values, which are related to the thermal disorder and are difficult to obtain, are determined here "as a bonus." Analysis of EXAFS spectra is considerably improved by calculating the Fourier transform of the function $k\chi(k)$. This leads to the EXAFS RDF defined in a way similar to that already described (Chapter 4). The important difference, however, is that the RDF is here measured relative to an identifiable central atomic species. This permits the separation of the contributions of the different pairs in the heteroatomic case. Figure 5.23 shows the radial correlation functions obtained for crystalline and vitreous GeSe relative to either Ge or Se atoms.[206]

The EXAFS technique has greatly benefited from the extremely high-intensity X-ray sources such as synchrotron radiation provided by high-energy electron storage rings (e.g. LURE in France). Using double crystal monochromators, these facilities provide X-ray beams with a 1–2 eV resolution and fluxes of 10^8–10^{12} photons s^{-1}. The data acquisition times are then of the order of minutes.

Fig. 5.23. Radial correlation functions for crystalline or amorphous GeSe; (*a*) relative to a Ge atom, (*b*) relative to a Se atom. After Ref. 206.

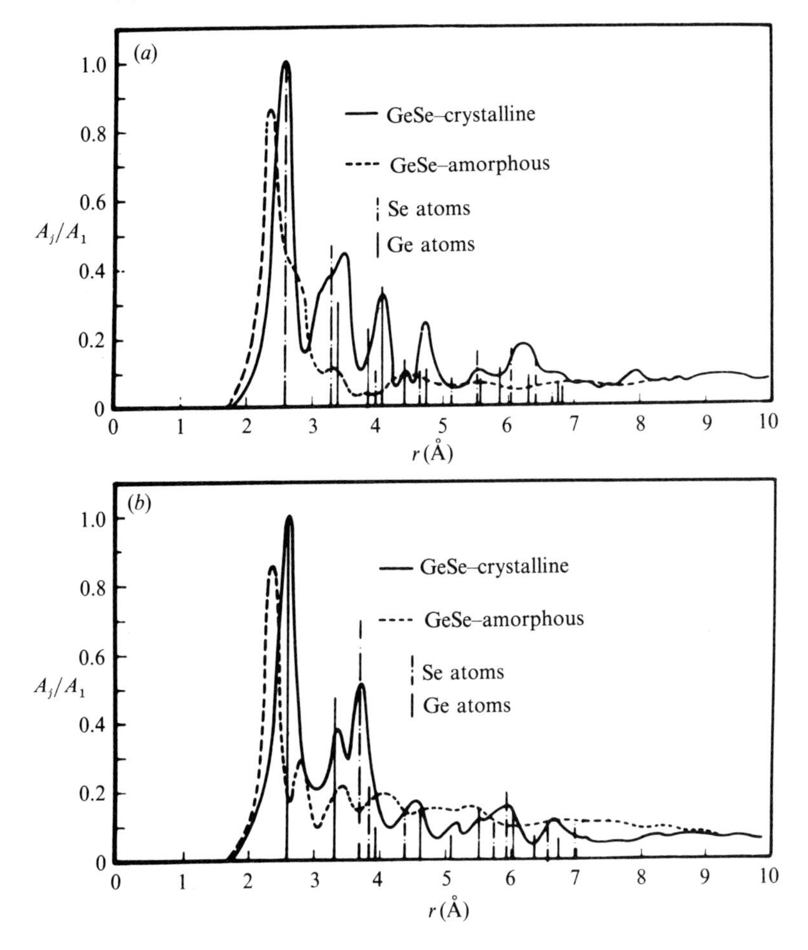

K edge EXAFS spectra are readily obtained for materials containing elements of atomic numbers, Z, between 19 (K) and 47 (Ag). For $Z < 19$, EXAFS spectra are severely limited due to high absorption rates for the low energies required. For $Z > 50$, resolution is limited because of K level life-time broadening. For elements between 56 (Ba) and 94 (Pu), L edge measurements may be used.

For practical application of these methods Refs. 207–210 should be consulted.

6 Phase separation in glasses

6.1 Immiscibility phenomena in glasses

Certain glasses consist of several non-crystalline phases of different compositions which form a complex heterogeneous structure. This is the result of immiscibility phenomena which have been studied mainly in the last twenty years.

Immiscibility in oxide systems, particularly those containing both SiO_2 and B_2O_3 (borosilicates), has been known for a long time. Greig[211] thoroughly studied immiscibility in molten silicates, especially in the binary systems SiO_2–alkaline earth oxides.

Glasses formed from such liquids are inhomogeneous, resulting in *emulsions* which lead to the phenomenon of opalescence or sharp separation into two layers. The initial unmixing of the liquid glass composition thus limits the range of homogeneous glass formation.

Most earlier studies focused on immiscibility above the liquidus – also called stable unmixing. In certain cases, however, a *homogeneous glass* can be obtained by cooling, which then, by appropriate thermal treatment below the liquidus, separates into a system of two glasses. One such case of metastable or sub-liquidus unmixing had been known for a long time and was even the source of an industrial process: the manufacture of "VYCOR" containing 96 % SiO_2.

In this process, a glass of a particular composition from the system SiO_2–B_2O_3–Na_2O is thermally treated in the softening range. The initial glass separates into two intimately mixed phases, one of which is almost completely pure SiO_2 glass, the other one being sodium borate glass. Eliminating the soluble borate phase by dissolving it in acid leaves behind a sort of SiO_2 sponge with very fine pores (~ 100 Å), which can be transformed into a compact glass through sintering at about 1000 °C, which is far below the fusion point of pure SiO_2, near 1800 °C.

Unmixing has been widely used in making opaque *enamels* which are glassy emulsions containing dispersed phases in the form of fine droplets – systems in which the unmixing is initiated by PbO or TiO_2.

In *opal glasses* opalescence is also caused by a phase separation phenomenon (certain opal glasses can, however, contain crystalline phases).

More recently, metastable immiscibility has been the object of numerous studies relating to glass ceramic processes (cf. Chapter 16) where unmixing can play a *precursor* rôle in catalyzed crystallization of glasses.

Phase separation by unmixing modifies the *texture* of a glass on a scale from tens to thousands of angströms according to the system and its thermal treatments. The associated modification of the physico-chemical properties of such systems can be important and in some cases has been systematically investigated. For example, the inclusion of a soluble phase within another insoluble phase explains the chemical durability of "PYREX" glasses (which are borosilicates). In contrast, it is important in other cases to avoid regions of metastable immiscibility to maintain glass homogeneity, in particular in optical applications. In the case of ordinary SiO_2–CaO–Na_2O glasses, the composition is generally outside the domain of immiscibility.

Note finally that for a time, *all* glasses were erroneously thought to be susceptible to unmixing (and that they were "micro-heterogeneous"). We now know that the phenomenon is confined to particular systems and well-defined compositions.

Phase separation has been studied mostly in silicate, borate and borosilicate glasses. Separation is also important for fluoroberyllates which have been studied systematically because they are isostructural with silicates and in chalcogenide systems where the range of unmixing is currently being investigated.

6.2 Thermodynamic studies

6.2.1 *The coexistence curve*

The separation of a liquid into two immiscible phases, or unmixing, is a well-known phenomenon. In a binary system A–B, the field corresponding to two liquids is delimited by the *coexistence curve* (Fig. 6.1) which defines the *miscibility gap* as a function of temperature. Depending on the system, the gap can have an upper or lower consolute point or constitute a closed domain. At a given temperature T a liquid M of global composition c separates into two liquids M_1 and M_2 of respective compositions c_1 and c_2 corresponding to the intersections with the coexistence curve.

The proportion of the two phases can be deduced by the well-known *lever rule*:

$$\text{Mass}\,(M_1)/\text{Mass}\,(M_2) = \overline{MM_2}/\overline{MM_1} = (c_2 - c)/(c - c_1)$$

Consider a binary system with an upper consolute temperature or critical temperature T_c (Fig. 6.2). At a temperature higher than T_c the molar free energy curve G as a function of composition must be concave upwards, otherwise a separation into two phases would result in a reduction of G. Given that:

$$\left(\frac{\partial G}{\partial T}\right)_p = -S$$

Fig. 6.1. Miscibility gap.

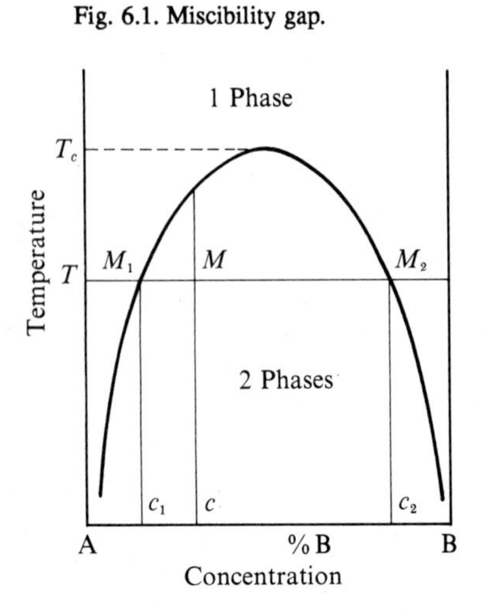

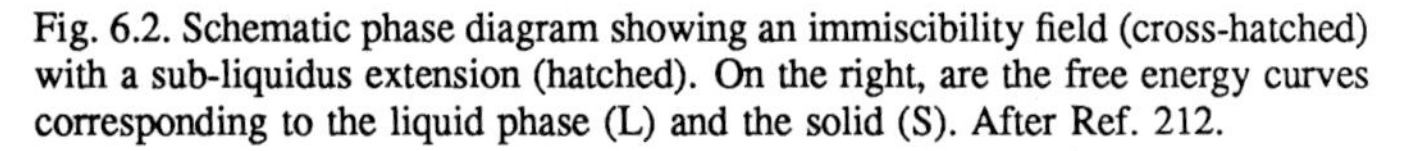
Fig. 6.2. Schematic phase diagram showing an immiscibility field (cross-hatched) with a sub-liquidus extension (hatched). On the right, are the free energy curves corresponding to the liquid phase (L) and the solid (S). After Ref. 212.

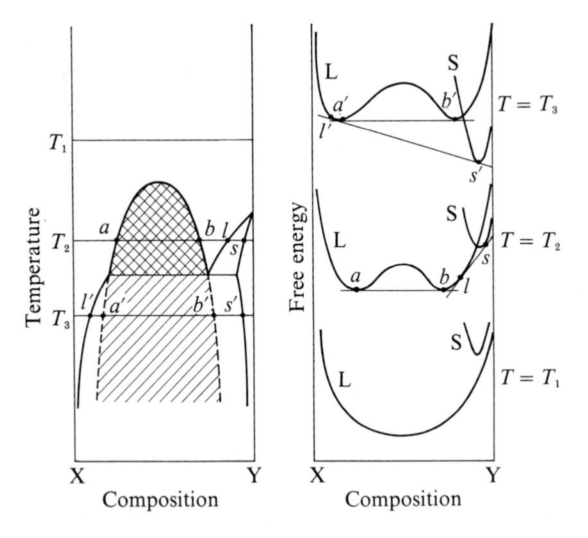

G increases in proportion to the entropy S as the temperature decreases. Since the entropy is lower for the pure components and is higher at the center of the composition interval, the G curve flattens as the temperature decreases and finally, a central portion develops a negative curvature.

If the composition M lies between the minima a and b, the free energy is minimized for a mixture of the two phases a and b, where the point m representing the mixture is located on the tangent common to both minima. Outside the intervals a b only one phase is stable.

6.2.2 *Stable and metastable unmixing*

Consider now the case where a *crystalline phase* can appear (Fig. 6.2). At temperature T_1 the liquid phase is stable for all compositions because the solid has higher free energy. When the temperature is lowered, the curve for the liquid goes up faster than the one for the solid because the entropy of the liquid is higher. At the melting point, the liquid curve catches up with that of the solid. At lower temperatures, the liquid–solid equilibrium corresponds on one side to the tangent $l - s$, and on the other side, the tangent $a - b$ corresponds to two coexisting liquid phases.

At the monotectic temperature, the liquidus shifts towards the other liquid phase, and at lower temperatures, the stable equilibrium is given by the tangent $l' - s'$. However, one can always define the tangent $a' - b'$ corresponding to a *metastable* equilibrium between two liquid phases.

This shows that the coexistence curve which defines a *stable* unmixing field bounded by the liquidus can be extrapolated to lower temperatures. The two branches thus extrapolated correspond to a region of *metastable* unmixing which extends the stable field previously studied.

If for kinetic reasons a crystalline phase does not appear, the system passes to a vitreous state and the metastable branches can be crossed reversibly.

This is particularly true in the SiO_2–RO and B_2O_3–RO systems where R is an alkaline earth element (Fig. 6.3). The range of the miscibility gap depends on R as shown in Fig. 6.4 for the B_2O_3–PbO system.

In certain glass-forming systems, the miscibility gap can be entirely subliquidus. This case corresponds to the situation represented in Fig. 6.5. The curve for the solid passes below the curve for the liquid before the portion of negative curvature develops.

The preceding reasoning leads to a definition of a completely metastable coexistence curve situated below the liquidus which then presents a characteristic flat portion with an inflection indicating a nearby unmixing immediately below. A typical example is given by the system SiO_2–Na_2O (Fig. 6.6).

With regard to unmixing, the two situations are identical and the field of unmixing can be considered as a whole, remembering that part can be metastable relative to phases susceptible to crystallization.

In ternary systems, unmixing surfaces are likewise defined which can be either partially or totally metastable. The metastable surfaces constitute an extension of the stable surfaces and can be estimated by extrapolation.

Fig. 6.3. Extent of the miscibility gaps in B_2O_3–RO systems. After Ref. 213.

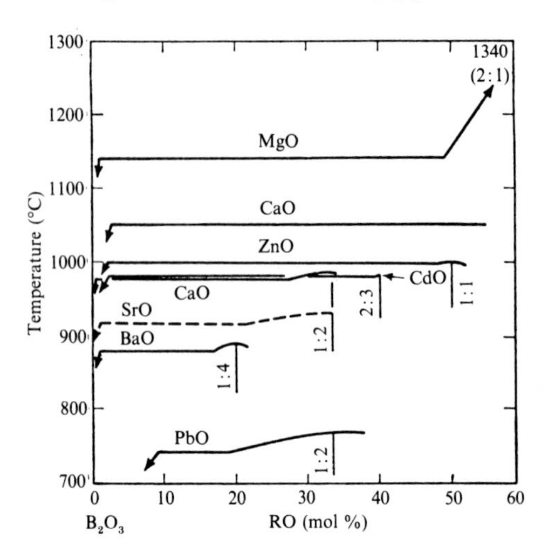

Fig. 6.4. Miscibility gap in the B_2O_3–PbO system (the horizontal dashed line indicates the liquidus). Curves of equal viscosity from 10^2 to 10^{13} dPa s are also shown. After Ref. 214.

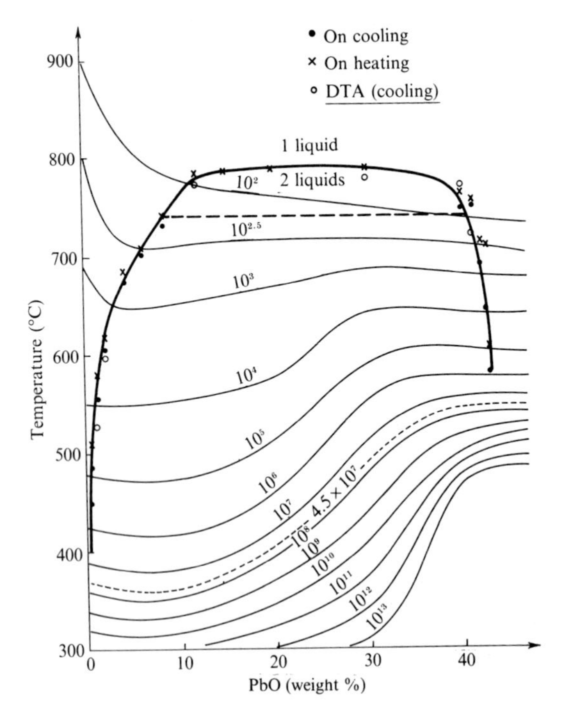

Fig. 6.5. Schematic phase diagram showing an entirely sub-liquidus immiscibility field (hatched). On the right are the corresponding free energy curves. After Ref. 212.

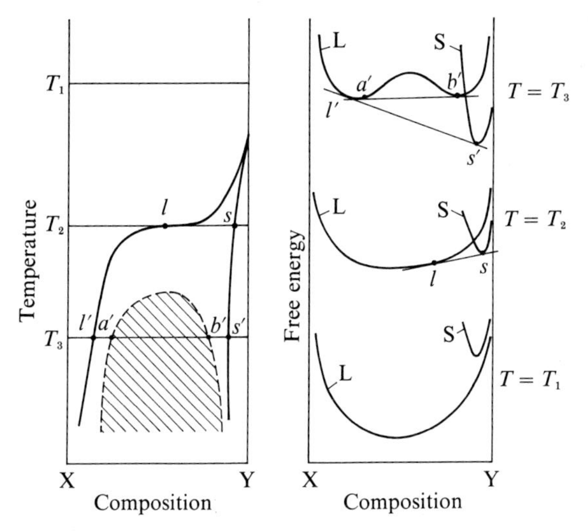

Fig. 6.6. Sub-liquidus coexistence curve in the system SiO_2–Na_2O. The spinodal (see Section 6.3.1) is shown by the dashed line. After Ref. 215.

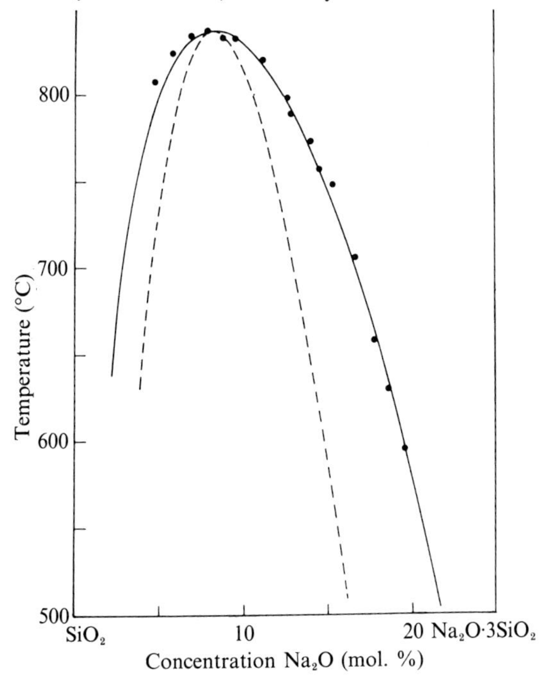

Knowledge of the binary unmixing curves can be a useful guide in the study of an unknown ternary system. Figure 6.7 shows an example of the SiO_2–Li_2O–Na_2O system and Fig. 6.8 the B_2O_3–PbO–Al_2O_3 system.

Fig. 6.7. Metastable immiscibility volume in the SiO_2–Li_2O–Na_2O system. After Ref. 216.

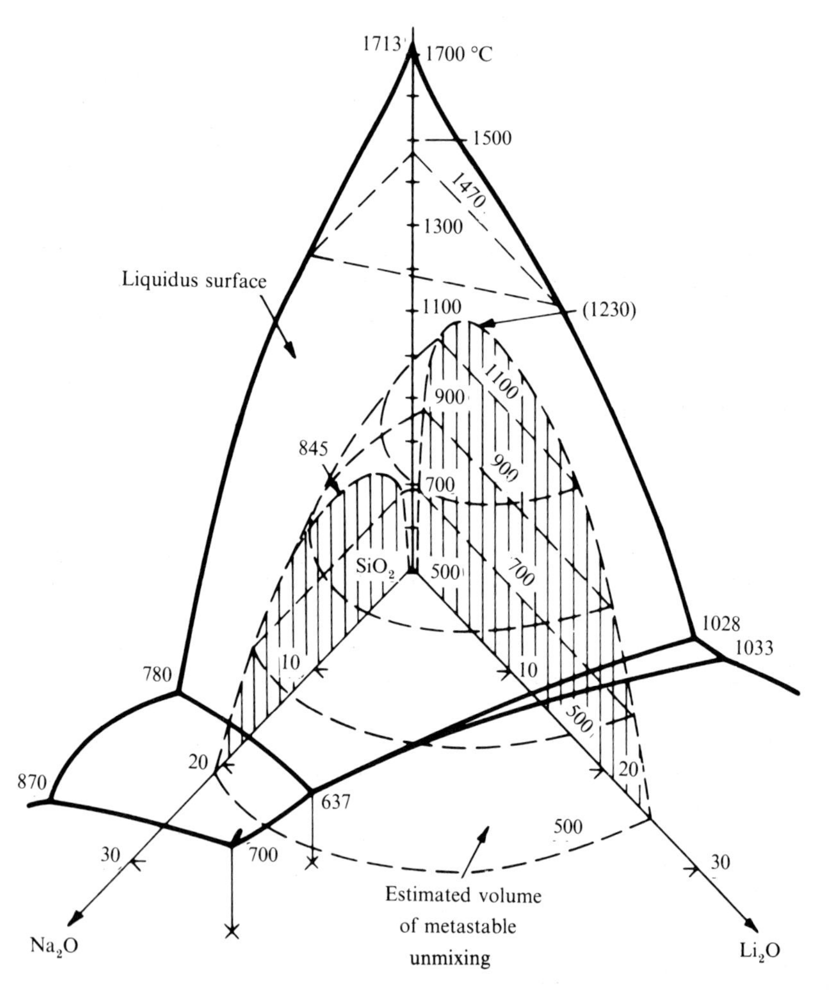

6.2.3 *Suppression of immiscibility*

The addition of a supplementary component (C) to a binary system (A–B) with a miscibility gap can have the effect of increasing or decreasing the tendency to immiscibility. The thermodynamic problem

Fig. 6.8. Immiscibility surface in the B_2O_3–PbO–Al_2O_3 system. After Ref. 217.

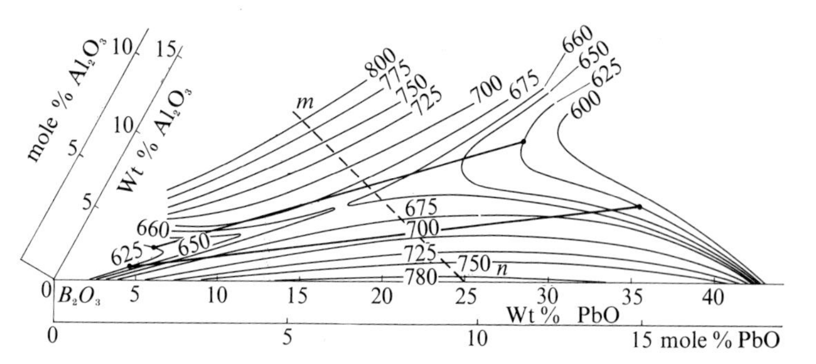

has been discussed by Prigogine and Defay (cf. Ref. 218). The variation of the critical temperature T_c can be calculated if the coexistence curves of the partial binary systems (A–C) and (B–C) are known.

In practice, the influences of certain components were determined experimentally. Thus, for example Al_2O_3 causes a regression of the immiscibility zone in silicates and borates. The addition of Al_2O_3 to the B_2O_3–PbO system causes a lowering of the miscibility gap, a large portion of the unmixing surface of the B_2O_3–PbO–Al_2O_3 ternary being metastable.

6.2.4 *The origin of immiscibility*

An explanation of immiscibility in liquid silicate or borate systems may be given by borrowing concepts from crystal chemistry. Warren and Pincus[219] and then Levin and Block[220] discussed the relations between the ionic force of the cations and the competition between the ions of the mixture seeking an optimal coordination. However, these considerations ignore the influence of temperature, a parameter essential in the complete thermodynamic treatment and which alone is able to account for the observed phenomena. These models are only applicable to very limited systems.

Phase separation is currently encountered in vitreous chalcogenide systems, but, apart from isolated cases, systematic studies have not yet been made.

6.3 **Kinetics of unmixing**

6.3.1 *Regions of stability: spinodal*

Consider a binary system susceptible to unmixing and the effect of an infinitesimal fluctuation of the composition Δc around the initial composition c_0.

The variation of the corresponding free energy will be:

$$\delta G = \frac{1}{2}\left[G(c_0 + \Delta c) + G(c_0 - \Delta c)\right] - G(c_0)$$

By expanding in a Taylor series one obtains:

$$\delta G = \frac{1}{2}\, G''(c_0)\Delta c_0^2$$

Two distinct cases arise depending on the sign of δG (which, in turn, depends on $G'' = \partial^2 G/\partial c^2$).

1. When $G'' > 0$, $\delta G > 0$; a small fluctuation increases G. The system is stable with respect to infinitesimal composition fluctuations.

2. When $G'' < 0$, $\delta G < 0$: every fluctuation, however weak, has the effect of reducing G. The system is unstable to any fluctuation, no matter how small.

The limit between the two cases corresponds to $\partial G^2/\partial c^2 = 0$. This condition defines the inflection points i and j of the free energy–composition curve, called *spinodal points*. In the temperature–composition diagram, the locus of the points corresponding to $G'' = 0$ is called a *spinodal*: this curve is necessarily interior to the coexistence curve and tangent to it at the critical point. The spinodal splits the unmixing field into two regions designated by I and II on Fig. 6.9. The behavior of the system in the course of unmixing differs according to which region it is in. III corresponds to the critical opalesence region (see Section 7.2.6).

6.3.2 *Modes of unmixing*

We will now study in detail the kinetics of phase separation in a system having a miscibility gap with an upper consolute point by considering the whole of the unmixing field which is either stable or metastable, i.e. assuming that eventual crystalline phases do not have time to precipitate.

The method consists of bringing a liquid into the two-phase field by cooling (quenching) in the form of a homogeneous glass and following its mode of decomposition in the course of isothermal treatment.

We saw that in region I the system is stable relative to small composition fluctuations but can become unstable relative to a mixture of two phases, i.e. a *finite* fluctuation is necessary to destabilize the initial homogeneous phase. In this context, consider the larger composition fluctuations shown in Fig. 6.10.

Let c_0 be the initial composition and c_β the composition of a phase β which tends to separate. If interfacial effects are neglected, it can

Fig. 6.9. Definition of the spinodal.

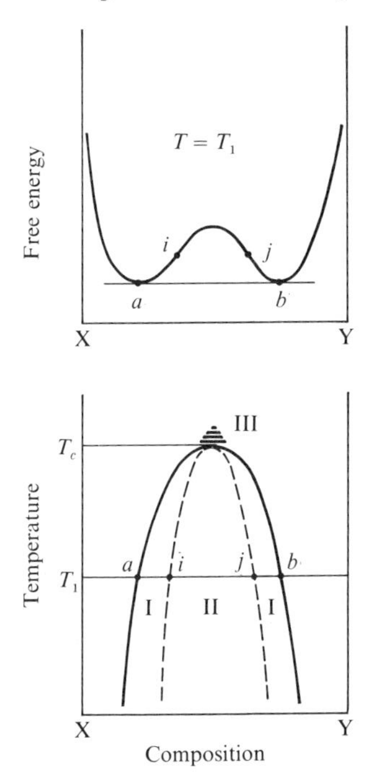

be shown that the variation of the free energy per atom of β is given graphically by the vertical segment $\overline{DE}$ between the tangent to the curve G at the initial composition point c_0 and the point on curve G for c_β. This corresponds to the driving force for the separation of the phase β (cf. Ref. 220).

Obviously, $\overline{DE}$ is positive up to c_e which corresponds to the intersection of tangent at c_0 with the curve G, and then becomes negative, reaching its greatest absolute value at c_β.

This shows that a finite fluctuation must go beyond composition c_e in order for the variation to become negative; the optimum composition being, of course, the equilibrium composition $c_\beta = c_b$. Therefore there exists a *thermodynamic barrier* which must be overcome before phase separation which occurs through *nucleation and growth*. For c_0 between c_a and c_i corresponding to the inflection point i, the system is *metastable* relative to composition fluctuations. The thermodynamic barrier ΔG_v^* corresponding to the maximum positive ΔG is obtained by tracing the tangent to curve G parallel to the tangent at c_0 (Fig. 6.10). This barrier disappears when $c_0 = c_i$; $\overline{DE}$ is then constantly negative (MD being tangent at the inflection point).

Fig. 6.10. Conditions for nucleation in a system undergoing phase separation.

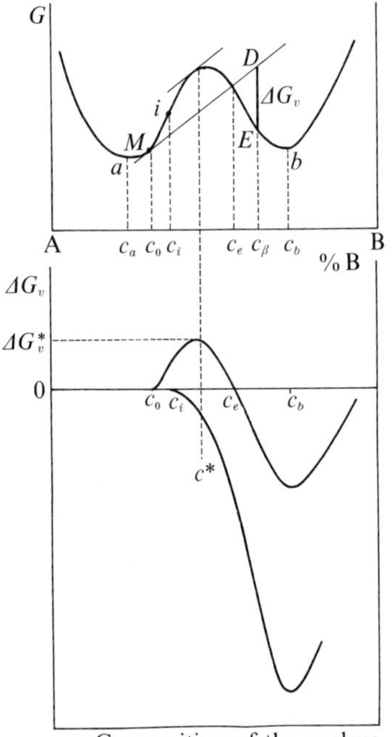

Between the spinodal points (region II) the system is unstable even to infinitely small fluctuations. The separation is accompanied by a constantly decreasing G and the system is controlled solely by diffusion. This process is called *spinodal decomposition* and will be studied in detail later (see Section 6.3.4).

6.3.3 *Phase separation by nucleation and growth*

The creation of a second phase of clearly different composition implies the formation of an interface. The phenomenon of phase separation is then controlled by the process of nucleation and growth (cf. Chapter 3).

(a) *Two-component system – Becker theory*

Assume that the nuclei have a constant composition corresponding to the stable phase c_b and that only their dimensions vary.

Classical theory[221] is easier to apply when there is no change in composition as in the case of the crystallization of a single-component system.

For a multicomponent system, the thermodynamic barrier can only be estimated if the detailed dependence of G on the composition is known. It is also necessary to make assumptions as to the variation of the surface energy and the diffusion barrier with composition and temperature.

The coexistence curve can be calculated from the free energy of the solution $\Delta G_m = \Delta H_m - T\Delta S_m$, where $\Delta G_m = G_{heterogeneous} - G_{homogeneous}$. Conversely, a model of ΔG_m can be adjusted to an experimental coexistence curve. A first approach uses a *regular solution model*. For a binary system, it is equivalent to stating:

$$\Delta G_m = \alpha c_1 c_2 + RT\,(c_1 \ln c_1 + c_2 \ln c_2)$$

where c_1 and c_2 are the fractions of the components 1 and 2 with $c_1 + c_2 = 1$ and α is a coefficient depending on the interatomic forces. The model leads to a coexistence curve which is symmetric relative to $c_1 = c_2 = 1/2$. The critical temperature $T_c = \alpha/2R$ can be used to determine α which plays the rôle of an adjustable parameter. In reality, the immiscibility curve is generally not symmetric and to make it symmetric, appropriate terminal components can be chosen to obtain a transformation of the abscissa. In the case of the system SiO_2–X_2O it leads, for example, to taking SiO_2 and $2SiO_2{\cdot}X_2O$ as terminal components.

In Becker's treatment the surface energy Δg_s of a nucleus is given by:

$$\Delta g_s = 2\alpha n_s$$

where n_s is the number of atoms per unit surface of the nucleus.

The model can be refined by introducing the activities $a_1 = \gamma_1 c_1$ and $a_2 = \gamma_2 c_2$ of the components 1 and 2 (γ_1 and γ_2 being the activity coefficients) and writing:

$$\Delta G_m = RT(c_1 \ln a_1 + c_2 \ln a_2)$$

The activities can be calculated from the liquidus curve of the equilibrium diagram. For example a_1^l, the activity of liquid component 1 in the liquidus composition is:

$$\ln a_1^l = \frac{\Delta H_F}{R}\left(\frac{1}{T_F} - \frac{1}{T_l}\right)$$

where ΔH_F is the heat of fusion of the component, T_F the melting temperature and T_l the liquidus temperature. The activities corresponding to other temperatures are calculated by assuming that the partial heat of fusion is constant for a given composition:

$$RT_l \ln \gamma_{1l} = RT \ln \gamma_{1T}$$

where T_l is the liquidus temperature, γ_{1l} the activity coefficient for component 1 at the liquidus and γ_{1T} the activity coefficient corresponding to T. The isothermal activity coefficient γ_2 for the other component is then obtained from the Gibbs–Duhem relation:

$$c_1 \, \mathrm{d} \ln \gamma_1 + c_2 \, \mathrm{d} \ln \gamma_2 = 0$$

Other models can be used, e.g. the *subregular solution model* in which the enthalpy ΔH_m is given by the expression:

$$\Delta H_m = (A_1 + A_2 c_1) c_1 c_2$$

where A_1 and A_2 are adjustable parameters from a given experimental coexistence curve.

The model of Van der Toorn and Tiedema[222] is an extension of the subregular model in which ΔH_m is expressed in the form of a polynomial in c_1 with adjustable coefficients calculated from the immiscibility curve.

The weak point in all these approaches is that the final result is very sensitive to the model adopted; most of the calculations are uncertain because of the associated character (or polymerization) of the systems studied which makes the choice of the terminal components difficult.

When ΔG_m is known, the position of the spinodal is obtained from the condition $\partial^2 \Delta G_m / \partial c^2 = 0$. The position of the spinodal can be located approximately from the coexistence curve by using the relation:

$$c_s - c = (c - c_c)/\sqrt{3}$$

obtained by expanding ΔG_m about T in a Taylor series. In this formula c and c_s are respectively the compositions of the coexistence curve and the spinodal for a temperature T and the critical composition c_c.

It can also be shown that near the critical point T_c, $c - c_c$ varies proportionally to $|T - T_c|^{1/2}$.

(b) *Borelius and Hobstetter–Scheil theory*

The Becker theory assumes that the composition of the embryos is constant and that they become nuclei by dimensional fluctuation. In the Borelius theory,[223] a group of atoms of constant size becomes stable by a composition fluctuation. The treatment neglects interfacial energies and the critical composition c^* of the embryo is obtained from the maximum of the ΔG_v curve (Fig. 6.10).

The mixed treatment of Hobstetter[224] and Scheil[225] combines the hypotheses of Becker and Borelius: the nucleus results from an optimization which includes size and composition at the same time. The interfacial energy Δg_s is considered and varies with the composition of the nucleus.

All these theories indicate that the mechanism of phase separation by nucleation is more probable in region I; however, it is not excluded in region II.

(c) *Hilliard–Cahn theory*

Cahn and Hilliard[226] developed a more general approach in which the free energy of the system shows spatial variations. The theory does not separate volume and surface contributions nor does it imply that the nucleus has a uniform composition. It is particularly appropriate for the analysis of phase separation phenomena in region II.

The Helmoltz free energy f per unit volume is expressed here as a function of the local concentration c (the mole fraction B in a binary solution A–B) and the derivatives of the concentration of the immediate environment. The free energy per unit volume of a solution of a uniform composition c is considered by means of a Taylor series expansion around $f(c)$. The total energy F is obtained by integration over the entire volume:

$$F = \int_v [f(c) + \kappa(\nabla c)^2 + \ldots] \mathrm{d}V \qquad (6.1)$$

The term $\kappa(\nabla c)^2$ represents the increase of f due to the concentration gradient and can be associated with the formation of an interface.

Note that the authors use the Helmoltz free energy F in their theory, but in condensed systems at normal pressure, the difference between F and G can be neglected.

Cahn and Hilliard have deduced the properties of a critical nucleus in the metastable region. At low supersaturations, the properties found are those of a classical critical nucleus. As supersaturation increases, the formation energy W^* becomes progressively smaller and tends towards zero at the spinodal. Simultaneously, the interface becomes more and more diffuse. Instead of resembling a droplet of a second phase, the nucleus begins to resemble a composition fluctuation. Near the spinodal, the energy formation of the nucleus varies as $(\Delta T_s)^{3/2}$ where ΔT_s is the difference in temperature $T - T_s$ relative to the spinodal.

6.3.4 *Phase separation by spinodal decomposition*

Consider equation (6.1) and develop $f(c)$ in a Taylor series around the average composition c_0:

$$f(c) = f(c_0) + (c - c_0)\frac{\partial f}{\partial c} + \frac{1}{2}(c - c_0)^2 \frac{\partial^2 f}{\partial c^2} + \ldots$$

taking into account the fact that:

$$\int_v (c - c_0)\mathrm{d}V = 0$$

The difference of the free energy ΔF between the initial homogeneous solution and the inhomogeneous solution is:

$$\Delta F = \int_v \left[\frac{1}{2} \frac{\partial^2 f}{\partial c^2}(c - c_0)^2 + \kappa(\nabla c)^2 \right] \mathrm{d}V$$

where it is assumed that the gradient coefficient $\kappa > 0$.

If $\partial^2 f / \partial c^2 > 0$, the solution is stable towards all infinitesimal fluctuations. If, however, $\partial^2 f / \partial c^2 < 0$ (region II), the solution will be unstable towards fluctuations for which the first term predominates.

The kinetics of initial phase separation can be obtained by considering the diffusion equation. The mobility M is defined as the quotient of the flux J_A (or J_B) of the components A (or B) divided by the gradient of the chemical potential (cf. Chapter 10):

$$J_B = -J_A = M\nabla(\mu_A - \mu_B)$$

the generalized diffusion equation is obtained (neglecting the higher-order terms):

$$\frac{\partial c}{\partial t} = \frac{M}{N_v} \left(\frac{\partial^2 f}{\partial c^2} \right) \nabla^2 c - \frac{2M\kappa}{N_v} \nabla^4 c \tag{6.2}$$

where N_v is the number of atoms per unit volume. This *linearized* equation is applicable to isotropic systems without elastic energy contributions.

The evolution of the composition as a function of time can be obtained by considering the Fourier components of the composition given by the expression:

$$c(\vec{r}, t) - c_0 = \left(\frac{1}{2\pi} \right)^3 \int_\beta A(\vec{\beta}, t) \exp(\mathrm{i}\vec{\beta} \cdot \vec{r}) \mathrm{d}^3 \vec{\beta} \tag{6.3}$$

where $c(\vec{r}, t)$ represents local composition at time t at a point in space defined by the position vector $\vec{r}$; c_0 is the average composition, $\vec{\beta}$ a vector in Fourier space with a magnitude $\beta = 2\pi / \Lambda$ and Λ "the wavelength" of the spatial component.

In order for expression (6.3) to satisfy the diffusion equation (6.2), the amplitude $A(\vec{\beta}, t)$ must be connected with the initial amplitude $A(\vec{\beta}, 0)$ by the exponential expression:

$$A(\vec{\beta}, t) = A(\vec{\beta}, 0) \exp[R(\beta) \cdot t]$$

which defines an *amplification coefficient* $R(\beta)$ given by the expression:

$$R(\beta) = -M\frac{\partial^2 f}{\partial c^2}\beta^2 - 2M\kappa\beta^4$$

The initial amplitude distribution $A(\vec{\beta}, 0)$ is obtained by analyzing the situation at the start. The sign of $R(\beta)$ then determines whether the amplitude of a particular component, for which the wave number is β, is increasing or decreasing.

All components with $R(\beta) > 0$ increase, i.e. $\partial^2 f/\partial c^2 + 2\kappa\beta^2 < 0$; this condition is satisfied inside the spinodal $\left(\partial^2 f/\partial c^2 < 0\right)$ but only for the components $\beta < \beta_c$ where β_c is a critical wave number defined by:

$$\beta_c^2 = -\frac{1}{2\kappa}\frac{\partial^2 f}{\partial c^2}$$

Figure 6.11 gives the form of the curve $R(\beta)$ as a function of β; $R(\beta)$ is maximum for $\beta_m = \beta_c/\sqrt{2}$. Because the curve has a sharp maximum, the components close to β_m will grow more rapidly in amplitude and finally dominate the others. The system will tend to amplify a narrow band of spatial frequencies centered on β_m. The "*spinodal wavelength*" is the name given to $\Lambda = 2\pi/\beta_m$.

An interdiffusion coefficient $\tilde{D}$ can be introduced , which is connected to M by the relation:

$$\tilde{D} = \frac{M}{N_v}\frac{\partial^2 f}{\partial c^2}$$

Fig. 6.11. Amplification factor $R(\beta)$.

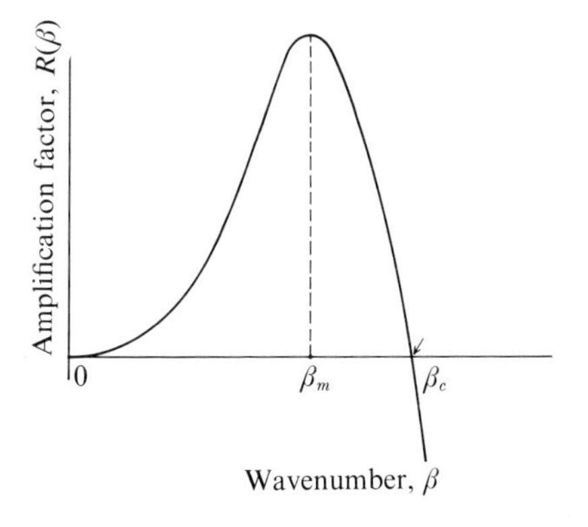

Since M is always positive, $\tilde{D}$ is negative inside the spinodal; it corresponds to a diffusion in a direction opposite to the concentration gradient (counter-current migration) as shown in Fig. 6.12.

The amplification factor can also be written:

$$R(\beta) = -\tilde{D}\beta^2 \left(1 - \beta^2/\beta_c^2\right)$$

The graph of $R(\beta)/\beta^2$ as a function of β^2 is linear and its intersection with the ordinate provides $\tilde{D}$.

Expressing $\partial^2 f/\partial c^2$ as a function of temperature, it can be shown[227,228] that β_m^2 or $1/\Lambda_m^2$ must vary proportionally to undercooling $\Delta T_s = T - T_s$ relative to the spinodal:

$$\frac{1}{\Lambda_m^2} = \alpha \Delta T_s$$

where α is a constant. This relation allows the determination of the spinodal temperature T_s by studying the variations of Λ_m as a function of the decomposition temperature. In the calculation, the temperature is generally assumed constant (isothermal treatment). Expressions have been obtained which give Λ_m during quenching processes at variable temperature.

6.3.5 *Morphology of unmixed phases*

The morphology of the phases resulting from an unmixing varies according to the mechanism of decomposition.

Figure 6.12 shows the evolution of the concentration of a component as a function of time. In the case of spinodal decomposition there is a continuous variation of the extreme compositions up to the point where equilibrium is attained; the initially diffuse interface tends to become sharper as the decomposition evolves. In contrast, for the classical nucleation and growth case (Fig. 6.12(*b*)), the second phase composition does not vary with time and the interface remains distinct throughout the process.

Cahn made an attempt to visualize the morphology of spinodal decomposition in the isotropic case: he superposed a set of spatial sinusoidal concentration waves with constant wavelength ($= \Lambda_m$) but with variable orientation, phase and amplitude. The computer-generated image of the structure shows a characteristic regularity for the evolving phases. Figure 6.13(*a*) corresponds to a planar section of such a mathematical model – note the similarity with the interconnected structures revealed by electron microscopy in the case of glasses (cf. Section 6.2).

Fig. 6.12. Evolution of a composition fluctuation as a function of time: (*a*) spinodal decomposition; (*b*) nucleation and growth. Here c_a and c'_a are the final (limiting) compositions of the separate phases. After Ref. 229.

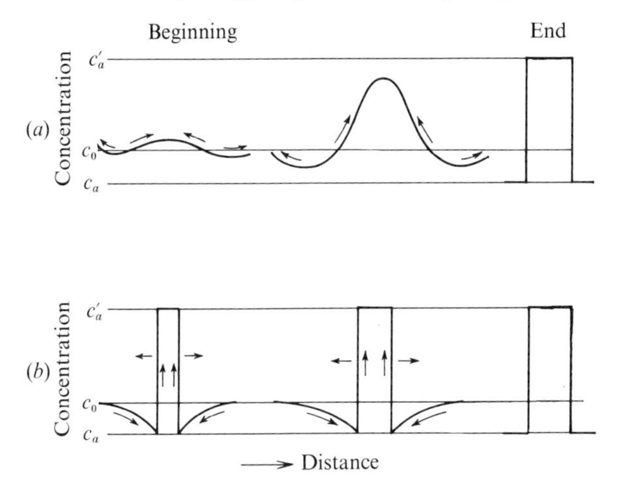

Fig. 6.13. (*a*) Plane section of a three-dimensional spinodal structure simulated on a computer by adding 100 random sinusoidal waves of wavelength λ. The points define the region where the concentration exceeds the average. After Ref. 227. (*b*) Plane section across a population of equal sized spheres distributed randomly in space. After Ref. 230.

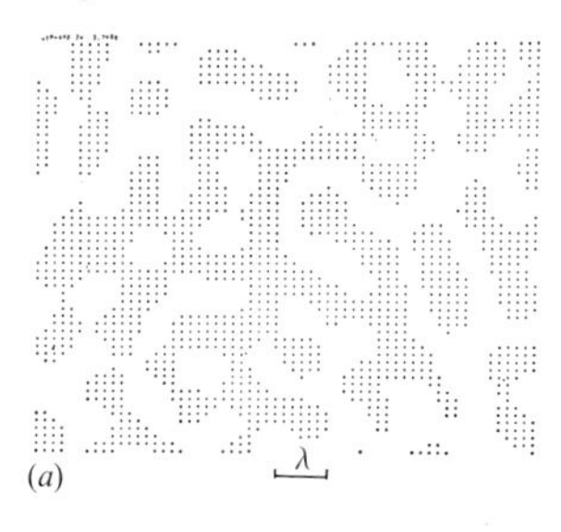

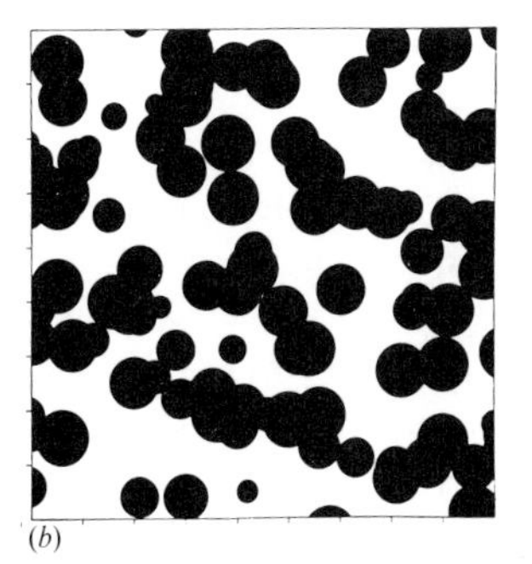

In the case of nucleation and growth, there is a tendency towards a random distribution of both the size and position of the phases (Fig. 6.14(*a*)). The particles tend to be spherical and have a low connectivity, whereas phases separated by spinodal decomposition are generally non-spherical ("sponge" structure) and have a high degree of connectivity (Fig. 6.14(*b*)).

Fig. 6.14. Transmission electron micrographs: (*a*) glass unmixed by nucleation and growth (98% B_2O_3, 2% PbO treated 0.5 h at 415 °C); (*b*) glass unmixed by spinodal decomposition. (76% B_2O_3, 19% PbO, 5% Al_2O_3 treated 1 h at 425 °C).

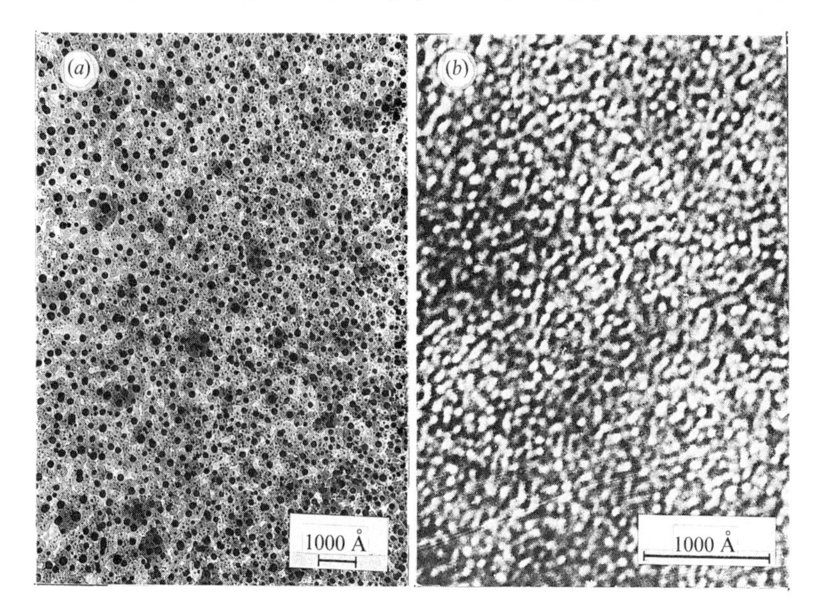

This phase connectivity was sometimes considered as a *criterion* for the spinodal decomposition mode. Haller[230], however, showed that similar textures can be generated by a nucleation mechanism: the spherical particles can bond together to form high connectivity regions if the volumetric density of nuclei centers is high. This is precisely the case for the central interior regions of the spinodal where the volumes occupied by each of the phases are approximately the same (no major phase). A computer simulation gave the diagram shown in Fig. 6.13(*b*).

Since the nucleation and growth mechanism cannot be excluded, even in the interior of the spinodal field, the connectivity criterion is not sufficient to identify spinodal decomposition. It is impossible to answer the question based on electron microscopy alone. Only a complete kinetic study can provide guidance on the nature of the operating decomposition process.[212]

6.4 Ripening phenomena

In the case of a saturated solution which undergoes phase separation, two main stages of development can be distinguished: the first, in which the system unmixes either by a nucleation and growth or spinodal mechanism is followed by a second in which a *rearrangement* of the phase geometry occurs. The latter stage in which the system tends to reduce its interfacial energy is called *coalescence* or *ripening*. The particles or domains have then attained appreciable dimensions and the degree of supersaturation of the matrix has fallen to a very low level. The smaller particles tend to redissolve and the larger particles grow at their expense. Fluctuation effects play a negligible rôle at this stage. The overall effect is growth of the precipitates (or interconnected structure) which reduces the total interfacial area and thus the surface energy of the phases present.

The process, called "Ostwald ripening", requires a diffusion of the solute from the regions close to the small particle to regions around larger particles; this implies that the solute concentration in equilibrium with a small particle is greater than that in equilibrium with a larger particle. This can be visualized qualitatively on a free energy diagram (Fig. 6.15) which represents equilibrium conditions between the matrix (α) and the precipitate (β) present in the form of large and small particles. Since the proportion of atoms situated on the interface increases as the size decreases, the average free energy per atom increases; thus, the curve for the small (β) particles is above the curve for the large particles. The common tangent construction shows that the equilibrium concentration in the matrix c_α, i.e. the *solubility*, is larger for the small particles. In a system containing particles of different sizes, concentration gradients are developed which promote the transport of solute from regions around the small particles towards those around large particles on which the solute precipitates.

The dependence of the solubility on the particle size follows the Thomson–Freundlich law (see Ref. 122 or 231 for a derivation):

$$RT \ln\left(\frac{a_2}{a_1}\right) = 2V_m\sigma\left(\frac{1}{r_2} - \frac{1}{r_1}\right)$$

where, in a binary system, a_1 and a_2 are the activities of component B in the matrix in equilibrium with particles of radii r_1 and r_2: V_m is the molar volume of the precipitated phase and σ the interfacial surface energy. In an ideal solution, or in a dilute solution obeying Henry's law, we have $a_1/a_2 = c_1/c_2$, where c_1 and c_2 are the solubilities of the two particles.

When r_1 increases indefinitely, c_1 tends towards the concentration c_∞ which corresponds to a plane surface, i.e. the solubility in the usual

Fig. 6.15. Free energy diagram showing the solubility of particles of different sizes.

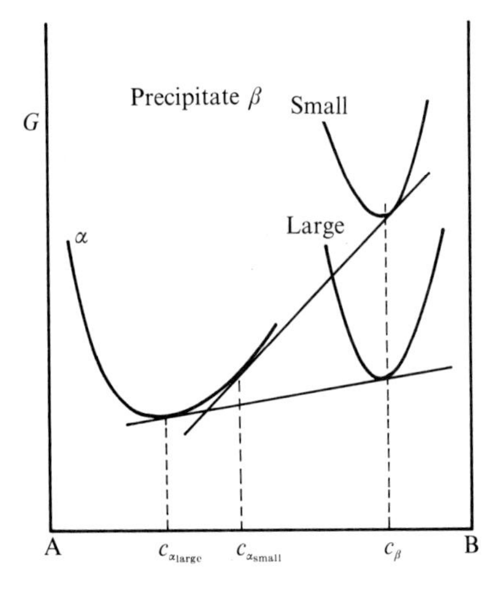

macroscopic sense. Thus, using a limited series expansion, for a particle of radius r_i, the solubility c_i is:

$$c_i = c_\infty \left(1 + \frac{2V_m\sigma}{RT}\frac{1}{r_i}\right)$$

If at a given instant the average concentration in the matrix at a point sufficiently remote from the particle is $\bar{c}$, it can be shown, by applying the diffusion equations for a spherical layer, that the concentration gradient at the interface is:

$$\left(\frac{\partial c}{\partial r}\right)_{r=r_i} = \frac{\bar{c} - c_i}{r_i}$$

In the steady state, the solute flux per unit of surface leaving from or arriving at the particles by volume diffusion is:

$$J_i = D\frac{\bar{c} - c_i}{r_i}$$

where D is the diffusion coefficient.

The sign of J_i depends on the sign of the difference $\bar{c} - c_i$, and J_i is zero for $c_i = \bar{c}$. Consider an ensemble of particles of variable radii r_i and seek in this population those of radius r^* which satisfy the condition $c_i = \bar{c}$, i.e. those with a solubility corresponding to the average concentration. They have a radius r^* given by the formula:

$$\bar{c} = c_\infty \left(1 + \frac{2V_m\sigma}{RT}\frac{1}{r^*}\right)$$

Under these conditions, all particles with radii $r_i < r^*$ dissolve and those with radii $r_i > r^*$ grow. The *critical* radius r^* thus represents the radius of those particles which at a given instant are in a stationary state when the concentration of the solute in matrix is $\bar{c}$, i.e. for a supersaturation equal to $\bar{c} - c_\infty$. Figure 6.16 shows the concentration profiles near particles for the three preceding cases.

Fig. 6.16. Schematic representation of the evolution of the concentration profile in the course of Ostwald ripening. Particles with a critical radius less than r^* dissolve and feed those of radius greater than r^*. After Ref 212.

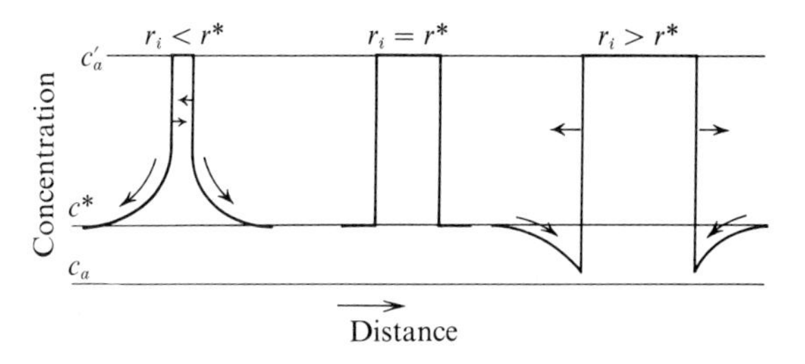

Lifshitz and Slyozov[232] and independently Wagner[233] resolved the mathematical problem for the evolution of the size distribution of a particle population under the preceding conditions. The kinetic equation which describes the evolution of the radius r_i is:

$$\frac{\mathrm{d}r_i}{\mathrm{d}t} = \frac{2\sigma}{RT} V_m^2 D c_\infty \frac{1}{r^*}\left(\frac{1}{r^*} - \frac{1}{r_i}\right)$$

The problem consists of solving the system of n equations of the preceding type for the interactions between all the particles under the conditions that the total mass of the precipitated fraction is conserved during the precipitation.

The theory furnishes the following results for the case where the process is assumed to be controlled by volume diffusion:

1. The critical radius r^* is identical with the average radius $\bar{r}$ of the population.

2. Regardless of the initial particle size distribution, the system tends asymptotically towards a certain universal distribution $f(z)$ of the reduced variable $z = r/r^*$ as shown in Fig. 6.17 and which is equal to zero for $z > 3/2$.

3. The average radius (critical) of the population increases with time following a cube law:

$$\bar{r}^3 - \bar{r}_0^3 = K(t - t_0)$$

where $\bar{r}_0$ is the average radius at time $t = t_0$ and K a constant defined by:

$$K = \frac{8}{9}\frac{\sigma}{RT}V_m^2 D c_\infty$$

4. because the fraction precipitated remains constant, the evolution of the number of particles is found to follow the law:

$$\frac{1}{N} - \frac{1}{N_0} = K'(t - t_0)$$

where N and N_0 are the number of particles at time t and t_0 and K' a constant depending on K and the total fraction precipitated.

Fig. 6.17. Distribution function $f(z)$ from the Lifschitz–Slyozov–Wagner theory. After Ref.214.

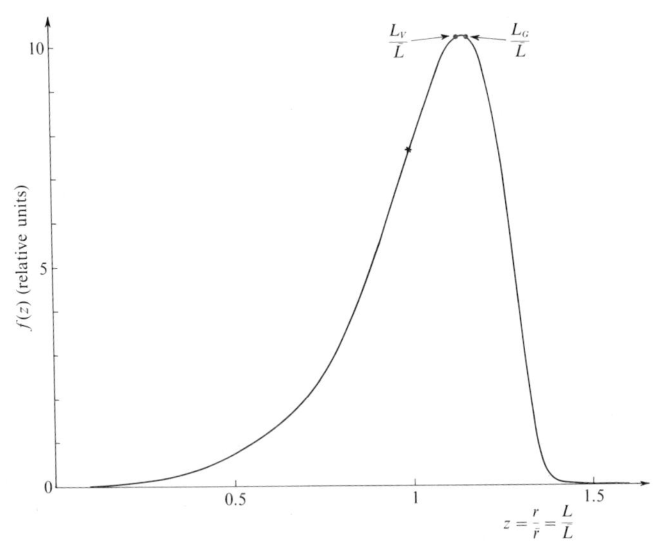

The two latter laws are particularly useful in practice and allow the Ostwald ripening to be detected.

The mechanism plays an important rôle in phase separation in glasses and where the phenomena of nucleation and growth lead to spherical precipitates under isotropic conditions.

The interconnected structures resulting, for example, from a spinodal decomposition also undergo a ripening which has the effect of increasing the size of the heterogeneities, but in this case, the preceding theory is not directly applicable because the precipitates with simple geometric forms cannot be identified. Instead, Haller's approach[230] can be used. This relates the *curvature* of the interface to the local solubilities to deduce the conditions for transport by local volume diffusion.

6.5 Experimental methods

6.5.1 *Determination of coexistence curves*

Immiscibility can be detected by optical techniques; if the phase dimension is of the same order of magnitude as the wavelength of the light, the liquid (or glass) becomes first opalescent (Tyndall effect), then opaque. However, apparently transparent and homogeneous glasses can already contain separated domains of several hundred angströms and to define their state precisely, electron microscopy or Small-Angle X-ray Scattering (SAXS) techniques must be employed.

(a) *Stable unmixing*

The determination can be made directly by placing the liquid in a light path while varying the temperature. Another approach is to quench from a series of temperatures and examine the samples in the solid state; this allows the unmixing temperature to be bracketed. By varying the composition, the curve (or surface) of unmixing can be determined.

(b) *Metastable (sub-liquidus) unmixing*

The method starts with samples unmixed by a previous thermal treatment. The minimum treatment temperature at which an opalescent sample again becomes transparent is then determined. Elongated samples or a series of samples placed in a thermal gradient furnace can be used to speed up the process.

(c) *Ternary diagrams – determination of tie lines*

In a binary diagram the composition of the intermixed phases is fixed at each temperature as soon as the coexistence curve is known. This is not the case for ternary diagrams where the direction of the tie lines varies with the composition of the coexistent phases and must be determined independently.

The ideal method is to analyze each separate phase chemically. This can be done systematically when the two phases have large differences in solubility facilitating their separation but this is usually not possible. Figure 6.18 from the B_2O_3–PbO system shows isolated spheres of the less soluble phase before and after their extraction from the matrix.

Fig. 6.18. (*a*) Unmixing of a glass: 75% B_2O_3, 25% PbO (wt.%). (*b*) Spheres extracted from the matrix of a 96% B_2O_3, 4% PbO glass treated 16 h at 415 °C after dissolution in water.

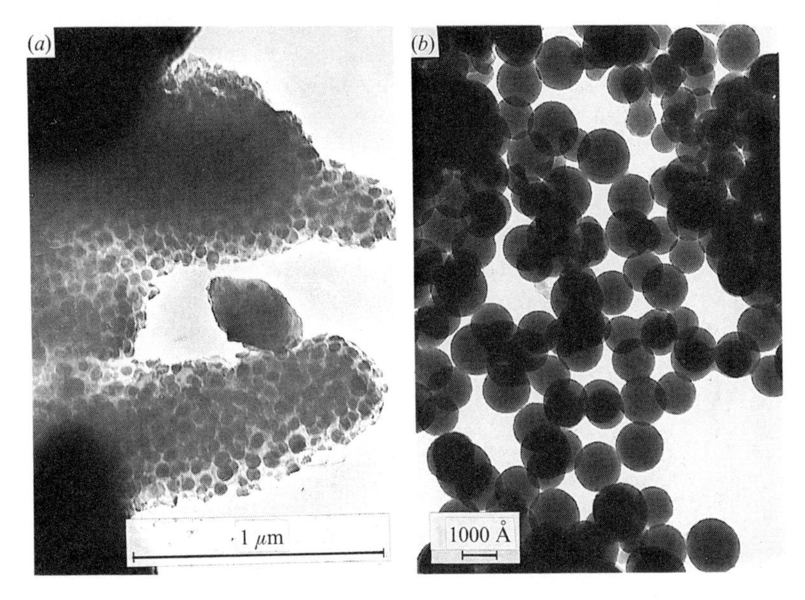

Direct analysis by electron beam microprobe can be attempted if the domains of the analyzed phase can be grown up to dimensions compatible with the dimensions of the analyzing beam ($\sim 1 \mu m^2$).

In general, each of the two vitreous phases resulting from unmixing has a different T_g. Since a given tie line corresponds to a defined pair of phases in variable proportions, each T_g must be constant along the tie line, thus allowing its approximate determination. This is called the "constant T_g" method by Mazurin *et al.*[234]

6.5.2 *Study of the textures resulting from phase separation*

Given the fineness of the precipitates, which are about 100–1000 Å, electron microscopy, SAXS (and more recently, Small Angle Neutron Scattering (SANS)), as well as light scattering find here a natural application.

(a) *Electron microscopy*

Transmission electron microscopy is currently used for the detection and characterization of phase separation. The *replica* method allows the observation of the phases present, occasionally after a moderate chemical attack on one of them to accentuate the relief (Fig. 6.19). However, it may sometimes present problems in the case of the finest structures of the order of 100 Å. Observation by direct transmission of thin blown films or drawn fibers is then preferred. Thinning by ion bombardment is a possible technique. The use of the Scanning Electron Microscope (SEM) now tends to replace the replica method, but its resolution is often insufficient and makes direct transmission methods indispensable for study of the finest structures. Scanning Transmission Electron Microscopy (STEM) can be advantageously used.

Fig. 6.19. Electron micrograph of an unmixed SiO_2-Na_2O glass after etching to enhance surface relief.

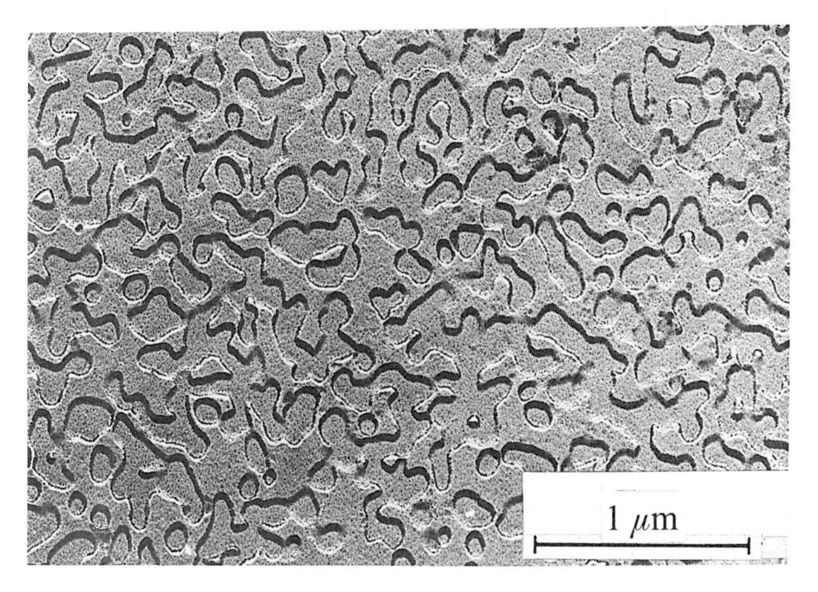

The problem becomes complicated when *quantitative* measurements are sought to follow phase separation kinetics. The replica or SEM methods provide plane sections and, to obtain *volume* distributions, statistical methods of quantitative metallography or *stereology*[235] must be applied. In the case of direct measurements by transmission, the difficulty lies in the assessment of the observed *volume*, given the thinness of the samples and vertical superposition effects of the particles. It is then possible to have recourse to stereoscopic methods.[236]

(b) *SAXS*

In contrast to microscopic methods, small-angle scattering of X-rays leads directly to quantitative measurements by providing various average statistics on the systems observed.

In the case of phase separation, the most productive areas of application have been:

1. the study of "dilute" systems;
2. spinodal decompositions.

1. *"Dilute" system of particles*. The theory of SAXS[237] leads to unequivocal and quantitative results if the system *is not dense*, i.e. if the particles are sufficiently separated from each other to avoid the effect of interparticle interferences.

Let us consider the simple case of scattering by a dilute system of N *identical* particles of volume v, of uniform electron density ϱ contained in a matrix with electron density ϱ' (monodisperse system).

The scattered intensity $I(Q)$ is:

$$I(Q) = I_e(\Delta\varrho)^2 N v^2 \left(1 - \frac{1}{3} R_G^2 Q^2 + \ldots\right)$$

with $Q = (4\pi \sin\theta)/\lambda$ (for a small scattering angle $\epsilon = 2\theta$, $Q = (2\pi\epsilon)/\lambda$). I_e is the intensity scattered by an electron, and $\Delta\varrho = \varrho - \varrho'$ is the difference in the electron densities of the particle and the matrix.

The form of the curve $I(Q)$ depends on the "radius of gyration" R_G of the particle defined from the density $\varrho(r)$ by the relation:

$$R_G^2 = \frac{\int_v r^2 \varrho(r)\, dv_r}{\int_v \varrho(r)\, dv_r}$$

For a homogeneous particle of density $\varrho(r) = \varrho$ and diameter L:

$$R_G = \frac{L}{2}\sqrt{\frac{3}{5}}$$

The *Guinier approximation* consists of replacing the limited series expansion by an exponential:

$$I(Q) = I_e(\Delta\varrho)^2 N v^2 \exp\left(-\frac{1}{3} R_G^2 Q^2\right)$$

The graph of ln $I(Q)$ as a function of Q^2 (or of ϵ^2) (Fig. 6.20) is, under these conditions, a straight line of slope α which is proportional to R_G^2 and allows the determination of L.

Fig. 6.20. Principle of SAXS.

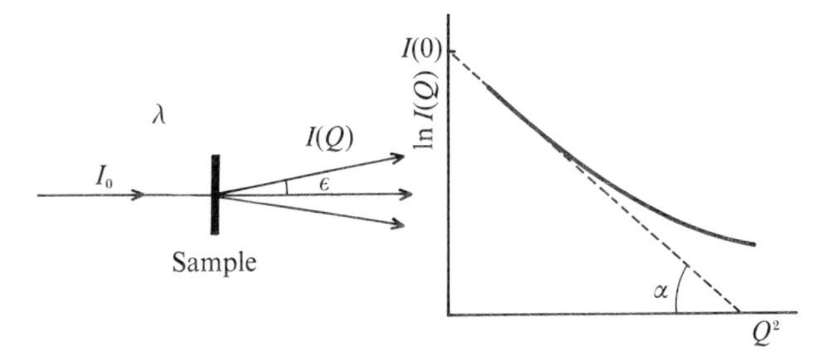

Also, the extrapolated value $I(0)$ for $Q \to 0$ is proportional to the number of particles and the *square of the volume* of a particle:

$$I(0) = I_e(\Delta\varrho)^2 Nv^2$$

Another significant value is the "integrated intensity" Q_0 defined by the expression:

$$Q_0 = \frac{1}{4\pi^2}\int_0^\infty Q^2 I(Q)\,\mathrm{d}Q$$

Q_0 is proportional to the number of particles and to their volume i.e. to the *total volume* of the precipitated phase:

$$Q_0 = I_e(\Delta\varrho)^2 Nv$$

The proportionality constants are eliminated forming the ratio:

$$I(0)/Q_0 = v = \frac{\pi}{6}L_v^3$$

which provides a direct evaluation of the volume.

The comparison of this value L_v with the value L deduced from R_G makes it possible to verify the hypothesis assumed for the *form* of the particle. $L_v \neq L$ if the particles deviate from spherical form.

The general case is of a particle population with various shapes and dimensions (polydisperse system); the Guinier plot is then no longer linear. The limiting tangent at the origin obtained by extrapolation defines a weighted radius of gyration $\bar{R}_G$ such that:

$$(\bar{R}_G)^2 = \frac{\sum_i R_i^2 v_i^2 p_i}{\sum_i v_i^2 p_i}$$

where R_i, v_i, and p_i are respectively the radius of gyration, volume and the fraction of particles of type i.

If, as is frequently the case in phase separation, the particles are all identical in shape (spherical) and differ only in size, the preceding expression becomes:

$$(\bar{R}_G)^2 = \frac{\sum_i R_i^8 p_i}{\sum_i R_i^6 p_i}$$

which defines a weighted diameter $\bar{L}_G$ of the population:

$$\bar{L}_G = 2\sqrt{\frac{5}{3}}\bar{R}_G$$

This diameter depends strongly on the fraction of large particles.

In the polydisperse case, the general equations become:

$$I(0) = I_e(\Delta\varrho)^2 N\overline{v^2}$$

$$Q_0 = I_e(\Delta\varrho)^2 N\bar{v}$$

$$\frac{I(0)}{Q_0} = \frac{\overline{v^2}}{\bar{v}} = \frac{\pi}{6}(\bar{L}_v)^3$$

where $\overline{v^2}$ and $\bar{v}$ are respectively the average values of the volume squared and the volume of the population. Note that in general $\overline{v^2}/\bar{v} \neq \bar{v}$.

By introducing a particle size distribution function $m(L)$ satisfying the condition:

$$\int_0^\infty m(L)\,\mathrm{d}L = 1$$

and defining the various average values by:

$$\overline{L^n} = \int_0^\infty L^n m(L)\,\mathrm{d}L$$

the values L_G and L_v respectively obtained from the preceding equations can be expressed as:

$$(\bar{L}_G)^2 = \overline{L^8}/\overline{L^6}$$

and

$$(\overline{L}_v)^3 = \overline{L^6} / \overline{L^3}$$

Both relations can be used to determine the parameters of a distribution function: Gaussian distribution, Lifschitz distribution etc.

It can be shown that for an ensemble of phases *bounded by a sharp interface*, when $Q \to \infty$ the expression $Q^4 I(Q)$ becomes a constant which depends upon the specific interfacial surface.[237]

This relation, called Porod's relation, enables the evaluation of the total particle surface and thus provides an additional characterization tool.

2. *Dense system, spinodal case.* In the case of *dense* systems, interference between the waves scattered by neighboring particles produces complex effects, and interpretation of the small-angle scattering curves does not generally lead to a useful unique solution.

However, there is a dense system which can be studied with unambiguous results and which is particularly interesting for the study of unmixing: spinodal decomposition. Since Cahn's theory is developed in reciprocal space, a direct correlation exists with small-angle scattering.

Relation (6.3) (Section 6.3.4) is the Fourier transform of the amplitude spectrum of the composition fluctuation in the solution. The spatial components are obtained from the inverse transformation:

$$A(\vec{\beta}, t) = \int_v [c(\vec{r}, t) - c_0] \exp(-\mathrm{i}\vec{\beta} \cdot \vec{r}) \, \mathrm{d}^3\vec{r}$$

the integration being extended over the volume v of the sample.

If, on the other hand, we calculate the X-ray scattering spectrum for a sample in which the composition distribution satisfies (6.3), the amplitude $B(\vec{Q}, t)$ of the scattered radiation is:

$$B(\vec{Q}, t) = \int_v \varrho(\vec{r}, t) \exp(-\mathrm{i}\vec{Q} \cdot \vec{r}) \, \mathrm{d}^3\vec{r}$$

where $\vec{Q}$ is the scattering vector and $\varrho(\vec{r}, t)$ the local electron density at time t related to the concentration c by the expression:

$$\varrho(\vec{r}, t) - \varrho_0 = \Delta f [c(\vec{r}, t) - c_0]$$

where Δf is the difference between the average atomic structure factors of the two final equilibrium phases and ϱ_0, the average electron density. For small values of $\vec{Q}$, Δf can be considered independent of $\vec{Q}$. Neglecting the (non-observable) intensity due to the contribution of ϱ_0, one obtains:

$$B(\vec{Q}, t) = \Delta f \int_v [c(\vec{r}, t) - c_0] \exp(-\mathrm{i}\vec{Q} \cdot \vec{r}) \, \mathrm{d}^3\vec{r}$$

which shows that the two expressions, $B(\vec{Q},t)$ and $A(\vec{\beta},t)$ are proportional provided that:

$$\vec{Q} = \vec{\beta}$$

an identity equivalent to the Bragg relation. The equality of the values of $\vec{Q}$ and $\vec{\beta}$ implies that

$$\frac{4\pi \sin\theta}{\lambda} = \frac{2\pi}{\Lambda} \quad \text{or} \quad \lambda = 2\Lambda \sin\theta$$

Here λ is the wavelength of the scattered radiation and Λ the wavelength of the spatial component of the concentration fluctuation producing scattering at the Bragg angle.

The scattered intensity $i(\vec{Q},t)$ equal to $B(Q,t) \cdot B^*(Q,t)$ is also proportional to $A(\vec{\beta},t) \cdot A^*(\vec{\beta},t)$ and it is seen that the scattering spectrum gives directly the distribution of the square of the amplitude of the components of the concentration fluctuation.

Since Λ is typically of the order of 100 Å, the angles θ for the X-ray wavelengths usually employed (λ = 1.54 Å) are small (< 1°). Thus Cahn's theory can be verified directly by SAXS without the necessity of a Fourier transformation of the results. Thus, studying the evolution of the spectrum of the spatial components of the composition is reduced to studying the SAXS spectrum by replacing $\vec{\beta}$ by $\vec{Q}$ in the corresponding expressions.

In the case of an isotropic system such as glass, if the theory is valid, the spectrum should evolve as a function of time as follows:

$$I(Q,t) = I(Q,0) \exp\left[2R(Q)t\right]$$

or

$$\ln I(Q,t) = \ln I(Q,0) + 2R(Q)t$$

the logarithms of the intensity components must increase linearly with time, the rate depending on the amplification coefficient. Given the theoretical behavior of $R(\beta)$ (Fig. 6.11), the formation of an intensity maximum in the scattering spectrum is expected.

6.6 Examples of studies of vitreous systems

6.6.1 *Region I*

(a) *Nucleation and growth*

The classical study of nucleation and growth is that of Hammel[238] who investigated the kinetics of phase separation in a glass of composition $76\,SiO_2{\cdot}13\,Na_2O{\cdot}11\,CaO$ which is near the immiscibility limit on the side of low SiO_2 content.

In this region, the phase separation gives rise to the formation of spherical particles. Their size distributions were obtained from electron micrographs on samples treated for variable times at temperatures below the unmixing temperature.

Knowing the number and size of the particles within a given time interval and their growth rate, it was possible to determine the time of formation by extrapolation and thus obtain the nucleation rate under steady state conditions.

Figure 6.21 shows the number of nuclei as a function of time for three thermal treatment temperatures. We can see that the linear variation is preceded by a transient and followed by a decreasing rate. Δg_v was obtained from the experimental miscibility gap by adjusting the parameters of the Lumsden model.[239] Δg_s was estimated as 4.6 erg cm^{-2} by studying the variation of the solubility temperature with the particle radius. The activation energy for diffusion $\Delta G'$ was the same as that for growth (94.5 kcal $mole^{-1}$). The transition time appears more in accord with the value of Hillig (cf. Section 3.2.4). Under these conditions, the calculations are in acceptable agreement with experiments but this success may be fortuitous because the results depend strongly on the model chosen for the calculation of Δg_v. Choosing the Van der Toorn–Tiedema model leads to relative variations which can reach 10^4.

(b) *Ostwald ripening*

Figure 6.22 shows a series of electron micrographs of glasses in the system B_2O_3–PbO containing 2 wt% PbO after thermal treatment for periods between 15 min and 16 h. It is seen that the number of particles decreases and their size increases.

Studies by SAXS[214] demonstrate that this is a typical Ostwald ripening. Figure 6.23 shows that the r^3 variation law is verified after a transient period. Figure 6.24 shows the evolution of the size-distribution curves of the particles calculated for the Lifschitz–Slyozov–Wagner distribution using averages from SAXS.

The electron micrograph of Fig. 6.25 shows very distinctly the depletion of the solute (rich in PbO) around the growing particle which appears as a halo around the particle, lighter than the matrix, schematically illustrated in Fig. 6.16.

Fig. 6.21. Number of nuclei as a function of time during the unmixing of a glass in the system $76\,SiO_2 \cdot 13\,Na_2O \cdot 11\,CaO$ treated respectively at (*a*) 601 °C, (*b*) 610 °C, and (*c*) 625 °C. After Ref. 238.

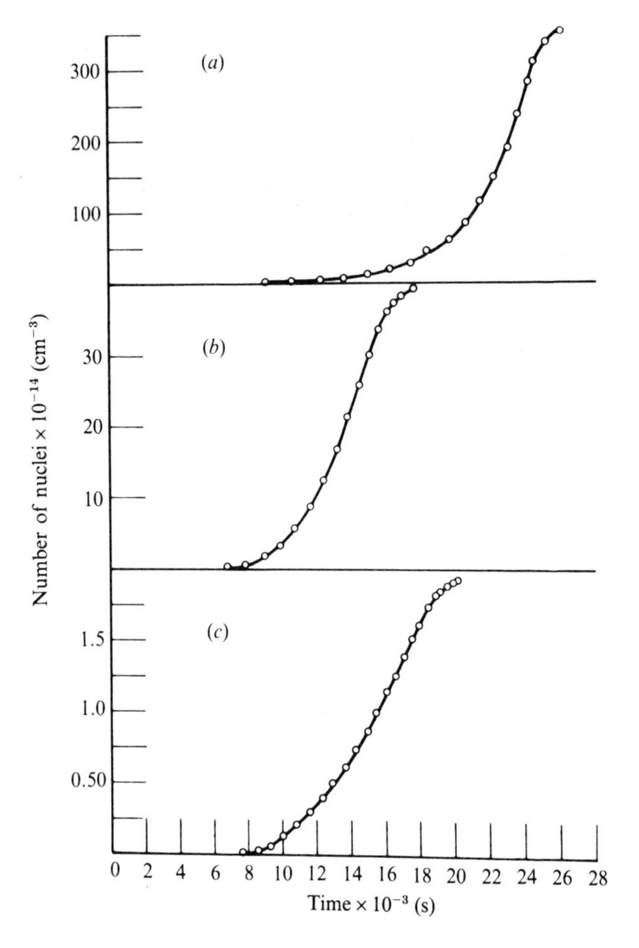

6.6.2 *Region II*

(a) *Spinodal decomposition*

Figure 6.26 shows the spectra obtained by SAXS for the system $76\,B_2O_3 \cdot 19\,PbO \cdot 5\,Al_2O_3$.[240] The glasses quenched from the liquid at 1150 °C were thermally treated at 450 °C. Two stages of decomposition can be distinguished. In the first, the logarithms of the intensity components vary proportionally with time (Fig. 6.27(*a*)). The amplification coefficients calculated from these results show that the curve $R(Q)$ is wider than predicted by Cahn's theory; for higher values of Q, $R(Q)$

Fig. 6.22. Electron micrographs of samples of glass B_2O_3–2% PbO showing Ostwald ripening. (*a*) untreated, (*b*) treated at 415 °C for 15 min, (*c*) 30 min, (*d*) 1 h, (*e*) 3 h, and (*f*) 16 h. After Ref. 214.

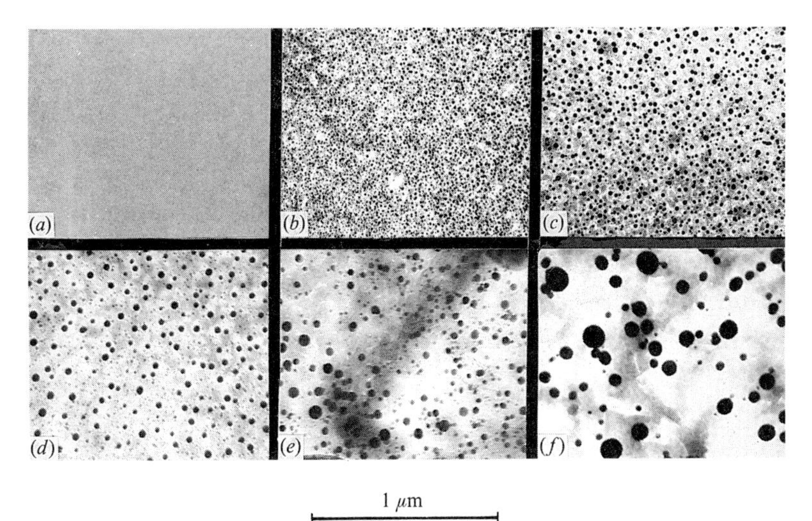

Fig. 6.23. Variation of L_G^3 during ripening of a sample of 99% B_2O_3,1% PbO glass thermally treated at the indicated temperatures. After Ref. 214.

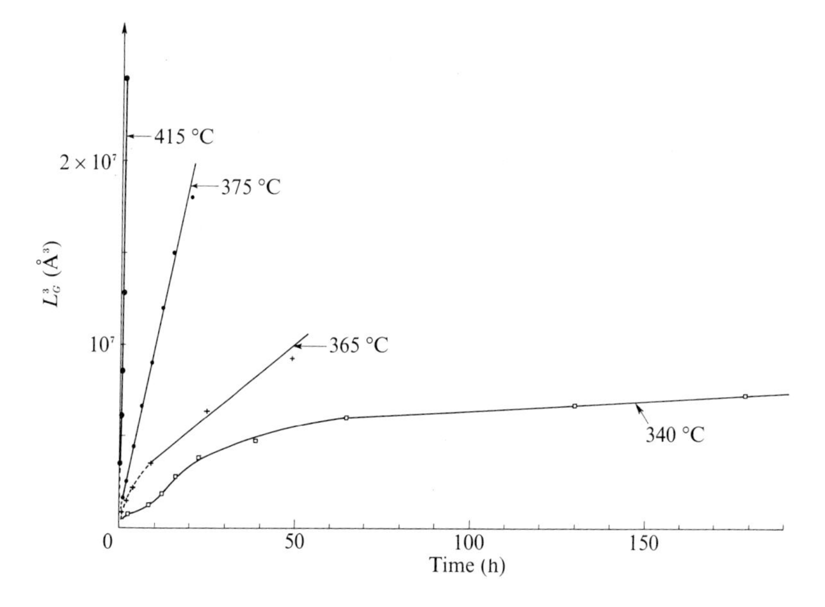

Fig. 6.24. Evolution of the size-distribution curves for particles in a 99% B_2O_3, 1% PbO glass thermally treated at 375 °C, calculated for a Lifschitz–Slyozov–Wagner distribution from SAXS data. The critical diameter L* is indicated by * on each curve. After Ref. 214.

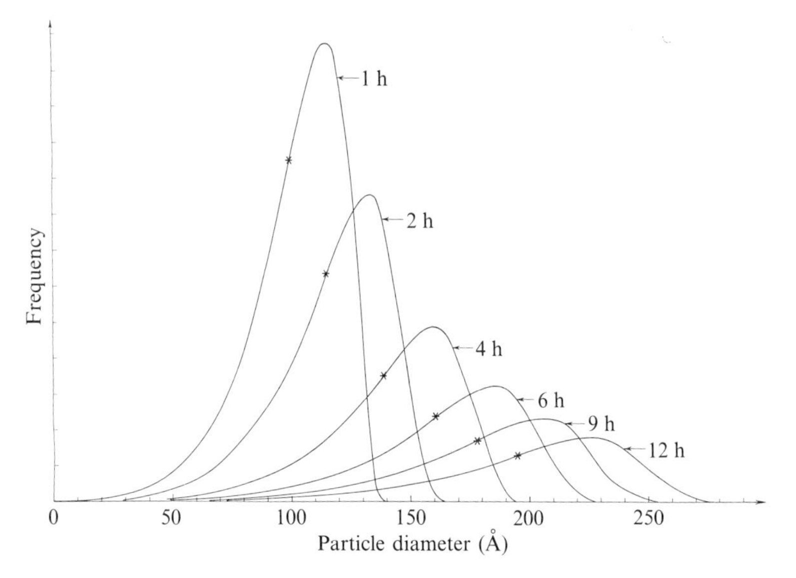

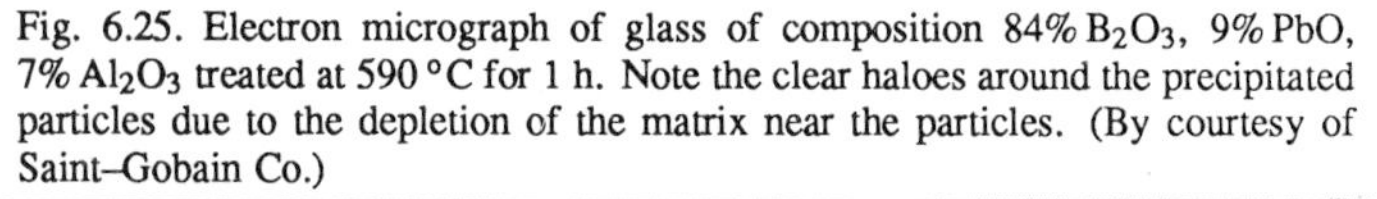

Fig. 6.25. Electron micrograph of glass of composition 84% B_2O_3, 9% PbO, 7% Al_2O_3 treated at 590 °C for 1 h. Note the clear haloes around the precipitated particles due to the depletion of the matrix near the particles. (By courtesy of Saint–Gobain Co.)

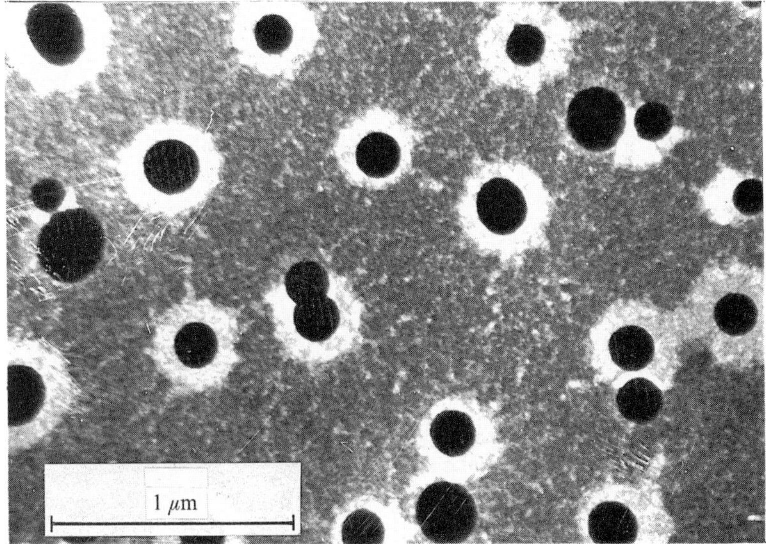

Fig. 6.26. SAXS spectra for a glass $76\,B_2O_3 \cdot 19\,PbO \cdot 5\,Al_2O_3$ (wt.%) quenched from 1150 °C and then held at 450 °C for the times indicated. After Ref. 240.

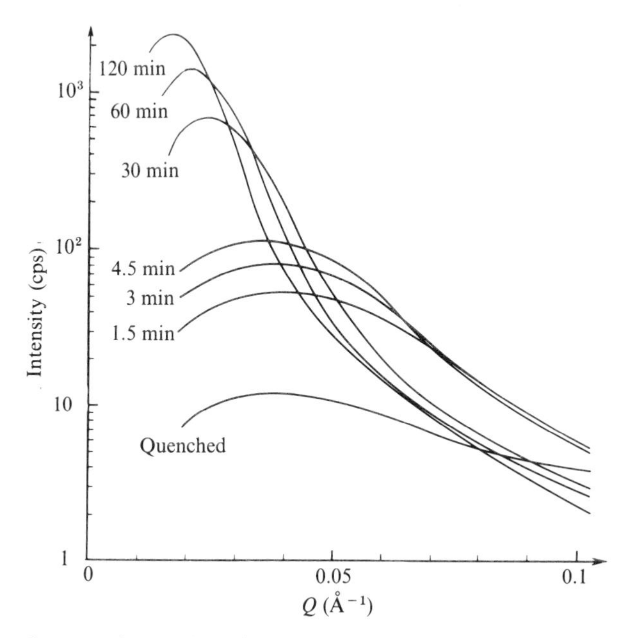

tends towards small positive values instead of becoming zero or negative (Fig. 6.27(*b*)). The graph of $R(Q)/Q^2$ as a function of Q^2 is not linear but hyperbolic (Fig. 6.27(*c*)). It shows that the linearized theory is only applicable to the very first moment of decomposition. The interdiffusion constant $\tilde{D}$ obtained from these measurements is $\tilde{D} = -3 \times 10^{-19}$ cm^2 sec^{-1}, and the "spinodal" wavelength $\Lambda = 130$ Å.

The same findings were made on the SiO_2–Na_2O system by Tomozawa, MacCrone and Herman[241] who recognized the necessity of introducing higher-order terms into the Cahn theory. Of course, this linear theory can only apply to the first moments of decomposition.

(b) *Ripening*

In the second stage, the intensity maximum of the spectrum shifts towards smaller values of Q. This corresponds to the *ripening* stage of the structure. It is still possible to define a wavelength Λ_p corresponding to the intensity maximum but in contrast to the spinodal Λ_m, Λ_p is function of time (Fig. 6.26); measurements[242] show that:

$$\Lambda_p^3 - \Lambda_{p0}^3 = \beta'(t - t_0)$$

where the initial wavelength Λ_{p0} corresponds to time t_0: β' is a constant depending on the thermal treatment temperature. This is similar to the Ostwald ripening law for a particulate system.

Fig. 6.27. Treatment of the spectra from Fig. 6.26. (*a*) Variation with time of two intensity components. (*b*) Amplification factor $R(Q)$. The dashed line is Cahn's theoretical curve. (*c*) Deviation of experimental $R(Q)/Q^2$ values from linear behavior. After Ref. 240.

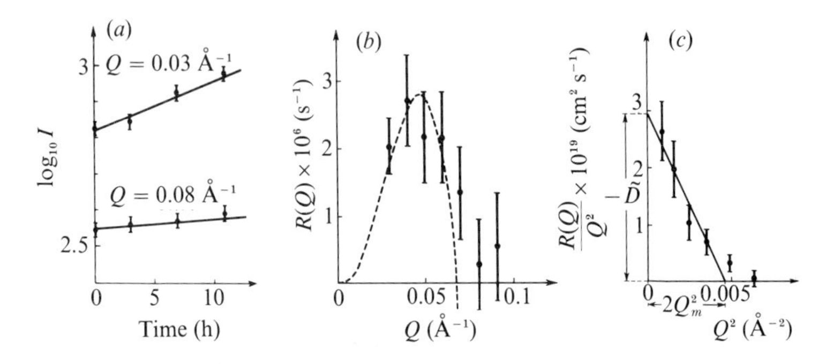

Fig. 6.28. Optical transform of an electron micrograph. After Ref. 243.

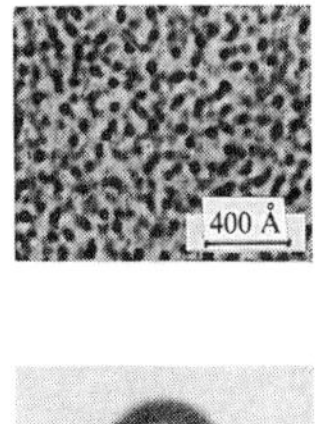

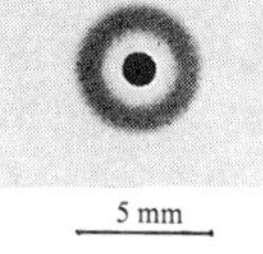

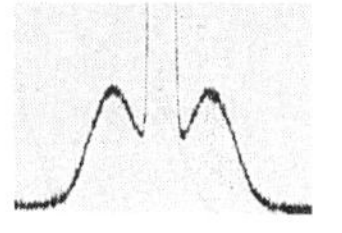

A correlation was established between the results from SAXS and those of electron microscopy by Fourier analysis of electron micrographs with *optical* means.[243] Thus, the scattering spectrum in *visible* light of a micrograph reduced to an appropriate scale and constituting a scattering "grating" provides a Fraunhofer spectrum. This gives a quantitative evaluation of the texture in reciprocal space quite analogous to the SAXS given by the real sample texture (Fig. 6.28). Λ_p is deduced from measurement of the ring diameter of the scattered spectrum. It is possible to use SANS in a manner similar to that for X-rays.[244] The first results, concerning the unmixing of a SiO_2–Na_2O glass verified Porod's law, showing that the system was already in the course of ripening.

The application of this method may be valuable in systems containing magnetic atoms; it is then possible to separate the magnetic heterogeneities from the usual "chemical" heterogeneities.[245]

7 Study of "medium-range order" in glasses

As we have seen in the preceding chapters, the structure of glass can be studied at several different levels. Diffraction and spectroscopic methods (Chapters 4 and 5) confirm the existence of *short-range* order, which, in the majority of cases, is identical or similar to that of the parent crystalline phases. This local order is known up to a distance of about 6–8 Å, in the most favorable cases although constant efforts are being made to increase this limit. As seen in Chapter 6, the *textures* resulting from *phase separation*, the finest of which are of the order of 30–50 Å are easily identified by small-angle scattering methods or electron microscopy.

Between these two limits is the zone corresponding to what could be called *medium-range order*[246] which is difficult to study even by the most sophisticated current methods of investigation. It is precisely in this region, typically ten to several tens of angströms in size, that the characteristic structural organization of glasses is found.

The variety of terms used, such as structural "micro-heterogeneity", structural "domains", local composition fluctuations etc. expresses the difficulty in quantitatively defining a structural disorder in intermediate regions.

Since the ideal method is direct observation, we shall begin by examining results furnished by electron microscopy.

7.1 High-resolution electron microscopy

Numerous electron microscopy studies of glass mention a microheterogeneous structure. Leaving aside textures originating from phase separation already discussed, we shall examine the case of simple glasses which are not susceptible to unmixing. Extremely thin samples of vitreous SiO_2, prepared *in situ* within the microscope to avoid surface modification by atmospheric attack, show a heterogeneous structure which can be described in terms of "domains" 30–50 Å in extent, Fig. 7.1.[246] These results were obtained with microscopes capable of 6–10 Å resolution. Technical progress in the construction of electron microscopes now allows atomic resolution of about 2 Å; point resolution currently reaches values of about 3 Å for the best equipment. It was even possible to detect isolated heavy atoms directly using a sophisticated scanning electron microscope.[247] It is thus interesting to consider how well a high-resolution electron microscope can provide information on a vitreous structure. To do so, the problem of image formation, i.e. the mechanism producing the observed contrast, must first be examined.

Fig. 7.1. The structure of vitreous SiO_2 observed in a transmission electron microscope. After Ref. 246.

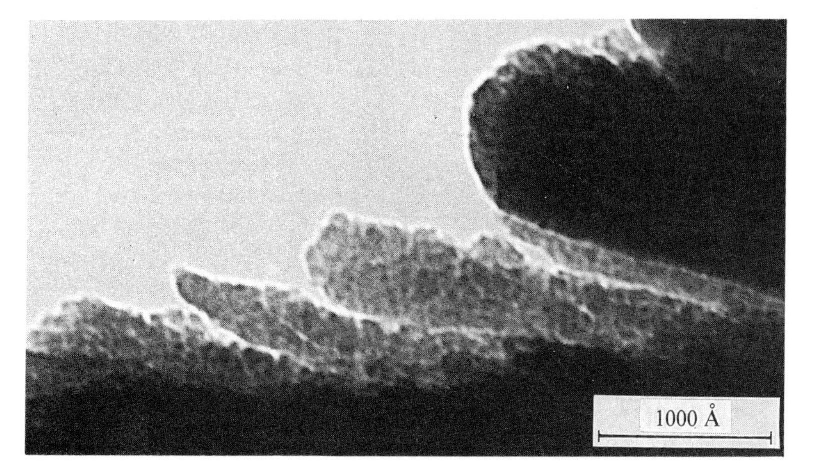

7.1.1 *Image formation theory*

We shall restrict the discussion to Transmission Electron Microscopes (TEM), because current scanning microscopes do not have sufficient resolving power to study disordered structures. TEM is analogous to a classical optical microscope using visible light. However, the glass lenses are replaced by magnetic ones the power of which can be adjusted with the lens excitation current. Two principal stages must be distinguished in the formation of an image.

In the first stage, the incident electron undergoes interaction with the object. The incident wave $\psi_0(x,y)$ is modified by the electrostatic potential of the object which results in a phase shift:

$$\frac{-\pi}{\lambda E}\Phi(x,y) = \sigma\Phi(x,y)$$

where E is the accelerating potential, λ the wavelength and $\Phi(x,y)$ the projection of the potential distribution on a plane normal to the incident beam. The sample, which is assumed to be thin enough that the amplitude is not affected, thus acts as a *phase object*. The amplitude distribution of the wave function on the exit face is then:

$$\psi(x,y) = \psi_0(x,y)\, q(x,y)$$

with

$$q(x, y) = \exp\left[-\mathrm{i}\sigma\Phi(x, y)\right]$$

In the second stage, the wave is collected by the objective lens (and all successive lenses) and transferred to the observation screen where an enlarged image of the object appears. The second stage is important because imperfections of the optical system introduce distortions and modifications of the transmitted information which affect the final image. In the discussion of image formation, it is enough to consider the action of the objective lens: the following lenses (projection and intermediate lenses) only amplify the image given by the objective. Figure 7.2 shows a simplified schematic of the ray paths during image formation by a single lens L. Three important planes are distinguished: the *object plane*, containing the object, the image *focal plane* and the *image plane*, the latter being coupled with the object plane by means of L.

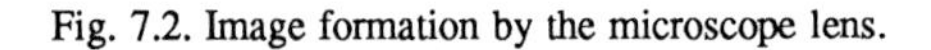
Fig. 7.2. Image formation by the microscope lens.

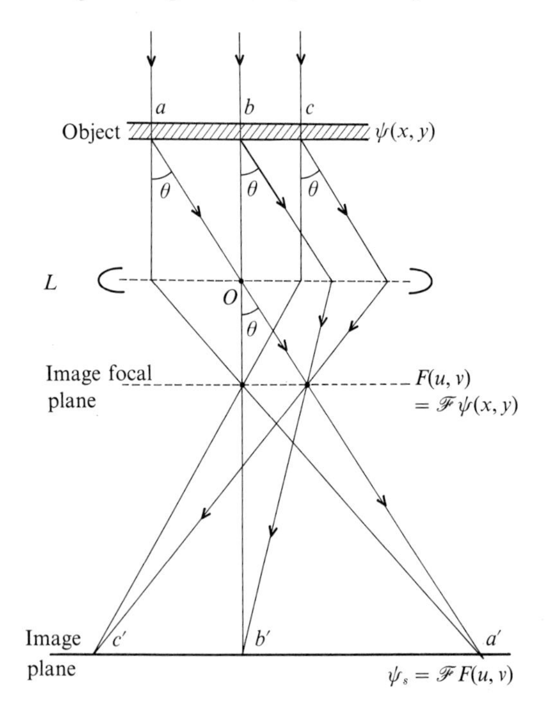

The process of image formation can be described by a generalized form of the Abbe theory (cf. e.g. Ref. 248). The lens focuses the directly transmitted non-scattered rays at a point located on the axis of the image focal plane. All radiation scattered at an angle θ is focused on the same plane in a position proportional to θ. Thus an amplitude distribution is formed in the focal plane having at each point an amplitude which corresponds to the sum of all scattered amplitudes at a given angle.

The corresponding intensity distribution constitutes the Fraunhofer spectrum of the sample. Formally, the amplitude distribution is the Fourier transform of the amplitude of the wave leaving the sample:

$$\begin{aligned} F(u,v) &= \mathcal{F}\left[\psi_0 q(x,y)\right] \\ &= \psi_0 \int\int q(x,y) \exp\left[2\pi \mathrm{i}(ux+vy)\right] \mathrm{d}x\, \mathrm{d}y \end{aligned}$$

where u and v are the angular variables locating the positions in the focal plane $u = x/f\lambda$ and $v = y/f\lambda$ (f is the focal distance). Then, the lenses transfer the rays from the focal to the image plane which in optical terms represents a second Fraunhofer diffraction process with interferences between different waves arriving at each point of the image plane. Mathematically , this corresponds to a second Fourier transformation:

$$\psi_s(x,y) = \mathcal{F}\left\{\mathcal{F}\left[\psi_0 q(x,y)\right]\right\} = \psi_0 q\left(-\frac{x}{M}, -\frac{y}{M}\right)$$

which reconstructs the transmission function q of the sample (inverted and enlarged by the magnification ratio M). This perfect reconstruction which assumes the presence of all the spatial components and the conservation of the correct phase relation in the transfer is not possible in practice. The presence of diaphragms eliminating higher-frequency components limits spatial resolution. The defocus effect and spherical aberration of the objective introduce additional phase changes. In this sense, the defocus, i.e. the focusing on a plane located at Δz from the exit face of the sample, introduces a dephasing: $\pi \Delta z \theta^2/\lambda$. The (unavoidable) third-order spherical aberration introduces a dephasing $\pi C_s \theta^4/2\lambda$ where C_s is the coefficient of spherical aberration. The dephasings combine and the wave amplitude is multiplied by $\exp\left[\mathrm{i}\gamma(\theta)\right]$ with:

$$\gamma(\theta) = \frac{\pi \Delta z \theta^2}{\lambda} + \frac{\pi C_s \theta^4}{2\lambda}$$

The objective lens then forms an enlarged image by recombining the different waves with appropriate phase change $\gamma(\theta)$. Two functional modes are used in practice: *bright field* and *dark field*.

In bright field (Fig. 7.3(*a*)) all electrons, scattered or not, contribute to the image. In the case of a perfect optical system, the image would not have any contrast. In a real system, the contrast is due to the function $\gamma(\theta)$. By defocusing, the observed contrast is enhanced but different spatial components will be modified in different ways, thus introducing effects which must be taken into account in the interpretation of the image.

In the case of axial illumination and thin samples, a Fourier component of periodicity $d = \lambda \cdot 2\sin(\theta/2)$ is transferred with a contrast $\sin\gamma(\theta)$. For amorphous samples, there is a continuous spectrum of scattered waves; some of these intensity components can, after *spatial filtering*, show a null or even reversed contrast effect which will depend on the defocus.

In practice, optimal defocus conditions are used which take advantage of the inflections in the *transfer function* $\sin\gamma$, allowing imaging of a particularly wide range of spatial periodicity in the object. The curve (Fig. 7.4) indicates the value of $\sin\gamma$ for a given value of the ratio d/d_0 where $d_0 = (C_s\lambda^3)^{1/4} \simeq 5.6$Å for 100 keV. For a given defocus, called the "Scherzer defocus," $\Delta z_1 = -1.2\sqrt{C_s\lambda}$, periodicities d between 4 and 10 Å, will be favored while for a larger defocus $\Delta z_2 = -1.85\sqrt{C_s\lambda}$, the periodicities $8 < d < 10$Å as well as $3.3 < d < 6$Å will be favored, the latter with reversed contrast.

For an even larger defocus, Δz_3, three enhanced bands are seen. Thus, according to the defocus value, different periodicities will be favored which will strongly affect the aspect of the image.

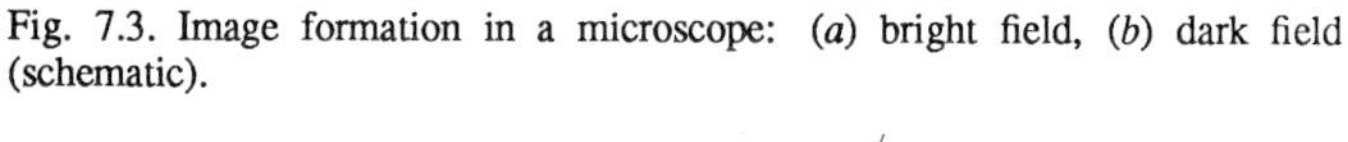

Fig. 7.3. Image formation in a microscope: (*a*) bright field, (*b*) dark field (schematic).

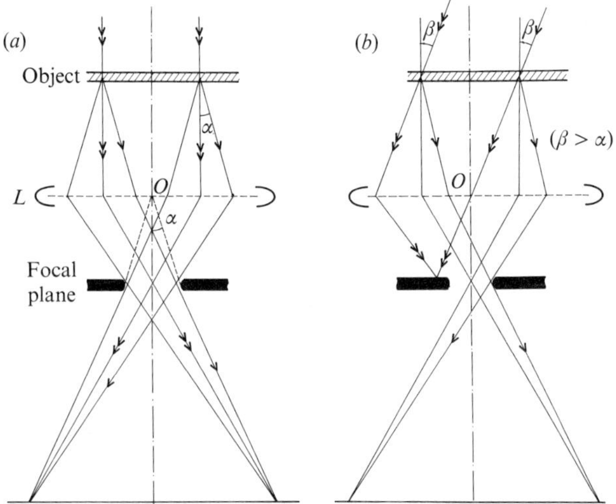

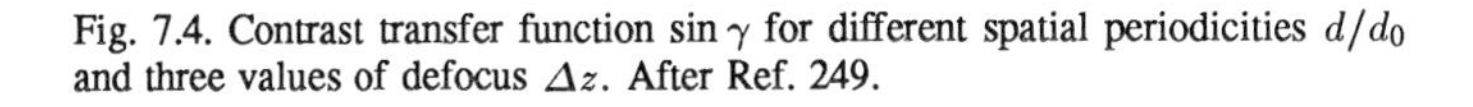
Fig. 7.4. Contrast transfer function $\sin\gamma$ for different spatial periodicities d/d_0 and three values of defocus Δz. After Ref. 249.

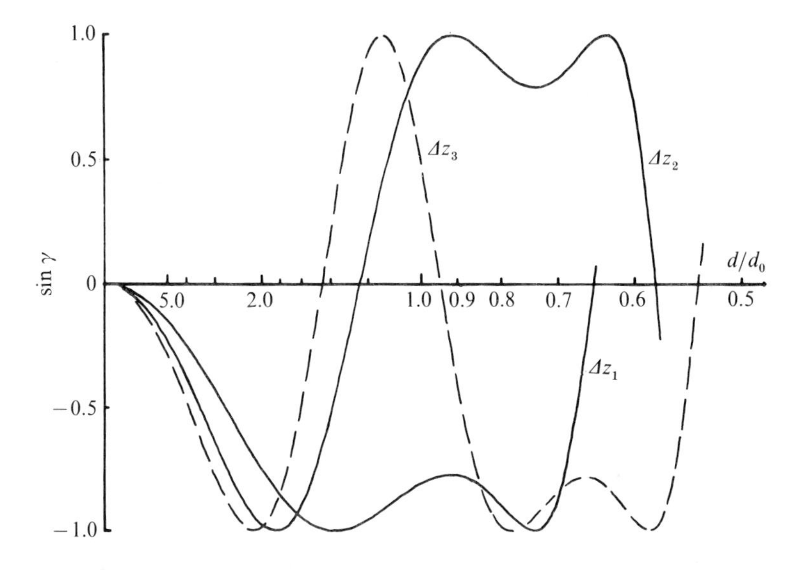

7.1.2 *Results of observations*

The nature of the "domains" frequently revealed by electron microscopy of amorphous substances has still to be determined. It would be particularly interesting to discover a higher level of local order.

In bright field, a *more highly ordered* region in an amorphous matrix can give rise to the local formation of a system of *fringes*. Numerous studies, namely on amorphous C or Ge, showed this possibility.[250–5]

Figure 7.5 shows the structure of a SiO_2 gel. Domains of about 100 Å are apparent while internal details are obscured by the "background noise" related to the image formation. It is possible to apply optical filtering techniques to the micrographs to remove non-significant details obscuring the image.[251] However, the appearance of a system of fringes does not constitute clear proof of order; in an amorphous system, such an effect can result from random superposition enhanced by the oscillatory character of the filter function $\sin\gamma$.

Moreover, it is important to remember that the observed image corresponds to the projection of the structure on a plane parallel to the plane of observation. A serious problem arises from *overlapping* and superposition of the atoms within the observed layer.

Fig. 7.5. High-resolution micrograph of a SiO_2 gel with part of the high-frequency "noise" suppressed by optical filtering. After Ref. 251.

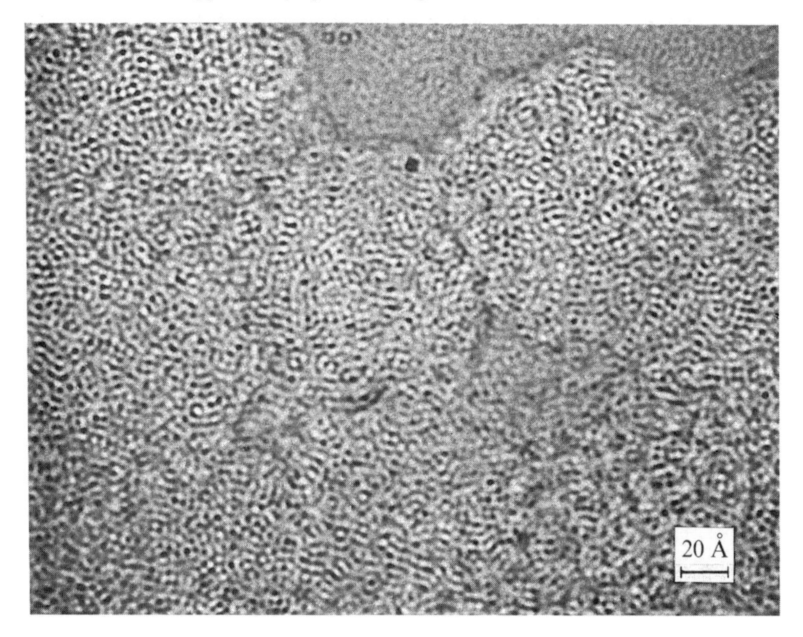

These superpositions can occasionally cause fringes (even in a totally amorphous structure), or render impossible the observation of fringes coming from an ordered portion because of averaging effects. To avoid such difficulties, *dark field* observation can be used (Fig. 7.3(*b*)). In this mode, all non-scattered (directly transmitted) electrons as well as a large part of the scattered electrons are blocked by a diaphragm appropriately placed in the focal plane. (In practice, the illuminating source is tilted to avoid non-alignment effects.)

Under these conditions, only beams corresponding to the selected scattering angles can contribute to the image formation. When these transmission conditions are fulfilled, an ordered region appears bright on a dark background (Fig. 7.6).

In order to improve the resolution in imaging, a large fraction of the scattered beams is used, introducing other kinds of random effects giving rise to bright points ("speckles") which can be confused with images coming from actual ordered portions. A method proposed[249] to minimize these random effects is dark field illumination using a beam in the form of a hollow cone. The speckles from separated atoms are then eliminated by an averaging effect and there is increased probability that the ob-

Fig. 7.6. Dark field high-resolution micrograph of vitreous SiO_2 After Ref. 252.

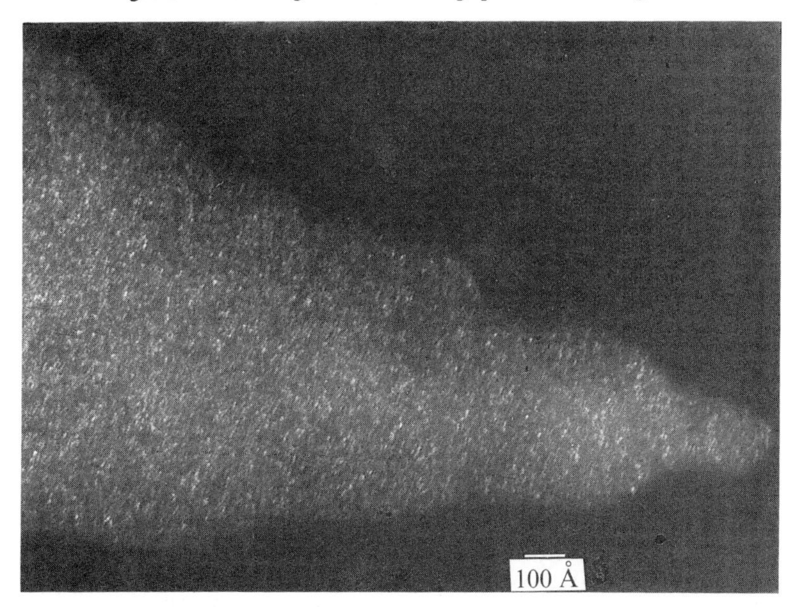

served intense spots come from significant structural correlations related to real short-range order. Another method of resolving the problem of superposition is to take stereoscopic images. For technical reasons (focus compensation and precise location of the same structure in the observed field), this method is rarely used at extreme magnifications.

7.1.3 *Image simulation*

Using the above approach, it is possible to *calculate a priori* the predicted image contrast from a given structural model (see later) and to compare the diagram thus obtained to the experimental micrograph.[253] The calculations made for amorphous Si and Ge[254] and vitreous SiO_2[255] allowed clarification of some unresolved issues. For instance, it became apparent that in dark field speckles can be generated by totally disordered structures. On the other hand, various disordered network models can give local fringes in bright field. A detailed comparison seems difficult because the models used for calculation are often thinner than actual specimens and not all instrumental aberrations are taken into account.

At the present time, a complete interpretation of a high-resolution electron micrograph of a non-crystalline structure has not yet been solved satisfactorily though significant progress has been made in recent years.[256]

In any case, sample thickness is the essential factor limiting the technique and attempts at the visualization and interpretation of an amorphous structure. Thicknesses as small as 10 Å were recommended but often cannot be technically realized. It is difficult to obtain self-supporting films less than 60 Å thick; however, discontinuous films of about 20 Å supported on crystalline graphite have been employed. The question arises whether the structure of such thin specimens is identical to that of bulk material. Surprisingly enough, it seems that the most useful results for vitreous material actually come from studies at *moderate resolutions*. Are the "domains" revealed in single-phase glasses simply due to structure fluctuations?

7.2 Fluctuations in glasses

Glasses can be considered as resulting from the freezing of a liquid and as such, they can contain local fluctuations which according to the case, may be density fluctuations or composition fluctuations or both.

7.2.1 *Density fluctuations*

Consider first the problem of density fluctuations in single component glasses. In any system, the local density $\varrho(\vec{r})$ of the material is not uniform on a microscopic scale but presents small local deviations relative to the average density ϱ_0:

$$\varrho(\vec{r}) = \varrho_0 + \Delta\varrho(\vec{r})$$

It is convenient to break up the fluctuations into their spatial Fourier components $\Delta\varrho_\kappa$:

$$\Delta\varrho_\kappa = \frac{1}{V}\int_V \Delta\varrho(\vec{r})\exp\left(-\mathrm{i}\vec{\kappa}\cdot\vec{r}\right)\mathrm{d}\vec{r}$$

where V is the volume of the system and κ is the spatial wavenumber.

For a system in equlibrium, the mean square value of the fluctuation is obtained from thermodynamics (see e.g. Ref. 257.)

$$\langle\Delta\varrho_\kappa^2\rangle = \frac{\varrho_0^2\,kT}{V}\beta_T \tag{7.1}$$

where k is Boltzman's constant, T the absolute temperature, β_T the isothermal compressibility and where $\langle\ \rangle$ designates the ensemble average for the system. Knowledge of β_T thus allows the evaluation of the fluctuation level present in the liquid at a given temperature.

For an equilibrium viscoelastic system, this expression can be split into a contribution from entropy fluctuations at constant pressure and one due to adiabatic pressure fluctuations.[258]

$$\frac{V\langle \Delta \varrho_\kappa^2 \rangle}{\varrho_0^2} = kT(\beta_T - \beta_s) + kT\beta_s$$

in which β_s is the adiabatic compressibility.

The second term can in turn be sub-divided into a contribution due to phonons and one due to local structural variations caused by relaxation. If the material has sufficient viscosity, the structural rearrangements are slow relative to the phonon frequency.

A high-frequency, adiabatic compressibility β_s^∞ is associated with the phonons where the complement β_s^r represents the relaxational compressibility:

$$\beta_s^r = \beta_s - \beta_s^\infty$$

Under these conditions:

$$\frac{V\langle \Delta \varrho_\kappa^2 \rangle}{\varrho_0^2} = kT(\beta_T - \beta_s) + kT\beta_s^r + kT\beta_s^\infty$$

This formula is of interest because it reveals the part relating to fluctuations due to structural variations.

For a glass which is a non-equlibrium system, the fluctuations are frozen-in at a temperature $\overline{T}_f$ in the transformation interval, when the time constant for equilibration becomes large relative to the cooling speed imposed. This part of the fluctuations is thus fixed and must be evaluated at a temperature $\overline{T}_f$ corresponding to the *fictive* temperature of the glass (cf. Chapter 2). The last term corresponds to the vibrational modes (phonons) which depend on the *actual* temperature of the specimen. Thus for a glass:

$$\frac{V\langle \Delta \varrho_\kappa^2 \rangle}{\varrho_0^2} = \left[k\overline{T}_f(\beta_T - \beta_s) + k\overline{T}_f\beta_s^r\right] + \left[kT\beta_s^\infty\right] \qquad (7.2)$$

The term $k\overline{T}_f(\beta_T - \beta_s)$ can be evaluated by using the thermodynamic relation:

$$\beta_T - \beta_s = \frac{\alpha_f^2 \overline{T}_f}{\varrho_0 C_{pf}}$$

where α_f and C_{pf} are respectively the thermal expansion coefficient and the specific heat at the fictive temperature $\overline{T}_f$. From the point of view of magnitude, the term $k\overline{T}_f\beta_s^r$ dominates and the other two each contribute about 8% of the total. To determine $\langle\Delta\varrho_\kappa^2\rangle$, it is thus necessary to know $\overline{T}_f$, β_s, and β_s^∞: $\overline{T}_f$ may be taken as the temperature where the system has a viscosity $\sim 10^{14}$ dPa s. However, as will be shown, $\overline{T}_f$ can be eliminated using light scattering measurements. The compressibility can be obtained from ultrasonic or hypersonic propagation measurements.

7.2.2 *Measurement of compressibility from ultrasonic propagation*

The propagation of an ultrasonic wave of frequency ω in a visco-elastic material with specific mass ϱ_0 is described by a complex longitudinal modulus $M(\omega)$ of which the real part $M'(\omega)$ is (cf. Chapter 9):

$$M'(\omega) = \varrho_0 v_L^2 \frac{1-m^2}{(1+m^2)^2} \qquad \text{with} \qquad m = \alpha v_L/\omega$$

where α is the absorption coefficient and v_L the ultrasonic velocity. For $\alpha v_L/\omega \ll 1$, $M' \simeq \varrho_0 v_L^2$

In general:

$$M'(\omega) = \frac{1}{\beta_s} + \left(\frac{1}{\beta_s^\infty} - \frac{1}{\beta_s}\right) \int_0^\infty \Phi(\tau)\frac{\omega^2\tau^2}{1+\omega^2\tau^2}\,\mathrm{d}\tau$$

where $\Phi(\tau)$ is the normalized distribution function of the relaxation times.

Defining an average relaxation time $\bar{\tau}$ by the relation:

$$\bar{\tau} = \int_0^\infty \tau\Phi(\tau)\,\mathrm{d}\tau$$

for low frequencies ($\omega\bar{\tau} \ll 1$):

$$M'(\omega) \to \frac{1}{\beta_s} = \varrho_0 v_0^2$$

and for high frequencies:

$$M'(\omega) \to \frac{1}{\beta_s^\infty} = \varrho_0 v_\infty^2$$

The dispersion of the sonic velocity is associated with the variation of $M'(\omega)$, $\bar{\tau}$ depending strongly on temperature; it can be shown that:

$$\bar{\tau} = \left(\eta_v + \frac{4}{3}\eta_s\right)\left(\frac{1}{\beta_s^\infty} - \frac{1}{\beta_s}\right)$$

where η_v and η_s are respectively the volume and shear viscosities (cf. Chapter 9).

Since the compressibilities vary only slightly with temperature and generally the ratio η_v/η_s is constant, $\bar{\tau}$ varies as η_s.

At sufficiently high temperatures $\omega\bar{\tau} \ll 1$ and the measured sound velocity is v_0, independent of ω. Similarly at low temperatures, $\bar{\tau}$ is much higher, so that for the same frequency, $\omega\bar{\tau} \gg 1$ and the measurement provides v_∞ (independent of the frequency). At intermediate temperatures, $\omega\bar{\tau} \sim 1$, the velocity lies between v_0 and v_∞ and depends on ω (Fig. 7.7). Caution is therefore necessary in the evaluation of compressibilities from ultrasonic measurements and it is well to verify that the indicated constant levels are actually attained.

Fig. 7.7. Dispersion relations of the ultrasonic velocity. After Ref. 258.

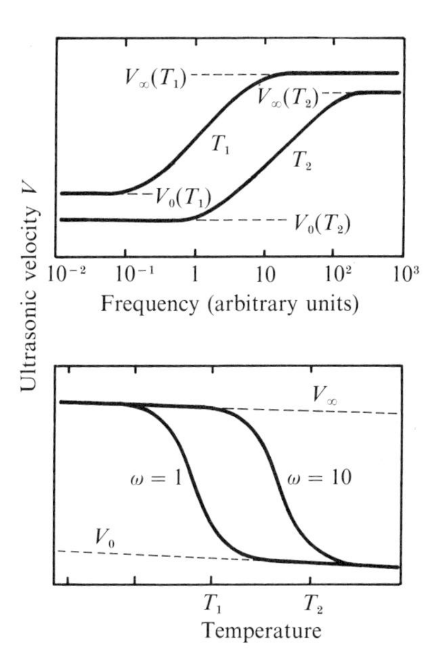

The propagation of hypersonic (thermal) waves in solids is another way to measure compressibility using light scattering. Schroeder[259] has published a useful article on this technique.

7.2.3 *Light scattering, Rayleigh and Brillouin effects*

In measurements of light scattering, the specimen is illuminated with monochromatic light polarized perpendicularly to the scattering plane with frequency ν_0, wavelength λ and intensity I_0. The intensity scattered by a volume V at an angle δ is:

$$I = \frac{I_0 \pi^2}{\lambda_0^4} V^2 (1 + \cos^2 \delta) \langle \Delta\epsilon_\kappa^2 \rangle$$

where $\Delta\epsilon_\kappa$ is a spatial Fourier component of the fluctuation $\Delta\epsilon(\vec{r})$ of the dielectric permittivity:

$$\Delta\epsilon_\kappa = \frac{1}{V} \int \Delta\epsilon(\vec{r}) \exp(-i\vec{\kappa} \cdot \vec{r}) \, d\vec{r}$$

The spectrum of scattered light (Fig. 7.8) contains the Rayleigh line, centered on the frequency ν_0 between a pair of Brillouin lines shifted by $\pm\nu_B$. These lines result from the presence of longitudinal phonons in the substance.

Fig. 7.8. Brillouin scattering spectrum (schematic). ν_0, the frequency of Rayleigh scattering; (L) Brillouin lines due to longitudinal phonons ($\nu_0 + \nu_B$) and (T) those due to transverse phonons ($\nu_0 + \nu_S$).

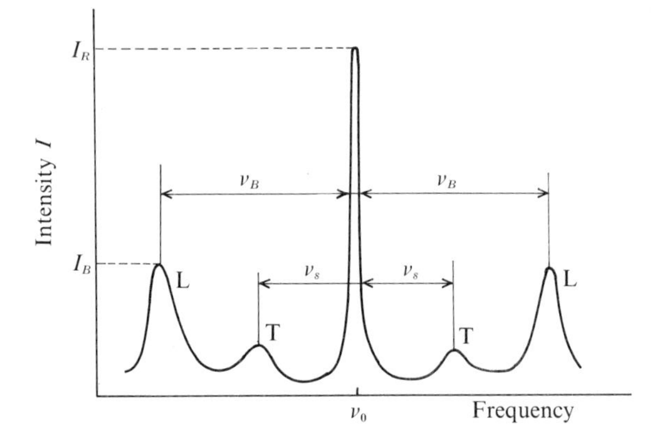

These hypersonic waves act like a grating scattering light which propagates with a velocity v; the frequency ν_B is equal to:

$$\nu_B = v\kappa / 2\pi$$

where the value of the wave vector κ associated with the fluctuation responsible for the scattering is given by the relation:

$$\kappa = \frac{4\pi n}{\lambda_0} \sin\left(\frac{\delta}{2}\right)$$

n being the index of refraction of the substance.

For a material capable of sustaining shear stresses, as is the case for a solid glass, there is, in addition, one pair of lines shifted by $\pm\nu_S$, corresponding to the presence of transverse phonons:

$$\nu_S = v_S \kappa / 2\pi$$

v_S being the transverse wave velocity.

The relation between the permittivity and density fluctuations is:

$$\langle \Delta\epsilon_\kappa^2 \rangle = \left(\frac{\partial\epsilon}{\partial\varrho}\right)_T^2 \langle \Delta\varrho_\kappa^2 \rangle + \left(\frac{\partial\epsilon}{\partial T}\right)_p^2 \langle \Delta T_\kappa^2 \rangle$$

the second term on the right hand side, containing the temperature fluctuations ΔT_κ, is, in general, negligible relative to the first term, hence:

$$\langle \Delta\epsilon_\kappa^2 \rangle \simeq \left(\frac{\partial\epsilon}{\partial\varrho}\right)_T^2 \langle \Delta\varrho_\kappa^2 \rangle$$

Light scattering experiments allow the measurement of the different terms in equation (7.2). It can be shown that the Rayleigh I_R and Brillouin I_B intensities are respectively proportional to the two terms between the square brackets in equation (7.2) with the same proportionality coefficient.

The ratio R_{LP} of the scattered intensities (I_R and I_B) constitutes the *Landau–Placzek ratio*:

$$R_{LP} = \frac{I_R}{I_B} = \frac{\overline{T}_f}{T}(\varrho_0 v_\infty^2 \beta_T - 1)$$

Measurement of R_{LP} allows the elimination of $\overline{T}_f$ and the level of the fluctuations χ is thus obtained:

$$\chi = \frac{V\langle \Delta\varrho_\kappa^2 \rangle}{\varrho_0^2} = (1 + R_{LP})\frac{kT}{\varrho_0 v_\infty^2}$$

Measurements by Schroeder[260] for vitreous SiO_2 at $T = 300\,\mathrm{K}$ provide the following values:

$$R_{LP} = 23.3 \qquad v_\infty = 5.92 \times 10^5\,\mathrm{cm\,s^{-1}}$$

which gives:

$$\chi = 1.29 \times 10^{-24}\,\mathrm{cm^3}$$

To have some idea of the level of the fluctuations present, for a 20 Å cube, the preceding value corresponds to a relative variation:

$$\frac{\sqrt{\langle \Delta \varrho_\kappa^2 \rangle}}{\varrho_0} = 0.012 \quad \text{or} \quad 1.2\%$$

The same calculation for B_2O_3[261] using the scattering results from Buccaro and Dardy[262] gives $\chi = 2.7 \times 10^{-24}$ cm^3 at 293 K.

7.2.4 *Results obtained by SAXS*

The quantity χ can be obtained independently from measurement of small-angle scattering of X-rays.

It has been shown (see Ref. 237) that the scattered intensity I_0 extrapolated to zero scattering angle is, for a scattering unit (e.g. the unit SiO_2):

$$I_0 = \frac{V}{N_v} \langle \Delta \varrho_e^2 \rangle$$

where $\langle \Delta \varrho_e^2 \rangle$ is the mean squared variation of the average electron density ϱ_{e0}, and N_v is the number of scattering units in a volume V. The electron density is proportional to the usual density:

$$\frac{\langle \Delta \varrho_e^2 \rangle}{\varrho_{e0}^2} = \frac{\langle \Delta \varrho^2 \rangle}{\varrho_0^2}$$

therefore:

$$I_0 = \frac{\chi N \varrho_0 \sum Z}{M}$$

where $\sum Z$ represents the summation of atomic numbers of the scattering unit, M its molar mass and N Avogadro's number.

The results of Weinberg[263] and Levelut and Guinier[264] lead to values of I_0 for vitreous SiO_2 which are in excellent agreement with results calculated from optical scattering measurements. If χ is taken equal to 1.29×10^{-24} cm^3, this corresponds, after the preceding formula, to a scattered X-ray intensity $I_0 = 25.7$ e.u./SiO_2 which is very close to the experimental results ($I_0 \sim 23$–26).

Thus when the fictive temperature of the glass is taken into account, the fluctuation level given by X-ray scattering is in perfect agreement with the results deduced from thermodynamic considerations and optical measurements.[246]

For vitreous GeO_2, Pierre and Uhlmann[265] measured $I_0 = 49 \pm 2.5$ e.u./GeO_2 which corresponds to $\chi = 1 \times 10^{-24}$ cm^3. For vitreous B_2O_3 the higher levels for χ calculated from optical scattering give the scattered X-ray intensity:[261]

$$I_0 = 49.7 \text{ e.u.}/B_2O_3$$

Measurements by Porai-Koshits[266] show that the scattered intensity remains essentially constant up to T_g and then increases proportionally to the temperature T (Fig. 7.9). In these measurements, the calibration is done relative to vitreous SiO_2 taken as an intensity standard ($I_0 = 25$e.u./SiO_2). It seems thus well established that in these simple glasses, the scattered intensity is due only to thermal density fluctuations.

Fig. 7.9. Scattered intensity $I(0)$ at zero angle per unit of composition:
× B_2O_3, vitreous and liquid, after Ref. 266.
• B_2O_3, vitreous and liquid, after Ref. 261.
+ GeO_2, vitreous, after Ref. 265.
△ SiO_2, vitreous, after Ref. 264.
(u.c. = units of composition – a stoichiometric unit defined by formula fixing composition.

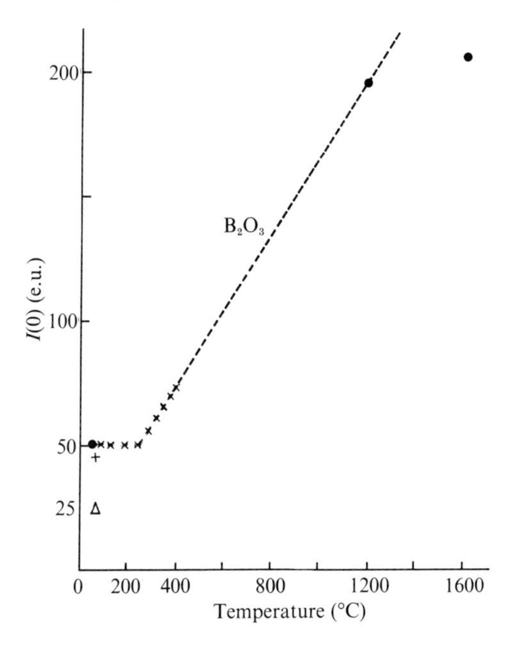

It is interesting to compare these results to those obtained for liquid B_2O_3 at higher temperatures.[261] Using equation (7.1) and the value

$\beta_T \sim \beta_s$ obtained by Bockris and Kojonen,[267] the values of I_0 expected at 1200 °C and 1600 °C were obtained. Figure 7.9 shows the excellent agreement with the results obtained by Porai-Koshits at least up to 1200 °C.

7.2.5 *Multi-component glasses – composition fluctuations*

In a system of several components, it is necessary to modify equations (7.1) and (7.2) to take into account composition fluctuations. In a binary system, a term proportional to:

$$\left(\frac{\partial \epsilon}{\partial c}\right)^2_{PT} \frac{k\overline{T}'_f}{N'} \left(\frac{\partial \mu}{\partial c}\right)^{-1}_{PT}$$

has to be added to the Rayleigh scattering. Here the concentration $c = N/N'$ is the mole ratio of solute to solvent, μ is the chemical potential and $\overline{T}'_f$ is a fictive temperature associated with the thermodynamic arrest of composition fluctuations. In general, $\overline{T}'_f \neq \overline{T}_f$. For $\overline{T}'_f$, a temperature corresponding to a viscosity $\sim 10^8$ P can be adopted.

The analysis becomes much more complex;[259] the R_{LP} ratio is equal to $R_\varrho + R_c$ where the two terms are associated respectively with fluctuations in density and composition.

One such evaluation has been made for the SiO_2–K_2O system.[258] It is remarkable that glasses containing more than 25% K_2O have an attenuation less than that for pure vitreous SiO_2. The attenuation due to fluctuations in glasses of composition $3\,SiO_2{\cdot}K_2O$ is about 2/3 that of pure SiO_2, a fact which might be used with advantage in the technology of optical conductors (cf. Chapter 12).

The results obtained by SAXS are less easily interpreted for multi-component systems. For a review of the work of Porai-Koshits and his colleagues who have studied such systems, see Ref. 268.

Thus light and X-ray scattering methods confirm the existence of fluctuations in glasses at a level which is in agreement with the results given by thermodynamics. In some instances the fluctuations can be particularly marked as in the case of supercritical fluctuations.

7.2.6 *Supercritical fluctuations – critical opalescence*

Zone III adjacent to a coexistence curve (Fig. 6.9) immediately above the consolution point is the site of particularly large fluctuations producing the phenomenon of *critical opalescence.*

These are equilibrium fluctuations the amplitude of which increases as the temperature approaches the critical temperature. At a constant temperature, the system is thus in equilibrium in contrast to the case of unmixing where the system evolves as a function of time below the coexistence curve.

This phenomenon, which is of great theoretical interest, has mostly been studied in systems of organic compounds and, more rarely, in some metal alloys after quenching to ambient temperature.

The first investigations at elevated temperatures of the supercritical region of an oxide system capable of becoming a glass were in the system B_2O_3–PbO–Al_2O_3. The immiscibility surface (Fig. 6.8) was determined in detail for studies of the kinetics of phase separation (cf. Chapter 6). Given that even a rapid quench can lead to a modification of the system by unmixing, particularly near the critical compositions, a direct study of the phenomena was made by SAXS at elevated temperatures.[269–71]

Figure 7.10 shows the SAXS spectra obtained for this system. Note that the intensity increases as the temperature decreases towards the critical temperature.

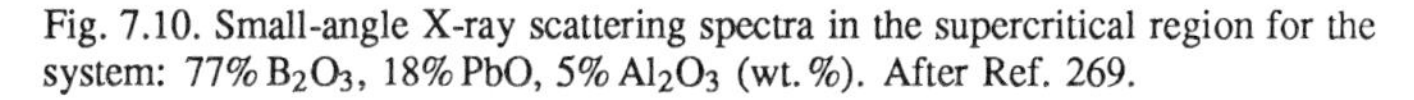

Fig. 7.10. Small-angle X-ray scattering spectra in the supercritical region for the system: 77% B_2O_3, 18% PbO, 5% Al_2O_3 (wt.%). After Ref. 269.

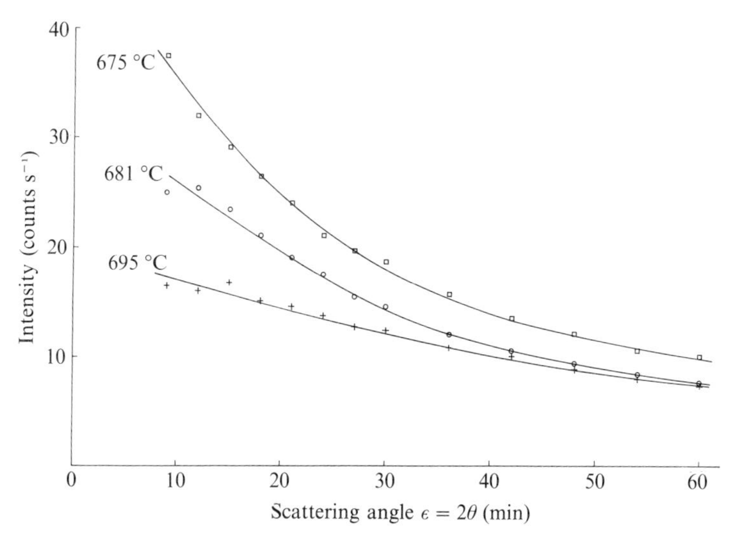

To interprete this phenomenon, Debye[272] defined a correlation function $c(r)$ by the relation:

$$c(r) = \langle \delta n_A \, \delta n_B \rangle / \langle (\delta n)^2 \rangle$$

where δn_A and δn_B are the local fluctuations of the number of electrons contained in the small elements of volume centered on the points A and B situated at a distance r from each other, and δn is the fluctuation for $r = 0$. (The $\langle \quad \rangle$ represents the average over the volume of the specimen.)

The scattered intensity $I(Q)$ is:

$$I(Q) = I_0 \int_0^\infty c(r) \frac{\sin Qr}{Qr} \, \mathrm{d}\tau \bigg/ \int_0^\infty c(r) \, \mathrm{d}\tau$$

with $Q = (4\pi \sin \theta)/\lambda$, I_0 the scattered intensity for angle $\theta = 0$ and $\mathrm{d}\tau = 4\pi r^2 \mathrm{d}r$, an elementary volume element.

Debye showed that if a correlation distance L_D is introduced by the expression:

$$L_D^2 = \int_0^\infty c(r) r^2 \, \mathrm{d}\tau \bigg/ \!\! \int_0^\infty c(r) \, \mathrm{d}\tau$$

the scattered intensity near the critical point can be obtained from the following equation:

$$\frac{1}{I(Q)} = \frac{1}{I_0} \left[1 + \frac{Q^2}{6} L_D^2 + \cdots \right]$$

The graph of $1/I(Q)$ as a function of Q^2, called the Ornstein–Zernike (OZ) plot is linear and its slope gives L_D^2.

Debye then defined a quantity l called the "range of molecular forces" by the second moment of the pair potential $\epsilon(r)$. For a single component system:

$$l^2 = \int r^2 \epsilon(r) \, \mathrm{d}\tau \bigg/ \!\! \int \epsilon(r) \, \mathrm{d}\tau$$

In a two-component system, $\epsilon(r)$ must be replaced by the expression:

$$\epsilon_{11}(r)/\omega_1^2 + \epsilon_{22}(r)/\omega_2^2 - 2\epsilon_{12}(r)\omega_1\omega_2$$

where ω_i is the molar volume of each component and $\epsilon_{ij}(r)$ the pair potentials.

For a temperature T near the critical point T_c, Debye showed that L_D and l are related by the expression:

$$L_D^2 = l^2 T_c/(T - T_c)$$

(where T and T_c are absolute temperatures). The graph of $1/L_D^2$ as a function of $\Delta T = T - T_c$ is then linear and from its slope l can be determined.

Figure 7.11 shows that the OZ plots corresponding to the spectra of Fig. 7.10 are reasonably linear; the weak curvature only begins for higher values of Q. The graph of $1/L_D^2$ vs T gives $l \sim 11$ Å, and its intersection with the T axis gives a precise value of T_c (Fig. 7.12).

Fig. 7.11. OZ plots corresponding to spectra from Fig. 7.10. After Ref. 269.

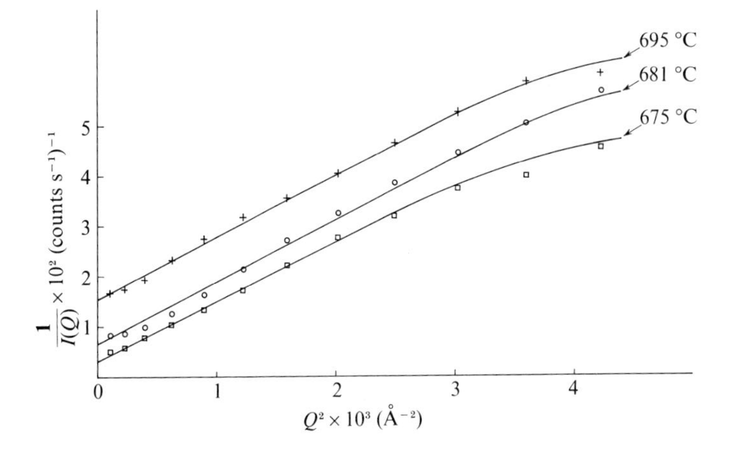

Fig. 7.12. Variation of the Debye correlation length L_D with temperature for the spectra of Fig. 7.10. After Ref. 269.

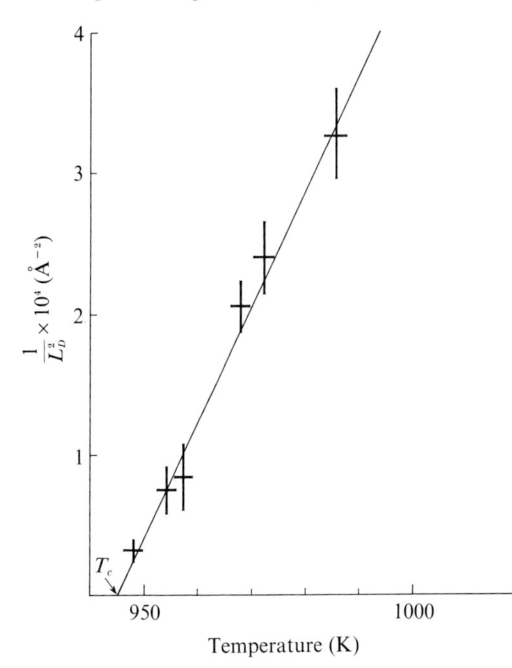

The same procedure was applied to a series of compositions corresponding to section mn of the coexistence surface of Fig. 6.8. To show the evolution of the supercritical fluctuations near the coexistence surface, the temperatures corresponding to Debye lengths of 30, 50 and 100 Å were calculated from plots of $1/L_D^2$ vs T. The results are given in Fig. 7.13 where lines for constant L_D are shown. (The section of the coexistence surface, marked CS, corresponds to $L_D \to \infty$.) This representation clearly reveals the topography of the supercritical region.

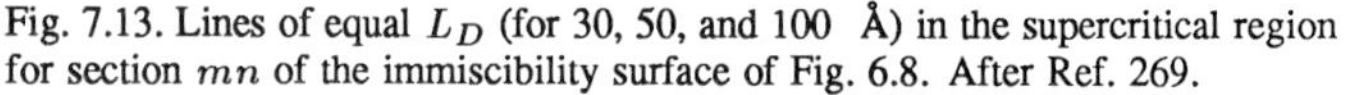

Fig. 7.13. Lines of equal L_D (for 30, 50, and 100 Å) in the supercritical region for section mn of the immiscibility surface of Fig. 6.8. After Ref. 269.

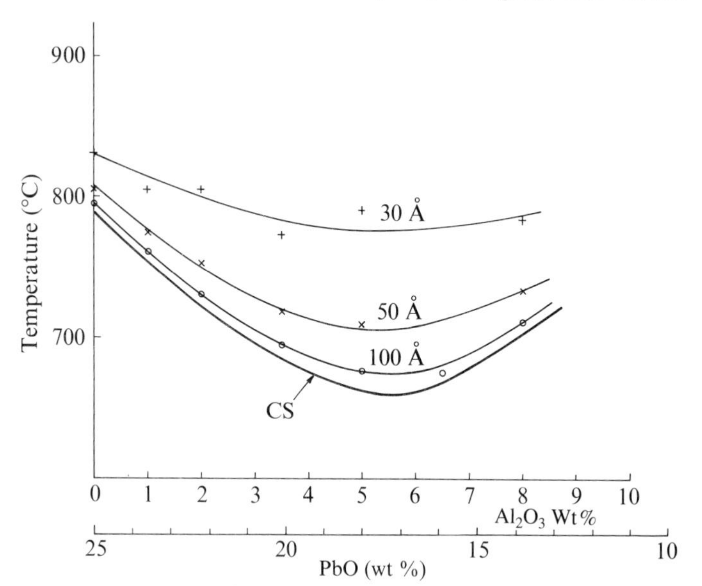

In the case of a structurally complex liquid, like the one considered here, it is difficult to give a molecular interpretation for l. This quantity must be related to the most prominent structural distances and in particular to the shortest interatomic distances between the atoms.

Assuming uniform distribution of atoms in the melted medium, the shortest interatomic distances $d_{(Pb-Pb)}$ and $d_{(Al-Al)}$ were calculated as a function of the composition. The experimental length l is of the same order of magnitude and close to the shortest available distance, d (Fig. 7.14). The change from $d_{(Pb-Pb)}$ towards $d_{(Al-Al)}$ occurs for a composition near the minimum of the coexistence surface.

Fig. 7.14. Comparison of the range of molecular forces l with the shortest interatomic distances $d_{(Pb-Pb)}$ and $d_{(Al-Al)}$ for section mn of Fig. 6.8. After Ref. 269.

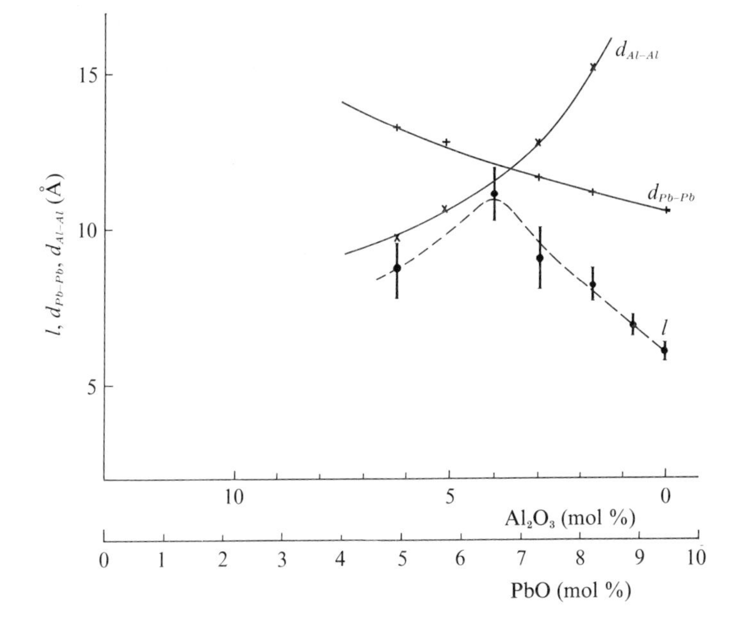

7.3 Modeling

The concept of short-range order in glass is quite well connected to parent crystalline structures. In the case of intermediate-range order, only diverse *hypotheses* are available which, in a more quantitative form, lead to *models* subject to verification.

Glasses being frozen liquids, the two classical points of view in the study of liquids are found in structural hypotheses for glasses. In the first, a liquid can be considered as a gas compressed to the point where the constitutive units come in contact producing a condensed phase essentially disordered at short and intermediate distances. According to the second, the liquid results from the fusion of a crystal and thus the structure of the liquid can be deduced by introducing a sufficient number of dislocations in the crystalline structure. This leads to an assembly of small quasi-crystalline domains with an elementary structure directly related to that of the crystal.

7.3.1 *Structural hypotheses*

Different hypotheses on the structure of glass have adopted similar points of view.

The *crystallite hypothesis* due to Lebedev is historically the oldest. The glass is considered as an assemblage of "microcrystallites" (Fig. 7.15(*a*)) too small to give sharp diffraction lines. However, a quick calculation of the line broadening shows that for this to be true, the crystallite size cannot exceed about 8 Å for SiO_2. It is difficult in this case to speak of "crystallites" when the size hardly exceeds that of the elementary unit cell of the crystalline component.

At the other extreme, the *perfectly disordered network hypothesis* of Zachariasen-Warren (Fig. 7.15(*b*)) postulates for oxide glasses the existence of a disordered arrangement of chains of glass-formers and a random distribution of network modifying cations. Nowhere is it possible to show a quasi-crystalline arrangement although local composition variations are allowable.

In general, all the proposed structural explanations start from one or the other of these viewpoints which certainly constitute the extreme positions enclosing a more complex reality which varies according to the nature of the glass considered.

The crystallite hypothesis, which mostly had partisans in the USSR, has been progressively replaced by the random continuous network theory, developed in the United States, which dominated most of the interpretations in latter decades. However, it appears that continued refinement of investigative methods has to some extent renewed interest in the non-random aspects of the structure.

Porai-Koshits proposed the hypothesis of a *paracrystalline* structure (Fig. 7.15(*c*)) derived from the concept of a *paracrystal* or a crystal with a randomly deformed lattice after Hosemann and Bagchi.[133]

Without speaking of a true ideal microcrystalline order, the possibility of a locally variable degree of order can be considered. A structure of "microdomains" visible in the electron microscope suggests an organization where the center of a domain would have a more ordered structure than the periphery.[273]

A localized progressive change from a disordered structure towards a more ordered one can be envisaged between the extremes of an assemblage of perfect crystallites and a perfectly disordered network. This is actually equivalent to the local formation of a paracrystalline network (deformed crystalline network) which may be stabilized by an occasional broken bond or an incorrect size ring.

This would reduce the difficulty encountered in the theory of perfect "microcrystallites" in which the mode of assembling the different microcrystallites is generally left unresolved or the existence of an "interstitial tissue" joining the organized domains is assumed. A degree of variable order from the center to the periphery of a domain could resolve these

Fig. 7.15. Hypotheses of intermediate-range order (schematic): (*a*) crystallites, (*b*) disordered network, (*c*) paracrystal.

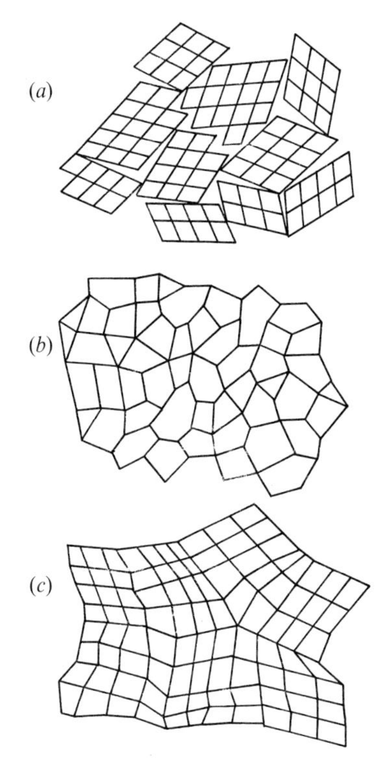

difficulties. Figure 7.16 shows an example of an SiO_2-type network with a progressive change from an ordered to a disordered structure. Scattering methods would not reveal such local order variations which do not appear in the average radial distribution function.

In general, the corresponding *models* can be arranged into two classes: (1) *disordered models* for which the most typical examples are *random close packing*, valid for metallic glasses, as well as the *random continuous network*, used for covalently bonded glasses; (2) the *partial order models*: based on microcrystallites, anti-crystalline polyhedral packing or a disordered dislocation distribution.

7.3.2 *Construction of models – general modeling principles*

The first disordered models were constructed by hand using balls and spikes to simulate atoms and bonds as for crystalline structures. However, the introduction of disorder calls for several comments. During the

Fig. 7.16. Example of a model showing a progressive transition from an ordered structure (left) to a disordered structure (right).

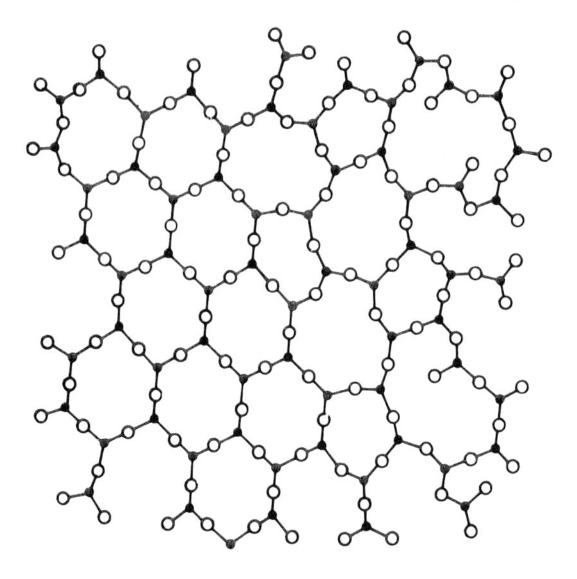

construction of these types of models, a general guiding principle must be followed (e.g. the structure must be built using a certain type of polyhedra). The steric constraints (space filling or the lack of flexibility of certain bonds) indicate to the builder the possible location of successive units.

But it is evident that the work remains largely subjective and that it is difficult in practice to obtain a perfectly disordered network. The subjectivity of the decision can be minimized by increasing the number of builders or by introducing a random system for decision making in the course of construction based, for example, on throwing dice or drawing lots.

In another class of models, the system is left to itself to physically find positions of equilibrium simulating the real situation.

In the simplest case, packing effects can be made use of to create a disordered mass of structural units; e.g. the spheres representing the ions or atoms placed and maintained in contact. *Random close packing models* operate on this principle. This corresponds to assuming an infinite repulsive potential at the contact of the units (hard, impenetrable spheres). A more realistic way is to use an interaction law between the elements of the model by introducing the interionic potential and to calculate the behavior of the model. The widespread use of modern means of calculation has allowed great progress in this modeling. Several avenues have been explored:

1. A model from one of the above methods may be used as a starting point and the energy of the system minimized by successive adjustments of the positions of the atomic centers to obtain the desired degree of optimization.

This *relaxation method* was employed for models made by stacking as well as those built by hand.

2. A model can be constructed entirely on a computer. Starting from an initial nucleus, the machine selects the coordinates of successive elements following various algorithms and they are accepted or refused according to the initial rules adopted. As the mass increases, repeated relaxation procedures are used as in 1.

3. "Monte Carlo methods" use successive and repetitive trials to adjust the model to obtain, for example, the best agreement between the radial distribution function of the model and that found experimentally by X-ray scattering.

4. A complete dynamic simulation of the liquid structure can be made on a computer by studying the displacements of the entire ion population subjected to given interaction laws. By reducing the level of thermal agitation, a "quench" of the system is induced which allows the frozen configuration to be visualized. This field of *molecular dynamics* offers the greatest possibilities for modeling. Its main handicap remains that it is impossible to operate with atomic populations greater than hundreds of units and that the calculation time remains very long.

Simulation of the dynamic behavior of a system can be obtained using a magnetic model.[274] Although essentially two-dimensional, such a model allows particularly striking visualization of the populations of more than 1000 units.

In spite of the diversity of approaches, the number of disordered models currently available is relatively small and the principal types will now be reviewed.

7.3.3 *Disordered models*

(a) *Random close packing*

Bernal[275] and Scott[276] independently determined the RDF of random close packing of hard spheres simulating a liquid.

They directly analyzed the structure developed by an assembly of identical spheres (steel ball bearings) held in contact in an elastic bag. This latter precaution was necessary to avoid the introduction of plane container walls which inevitably cause the beginning of regular organization of the balls or the "crystallization" of the system. Such a regular stacking propagates towards the interior of the volume for several tens of layers thus disturbing the desired disorder

To hold the system in place, Bernal and Masson[277] injected ink into the container and then removed the excess liquid. This glued the balls together and marked the points of contact by an accumulation of liquid between the balls through capillary action. The system of balls was then placed on a measuring table and as each ball was removed, the three coordinates of the center were obtained as well as the number and angular distribution of the contact points visible on the balls.

Figure 7.17 represents the radial density function obtained by Bernal for a system of 3000 balls and by Scott for a system of 20 000 balls. External layers were eliminated to avoid perturbations effects from the surface. The distribution of coordination numbers obtained by analysis of the contact points shows that the number of neighbors in contact is basically variable ranging from 5 to 12 with an average close to 8. The packing factor of such a structure is 0.637 which is to be compared with the packing factor of denser compact crystalline structures.

It is clear that the distribution function is only part of the rich structural information contained in the model. In particular, higher-order coordination spheres can be obtained by computer analysis of the coordinates obtained from the model.

Study of random close packing shows that the structure is composed of five types of polyhedra (Fig. 7.18). These polyhedra, called *canonical* or *Bernal polyhedra*, are the smallest polyhedra fulfilling the conditions that all edges be approximately the same length and that an atom could not be added to the center without increasing the edges by more than 10%.

Alternatively, to define a system of geometric neighbors from a given center, the concept of "Voronoi polyhedra" or "Wigner-Seitz cells" can be introduced.

By considering the median planes and the vectors r_{ij} which join an atom i to its neighbors j, it is possible to define the smallest polyhedron which contains all the points in space which are closer to a given point in the network than to any other point in the network. (The equivalent in two dimensions is called the Dirichlet polygon, Fig. 7.19.)

It is evident that the system of Voronoi polyhedra is space-filling and allows the study of stacking geometry. This representation is important because in the evolution of most of the thermodynamic properties, only the nearest neighbors are taken explicitly while the rest of the structure can be considered as a continuum.

Bernal produced such a model by compressing together an irregular pile of plasticene spheres previously covered with talc to enable their separation for analysis. The average number of faces per polyhedron was

Fig. 7.17. Radial density function for a system of hard spheres. After the results of Bernal[275] [○] and Scott.[276] [▽]. (The continuous line is for liquid Ar and the dashed line corresponds to average (macroscopic) density.)

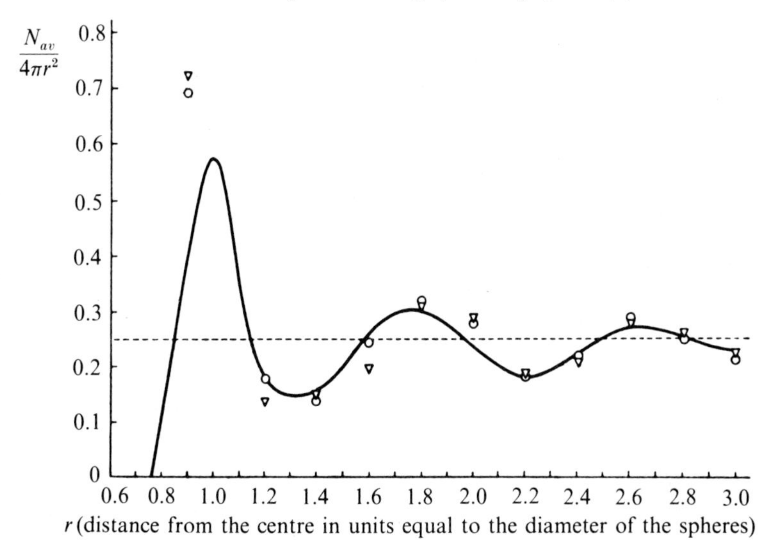

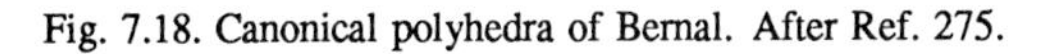
Fig. 7.18. Canonical polyhedra of Bernal. After Ref. 275.

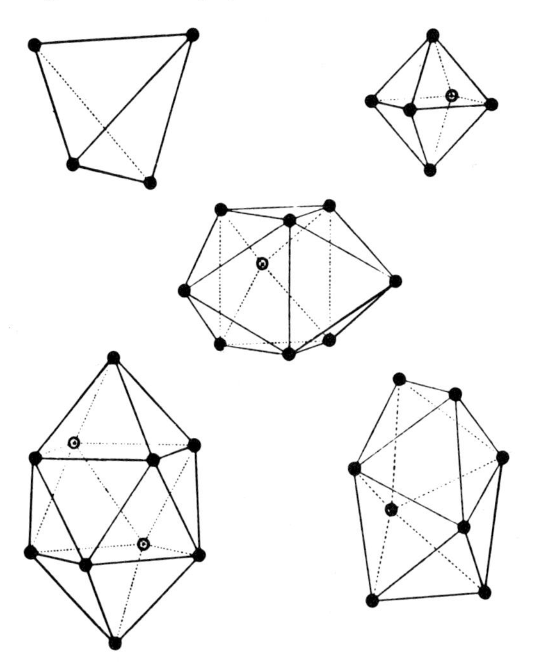

Fig. 7.19. Dirichlet polygon.

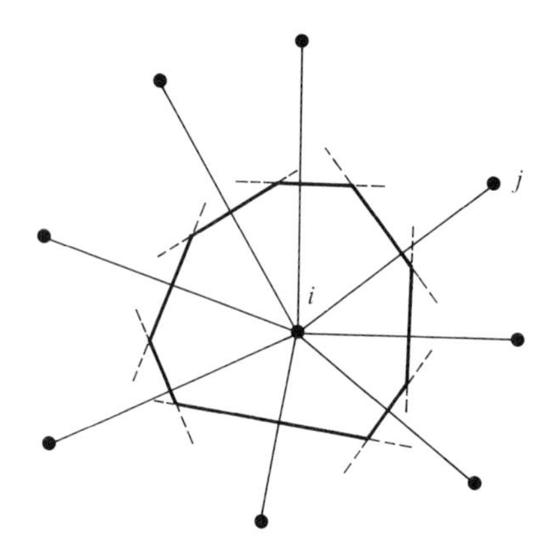

13.6 and the average number of edges per face was 5.16. This conforms to the *Euler relation* between the number of vertices V, edges E and faces F of a polyhedron:

$$V - E + F = 2$$

Since there are three edges for each vertex: $3V = 2E$. The average number of edges for each face is: $2E/F$, and for $F = 13.6$, this gives 5.1 which is in good agreement with experiment. There is a predominance of *pentagonal* faces. It can be shown that this is related in turn to a predominance of five element rings leading to the formation of a disordered stacking. This confirms the theoretical notion that a five-fold symmetry cannot produce an extended regular three-dimensional arrangement.

This model was improved by Finney[278] and Bennett.[279] It applies well to liquid metals and has been refined by introducing Lennard–Jones or Morse potentials in place of the hard-sphere interactions. In the case of metal–metalloid glasses, it is thought that the metalloid atoms fill holes formed by the disordered stacking of the metal atoms and that the space thus created could be as much as 20% of the volume.[280] However, recent measurements cast doubt on these assertions[281] because "interstitial" atoms can significantly influence their own environment.

A more realistic approach uses models with spheres of several sizes. Sadoc, Dixmier and Guinier proposed such models with 500 spheres of two sizes[282] with the additional condition that the metalloid atoms must be surrounded by nine metallic atoms. Analogous methods were proposed by Boudreaux and Gregor in a study of amorphous alloys: $Fe_{73}P_{27}$ and

$Fe_{86}P_{14}$. Models imposing chemical homogeneity were then optimized by minimizing the energy.[283]

Bernal's model is particularly important because it proves the possibility of the existence of a continuous and homogeneous geometric structure without any pseudocrystalline arrangements. Bernal thus laid the foundations for a statistical geometry of disordered systems.

(b) *Models of a continuous random network*

This type of model assumes that the local environment of an atom is determined by the nature of the bonds which impose the formation of stable groups.

This is the case for amorphous Si or Ge where tetrahedral coordination is preserved, or for the glasses SiO_2, B_2O_3, or As_2S_3 which are made up of the stable units (SiO_4), (BO_3) or (AsS_3) respectively. It is the arrangement of these preceding units which introduces the disorder implying variability of bridging bond angles.

Current models relate to the linking of *tetrahedral* groups. The model of Bell and Dean[284] for vitreous SiO_2, was constructed manually from rods and balls and contained 188 tetra-coordinated units with a total of 614 atoms (Fig. 7.20). The disordered linking of the (SiO_4) groups, assumed rigid, satisfies Zachariasen's rules. While the construction attempted to produce a totally interconnected network, the criterion of disorder was purely visual. Without attempting to introduce a pre-established bond angle distribution, an average bond angle near 160° was selected with the variability of this angle resulting from the elastic tensions appearing as the model advanced. Determination of the atomic coordination was done photogrammetrically using stereoscopic pairs and the classical methods of cartography. Analysis of the model using a computer allowed the construction of partial RDFs relative to the pairs Si–Si, Si–O and O–O (Fig. 7.21). The average angle of the bridges is $\overline{\theta} = 153°$ and the approximate density (calculated for an internal portion consisting of 72 SiO_2 units) is 1.99 g cm^{-3} compared to the experimental value of 2.20 g cm^{-3} for vitreous SiO_2.

To complete the statistical study of the model, the proportions of the rings of different orders ($n < 9$) were determined, a ring being defined as a closed path passing only once through n different Si atoms *via* Si–O–Si bridges. The model contains significant proportions of five- and seven-member rings.

The model was used to calculate the vibration spectra of vitreous SiO_2 and isostructural glasses GeO_2 and BeF_2 as well as to evaluate the electronic spectra. The complete table of atomic coordinates for the model is contained in Ref. 284.

Fig. 7.20. Bell and Dean model of vitreous SiO_2. After Ref. 285.

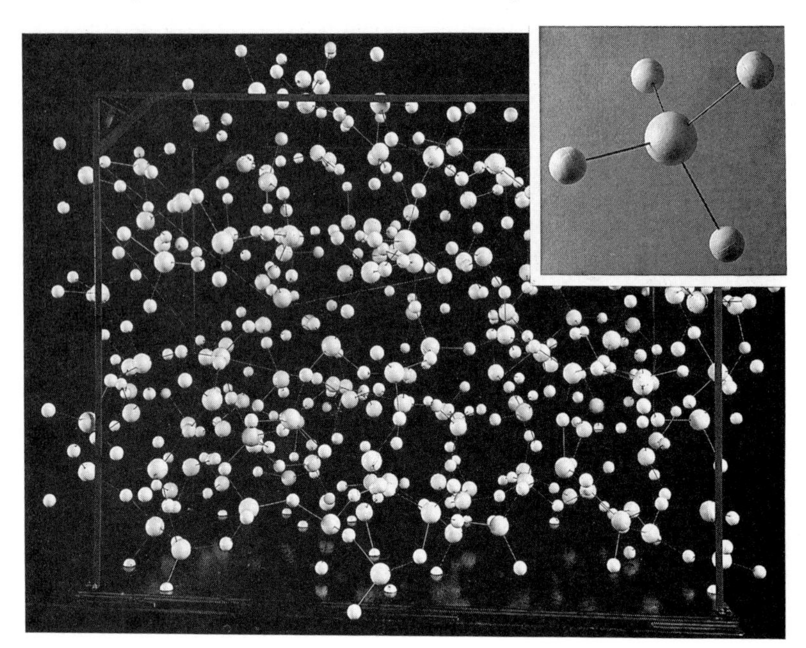

A similar model constructed by Evans and King;[286] did not reproduce the experimental RDF as well. Evans and Teter[287] studied the influence of different model parameters on the resulting density.

Note that currently there is not a comparable model based on tri-coordinated units which could, for example, be applied to a study of B_2O_3. For the latter glass it would be particularly interesting also to have a model based on larger units of three-connected triangles forming a plane ring to test the validity of the interpretations involving boroxol units.

Polk[288] manually constructed another disordered tetrahedral model consisting of 440 tetrahedral units bonded by elastic joints. The construction tended to exclude unsatisfied "dangling" bonds. The model was originally constructed to represent the structure of amorphous Si and Ge, but has been used in general studies of glasses based on coordination 4. Figure 7.22 shows the corresponding radial distribution histogram. This model was extended to 519 atoms by Polk and Boudreaux[289] and improved by Steinhardt *et al*[290] who minimized the elastic energy using a computer.

Fig. 7.21. Histogram of the distributions of interatomic distances for the model of Fig. 7.20. After Ref. 284.

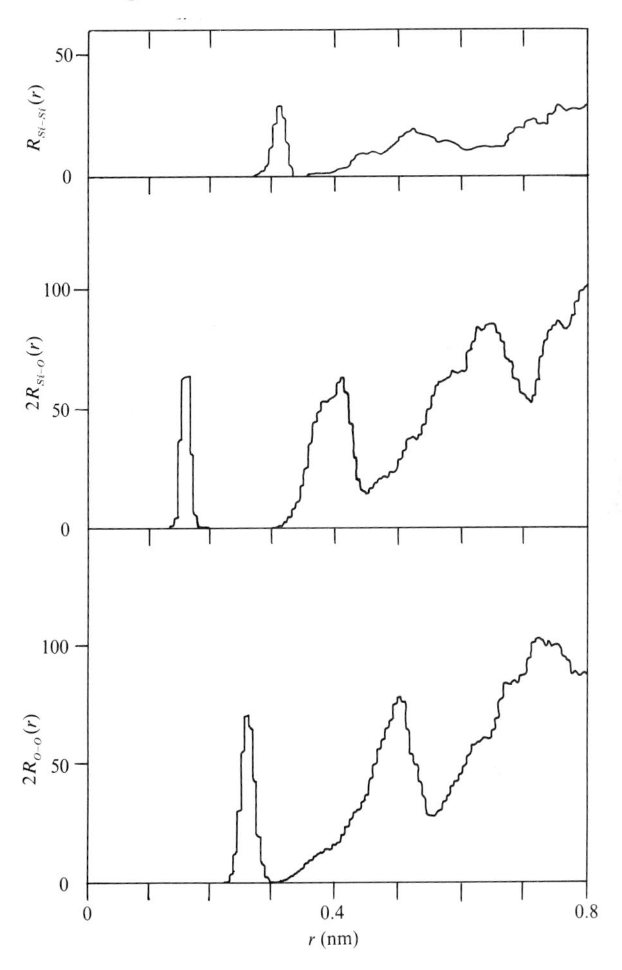

The disordered tetrahedral models constructed directly on a computer by Shevchik and Paul[291] and Henderson and Herman[292] do not, however, exhibit the small degree of distortion and absence of unsatisfied bonds characterizing the manually constructed Polk model.

The disordered tetrahedral model represents the structure of amorphous Si and Ge well. Because this model has a significant number of rings with an odd number of members, five or seven, in addition to the six-member rings (the only ones present in crystalline Si or Ge), the structure of these

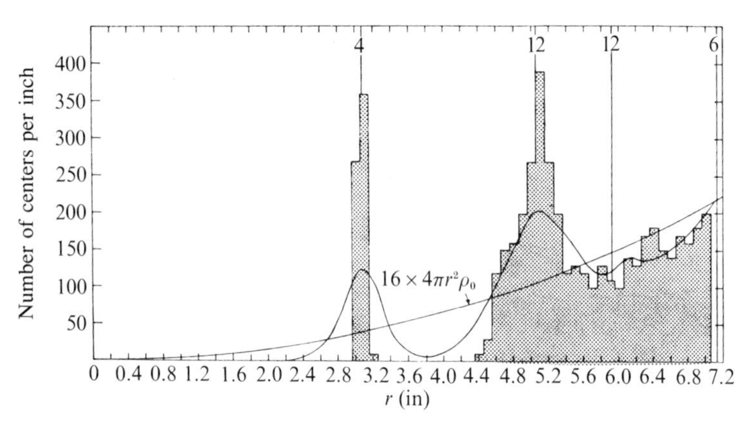

Fig. 7.22. Histogram of Polk's disordered tetrahedal model. After Ref. 288.

elements in the amorphous state is incompatible with "crystallite" models and can only be interpreted on the basis of a random disordered network.

Monte Carlo modeling methods have been applied by Renniger, Rechtin and Averbach[293] to study the configurations of chalcogenide glasses, especially those of the As–Se system. Starting from a quasi-random arrangement, the Monte Carlo method consists of progressively obtaining agreement between the calculated and experimental radial distribution curves. The final result (see e.g. Fig. 7.23) required 10^4 successive "trials" in which the structure was optimized by successive adjustments within the restrictions imposed on the most probable coordination (3 for As and 2 for Se) and the probability of persistence of the different bonds, Se–Se, As–As and As–Se, as a function of composition. The models consisted of 140–150 atoms. As the composition changes, the model shows gradual growth of interconnected chains from a mixture of chains and rings corresponding to pure Se up to about 40 % As.

For compositions richer in As, the structures progressively degrade. However, the results are unclear; equally good agreement can be obtained on the one hand for a ramdomly disordered model and on the other hand for a perturbed crystal type network.

7.3.4 *Ordered models*

These models postulate either the existence of crystalline micro-domains separated from each other by a poorly defined transition zone, or micro-domains with a structure which hinders any extended growth (anti-crystalline groups).

Fig. 7.23. Application of the Monte Carlo method to a glass of composition 40 As·60 Se: (*a*) radial distribution curve adjusted to the experimental curve; (*b*) a configuration of the adjusted model. After Ref. 293.

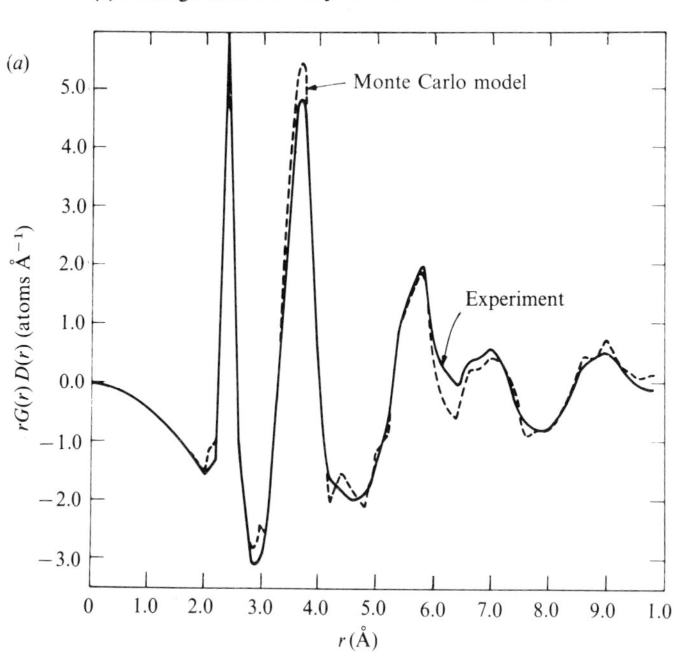

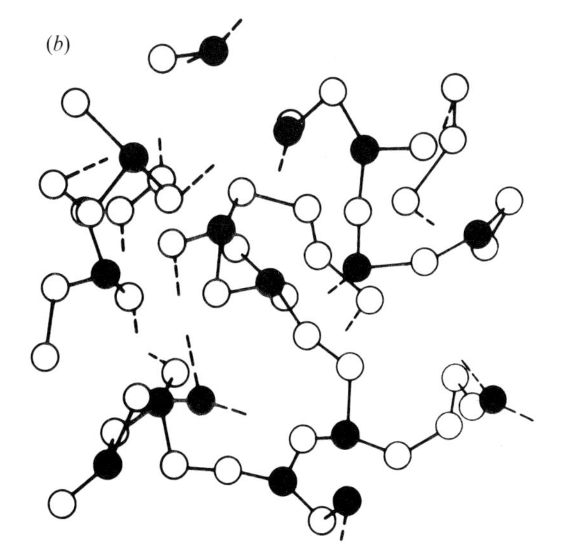

(a) *Quasi-crystalline models*

Most of the interpretations of diffraction or spectroscopy experiments are, in fact, made by assuming short-range order identical to that of a parent crystalline phase (cf. Chapter 4 and Chapter 5). To investigate the extent of this order, it is necessary to use a more quantitative model taking into account effects in the peripheral zone.

The Leadbetter and Wright[294] method progressively reduces to zero the correlations between the atomic positions in the crystal beyond a correlation length L. Thus L becomes a parameter specifying the extent of the short-range order. Moreover, the average for all angular orientations of the model is introduced into the calculation. Application to the interpretation of X-ray diffraction results for BeF_2 indicates a length $L \sim 11$ Å beyond which no order is detectable, the structure being best approximated by that of α-cristobalite.

(b) *"Anti-crystalline" groups*

In 1952, Frank[295] had already drawn attention to the possible existence of local non-crystalline order in simple liquids. The aggregates which possess five-fold symmetry cannot produce an extended network and their growth would therefore be self-limited.

(c) *"Vitrons"*

This idea has been taken up by Tilton[296] who explored the possibility of constructing the vitreous SiO_2 network using units based on a pentagonal dodecahedron. These structures (Fig. 7.24) called "vitrons" can be built up to a limited size beyond which the stresses prevent the construction of a more extended network. The same idea was developed by Robinson.[297] Difficulties in these models come from the need to imagine a disordered "connecting tissue" between the vitrons. The presence of such structures in glass has not yet been proved.

(d) *"Amorphons"*

An extensive study of groups with icosahedral symmetry was carried out by Hoare and Barker[298] who found a considerable variety of structures with a local symmetry, called "amorphons", consisting of 50–1000 atoms. They are also growth self-limited like the "vitrons" of Tilton. Figure 7.25 shows an example of such an aggregate consisting of 115 atoms.[299] Although the formation of such *isolated* structures has been established by computer simulation of an aggregation process from a super-cooled phase,[300] their presence in an extended system remains questionable. New interest in this concept arose after the discovery of icosahedral phases in Al_6Mn, Al–Mn–Si and other alloys prepared by

Fig. 7.24. A network constructed from dodecahedral units (vitrons).

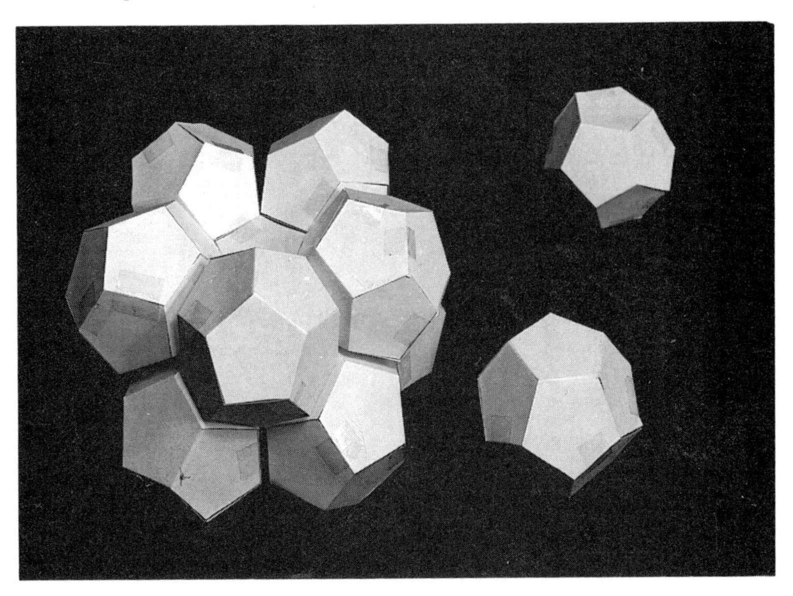

rapid quenching of melts.[301] These *quasi-crystals* follow models based on three-dimensional *Penrose tilings*.[302]

(e) *"Polytetrahedral" models*

Another type of model possessing a locally limited order was proposed by Gaskell[303] who studied the networks obtained by stacking of tetrahedral units in such a manner that {111} faces were in contact. Such networks, called "polytetrahedra" by the author, show pronounced alignments viewed at certain angles (Fig. 7.26) which Gaskell relates to the "fringes" observed in high-resolution electron microscopy (cf. Section 7.1.2). However, such models do not seem to give satisfactory agreement with experimental radial distribution curves.

7.3.5 *Local aggregates and vitreous transition*

According to Tammann[1], the vitreous transition is due to the formation of "domains" first growing at random in a supercooled liquid matrix, then interfering in the neighborhood of T_g producing a relatively rigid interconnected structure. Tammann's hypothesis does not explain why such "domains" do not crystallize during growth nor why the later stages of transition are particularly sudden. Hoare and Barker[298] think that Tammann's "domains" could in fact be "amorphons."

Fig. 7.25. "Amorphon" of 115 atoms obtained by computer optimization. After Ref. 299.

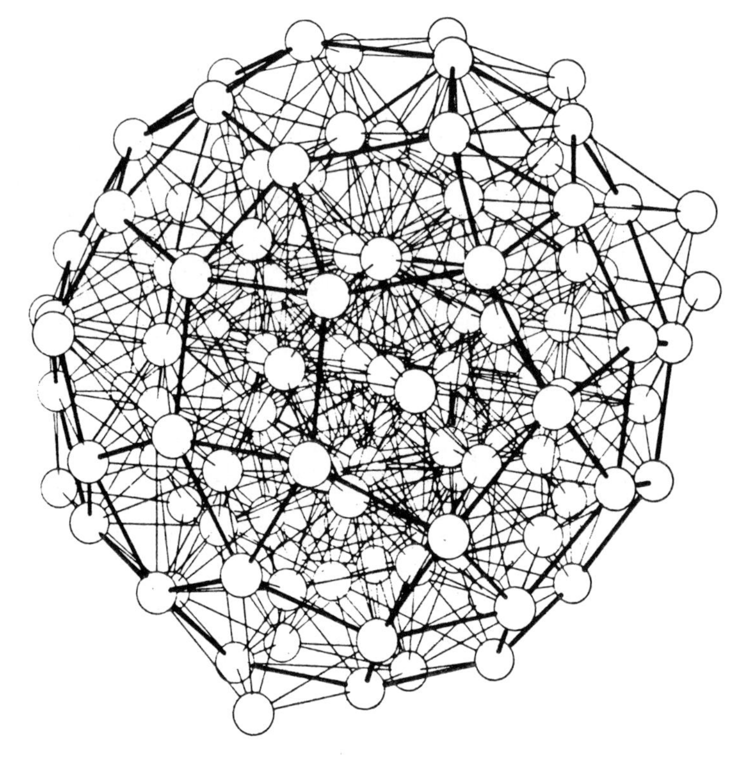

With such a structure, these groups are unlikely to crystallize and, their size being self-limited, solidification can only continue by a multiplication of their number. The dimension of an "amorphon" relative to a "monomer" unit is large enough for a "connective tissue" to be distinct. This undifferentiated tissue remains relatively mobile and allows the relaxation of constraints up to an advanced stage of solidification. The abrupt nature of the transition would be due to a rapid disappearance of the interconnecting tissue because of multiplication of the centers.

7.3.6 *Dynamic models*

(a) *Use of molecular dynamic methods*

Computer simulation of the movements of an atomic or ionic population representing a liquid allows frozen-in states corresponding to a glass to be easily obtained. However, for practical economical reasons, the simulated freezing speeds of about 10^{11} K s^{-1} are greater than freez-

Fig. 7.26. Polytetrahedral network models of Gaskell. (*a*) 280 atoms; (*b*) 798 atoms. After Ref. 303.

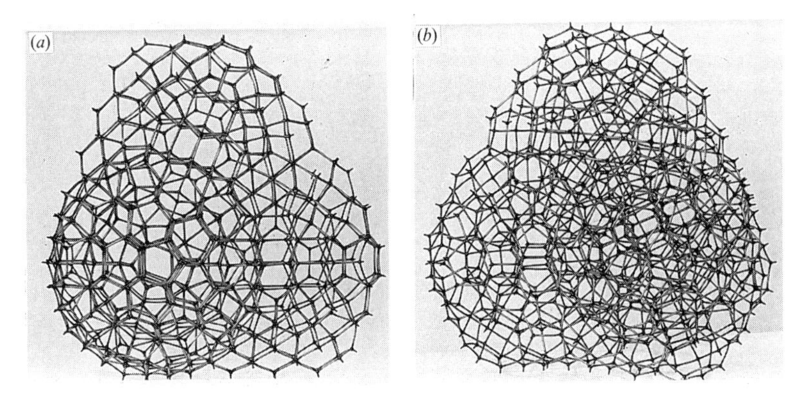

ing speeds obtained by hyperquenching (splat cooling). In other words, the structure of simulated glasses corresponds to very high "fictive temperatures".

The study of a group of 500 hard spheres by molecular dynamics leads to a distribution function identical to that obtained by Bernal.

Woodcock, Angell and Cheeseman[304] studied sphere systems obeying to the Lennard–Jones potential, ionic systems (M^+, X^-) as well as network-forming systems which can simulate vitreous SiO_2.

In spite of certain deviations which could be corrected by refining the potential used for simulation, agreement with experiment was satisfactory. The main point gained from the simulation experiment is that the formation of continuous tetrahedral networks does not imply the existence of directional bonds between atoms.

Computer simulation of the (K^+, Cl^-) system[305] showed that during fusion, the average coordination number decreases from 6 to 4, which conforms to the results obtained by X-ray scattering in such systems (cf. Chapter 4). Moreover, the existence of a "vitreous transition" during simulated ultrarapid "quenching" was demonstrated (Fig. 7.27).

(b) *Dynamic magnetic models*

It is possible to simulate the behavior of ionic populations using a dynamic magnetic model.[274] A group of floats with bar magnets at their centers simulating the ions is placed on the surface of water in a container. The movement of thermal agitation is simulated by the movement of the water surface and it is possible to follow the evolution of the configurations thus produced.

Fig. 7.27. Computer simulation of the fusion of a KCl crystal. (*a*) Variation of the internal energy U with temperature showing fusion at T_f and a "vitreous transition" at T_g. (*b*) Variation of the corresponding average coordination numbers. After Ref. 305.

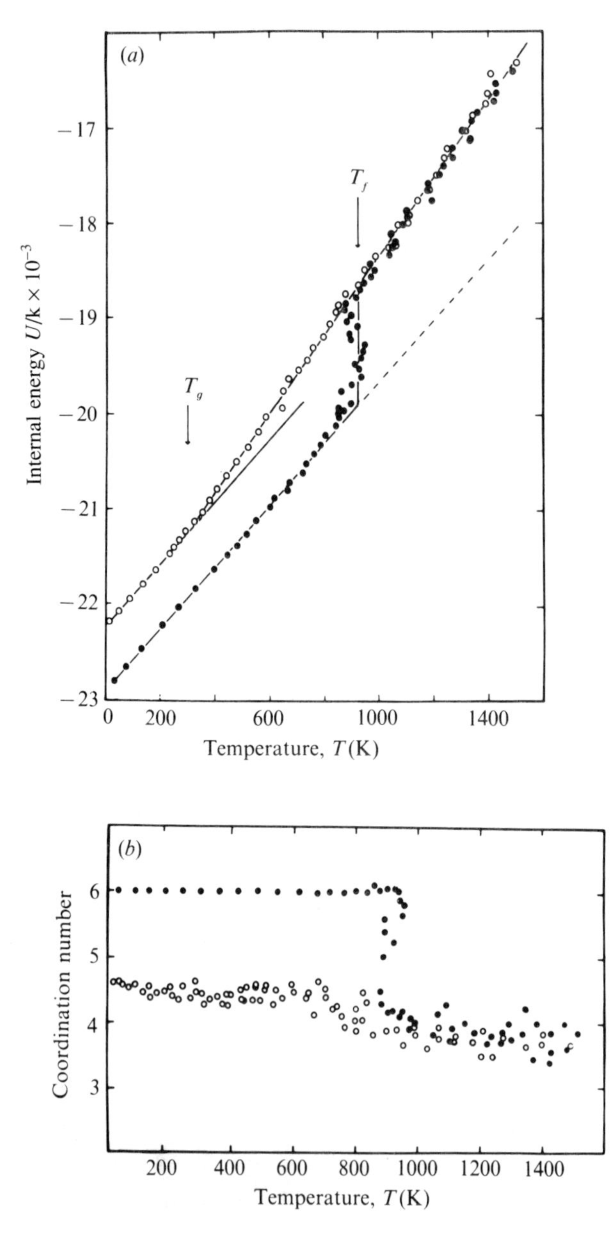

Fig. 7.28. Fusion of a two-dimensional "crystal" C^+–A^- observed on a dynamic magnetic model: (*a*) non-perturbed crystal; (*b*) appearance of dislocations, (*c*), (*d*) formation of a paracrystalline mass. The corresponding diffraction spectra can be obtained by optical transforms: (*e*) crystal, (*f*) liquid. After Ref. 274. Note: The cations C^+, tinted red in the original model, do not reproduce well here.

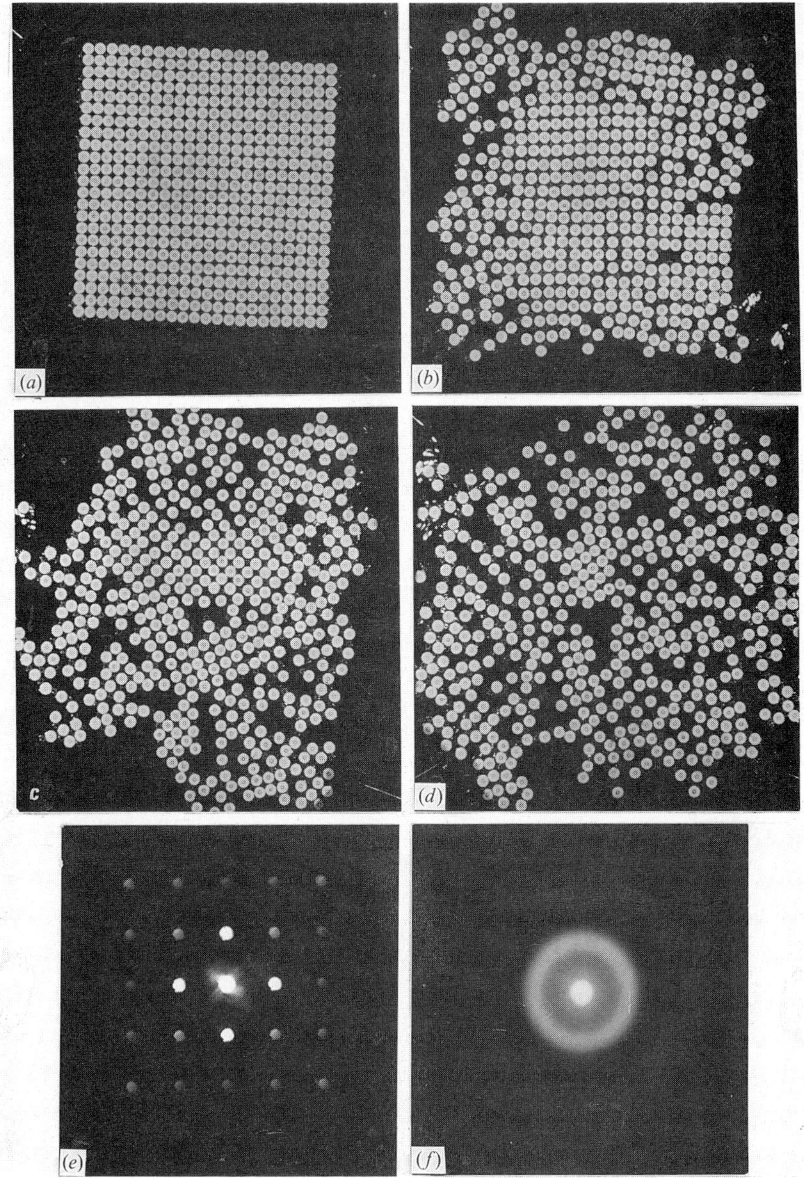

Although two-dimensional, this analogous arrangement is very valuable and permits a particularly striking visualization of cooperative movements during fusion (Fig. 7.28) and solidification of disordered structures.

Quantitative measurements can be made from cine-recordings of the evolving model, and these allow allow frame by frame analysis of the configurations. The influence of different parameters (size of ions, etc.) on coordination distributions and their "life-times" can be studied.[274]

8 Classification of glasses

8.1 Natural glasses

Substances rarely exist in the vitreous state in nature. In general, vitrification only occurs when molten lava reaches the surface of the Earth's crust and is cooled rapidly. Among the *volcanic rocks* which may contain higher or lower proportions of vitreous phases, *obsidians*, which are natural glasses with a composition close to that of current industrial glasses, are particularly notable. Obsidians are colored black, gray or reddish-brown by impurities (Fe, Mn, etc.) and contain less than 3 % water. *Pechsteins* are colored black, blue, green or red and contain more than 3 % water. Obsidians are saturated with volcanic gas retained in solution during quenching. On heating to 900–1000 °C the dissolved gas is released, resulting in a swelling of the semi-viscous material and the formation of a sponge-like rock called *pumice*. Pechsteins lose water at about 200–300 °C; they may have been formed by hydration of obsidians. The rapid cooling of volcanic glasses often results in internal stresses which make them birefringent. When the stresses are released, spherical inclusions with multiple overlapping layers (perlitic rocks) may appear.

Obsidians and pechsteins almost always contain microscopic crystalline inclusions (feldspars, pyroxenes or mica). They can be found on the Lipari Islands, in Iceland, Japan and Hawaii. In Hawaii, obsidian is often present in a fibrous form (rock wool) drawn during lava outflow.

Natural glasses can also be formed by the sudden temperature increase following strong shock waves, e.g. meteor impacts. The kinetic energy of the projectile is partially converted into heat which melts the rock at the point of impact; the rapid cooling which follows gives rise to the formation of glasses called *tectites*.[306] It is probable that the blocks of almost pure SiO_2 found in the Libyan desert come from meteor impacts. The same processes occur outside the planet Earth. The Apollo mission revealed that the lunar surface contains a large proportion of glass in the form of tiny balls created by fusion and rapid cooling of lunar rocks under meteor impact. Meteorites found on earth sometimes contain vitreous particles.

Biological processes can, in some rare cases, lead to glass formation; e.g. the skeleton of some deep water sponges (*monoharpis*) consists of a large rod of vitreous SiO_2.[307]

Apart from the preceding cases, all glasses are essentially artificial products. Topics on glass in planetary and geological phenomena are assembled in Ref. 308.

8.2 Artificial glasses

Many substances can be used to form glasses (cf. Chapter 1), but only some of them offer practical value. In the following we will take an overview of the main types of glasses and their principal characteristics. We will ignore organic polymers (plastic materials) which are amply treated in other works and need not be examined in this context.

Among inorganic glasses, oxide glasses, and above all silicates, are the most important ones, representing more than 95 % of the tonnage of industrial glass products. Rawson's excellent work[7] contains detailed descriptions of numerous glass systems, but does not include metallic ones.

8.2.1 *Oxide glasses*

(a) *One-component glasses (network-formers)*

• SiO_2. Silica is the glass-former *par excellence*. Vitreous SiO_2, also incorrectly called "fused quartz", is a typical glass with a structure already described, based on (SiO_4) tetrahedra forming a disordered three-dimensional network.

SiO_2 glass is technically important because of its excellent chemical resistance (except to HF and alkali) and its small expansion coefficient ($\sim 0.5 \times 10^{-6}$ K^{-1}) which gives it a very good thermal shock resistance.

SiO_2 is highly transparent to UV and is therefore used as an envelope for Hg vapor lamps. Its optical properties depend on impurities (OH^-, alkali metal ions, transition metal ions, Al^{3+} and Ge^{4+}, where the latter can introduce color centers). SiO_2 can be obtained in different ways: by fusion of quartz and $SiCl_4$ vapor phase hydrolysis (cf. Chapter 1); the different commercial varieties are mainly distinguished by their OH^- content and impurity level. For special applications (optical fibers), ultrapure SiO_2 has been produced (cf. Chapter 17).

Refs. 309 and 310 contain a large body of physical chemical data on SiO_2 glass and the crystalline allotropic varieties of silica.

• B_2O_3. Anhydrous boric oxide becomes vitreous on cooling and it is very difficult to obtain the crystalline form. The structure of B_2O_3 glass is based on triangular (BO_3) units. Their structural arrangement is not entirely clear even though the presence of boroxol units consisting of an association of three (BO_3) units is generally assumed. Because it is hygroscopic, B_2O_3 is never used alone but is incorporated in numerous industrial glasses.

A recent monograph[311] contains a large amount of information on the structure of vitreous B_2O_3, and on borate glasses in general.

• P_2O_5. Vitreous phosphoric anhydride does not have sufficient hygroscopic resistance to be used alone. Its structure is based on distorted tetrahedra (PO_4), one of the four bonds being a double bond, with the result that only three bonds participate in network formation.

• GeO_2 Germanium oxide forms a glass isostructural with SiO_2, based on (GeO_4) tetrahedra and is mainly of theoretical interest.

(b) *Glasses formed by a combination of several glass-formers*

Glasses in the systems $SiO_2-B_2O_3$, SiO_2-GeO_2, $SiO_2-P_2O_5$, and SiO_2-TiO_2 have been prepared either by direct fusion or by vapor phase hydrolysis of the corresponding halide compounds.[312] They were developed for optical fiber technology where it is necessary to have ultrapure glasses with refractive indices close to that of vitreous SiO_2.

Certain glasses in the SiO_2-TiO_2 series have a thermal expansion coefficient close to zero. They find application in the manufacture of rigid and thermally stable telescope mirrors.

The structure of these glasses is still largely unknown; in particular, it is not definitively established if mixed chains like

$$\begin{array}{ccccc} | & & | & & | \\ -Si & - O - & B & - O - & Si- \\ | & & | & & | \end{array}$$

form or if there are micro-domains richer in one of the components.

(c) *Glasses formed by a combination of glass-formers and other oxides*

The large majority of the usual oxide glasses are in this group. The addition of other oxides allows the properties of the resulting diverse glasses to be modified within wide limits.

The fine structural details of multicomponent glasses are still practically unknown and the various methods previously discussed only permit a fragmentary view or the study of certain particular cases. For example, the distribution of modifying cations in the network "holes" is unknown even though there are indications that they are not distributed at random as in the Zachariasen model. Limited information on the coordination of cations other than modifiers was obtained by various methods employing spectroscopic probes (cf. Chapter 5). Note, however, the difference between *technological glasses* which may have up to 10–20 significant components (either intentional or coming from impurities) and *model glasses* with two, three or four components, their numbers being intentionally reduced for the purpose of simplification and possibility of interpretation. Thus for example, the binaries SiO_2-X_2O or SiO_2-RO where X is an alkali metal and R an alkaline earth are studied systematically in an attempt to elucidate the influence of different oxides even though such glasses do not have practical applications. Likewise, a study of the ternary diagram SiO_2-Na_2O-CaO provides valuable information on industrial glasses which, with the addition of numerous secondary components, derive from this system.

It is usual to classify the different glasses according to the principal glass-former (or formers) i.e. glasses are called silicates, borates, germanates, borosilicates (or silicoborates), boroaluminates, etc.

Qualitatively, different oxides in glass compositions influence the properties of the glasses as follows:

- Network-forming oxides

SiO_2 reduces the thermal expansion coefficient (thus improving thermal shock resistance), raises working temperature and improves mechanical strength.

B_2O_3 (at less than 15 %) reduces the expansion coefficient. It reduces the viscosity η at high temperatures and raises it at low temperature (the glass becomes "shorter") cf. Section 9.2.2. It improves resistance to mechanical abrasion.

P_2O_5 raises the UV and reduces the IR transmissions; it reduces the chemical resistance of the glass.

- Network-modifying oxides

The general property of oxide modifiers, which depolymerize the network-forming oxides, makes them act as *fluxes* thus reducing the high melting temperature of pure SiO_2 ($\sim$ 1800 °C).

Na_2O is the oxide most used. Its introduction lowers viscosity, increases the expansion coefficient and electrical conductivity, and reduces the chemical resistance (SiO_2-Na_2O glasses are water soluble).

Li_2O is the most active "flux"; it increases the tendency for devitrification by reducing viscosity.

K_2O is the "flux" acting on viscosity to "lengthen" glass (i.e. to increase its working range).

Among the alkaline earth oxides, CaO and MgO are important in practice.

CaO is an essential ingredient in industrial glasses. It improves the chemical resistance of SiO_2-Na_2O glasses by greatly reducing their solubility.

MgO (< 4%) improves the viscosity characteristics around 800 °C (the glass becomes "longer").

BaO acts in a similar manner to CaO (it also raises the refractive index); it is used in optical glasses (barium crowns).

- Intermediate oxides

Al_2O_3 "lengthens" the glass, i.e. increases its working range, improves the mechanical and chemical resistance, increases the diffusion coefficient of alkali metal ions (the reason for its use in chemically tempered glasses) and reduces the tendency for unmixing.

ZnO increases the hardness of silicate glasses.

CdO in high concentrations (30–60 %) is used in the composition of glasses for thermal neutron shields.

PbO increases the index of refraction (used in so called "crystal" glass) and increases electrical resistivity. In high concentrations (40–80 %) it is used in X-ray shielding glasses; it lowers T_g and enhances phase separation, thus its use as an opacifier in certain enamels.

TiO_2 increases the index of refraction, enhances phase separation (used in glass ceramics) and improves acid resistance.

ZrO_2, like TiO_2, promotes glass devitrification and phase separation.

Transition metal oxides and rare earth oxides are used in colored glasses (art glasses and special optical glasses).

Silicate glasses are the most important and constitute most of the glass produced for usual applications (construction, transportation, lighting).

Typical compositions of several types of glasses are shown in Table 17.1. The fabrication processes of thees glasses are described in Chapter 17.

Borate glasses, which are soluble in water, have only theoretical interest. They have often been studied because their melting temperature is lower than that of the silicates, which makes them more accessible to numerous experimental techniques.

Borosilicate glasses of the $SiO_2-B_2O_3-Na_2O$ system are mainly used in chemistry (Pyrex, Vycor) because of their good chemical resistance and low expansion coefficient (resistance to thermal shock).

Boroaluminates of the $CaO-B_2O_3-Al_2O_3$ system, "Cabal" glasses without alkali metal ions, have high resistivity and are used in electrical applications.

Phosphate glasses are soluble in water and do not have practical applications but have been studied extensively by chromatography which allows the separation and analysis of various anions constituting the network.

Germanate glasses are mostly of theoretical interest; since they are isostructural with silicates, they are studied for comparative purposes.

Vanadate glasses have semi-conductive properties. Titanate and molybdate glasses, which can be prepared in narrow composition ranges, are currently only of academic interest.

Telluride glasses have interesting IR transmission properties similar to *aluminate* glasses.

The detailed structure of these different glasses is not known, at most only the coordination of the glass-forming cation has been determined in some cases.

8.2.2 *Halide glasses*

Before 1975 only a few halide systems had been studied for their glass-forming tendencies.[313] Beryllium fluoride, BeF_2 is a glass network-former which may be considered as a weakened model of SiO_2; its structure is based on (BeF_4) tetrahedra. BeF_2 glass and vitreous fluoroberyllates – which are isostructural with silicates – were, until lately, only of theoretical interest.[314,315] More recently, however, these glasses have become particularly important for their exceptional optical characteristics which make them the best candidates for high-power lasers for thermonuclear fusion applications.[316] Mixed oxyfluoride glasses such as fluoroborates and fluorophosphates have also been prepared for these applications. Their structure has not yet been studied in detail.[317]

Fluorozirconate glasses which were discovered in France in 1974[318] deserve particular attention. In these glasses ZrF_4 is the primary constituent (more than 50 %) associated with BaF_2 (about 30 %) and other additional fluorides such as LnF_3 (Ln = lanthanides) and AlF_3 (Fig. 8.1). These glasses are of great interest because of their optical transmission properties (Fig. 8.2) which make them good candidates for optical fibers[319,320] (cf. Chapter 18). The domain of glass formation is generally quite restricted. One of the best compositions with a low tendency towards devitrification is: $57\,ZrF_4 \cdot 34\,BaF_2 \cdot 5\,LaF_3 \cdot 4\,AlF_3$.

The structure of fluorozirconate glasses is based on a collection of ZrF_6, ZrF_7 and ZrF_8 polyhedra which have a tendency to polymerize, sharing corners or edges. More precisely, it is the edge-sharing dimer Zr_2F_{13} which by corner-sharing, generates disordered networks.[321–3]

Subsequently, many other families of fluoride glasses were discovered, based on heavy metal fluorides. The conditions of glass-formation are here very severe and at least three components are necessary. The

Fig. 8.1. Glass-formation region for various ternary ZrF_4 based glasses.

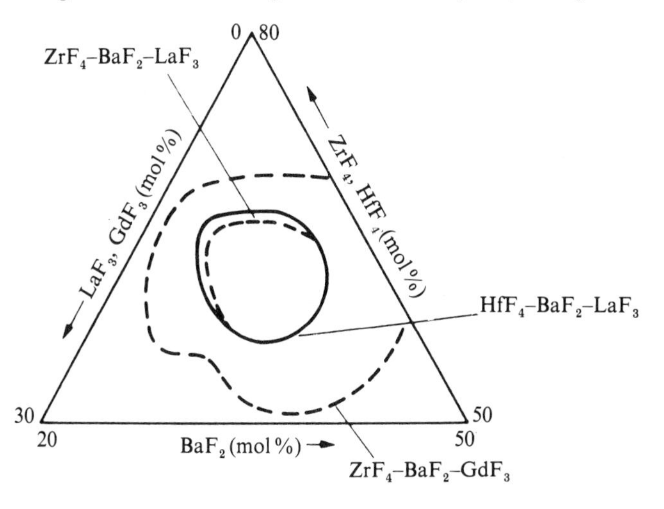

Fig. 8.2. Transmission characteristics of some fluoride glasses compared to vitreous SiO_2.

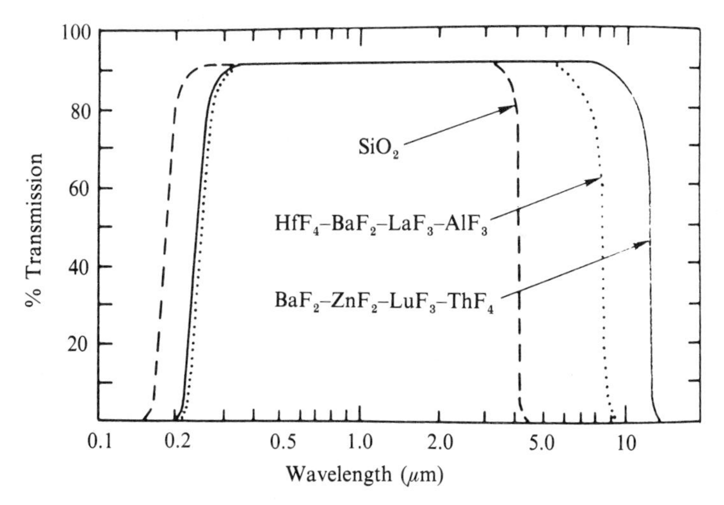

four-component glass $16\,BaF_2 \cdot 28\,ThF_4 \cdot 28\,YbF_3 \cdot 28\,ZnF_2$ and the five-component glass $30\,BaF_2 \cdot 30\,InF_3 \cdot 20\,ZnF_2 \cdot 10\,YbF_3 \cdot 10\,ThF_4$ are particularly promising.[324]

Other fluoride glasses[325] are based on transition metal fluorides combined with AlF_3 or GaF_3 and PbF_2. Some of these glasses which contain a high proportion of paramagnetic ions, show interesting magnetic properties. A good example is the composition $PbMnFeF_7$. Their devitrification tendency is higher than that of ZrF_4 based glasses and is accelerated by moisture.

Glasses based on $ZnCl_2$ also have interesting optical properties in a large spectral interval from UV up to 12 μm region in the IR, but their very great hygroscopicity prevents their practical application.

The AlF_3 based multicomponent glasses[326] require a rapid quench from the melt to prevent devitrification, which limits their potential value.

Oxygen-free glasses of the series HF–RF_n where R is an element of valence 2 or 3 so far only have theoretical interest.[327]

Refs. 328–330 constitute valuable reviews on the majority of halide glasses.

8.2.3 *Chalcogenide glasses*

These glasses are based on elements from group VI (S, Se, Te) combined with elements from group IV (Si, Ge) and group V (P, Si, Sb, Bi).[331,332] . These glasses, which do not contain oxygen, are interesting for their IR optical transmission (Chapter 12) and electrical switching (Chapter 11) properties.

The lighter elements in each group (of number N) have crystalline structures where each atom has 8–N neighbors. Thus Si and Ge have coordination 4, P and As coordination 3, and S and Se coordination 2. In all these elements the structure is determined by the character of the predominant covalent bonds. The metallic character increases with the atomic number passing towards structures characterized by higher coordination numbers. S and Se alone form glasses consisting of a mixture of polymer rings and chains. Introducing traces of halogen (e.g. I_2) ruptures the chains and shortens them. P exerts the opposite effect. Vitreous Se possesses photoconductive properties and is used in photocopiers (xerography).

Thermally more stable glasses with a wider range of properties are obtained by combination of elements from groups IV and V. The concepts of glass-formers and network-modifiers developed for the oxide glasses do not seem to be applicable to complex chalcogenide glasses. The rupture of bridging bonds does not always occur and it is more useful to consider most of these glasses as copolymers of the constituent elements.

Vitreous As_2S_3 has a good IR transmission up to 12 μm and its chemical stability is good compared to other transmitting materials. It is in industrial production. As_2S_3 and vitreous As_2S_3 have a structure similar to that of yellow arsenic.

As_2S_3 binary glasses with a composition between S and As_2S_3 can be considered as copolymers of the As_2S_3 groups, S_8 rings and S_∞ chains. For the ratio As/S $> 2/3$ the groups As_4S_4 can break the As_2S_3 network.

Bonds between Ge or Si and the chalcogenides are the strongest among those combinations still able to ensure glass-formation within wide limits. The corresponding glasses are technologically important because of their relatively high T_g.

The Ge–As–Si glasses have opto-acoustic applications[333] in which they are used as modulators and deflectors for IR rays because of their high refractive indices and low acoustic losses. Ge–As–Se glasses have similar uses.[334] The glass $Ge_{33}As_{12}Se_{55}$ has a high softening temperature (450 °C) which gives it technical significance as a candidate for optical windows in rocket guidance.

The Si–As–Te system has the largest composition field for glass-formation and contains glasses with the highest softening temperatures of which the glass $Si_{35}As_{25}Te_{40}$[335] is an example.

In a general way, the local structure of these glasses can be described either as a disordered "alloy" of the elements or as an arrangement of ordered chemical units. In the first hypothesis, a binary system would be a mixture of the two elements with the valence corresponding to the 8–N rule: this model is called a *disordered covalent network*.[336] The second hypothesis assumes the formation of coordination polyhedra obeying the 8–N rule but with the nearest neighbor atoms being dissimilar.

Taking the Ge–Se system as an example, the disordered covalent model allows the existence of the bonds Ge–Ge, Ge–Se and Se–Se for all compositions while the *ordered chemical model* only allows Ge–Se and Se–Se bonds. For compositions rich in Se, the Ge atoms have a coordination of 4 forming bridges between the chains. This is called the *cross-linked chain model* when the 3 or 4 coordinated element is a minor constituent. It has some similarity with the Zachariasen–Warren model for oxides. The cross-linked chain model appears appropriate for glasses containing Ge and As which are rich in S or Se. On the other hand, for compositions rich in Ge or As, the ordered chemical model seems to be preferable.

8.2.4 *Metallic glasses*

Metallic glasses, which are obtained by ultrafast quenching of liquid metal alloys, have shown a very rapid development because of their interesting properties which earned them the name "materials of the century" in the USA and "dream materials" in Japan. There is a vast literature on the subject which has evolved rapidly during the 1980s and only a brief summary will be attempted here.

Broadly speaking, metallic glasses may be sub-divided into two main classes:

1. Metal–metalloid alloys.

$(M_1, M_2, \ldots)_{80}$ $(m_1, m_2, \ldots)_{20}$ where M_i is a transition metal: Au, Pd, Pt, Fe, Ni, or Mn and m_i a metalloid: Si, Ge, P, C, or B. The 80/20 ratio is approximate.

2. Metal–metal alloys in approximately 50 % ratio.

One of the metals is generally a transition metal as in:

$Mg_{65}Cu_{35}$ $Au_{55}Cu_{45}$ $Sn_{90}Cu_{10}$
$Zr_{72}Co_{28}$ $Zr_{50}Cu_{50}$ $Ni_{60}Nb_{40}$

Many other compositions exist which are produced on an industrial scale.*

The main applications come from the fact that a variety of amorphous alloys can be made which have one or several unique characteristics, namely: extremely low magnetic losses, zero magnetostriction, high mechanical strength and hardness, radiation resistance, high chemical corrosion resistance or interesting catalytic properties. The soft magnetic properties which are the consequence of the lack of magnetocrystalline anisotropy make these materials ideal substitutes for core transformer steels (losses are 20 times less in the laboratory and about 3 times less in practice). Extremely low coercivities and square loop characteristics require lower excitation power and result in great energy saving.

Glassy metals are used in numerous electronic devices: cores in moving magnets, recording cartridges (replacing Permalloy), amorphous heads for audio and computer tape recording, high-frequency power transformers, magnetic shielding, etc.

In thin film form, amorphous materials are useful for thermomagnetic recording media; present materials are suitable for audio and video digital recording applications.

The chemical stability of metallic glasses in corrosive environments is linked with the absence of heterogeneities such as grain boundaries, dislocations or the presence of second phases. They are therefore more chemically resistant than their crystalline counterparts.

Mechanical properties are characterized by the absence of extreme brittleness compared to oxide glasses. They exhibit remarkable strength, approaching one half of theoretical strength, compared to strength levels 10^{-3}–10^{-4} of theoretical for crystalline counterparts. Plastic deformation occurs at high stress levels before rupture.

The main limitation of these amorphous alloys is the fact that they can

* The first series were commercialized by Allied Corp. (USA) under the trade name METGLAS. Other producers followed: Vacuum-schmelze (Hanau, Germany) and, in Japan, Hitachi Metals, Nippon Amorphous Metals Co. (Tokyo) and Nippon Steel Corp. (Kawasaki City). Smaller quantities and finished products may be obtained from United States firms, e.g. Arnold Engineering (Marengo, IL); Magnetic Metals (Camden, NJ); Magnetic Inc. (Butler, PA); and SGL Electronics (NJ)

only be obtained in the form of wires or thin ribbons by rapid quenching of a liquid alloy stream in contact with a spinning cooled metallic wheel.

Attempts to produce large pieces were made by means of extra rapid compaction of fine glass powders using high-pressure shock waves produced by explosions (shock-compaction). This ensures that devitrification of the glassy particles, which would occur during normal sintering procedures, is avoided.

Methods of preparation and properties of metallic glasses are described in Refs. 337–45. Ref. 346 gives a comprehensive review of technological applications and Ref. 76 the present status in Japan.

8.3 Property–composition relations

By virtue of their non-crystalline structure, glasses have a number of characteristic properties. The response to temperature is typical. They do not possess a distinct melting point, but on elevation of the temperature, they progressively soften and transform into a liquid which is less and less viscous. Their characteristic mechanical properties are progressively modified and they pass without discontinuity from an elastic solid to a viscous liquid. Since the liquids are *isotropic* and *homogeneous* (in the absence of phase separation), their composition, like those of solutions, can vary over wide limits.

As non-equilibrium solids, they possess a variable excess stored energy – hence the importance of the thermal history and the conditions of stabilization for a precise definition of certain of the properties. These properties are of two principal types: those which are related to the structure and those which depend only on the composition. Among those sensitive to the structure and to impurities are transport properties (electrical and thermal conductivity) and the various "losses" (dielectric, viscoelastic). The properties not related to structure are those which depend in the first place on the nature of the atoms and interactions with nearest neighbors. In this category are the density, elastic constants, specific heat, dielectric permittivity etc. These properties are often less sensitive to long-range order and remain close to the values for crystals of similar compositions.

For properties insensitive to structure, *additive relations* have been proposed which treat the glass as a *mixture* of components contributing independently to the property. This technique is common in the case of *oxide glasses* where for practical reasons one often seeks to predict the properties of a complex glass from its composition by calculation.

The property X is expressed for example in the form:

$$X = \sum_i C_i X_i$$

Table 8.1. *Examples of factors X_i for the calculation of glass properties from weight fraction composition.*

Oxide	Specific volume $V \times 10^6$ ($m^3\ kg^{-1}$)	Thermal expansion coefficient $\alpha \times 10^8 (K^{-1})$	Young's modulus $E \times 10^{-8}$ ($N\ m^{-2}$)	Heat capacity C_p ($J\ kg^{-1}K^{-1}$)	Thermal Conductivity $K_T \times 10^3$ ($W\ m^{-1}K^{-1}$)
SiO_2	4.35	2.67	7.0	8.00	12.85
B_2O_3	5.26	0.33		9.51	6.66
P_2O_5	3.92	6.67		7.96	
Li_2O		6.67			
Na_2O	3.85	33.3	6.1	11.2	−5.40
K_2O	3.57	28.33	4.0	7.78	2.44
MgO	2.63	0.33	4.0	10.21	24.79
CaO	3.03	16.67	7.0	7.96	13.27
BaO	1.43	10.0	7.0	2.51	1.93
Al_2O_3	2.44	16.67	18.0	8.68	15.57
TiO_2		13.67			
ZnO	1.69	6.0	5.2	5.22	8.46
PbO	1.04	13.0	4.6	2.14	3.18

where C_i are the weight fractions expressed in % of each component ($0 < C_i < 100$) and X_i are the "factors" representing the contribution to the property X of a given component. A formula of the type $X = A + \sum C_i X_i$ or an even more complicated expression (logarithmic etc.) ia often used.

Table 8.1 gives the factors X_i for the calculation of several properties of glasses based on the indicated oxides. Extensive compilations of such coefficients may be found in the publications of Morey[3] and Scholze.[8]

It is clear that glass cannot be considered as an *aggregate* of the different oxides which lose their individuality in the glass-formation, but rather as a *complex solid solution.*

In reality, such expressions are only interpolation formulae since X_i coefficients are obtained from extensive glass series measured in different laboratories. Often, moreover, the validity of the coefficients X_i is limited to well-defined intervals or only applicable to a particular glass type, e.g. borosilicates or PbO containing glasses, etc.

Babcock[17] tried to make the additive method more systematic by using a group of X_i factors which differ for each primary crystalline phase field in the equilibrium phase diagram. This approach is based on the hypothesis that the liquid structure (and thus the resulting glass) is related to the occurrence of a particular crystalline phase and thus contains the "sub-structures" present in the crystal.

Linear interpolation formulae are the best approach to reality in the fields thus delineated. Numerous examples of computer assisted numerical calculations can be found in Babcock's book.

The method has been tested on the $SiO_2-TiO_2-Na_2O$ system by assuming the phase diagram to be unknown. Calculations showed the existence of seven contiguous groups within which the linear approximations of the glass properties were best followed and corresponded to composition regions of the phase diagram. For glasses in the SiO_2-Na_2O-CaO system, the method leads to distinguishing ten different types of glasses.

The additive formulae are often of technological value, but do not reveal anything about the nature of the phenomena. For the scientist, however, relations which depend on the structure are the most interesting.

9 Rheological properties of glasses

Rheology is the science of the *deformation* and *flow* of solids and fluids under the action of mechanical forces. The rheological behavior of glasses is complex; at temperatures below T_g, glasses behave like *elastic solids* while at sufficiently high temperatures they have the properties of viscous *liquids*.

The transition from one behavior to the other is continuous; in contrast to crystalline solids, no true *melting point* is observed but instead a progressive softening. In the intermediate region, which is particularly important for applications, glasses exhibit *viscoelastic* behavior.

9.1 Concepts of rheology

9.1.1 *Solids – definition of deformation*

Deformation of a solid under the action of a force is called *elastic* if it is completely reversible, i.e. the solid returns to its original form when the exterior force is removed. Irreversible deformation i.e. *plastic* deformation or *flow* leaves a permanent deformation. The *deformation* is defined analytically as the relative change in dimension and depends on the applied force per unit area or *stress*.

9.1.2 *Ideal elastic solid – definition of the moduli*

As long as it remains small, the deformation is proportional to the stress. This is known as Hooke's law.

The ratio stress/deformation is called the *elastic modulus*; for an isotropic solid (which is the case for a glass), the following moduli are defined according to the mode of stress application.

(a) *Longitudinal stress* (Fig. 9.1(*a*))

A prismatic solid of length l_0 subjected to an axial stress $\sigma_a = F/A$ undergoes a deformation $\epsilon = \Delta l/l_0$.

Young's modulus is defined by the relation:

$$E = \sigma_a/\epsilon$$

In the case where the lateral faces are free, the solid undergoes a simultaneous change in its transverse dimension $\Delta d/d_0$. The ratio:

$$\mu = \frac{\Delta d}{d_0} \Big/ \frac{\Delta l}{l_0}$$

is called *Poisson's ratio*. The inverse of E is the tensile compliance.

Fig. 9.1. Definition of the elastic moduli: (*a*) axial loading; (*b*) tangential loading; (*c*) isotropic pressure.

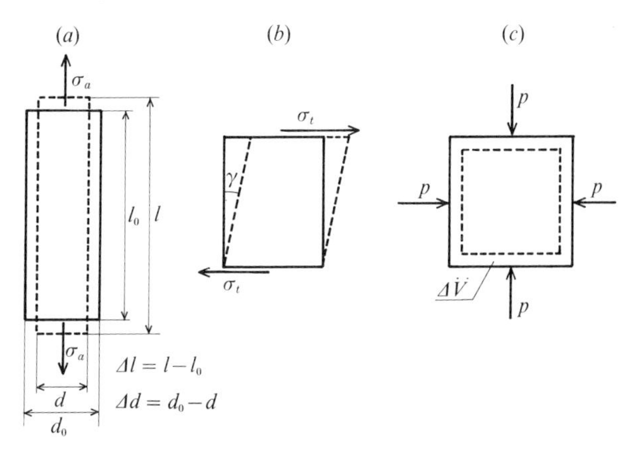

(b) *Tangential loading* (Fig. 9.1(*b*))

Under a tangential load (shear) $\sigma_t = F/A$, the solid suffers an angular deformation $\tan\gamma \sim \gamma$ proportional to σ_t. The *shear modulus* (shear or torsion) G is defined by the relation:

$$G = \sigma_t/\gamma$$

The inverse of G is the shear compliance $J = 1/G$.

(c) *Isotropic pressure* (Fig. 9.1(*c*))

Under the action of isotropic pressure p, the solid undergoes a relative volume change, $-\Delta V/V$. The *bulk modulus* K is defined by the relation:

$$K = -p\bigg/\frac{\Delta V}{V}$$

The inverse of K is the *compressibility* $\beta = 1/K$.

It can be shown that E, G, K, and μ are related as follows:

$$E = 2(1+\mu)G \qquad \text{and} \qquad E = 3(1-2\mu)$$

Two of these parameters are sufficient to characterize the response of an isotropic solid.

Units: The moduli E, G, and K are expressed in N m^{-2} or Pa. The number μ is dimensionless.

9.1.3 *Fluids – viscous flow*

The frictional resistance of a fluid in opposition to a force tending to make it flow is characterized by its *viscosity*.

(a) *Behavior in shear. Ideal viscous fluid*

The *coefficient of viscosity* is classically defined by considering a flow represented by Fig. 9.2. A plane plate of area A is displaced with a velocity V_0 parallel to a similar stationary plate at a distance z_0. For low velocities, flow between the plates is *laminar* and flow velocity in the fluid is proportional to z:

$$\frac{V}{V_0} = \frac{z}{z_0}$$

where the movement can be considered as a shear of superposed layers. Under these conditions, the velocity gradient $\mathrm{d}V/\mathrm{d}z$ which is identical to the shear rate, $\dot{\gamma} = \mathrm{d}\gamma/\mathrm{d}t$ is constant. To maintain the movement, it is necessary to apply a force F on the moving plate, i.e. a shear loading $\sigma_t = F/A$ such that:

$$\sigma_t = \eta\dot{\gamma}$$

The viscosity coefficient η is thus defined as the proportionality constant between the shear stress and the *velocity* of angular deformation $\dot{\gamma}$.

Fig. 9.2. Definition of the viscosity coefficient.

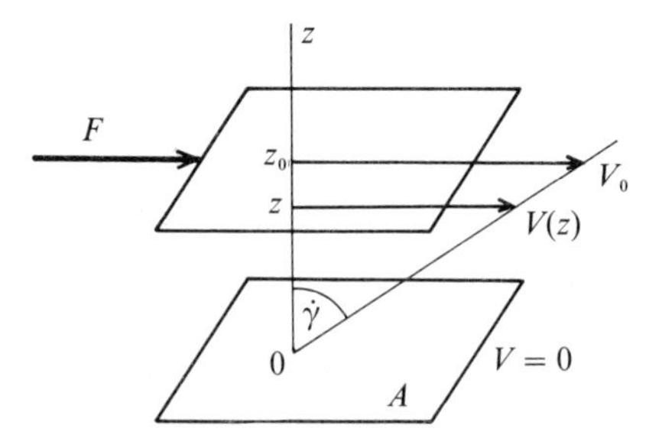

An ideal fluid, which is characterized by a linear relationship, is called a *Newtonian fluid*, and plays a rôle in viscosity analogous to that of a *Hookean solid* in elasticity.

Units: The viscosity coefficient has dimensions $(ml^{-1}t^{-1})$. The unit of measurement is the poiseuille (Pl) or Pa s.

In the CGS system the viscosity unit is the *poise*.

1 *poise* = 1 dPa s.

(b) *Behavior in tension*

The viscosity coefficient η plays a rôle in viscous flow analogous to that of the shear modulus G previously defined for elastic deformation of a solid.

In the case where a substance is stressed longitudinally, to find the equivalent of Young's modulus it can be assumed that the viscous flow occurs at *constant volume*. In these conditions $\mu = 0.5$ and $E = 3G$. The rate of extension $\dot{\epsilon}$ produced by a uniaxial tension σ_a in the case of viscous flow will be given by:

$$\sigma_a = 3\eta\dot{\epsilon}$$

The viscosity coefficient in extension is three times higher than that in shear. This formula finds application in the dilatometric determination of the viscosity coefficient of glass. One of the characteristics of linear viscosity is that it allows the *uniform drawing* of a glass softened by heating without formation of constrictions. In effect, if A is the cross-section of a rod of length L subjected to an extension force F, the axial stress being $\sigma = F/A$, for a deformation at constant volume:

$$\dot{\epsilon} = \dot{L}/L = -\dot{A}/A$$

from which it is seen that:

$$\dot{A} = -\dot{\epsilon}A = -\frac{\sigma_a A}{3\eta} = -\frac{F}{3\eta}$$

Since F is constant along the length of the rod, all sections decrease at the same rate regardless of their initial value. A section initially smaller would thus not have a tendency to greater restriction (necking).

9.1.4 *Comparison of the behavior of a Hookean solid and a Newtonian liquid – rheological models.*

Figure 9.3(a) shows the linear relation between the stress σ and the deformation γ of an ideal (Hookean) solid. As a function of time (Fig. 9.3(c)), the elastic deformation follows instantly upon the application of the stress and disappears when the stress is removed. It is said that the *recovery* is immediate and total. The mechanical rheological model is a "spring" subjected to a stress σ (Fig. 9.3(*b*)).

For a Newtonian fluid (Fig. 9.4), the relation between stress σ and the deformation rate $\dot{\gamma}$ is linear. As a function of time the deformation γ increases linearly and is not recovered when the stress is released (irreversible). The mechanical model of this behavior is a "dashpot", a piston moving freely in cylinder filled with liquid.

Fig. 9.3. Hookean solid: (*a*) linear behavior, (*b*) rheological model (spring), (*c*) behavior as a function of time.

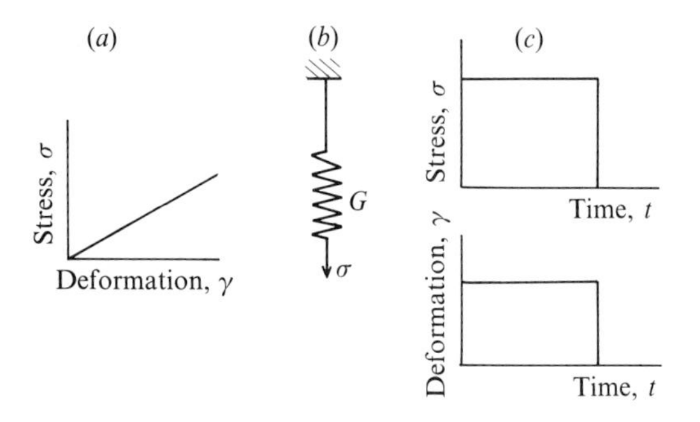

Fig. 9.4. Newtonian fluid: (*a*) linear behavior, (*b*) rheological model (dashpot), (*c*) behavior as a function of time.

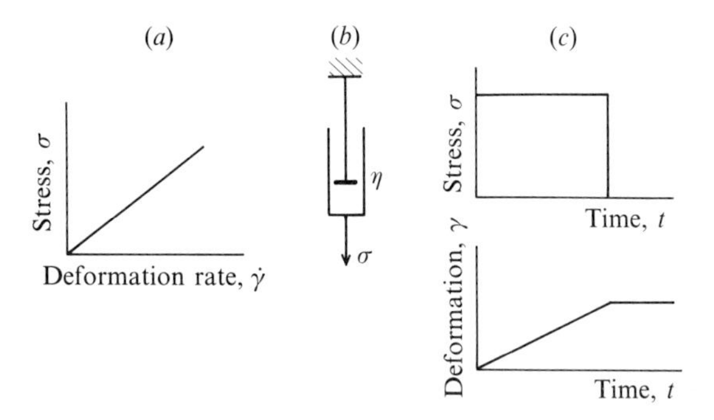

The rheological behavior of real viscous fluids can be represented by combinations of the ideal cases. *

* Although the "spring" and "dashpot" are subjected to longitudinal action, they can serve as a mathematical model for any type of deformation. In the following, equations will be developed for the case of the shearing deformation γ. The reasoning can be readily adapted for longitudinal or for isotropic deformation by using relations between the moduli.

(a) *Maxwell fluid*

By combining a spring (Hookean solid) and a dashpot (Newtonian fluid) in series, the model of a viscoelastic material called a *Maxwell fluid* is obtained (Fig. 9.5(*a*)).

If γ_1 and γ_2 are respectively the deformation of the spring and the dashpot and γ is the total deformation, the following relations hold:

$$\sigma = G\gamma_1 \qquad \sigma = \eta\dot{\gamma}_2 \qquad \gamma = \gamma_1 + \gamma_2$$

Eliminating γ_1 and γ_2, the differential equation for the system is obtained:

$$\frac{\sigma}{\eta} + \frac{\dot{\sigma}}{G} = \dot{\gamma} \tag{9.1}$$

Two tests are used to study the behavior of this model:

Fig. 9.5. Maxwell fluid. (*a*) rheological model, (*b*) flow, (*c*) relaxation.

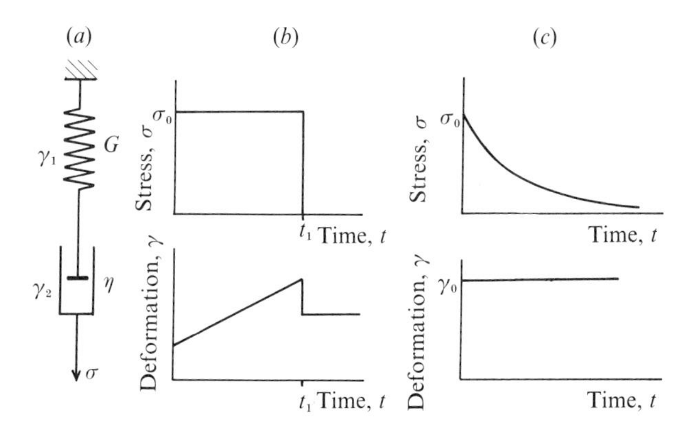

1. Flow:

The system is subjected to a constant stress and the resulting deformation is followed as a function of time. The solution of the equation for $\sigma = \sigma_0$ at $t = 0$ constitutes the *flow integral* $\gamma(t)$. It can be shown that:

$$\gamma(t) = \sigma_0 \left(\frac{1}{G} + \frac{t}{\eta} \right) \tag{9.2}$$

The deformation $\gamma(t)$ consists of an instantaneous elastic component σ_0/G due to the spring and a viscous component $\sigma_0 t/\eta$ due to the dashpot.

When the stress is removed at $t = t_1$ the elastic component instantly disappears while the viscous component remains (Fig. 9.5(*b*)).

2. Relaxation:

In contrast, if after applying a stress σ_0 to the model the level of deformation is maintained constant, the stress relaxes exponentially. The solution to equation (9.1) constitutes the *relaxation integral*:

$$\sigma(t) = \sigma_0 \exp\left(-\frac{t}{\tau}\right) \tag{9.3}$$

where the parameter $\tau = \eta/G$, called the *relaxation time*, characterizes the evolution of the stress which tends toward zero as time increases (Fig. 9.5(*c*)). For $t = 0$, $\sigma(0) = \sigma_0 = G\gamma_0$ where γ_0 is the instantaneous (elastic) deformation. At this moment, if the load is applied instantaneously, the viscous component is zero. As time passes and σ decreases, there is an irreversible transfer from the elastic component to the viscous component and only the latter remains at the end of the test.

(b) *The Kelvin–Voigt solid*

By coupling a spring and a dashpot in parallel, the Kelvin–Voigt solid is obtained (Fig. 9.6(*a*)). The viscoelastic behavior of this model can be obtained by taking the sum of the stresses σ_1 in the spring and σ_2 in the dashpot equal to the applied stress σ to produce the deformation γ.

$$\sigma_1 = G\gamma \qquad \sigma_2 = \eta\dot{\gamma} \qquad \sigma = \sigma_1 + \sigma_2$$

The elimination of σ_1 and σ_2 among the three relations leads to the equation which describes the model:

$$\sigma = G\gamma + \eta\dot{\gamma} \tag{9.4}$$

Study of the behavior of the model can be carried out using the two preceding standard tests again.

1. Flow:

Applying a constant stress σ_0 at $t = 0$, the deformation produced is obtained from the solution of equation (9.4). The flow integral is:

$$\gamma(t) = \frac{\sigma_0}{G}\left[1 - \exp\left(-\frac{t}{\tau}\right)\right] \tag{9.5}$$

where $\tau = \eta/G$ is called the *retardation time* of the model. For $t = 0$, the deformation is zero: as time increases, it tends in a uniform way toward a finite limit $\gamma_0 = \sigma_0/G$ (Fig. 9.6(*b*)).

Fig. 9.6. Kelvin–Voigt solid: (*a*) rheological model, (*b*) flow and recovery, (*c*) relaxation.

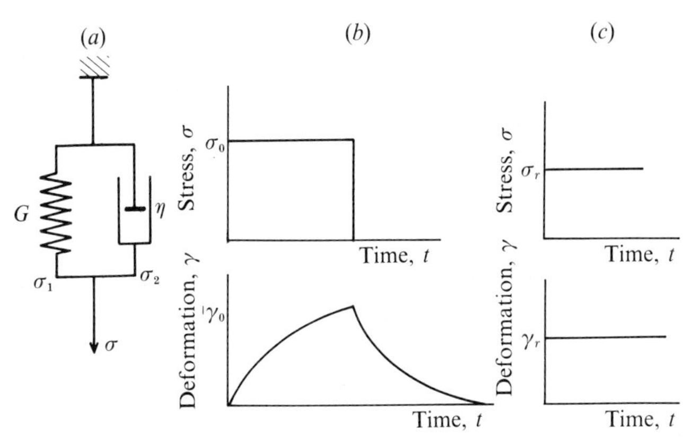

In the Kelvin model, with sudden load application followed by flow under constant load, it is the dashpot which first takes all the load and then transfers it slowly and progressively to the spring. In due course, $\dot{\gamma} \to 0$, $\sigma_2 \to 0$ and $\sigma_1 \to \sigma$.

When the stress is removed at $t = t_1$, the system progressively recovers its original configuration. The curve followed by $\gamma \to 0$ is called the *recovery* curve.

2. Relaxation:

Maintaining γ constant and equal to γ_r, the stress remains constant and equal to $\sigma_r = G\gamma_r$ (Fig. 9.6(*c*)). It is seen that there is a certain reciprocity between the Maxwell model (transfer of deformation) and the Kelvin model (transfer of stress)

9.1.5 *Viscoelastic functions: relaxation modulus and flow compliance*

The idea of the elastic modulus is generalized by introducing the *time dependent modulus*. Thus, the *relaxation modulus* $G(t)$ is defined by the relation:

$$G(t) = \frac{\sigma(t)}{\gamma_0} \tag{9.6}$$

where $\sigma(t)$ represents the response of the system to a constant deformation $\gamma = \gamma_0$ applied at time $t = 0$ and maintained for a long time (relaxation test).

Likewise, the compliance J is generalized by introducing the *flow com-*

pliance $J(t)$ defined by the relation:

$$J(t) = \frac{\gamma(t)}{\sigma_0} \tag{9.7}$$

where $\gamma(t)$ is the deformation resulting from the application of stress σ_0 at $t = 0$ and holding it constant with time (flow test).

In the general case, $J(t)$ and $G(t)$ vary as shown in Fig. 9.7. On a logarithmic time scale $J(t)$ and $G(t)$ give curves with a sigmoidal shape. For $t = 0$, $J(0) = J_u$ and $G(0) = G_u$, *instantaneous* or *non-relaxed* values. For $t \to \infty$, $J(\infty) \to J_R$ and $G(\infty) \to G_R$, *relaxed* values. The relations $G_u = 1/J_u$ and $G_R = 1/J_R$ are valid, but for all intermediate times:

$$G(t) \neq \frac{1}{J(t)}$$

In the case of a *solid*, J_R and G_R have finite values while for a *liquid* $J_R \to \infty$ and $G_R \to 0$.

Fig. 9.7. Variation of: (*a*) the flow compliance $J(t)$ and (*b*) the relaxation modulus $G(t)$ (schematic).

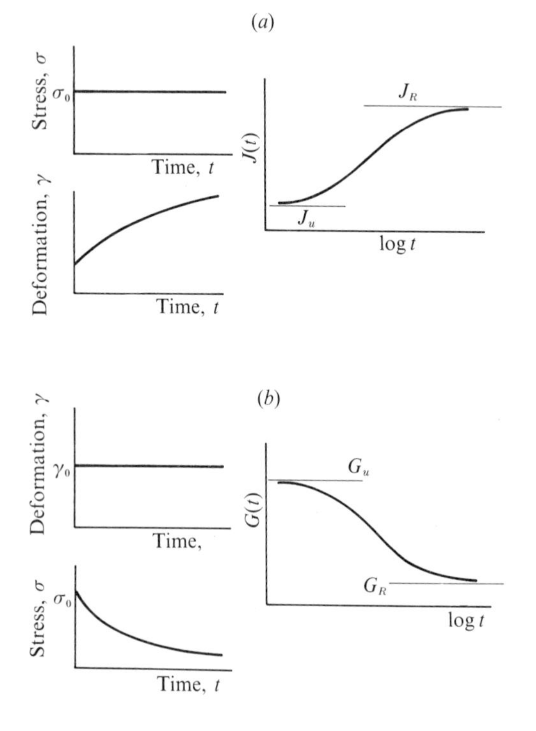

The functions $G(t)$ and $J(t)$ for Maxwell and Kelvin–Voigt models are assembled in Table 9.1.

9.1.6 *Generalized models*

The preceding simple models are not adequate to reproduce *quantitatively* the viscoelastic behavior of a real material. In order to follow more accurately the variations of $G(t)$ and $J(t)$, more complex models are used which are formed by combining several different Maxwell or Kelvin–Voigt models.

(a) *Generalized Maxwell model*

By combining in parallel m Maxwell elements with parameters (G_i, n_i) the *generalized Maxwell model* is obtained (Fig. 9.8). The additivity of the stresses in the individual branches of the model leads to a simple expression for the modulus which is equal to the sum of the moduli of the constituent elements.

$$G(t) = \sum_i G_i(t)$$

or

$$G(t) = \frac{\sigma(t)}{\gamma_0} = \sum_i G_i \exp\left(-\frac{t}{\tau_i}\right) \tag{9.8}$$

where τ_i are the *relaxation times* of the constituent elements, defined by

Fig. 9.8. Generalized Maxwell model

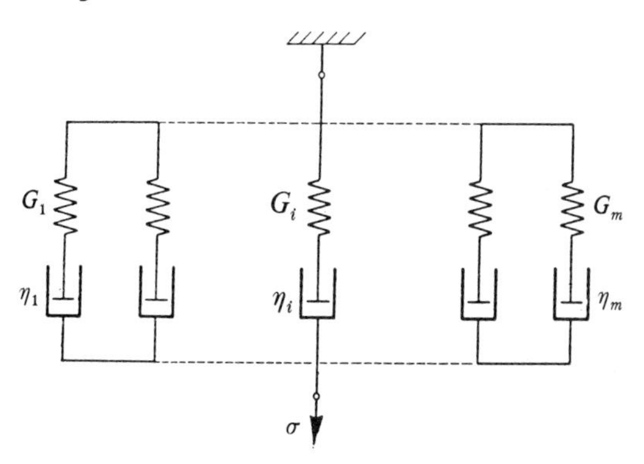

Table 9.1. *Relaxation modulus and flow compliance for simple rheological models.*

	$G(t)$	$J(t)$	G'	G''	J'	J''
Maxwell fluid	$G \exp - \left(\frac{t}{\tau}\right)$	$\frac{1}{G} + \frac{t}{\eta}$	$G \frac{\omega^2\tau^2}{1+\omega^2\tau^2}$	$G \frac{\omega\tau}{1+\omega^2\tau^2}$	$\frac{1}{G}$	$\frac{1}{G\omega\tau}$
	$(\tau = \eta/G)$					
Kelvin–Voigt solid	G	$\frac{1}{G}\left(1 - \exp - \left(\frac{1}{\tau}\right)\right)$	G	$G\omega\tau$	$\frac{1}{G\left(1+\omega^2\tau^2\right)}$	$\frac{\omega\tau}{G\left(1+\omega^2\tau^2\right)}$
		$(\tau = \eta/G)$				

$\tau_i = \eta_i/G_i$. Therefore a simple approximation for the relaxation modulus of a real material consists of taking a sufficient number of terms, i.e. by introducing a *distribution of relaxation times* τ_i.

The expression for the compliance for this model has a less simple form and the representation of $J(t)$ by the following model is preferred.

(b) *Kelvin chain*

By associating in series n Kelvin models with parameters (G_i, η_i) and a Maxwell model (to preserve the general applicability of the model to a fluid) (Fig. 9.9), a *Kelvin chain* is obtained.

$$J(t) = \sum_i J_i(t)$$

or

$$J(t) = \frac{\gamma(t)}{\sigma_0} = \frac{1}{G} + \frac{t}{\eta} + \sum_i \frac{1}{G_i}\left[1 - \exp\left(-\frac{t}{\tau_i}\right)\right] \qquad (9.9)$$

Fig. 9.9. Kelvin chain.

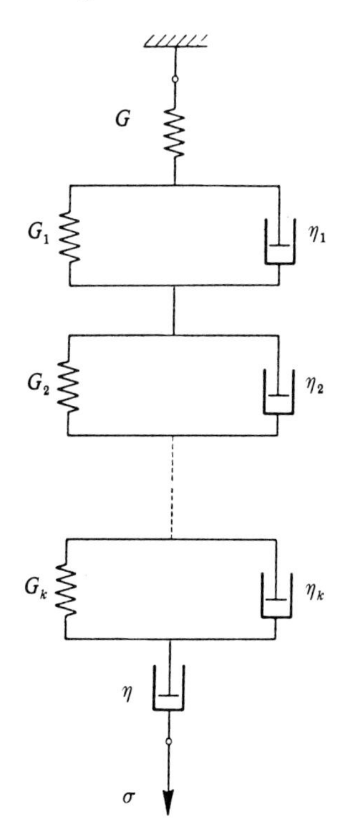

where τ_i are the *retardation times* defined by $\tau_i = \eta_i G_i$

Figure 9.10 shows that the total deformation $\gamma(t) = \sigma_0 J(t)$ is composed of three terms:

– a term (σ_0/G) corresponding to the *instantaneous elastic deformation* γ_i;

– a term $(\sigma_0\, t/\eta)$ proportional to time and corresponding to the viscoelastic deformation γ_v;

– a term composed of several exponentials and which corresponds to the *delayed or anelastic deformation* γ_d expressed by a *distribution of retardation times* τ_i.

Upon removal of the stress at time $t = t_1$, γ_i disappears immediately, γ_v remains, and γ_d decreases progressively toward zero (Fig. 9.10).

The modulus $G(t)$ would have a form analogous to that of the generalized Maxwell model, but the coefficients would have more complex analytical forms.

Either of the models can serve for the purpose of reasoning, thus the first can be used for the discussion of $G(t)$ and the second for $J(t)$.

(c) *Continuous spectra*

In the limit, one can consider models with an *infinite* number of elements and replace a *discrete distribution* of relaxation times (G_i, τ_i) by

Fig. 9.10. Generalized model. Deformation $\gamma(t)$ corresponding to stress $\sigma(t)$: γ_i, instantaneous elastic deformation; γ_v, viscous deformation: γ_d, delayed (anelastic) deformation.

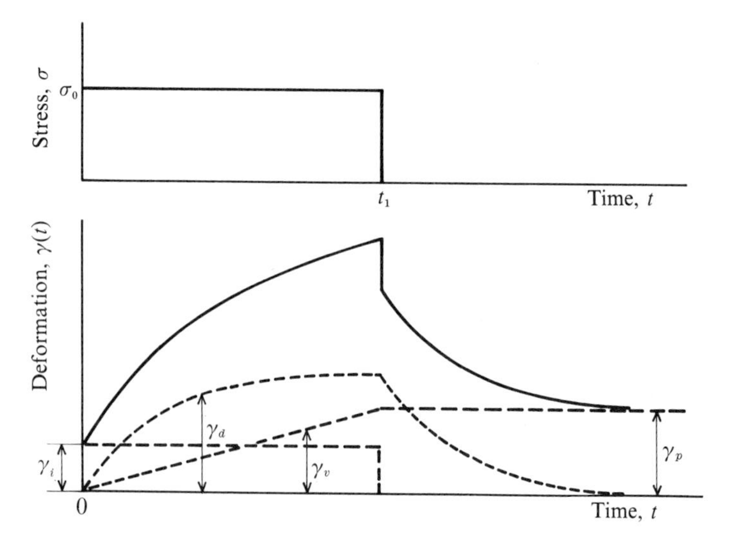

a *continuous distribution* $G(\tau)$, where $G(\tau)\,\mathrm{d}\tau$ represents the contribution to the modulus of elementary couples with relaxation times between τ and $\tau + \mathrm{d}\tau$.

Equation (9.8) is thus generalized and becomes:

$$G(t) = \int_0^\infty G(\tau) \exp\left(-\frac{t}{\tau}\right) \mathrm{d}\tau \tag{9.10}$$

where $G(\tau)$ is the *distribution of relaxation times* of the model.

Introducing a logarithmic time scale,which is easier to apply, the distribution is usually defined by the function:

$$H(\tau)\,\mathrm{d}\ln\tau \quad \text{with} \quad H(\tau) = \tau G(\tau)$$

and equation (9.10) becomes:

$$G(t) = G_R + \int_{-\infty}^{+\infty} H(\tau) \exp\left(-\frac{t}{\tau}\right) \mathrm{d}\ \ln\tau \tag{9.11}$$

where the constant term or *relaxed modulus* G_R has been separated. This represents the discrete contribution to the spectrum for $\tau \to \infty$ in the case of viscoelastic solids (for viscoelastic liquids, $G_R = 0$).

Likewise, generalizing the Kelvin–Voigt model (Equation (9.9)), a continuous distribution of *retardation times* $L(\tau)$ is introduced:

$$J(t) = J_u + \frac{t}{\eta} + \int_{-\infty}^{+\infty} L(\tau) \left[1 - \exp\left(-\frac{t}{\tau}\right)\right] \mathrm{d}\ \ln\tau \tag{9.12}$$

In this formula, the *instantaneous elastic compliance* or *non-relaxed* J_u has been separated, corresponding to the discrete contribution $\tau = 0$ as well as the flow term t/η.

The distribution functions $L(\tau)$ and $H(\tau)$ can be obtained from creep or relaxation experiments.

9.1.7 *Dynamic experiments*

(a) *Complex compliance and modulus*

Viscoelastic behavior is often studied with a stress or deformation which varies periodically with time. Applying a sinusoidal stress of frequency ω:

$$\sigma = \sigma_0 \sin\omega t$$

Fig. 9.11. Definition of the complex compliance J^* and the complex modulus G^*.

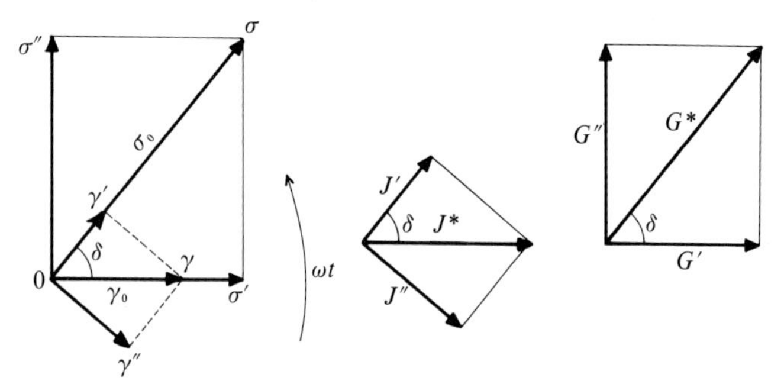

a deformation γ is observed which is lagging with a phase angle δ:

$$\gamma = \frac{\sigma_0}{G} \sin(\omega t - \delta)$$

Since $\sin(\omega t - \delta) = \sin \omega t \cos \delta - \cos \omega t \sin \delta$, the deformation consists of two components: an amplitude $(\sigma_0/G)\cos\delta$ in phase with the stress and another amplitude $(\sigma_0/G)\sin\delta$ lagging 90° out of phase (Fig. 9.11).

Using complex notation:

$$\sigma^* = \sigma_0 \exp \mathrm{i}\omega t$$

$$\gamma^* = \gamma_0 \exp[\mathrm{i}(\omega t - \delta)]$$

The *complex compliance* J^* is then defined by the relation:

$$\gamma^* = J^* \sigma^* = (J' - \mathrm{i}J'')\sigma^*$$

where J' and J'' are respectively the real and imaginary components of the compliance J^*:

$$J' = |J| \cos\delta \qquad J'' = |J| \sin\delta \quad \text{where} \quad |J| = \frac{\gamma_0}{\sigma_0}$$

Similarly the complex modulus G^* is defined by the relation:

$$\sigma^* = G^* \gamma^* = (G' + \mathrm{i}G'')\gamma^*$$

The components of G^* are:

$$G' = |G| \cos \delta \qquad G'' = |G| \sin \delta \quad \text{where} \quad |G| = \frac{\sigma_0}{\gamma_0}$$

Since:

$$G^* = \frac{1}{J^*} = \frac{(J' + \mathrm{i}J'')}{|J|^2}$$

the result is:

$$G' = \frac{J'}{|J|^2} \qquad G'' = \frac{J''}{|J|^2}$$

The ratios

$$\frac{J''}{J'} = \frac{G''}{G'} = \tan \delta$$

define the *loss angle* δ.

Substituting σ^* and γ^* into the differential equations for the different models gives J^* and G^* from the defining relations. For example, for the Maxwell fluid:

$$\frac{\sigma^*}{\eta} + \mathrm{i}\omega\frac{\sigma^*}{G} = \mathrm{i}\omega\gamma^*$$

whence

$$G^* = \frac{\sigma^*}{\gamma^*} = G\frac{\mathrm{i}\omega t}{1 + \mathrm{i}\omega t}$$

The components G' and G'' (or J' and J'') are functions of the frequency ω. Table 9.1 lists the results for the Maxwell and Kelvin–Voigt models.

Introducing the time distributions for the relaxation times $H(\tau)$ or retardation times $L(\tau)$, the following components are obtained:

$$G' = G_R + \int_{-\infty}^{+\infty} \frac{H(\tau)\omega^2\tau^2}{1 + \omega^2\tau^2} \,\mathrm{d}\ln\tau$$

$$G'' = \int_{-\infty}^{+\infty} \frac{H(\tau)\omega\tau}{1 + \omega^2\tau^2} \,\mathrm{d}\ln\tau$$

$$J' = J_u + \int_{-\infty}^{+\infty} \frac{L(\tau)}{1+\omega^2\tau^2} \,\mathrm{d}\ln\tau$$

$$J'' = \frac{1}{\omega\eta} + \int_{-\infty}^{+\infty} \frac{L(\tau)\omega\tau}{1+\omega^2\tau^2} \,\mathrm{d}\ln\tau$$

Figure 9.12 gives the form of the variations with frequency for the dynamic components.

One can also introduce the *normalized* spectra of relaxation Φ and retardation ψ defined as follows:

$$\Phi(\tau) = \frac{H(\tau)}{G_u - G_R} \qquad \psi(\tau) = \frac{L(\tau)}{J_R - J_u}$$

Fig. 9.12. Form of the variations of the components J', J'', G', G'' and $\tan\delta$ with frequency ω (schematic).

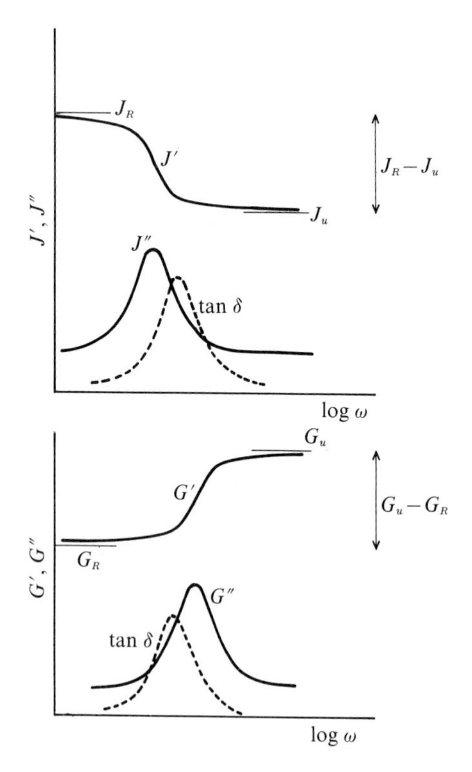

with

$$\int_{-\infty}^{+\infty} \Phi(\tau)\,\mathrm{d}\ln\tau = 1 \qquad \int_{-\infty}^{+\infty} \psi(\tau)\,\mathrm{d}\ln\tau = 1$$

In these conditions the preceding general relations can be written in a form which makes apparent the relaxed (index R) and the non-relaxed (index u) values. For example:

$$G(t) = G_R + (G_u - G_R) \int_{-\infty}^{+\infty} \Phi(\tau) \exp\left(-\frac{t}{\tau}\right) \mathrm{d}\ln\tau$$

$$J(t) = J_u + (J_R - J_u) \int_{-\infty}^{+\infty} \psi(\tau) \left[1 - \exp\left(-\frac{t}{\tau}\right)\right] \mathrm{d}\ln\tau$$

and the analogous expressions for the components G', G'', J' and J''.

This formulation which is analogous to dielectric relaxation (cf. Chapter 11) is less frequently used in viscoelasticity because the normalization factors are often inaccessible experimentally.

(b) *Dissipated energy*

The significance of J'', G'' and $\tan\delta$ is made apparent by calculating the energy W absorbed by the specimen in the course of a quarter cycle from $t = 0$ to $\pi/2\omega$.

$$W = \int_0^{T/4} \sigma \frac{\mathrm{d}\gamma}{\mathrm{d}t}\,\mathrm{d}t$$

Substituting $\sigma = \sigma \sin\omega t$ and $\gamma = \gamma_0 \sin(\omega t - \delta)$ this becomes:

$$W = \sigma_0\gamma_0 \left(\frac{\cos\delta}{2} + \pi\frac{\sin\delta}{4}\right)$$

The second term represents the mechanical energy dissipated in a quarter cycle, the first is the stored elastic strain energy W_e.

For a complete cycle, the energy dissipated is:

$$\Delta W = \pi\sigma_0\gamma_0 \sin\delta = \pi\sigma_0^2 J'' = \pi\gamma_0^2 G''$$

and the *specific loss* is:

$$\frac{\Delta W}{W_e} = 2\pi \tan \delta$$

(c) *Dynamic viscosity*

The loss effect can also be described by the ratio of the stress in phase with the deformation speed and the deformation speed. This corresponds to the real part η' of a *complex viscosity* $\eta^* = \eta' - i\eta''$, defined by $\eta^* = G^*/i\omega$. The components of η^* are thus related to those of G^* by the relations:

$$\eta' = \frac{G''}{\omega} \qquad \eta'' = \frac{G'}{\omega}$$

Whence in the general case:

$$\eta' = \int_{-\infty}^{+\infty} \frac{H(\tau)\tau}{1+\omega^2\tau^2} \mathrm{d} \ln \tau$$

For a viscoelastic liquid, the viscosity η for steady flow is obtained by setting $\omega = 0$ in the preceding equation which leads to the relation of Ferry:[347]

$$\eta = \int_{-\infty}^{+\infty} \tau H(\tau) \mathrm{d} \ln \tau$$

9.1.8 *Boltzmann superposition principle. Relations between the various viscoelastic functions*

Linear rheology is founded on Boltzmann's principle of superposition. If a stress σ_A is applied at time $t = 0$ and an additional stress σ_B at time $t = t_1$, the resulting deformation at time t is:

$$\gamma(t) = \sigma_A J(t) + \sigma_B J(t - t_1)$$

For a series of increases, $\Delta\sigma_i$ at times u_i:

$$\gamma(t) = \sum_{u_i=-\infty}^{t} \Delta\sigma_i J(t - u_i)$$

and, in the limit:

$$\gamma(t) = \int_{-\infty}^{t} \left(\frac{\mathrm{d}\sigma}{\mathrm{d}u}\right) J(t-u)\,\mathrm{d}u$$

The final deformation can be calculated knowing $\sigma(t)$ and the function $J(t)$.

Using the same approach for the modulus $G(t)$, the resulting stress $\sigma(t)$ of a series of deformations $\Delta\gamma_i$, applied at times u_i gives at the limit:

$$\sigma(t) = \int_{-\infty}^{t} \left(\frac{\mathrm{d}\gamma}{\mathrm{d}u}\right) G(t-u)\,\mathrm{d}u$$

which permits the calculation of the stress knowing $\gamma(t)$ and $G(t)$. As a consequence of the Boltzmann superposition principle, the functions $G(t)$ and $J(t)$ are not independent; it can be shown[347] that they are related by the equations of convolution:

$$\int_{0}^{t} G(u)\,J(t-u)\,\mathrm{d}u = t$$

$$\int_{0}^{t} J(u)\,G(t-u)\,\mathrm{d}u = t$$

which allows the definition of one function when the other is known. Likewise, relations exist between transient and dynamic functions, allowing the correlation between both stress types to be determined. The components of the complex dynamic modulus can be obtained from the relaxation modulus by Fourier transforms:

$$G'(\omega) = G_R + \omega \int_{0}^{\infty} [G(t) - G_R]\sin\omega t\,\mathrm{d}t$$

$$G''(\omega) = \omega \int_{0}^{\infty} [G(t) - G_R]\cos\omega t\,\mathrm{d}t$$

Likewise, $G(t)$ can be obtained from G' or G'' by the inverse transformations:

$$G(t) = G_R + \frac{2}{\pi} \int_0^\infty \left[(G' - G_R)/\omega\right] \sin \omega t \, d\omega$$

$$G((t) = G_R + \frac{2}{\pi} \int_0^\infty (G''/\omega) \cos \omega t \, d\omega$$

Analogous relations exist between $J'(\omega)$ and $J''(\omega)$ and $J(t)$. The work of Ferry[347] which gives numerous practical calculation methods may be consulted for further details.

9.1.9 *Practical determination of viscoelastic moduli*

The complete determination of a viscoelastic modulus, although theoretically possible, suffers from experimental difficulties. For example, in the transient experiments only a small part of the function $G(t)$ can be explored in a physically practical time interval of several seconds to several hours. Similarly, $G^*(\omega)$ can only be determined in practice in a restricted frequency interval. To cover a more extended interval, several different measurement techniques are often used together (cf Section 9.3.1).

All the moduli are temperature dependent, and it is possible to use experimental measurements made at different temperatures to reconstruct a more complete $G(t)$ or $G^*(\omega)$ function. For this purpose, the principle of time–temperature equivalence is used as will be shown in an example.

(a) *Principle of time–temperature equivalence – reduced curves*

At a temperature T_0, $G(t)$ is governed by a distribution $\Phi_{T_0}(\ln \tau)$ of relaxation times. Assume that all the relaxation times are shifted along the $\ln \tau$ axis by a quantity $\ln a_T$ while the temperature passes from T_0 to T (Fig. 9.13(*a*)). The old distribution $\Phi_{T_0}(\ln \tau)$ is related to the new distribution $\Phi_T(\ln \tau)$ by the relation:

$$\Phi_T(\ln \tau) = \Phi_{T_0}\left(\ln \frac{\tau}{a_T}\right)$$

giving e.g. for $G(t)$:

$$G(T, t) = G\left(T_0, \frac{t}{a_T}\right)$$

Fig. 9.13. Principle of time-temperature equivalence: (*a*) shift of relaxation times with temperature; (*b*) determination of $G(t)$ from partial curves.

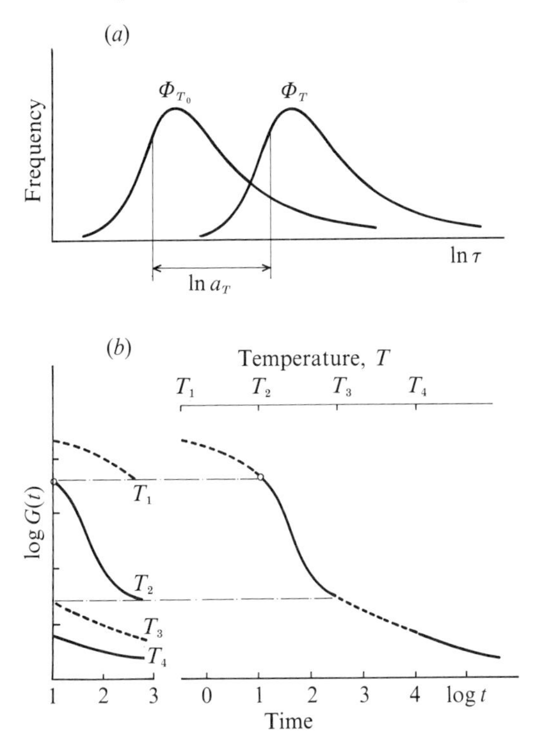

The effect of changing the temperature is the same as multiplying the time scale t by a_T or adding a factor to $\ln t$.

Consider as an illustration the determination of $G(t)$ (Fig. 9.13(*b*)). At temperature T_1, for a time interval limited to about 1000 s, the value of $G(t)$ is divided by about 10. To determine $G(t)$ at longer times, two methods can be used; either the observation time is increased, which rapidly becomes prohibitive, or the temperature is increased while keeping the time interval constant at 1000 s, and the time–temperature equivalence principle is applied. The left of Figure 9.13(*b*) shows determinations made at increasing temperature T_2, T_3, T_4. It is apparent that a horizontal displacement of the curve relative to T_2 will bring into coincidence the portions relative to T_1 and T_2 in the overlap region. Operating in the same way by translation of the curves T_3 and T_4, the *reduced curve* (master curve) is progressively constructed relative to a reference temperature (T_2 in Fig. 9.13(*b*)).

(b) *Williams–Landel–Ferry equation*

Experimentally, it is found that in most cases the displacement factor $\ln a_T$ is a simple function of T; a_T follows the Arrhenius relation:

$$\ln a_T = \frac{\Delta H}{R}\left(\frac{1}{T} - \frac{1}{T_0}\right)$$

where T_0 is the reference temperature ($a_T = 1$) and ΔH is the activation enthalpy for relaxation. a_T also follows an empirical expression, called the Williams–Landel–Ferry (WLF) equation:

$$\ln a_T = -\frac{C_1(T - T_0)}{C_2 + (T - T_0)}$$

where C_1 and C_2 are "universal" constants ($C_1 = 17.4$; $C_2 = 51.6$), T_0 being a reference temperature often taken equal to T_g determined by dilatometry. In reality, the "constants" C_1 and C_2 vary according to the glass considered. In the construction of the reduced curves it is equally necessary to take into account the variations of G related to variations in the density with temperature which lead to small vertical readjustments of the partial curves before the shift $\ln a_T$. For details of the different techniques consult Refs. 347 and 348.

It can be shown (see e.g. Ref. 347) that a_T represents the ratio of the viscosity η at temperature T to that relative to a reference temperature T_0:

$$a_T = \frac{\eta(T)}{\eta(T_0)}$$

From the Doolittle equation for viscosity (cf. Chapter 2):

$$\ln \eta = \ln A + B\left(\frac{V - V_f}{V_f}\right)$$

where A and B are constants, V is the total volume of the system and V_f the free volume. Assuming that $f = V_f/V$ varies linearly above T_g:

$$f = f_g + \alpha_f(T - T_g)$$

α_f being the thermal expansion coefficient; it is easily shown that:

$$a_T = \frac{-B(T - T_g)}{f_g[f_g/\alpha_f + (T - T_g)]}$$

which has the same form as the WLF equation, the reference temperature being T_g. It can also be shown that the WLF equation follows from that of Adams–Gibbs (cf. Chapter 2).

9.2 The viscosity of glasses

9.2.1 *Temperature dependence – microscopic model*

For a given composition, the variation of η with temperature is very large. For a classical soda–lime–silica (Na_2O–CaO–SiO_2) glass the ratio of viscosities between ordinary temperature and high temperature is about 10^{18} (Fig. 9.14).

Fig. 9.14. Variation of the viscosity of an industrial soda-lime-silica glass as a function of temperature. The fixed points and the different regions for technical operations are indicated.

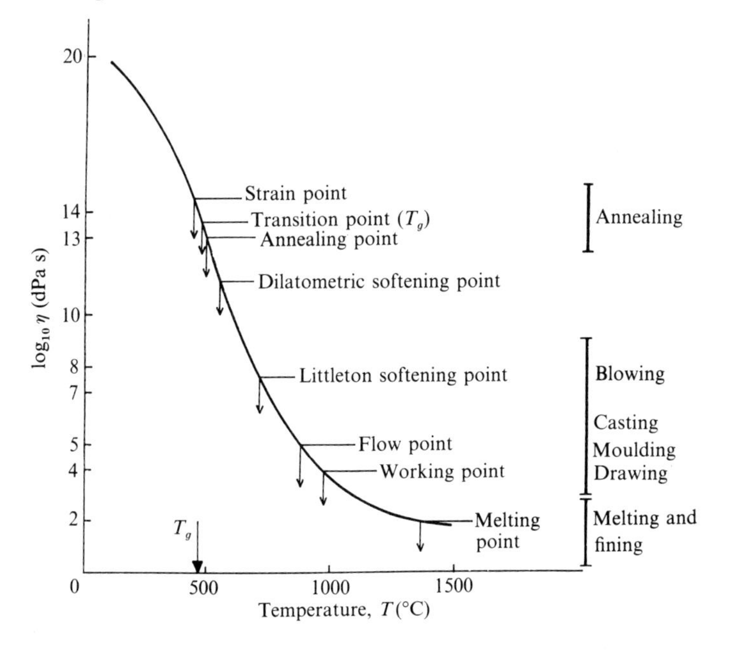

Viscous flow can be considered as an activated process dominated by a transition state of higher energy.[348] Consider (Fig. 9.15) two molecular layers separated by a distance λ_1 and assume that one slides relative to the other under the action of an applied force per unit area τ_t. If Δu is the difference in velocity of the two layers, by definition:

$$\eta = \sigma_t \frac{\lambda_1}{\Delta u}$$

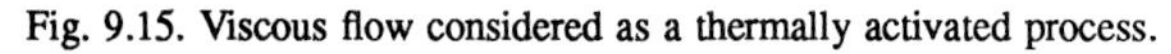

Fig. 9.15. Viscous flow considered as a thermally activated process.

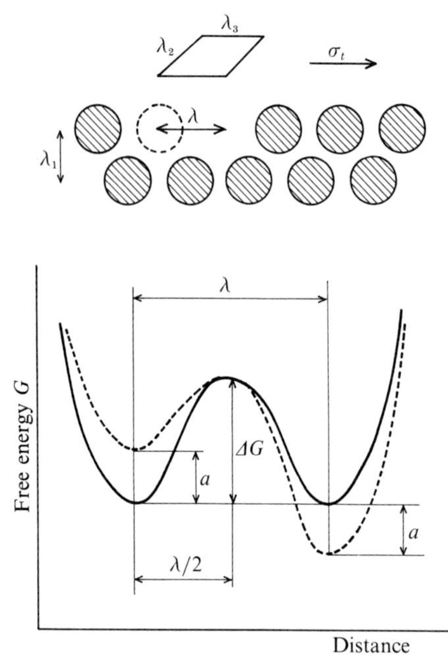

Assume that in the course of movement the molecules in the same layer pass from one equilibrium position to another separated by a distance $\lambda \sim \lambda_3$ between molecules in the direction of movement. The distance between the adjacent molecules perpendicular to the movement being λ_2, the applied force on one molecule is $\sigma_t \lambda_2 \lambda_3$ and the work done for a translation $\lambda/2$ is:

$$a = \frac{1}{2}\sigma_t \lambda_2 \lambda_3 \lambda$$

The potential barrier ΔG will be lowered by a in the direction of movement and raised by the same amount in the opposite direction when the stress σ is applied.

The classical analysis of an activated process consists of considering the jump frequencies in the direction of movement:

$$\omega_1 = \omega \exp(a/kT)$$

and in the opposite direction:

$$\omega_2 = \omega \exp(-a/kT)$$

where ω is the jump frequency of the thermally activated process:

$$\omega = \nu \exp(-\Delta G/kT)$$

The net rate of flow Δu is then:

$$\Delta u = \lambda(\omega_1 - \omega_2) = 2\lambda\omega \sinh\left(a/kT\right)$$

Taking $\lambda \sim \lambda_2$ and $\lambda_1\lambda_2\lambda_3$ approximately equal to the volume V_0 occupied by a molecule (the flow volume), then:

$$\eta = \frac{\sigma_t \exp(\Delta G/kT)}{2\nu \sinh(\sigma_t V_0/2kT)}$$

For low stresses, $\sigma_t V_0 \ll 2kT$ and the preceding expression becomes:

$$\eta \simeq \frac{kT}{\nu V_0} \exp\left(\frac{\Delta G}{kT}\right) = \eta_0 \exp\left(\frac{\Delta G}{kT}\right)$$

The theory thus leads to an Arrhenius dependence for η:

$$\log \eta = A' + \frac{B'}{T}$$

where A' and B' are constants. It is found experimentally that glass viscosity coefficients only follow this type of relation over limited temperature intervals. However, generally the graphs $\log \eta$ vs. $1/T$ are curved (Fig. 9.16) and a relation with three parameters, A, B, T_0, called the Vogel–Fulcher–Tamman equation is preferred:

$$\log \eta = A + \frac{B}{T - T_0}$$

It is found that an equation of this type represents remarkably well the viscosity of glasses over an extended temperature range. The three constants of this empirical equation are calculated in practice from three viscosity measurements e.g. at three fixed points (cf. Sections 9.9.2–9.2.3).

The theories of "free volume" or of the "cooperative rearrangements" discussed in Chapter 2 also lead to equations of the preceding form. However, all the current mathematical models apply rather poorly to associated liquids or covalently bonded networks as is the case for the usual oxide glasses. It is thus necessary to proceed to experimental determinations to evaluate the variations of η with temperature and composition.

Fig. 9.16. Comparison of the viscosity of different glasses: (*a*) SiO_2, (*b*) soda–lime–silica, (*c*) B_2O_3, (*d*) As_2S_3, (*e*) KNO_3–$Ca(NO_3)_2$, (*f*) Se, (*g*) glycerol.

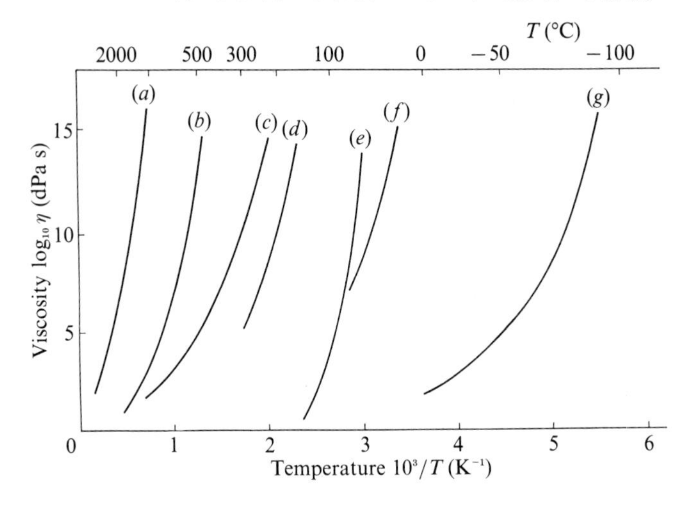

9.2.2 *Fixed points*

Since, in practice, certain intervals of viscosity have particular importance, a series of *standard viscosity values* has been adopted (Fig. 9.14). A given glass is then characterized from the point of view of its viscous behavior by the *temperatures* which correspond to these standard values and which are called *fixed points*. Table 9.2 lists the usual standard values of the fixed points. To retain the traditional numerical values of viscosity at the fixed points which were defined using the *poise* unit (CGS), the viscosities are given in dPa s. The higher viscosity values cannot usually be defined with precision. This is due to the experimental measurement techniques which will be discussed later.

The viscosity of $\sim 10^2$ dPa s corresponds to that of a liquid glass undergoing melting and fining. Of particular importance from a technical point of view is the interval 10^4–10^8 dPa s which controls the glass-forming operations. According to the extent of the temperature interval corresponding to these viscosities, the term "long" glass or "short" glass is used. "Long" glasses are better suited to manual blowing processes where precise control of temperature is difficult, while "short" glasses are amenable to mechanized processes which require a shorter working time.

The transition temperature T_g constitutes a particularly important fixed point from a theoretical view point, corresponding to a viscosity near 10^{13} dPa s; its significance has been considered in Chapter 2.

Table 9.2. *Definitions of fixed points.*

Designation	η (dPa s)
melting point	10^2
working point	10^4
sink point	$10^{4.22}$
flow point	10^5
softening point (Littleton's point)	$10^{7.8}$ (or 4.2×10^7)
dilatometric point	$\sim 10^{11.3}$
annealing point	$\sim 10^{13}$
transition point	10^{13}–$10^{13.6}$
strain point	$10^{14.5}$ (or 3.2×10^{14})

The *annealing point* and *strain point* temperatures are related to the *relaxation time* of internal stresses. At the annealing point the internal stresses are relaxed in about 15 min while at the strain point this stress relief requires about 2–3 h. The anneal point is the temperature which corresponds to the beginning of slow cooling in the course of the annealing operation whereas the strain point is the temperature below which the rate of cooling can be increased without the risk of introducing appreciable residual thermal stresses. The viscosity of a solid glass at ambient temperature is about 10^{19}–10^{20} dPa s.

9.2.3 *Measurement methods*

Given the large viscosity intervals to be covered, appropriate methods must be used for each specific domain. The following methods (Fig. 9.17) can be distinguished, in the order of increasing viscosities.

(a) *Falling ball viscometer*

For viscosities between 10–10^7 dPa s, methods based on Stokes' law are used. The Stokes equation $f = 6\pi r\eta V$, gives the drag force f which is exerted on a sphere of radius r moving with a velocity V in a fluid of viscosity η. This constitutes the principle of the falling ball viscometer (Fig. 9.17(*a*)).

If ϱ_S and ϱ are the respective densities of the sphere and the fluid at the temperature of the experiment, then:

$$\eta = \frac{2}{9}\frac{(\varrho_S - \varrho)}{V}r^2 g$$

Fig. 9.17. Methods of viscosity measurement (schematic): (*a*) falling ball viscometer, (*b*) counterweighted ball viscometer, (*c*) rotation viscometer, (*d*) sink point measurement, (*e*) free fiber elongation, (*f*) loaded fiber elongation, (*g*) creep deformation of a solid, (*h*) penetrometer, (*i*) torsion of a tube.

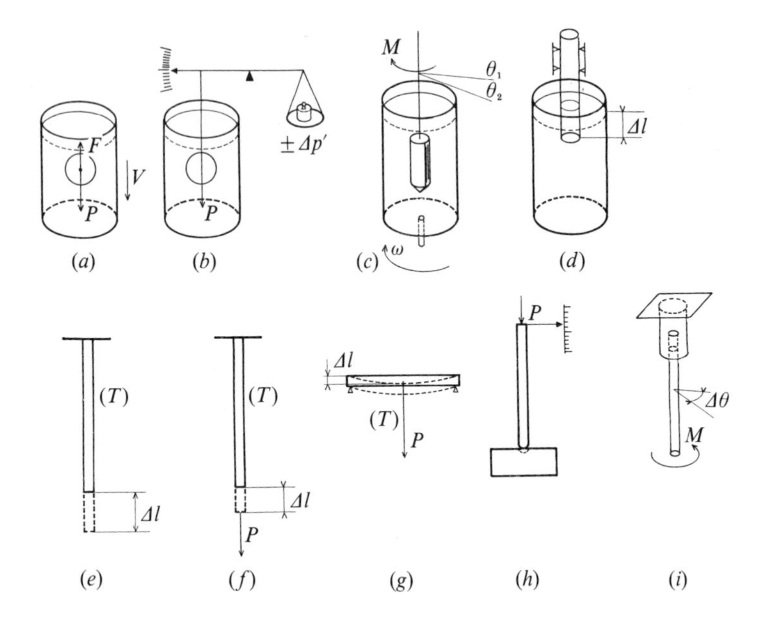

where g is the gravity constant. To slow the movement or even reverse its direction, the weight of the sphere P can be partially equilibrated by a variable counterweight $p' \pm \Delta p'$ (Fig. 9.17(*b*)).

From the damping coefficient of the oscillations, viscosities of less than 10 dPa s can be measured.

(b) *Rotation viscometer*

In this method one measures the couple exerted on a cylinder immersed in the fluid contained in a coaxial rotating crucible (or alternatively, the couple exerted on the crucible by a rotating cylinder) (Fig. 9.17(*c*)).

The viscosity is calculated from the relation:

$$\eta = C\frac{M}{\omega}$$

where M is the measured torsion couple, ω the angular velocity and C a constant depending on the geometry of the apparatus. For a cylinder of radius r immersed to a depth l in a coaxial container of radius R:

$$C = \frac{1}{4\pi l}\left(\frac{1}{r^2} - \frac{1}{R^2}\right)$$

The method is applicable between 10 and 10^4 dPa s. For higher viscosities (up to 10^7 dPa s) the aperiodic return method can be used. The time Δt is measured for a cylinder displaced from its equilibrium position ($\theta = 0$) to pass from θ_1 to θ_2. In these conditions, for $\theta_1 > \theta_2$:

$$\eta = C' \, \Delta t \Big/ \ln\left(\frac{\theta_1}{\theta_2}\right)$$

where C' is a constant for the apparatus obtained by calibration.

(c) *"Sink point" measurement*

The time necessary for a solid to sink to a given depth in a viscous medium can be used to measure the viscosity. To measure the *sink point* defined as $\eta = 10^{4.22}$ dPa s, a cylindrical metal bar of radius r and mass m is used (Fig. 9.17(*d*)). The time t required for the bar to sink a distance l is related to the viscosity η by:

$$t = a\pi r l^2 \eta / mg$$

where a is a constant. The temperature to obtain a standard penetration is found by trial and error.

In the case of an ordinary glass, for a bar 20 cm long and 0.5 cm in diameter made from 80 Pt–20 Rh alloy, $a = 16.94\,\text{cm}^{-1}$. A penetration of $l = 2\,\text{cm}$ in $t = 2\,\text{min}$ corresponds to a viscosity $\eta = 10^{4.22}$ dPa s which is defined as the "sink point."[349]

(d) *Fiber elongation viscometers*

To evaluate higher viscosities from 10^7 to 10^{15} dPa s, the rate of elongation of a solid fiber under a tensile stress is measured (Fig. 9.17(*e*)). If F is the force acting on a fiber of diameter d and length l, the viscosity η is obtained from the relation (cf. Section 9.1.3(b)):

$$\frac{4F}{\pi d^2} = 3\eta \frac{1}{l}\frac{\mathrm{d}l}{\mathrm{d}t}$$

This type of measurement is used for the determination of three important fixed points: the softening point, the annealing point and the strain point.

1. Measurement of the softening point temperature or the "Littleton point" ($\eta = 10^{7.6}$ dPa s).

This method uses the extension of a fiber under its own weight. The standard method[350] determines the temperature at which a fiber 0.55–0.77 mm in diameter and 23.5 cm long extends under its own weight at a rate of 1 mm min^{-1} when the upper 10 cm are heated in a standard furnace at a rate of approximately 5 °C min^{-1}.

In practice this method is used to measure differences in viscosity for a particular type of glass composition but it does not give the same viscosity level for all glasses. This is due to the fact that the applied force F equal to the weight of the fiber is reduced by the surface tension force γ which tends to shorten the fiber. The results depend on γ which varies according to the type of glass. In practice the Littleton temperature can be measured to ±2 °C but the corresponding viscosity values can vary from $10^{7.5}$ to $10^{8.1}$ dPa s depending on the composition; the value $10^{7.6}$ dPa s corresponds to ordinary soda–lime–silica glasses.

2. Measurement of the annealing point and the strain point.

By loading the fiber with a weight at its lower end, higher viscosities in the range 10^{8}–10^{15} dPa s can be measured (Fig. 9.17(*f*)). By following the variation of the elongation rate $\mathrm{d}l/\mathrm{d}t$ with temperature, the *annealing point* or *annealing temperature* defined by $\eta = 10^{13}$ dPa s can be determined. As shown in Fig. 9.18, by extrapolation to a rate which is less by a factor of 0.0316, (1.5 powers of ten lower on a logarithmic scale), the *strain point* $\eta = 10^{14.5}$ dPa s is obtained.

(e) *Various methods based on deformation of solids*

1. The deformation of a rod supported at its ends and loaded in the center (Fig. 9.17(*g*)) can be used to estimate viscosities from 10^{8} to 10^{15} dPa s.

2. The *dilatometer* permits the determination of not only T_g (cf. Chapter 2), but also of the *dilatometric softening* temperature which corresponds to the final hook in the expansion curve (Fig. 2.12). This point is reached when the specimen begins to flow (collapse) under the mechanical action of the dilatometer spring and is the temperature corresponding to a viscosity $\sim 10^{11.3}$ dPa s. Obviously this may vary with the type of apparatus (elastic constant of the spring) and has only a comparative value.

3. The *penetration viscometer*, effective between 10^{9} and 10^{13} dPa s, measures the penetration time of a weighted hemispherically tipped rod (Fig. 9.17(*h*)). The viscosity is calculated from the formula:

$$\eta = \frac{KPt}{\sqrt{Rl^3}}$$

Fig. 9.18. Principle of the determination of the strain point from the annealing point.

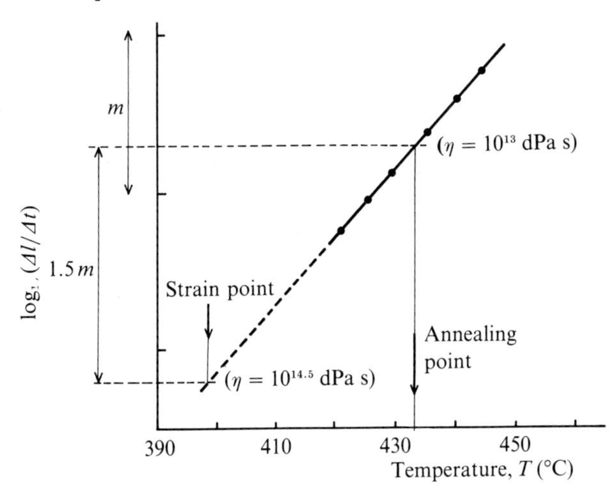

where t is the time for penetration l of the sphere of radius R under load P; K is an experimentally determined constant.

4. The torsion of a tube-shaped specimen allows the measurement of viscosities between 10^{13} and 10^{20} dPa s. From the applied moment M and the angular torsional velocity ω, the viscosity η is obtained by the formula:

$$\eta = \frac{2LM}{\pi\omega(R^4 - r^4)}$$

where R and r are respectively the exterior and interior radii and L is the length of the tube.

The different measuring apparatus can be standardized with glasses of known viscosity and *viscosity standards* are sold by the the National Institute of Standards and Technology, Washington DC for this purpose.

The establishment of a complete viscosity curve is very laborious. Often in practice only three fixed points are measured to determine the three coefficients of the Vogel–Fulcher–Tammann equation which allows extrapolation over an extended temperature interval.

9.2.4 *Composition dependence*

The dependence of glass viscosity on composition has only been studied empirically. In the case of vitreous SiO_2, the addition of modifying oxides (fluxes) strongly reduces the viscosity, and this has practical use in glassworking. For oxide glasses, the addition of Al_2O_3 increases the viscosity at all temperatures while CaO and B_2O_3 lower the viscosity at high temperatures and raise it at low temperatures. Figure 9.19 shows as an example the variations of viscosity in the SiO_2–R_2O systems at different temperatures.[351]

Fig. 9.19. Effect of modifying oxides R_2O on the viscosity of SiO_2. After Ref. 351.

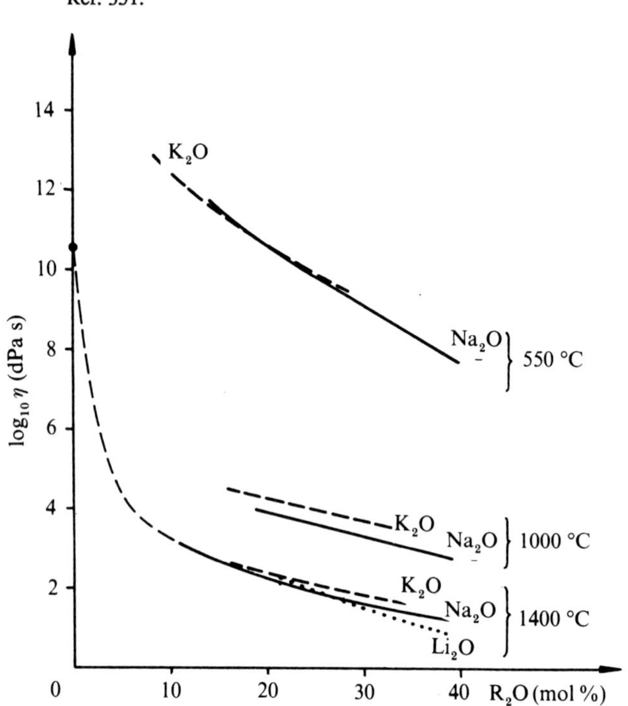

The introduction of OH^- or F^- ions lowers the viscosity of silicate glasses. These ions (mineralizers) play an important rôle in geological processes of magma crystallization.

Refs. 3 and 8 contain compilations of results as well as empirical interpolation formulae. Figure 9.20 gives viscosity curves for several industrial glasses.

9.3 Elastic and viscoelastic properties of glasses

Solid glass is a perfect Hookean solid; its deformation is proportional to the stress up to fracture which occurs suddenly without plastic deformation. This is called *brittle fracture* and will be the object of detailed study in Chapter 14.

Fig. 9.20. Variation of viscosity with temperature for several industrial glasses: (*a*) lead silicate, (*b*) soda-lime-silica; (*c*) borosilicate, (*d*) alumino-silicate, (*e*) SiO_2.

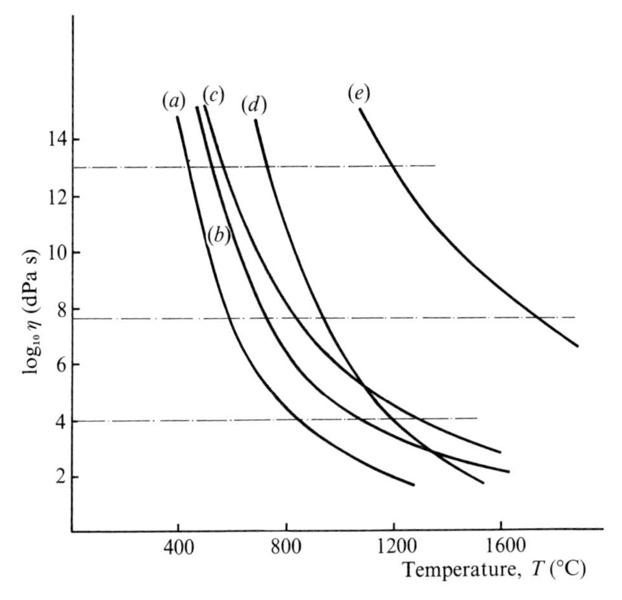

9.3.1 *Measurement of the elastic moduli and mechanical losses*

Young's modulus of glass is seldom obtained by direct tensile tests. Usually *dynamic* measurements of longitudinal or transverse resonant vibrations of prismatic specimens are used.

(a) *Forced oscillations*

In resonant techniques using the Cabarat elastometer,[352] a sample in the form of a bar is excited electrostatically to a resonance frequency ν in fundamental transverse flexure mode. The specimen is supported in vacuum by fine wires placed near the nodal points (Fig. 9.21(*a*)). If L is the length and b the width of the bar of density ϱ:

$$\nu = \frac{3.56}{\sqrt{12}}\sqrt{\frac{E}{\varrho}}\frac{b}{L^2}$$

Measurement of ν allows the determination of Young's modulus, E. Likewise the shearing modulus G can be obtained by measuring the vibrations of a specimen excited in torsion.

These methods also provide an evaluation of the mechanical loss for the oscillation frequency by measurement of "Q^{-1}" of the system. For a

Fig. 9.21. Measurement of the elastic modulus and internal friction: (*a*) resonant vibration of a specimen, (*b*) torsion pendulum, (*c*) ultrasonic pulse-echo method.

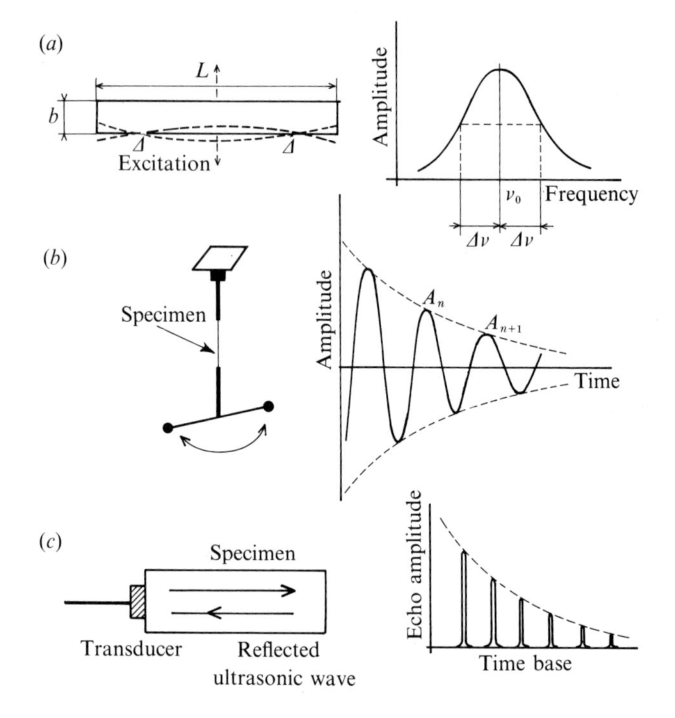

body placed into forced oscillation near resonance by a force of constant amplitude and variable frequency it can be shown that:

$$\frac{\Delta W_e}{W_e} = 2\pi \tan \delta_e = \frac{2\pi}{\sqrt{3}} \frac{\Delta \nu}{\nu_0} = 2\pi Q^{-1}$$

where ν_0 is the resonant frequency and $2\Delta\nu$ the full width at half maximum i.e. the difference between frequencies for which the amplitude is equal to half the resonant amplitude.

(b) *Damped oscillations*

The mechanical loss can be evaluated from the *logarithmic decrement* Λ of *damped* oscillations. Λ is defined as the ratio of the successive amplitudes of a body placed in oscillation and then released:

$$\Lambda = \ln\left(\frac{A_n}{A_{n+1}}\right)$$

where A_n and A_{n+1} are two successive amplitudes.

When $A_n/A_{n+1} \sim 1$:

$$\Lambda \sim \frac{1}{2}\frac{A_n^2 - A_{n+1}^2}{A_n^2} = \frac{1}{2}\frac{\Delta W}{W_e}$$

the square of the amplitude being proportional to the stored energy. Hence:

$$\Lambda = \pi \tan \delta_e$$

The classic *torsion pendulum* method (Fig. 9.21(*b*)) consists of the observation of a specimen in the form of a fiber of diameter d and length l carrying a movable load with moment of inertia I. The oscillations are recorded optically using the classical rotating mirror method of Poggendorf. Measurement of the natural frequency of oscillation ν permits the measurement of a quantity proportional to $\sqrt{G}$:

$$\nu = \frac{1}{2\pi}\sqrt{\frac{\pi G d^4}{32\, lI}}$$

and the measurement of Λ gives $\tan \delta_e$

(c) *Ultrasonic wave propagation*

The propagation of longitudinal waves in a medium in which the lateral dimensions are large relative to the wavelength allows the calculation of the uniaxial compression modulus, M^*:

$$M^* = M' + \mathrm{i}M''$$

where M' and M'' are respectively the real and imaginary parts of the complex modulus M^*. The parameters M' and M'' are related to the attenuation and velocity by the relations:

$$M' = \varrho v_L^2 \frac{1 - r^2}{(1 + r^2)^2}$$

$$M'' = 2\varrho v_L^2 \frac{r}{(1 + r^2)^2}$$

where v_L is the propagation velocity of longitudinal waves expressed in centimeters per second, ϱ the density of the material at the temperature considered and r is equal to $\alpha\lambda/2\pi$ where λ is the wavelength and α the attenuation.

Likewise the shear wave permits the measurement of the complex modulus:

$$G^* = G' + G''$$

The relations are identical to those cited above, α_r and v_r now being the attenuation and the velocity of the transverse waves. The bulk compressive modulus K^* is related to the two moduli defined above by the relation:

$$K^* = M^* - \frac{4}{3}G^*$$

In the pulse-echo method, a specimen is coupled to an ultrasonic generator and measurements are made of the attenuation of propagating pulses which are subjected to repetitive reflections from the free face (Fig. 9.21(*c*)).

9.3.2 *Results*

(a) *Elastic moduli*

Table 9.3 shows values for the moduli E and G for a number of glasses.

The elastic modulus is higher for glass with a more rigid structure. In the case of oxide glasses, it is reduced in proportion to the modifying oxides which weaken (depolymerize) the network. Research on high-modulus glasses, which are needed for the fabrication of plastic composite reinforcement, has led to compositions containing oxide ions with strong ionic forces (BeO, MgO, La_2O_3, etc.). Even with this approach, the moduli obtained do not exceed 15×10^{10} N m^{-2} and cannot match other materials as indicated in Table 9.3. Present attempts are centered rather towards the incorporation of *nitrogen* into silicate and silica-aluminate glasses. The N atom can bond with three Si atoms (instead of two for O) which produces a cross-linking of the network and increases Young's modulus and the hardness of the glass.[353]

Poisson's ratio μ is between 0.2 and 0.3 for glasses; the values of E and G are of the same order of magnitude. The compressibility β being very small, the modulus K is very large; for a theoretically incompressible liquid, $\mu = 0.5$ and K is infinite.

(b) *Viscoelastic functions in the transformation interval*

In the case of glasses, the viscoelastic behavior is most important in the transformation interval. Since, in this region, the glass is always susceptible to structural rearrangement through the phenomenon of stabilization, it is necessary to separate viscoelastic phenomena by using *stabilized* glasses, otherwise it is necessary to take into account the

Table 9.3. *Elastic moduli of glasses compared to those for other materials.*

Material	Young's modulus $E\,(\mathrm{N\,m^{-2}})$	Shear modulus $G\,(\mathrm{N\,m^{-2}})$	Poisson's ratio μ
Vitreous SiO_2	7.2×10^{10}	3.15×10^{10}	0.17
Soda-lime-silica glass ($75\,SiO_2, 17\,Na_2O, 9\,CaO$)	7.13×10^{10}	2.93×10^{10}	0.218
High modulus glass ($40.7\,SiO_2, 7.2\,Al_2O_3$, $26.7\,MgO, 25.4\,BeO$)	13.68×10^{10}		0.267
Pyrex	6.85×10^{10}		
Sintered Al_2O_3	36×10^{10}		
Dense SiC	46.5×10^{10}		
Cu	13.2×10^{10}	4.9×10^{10}	0.343
Pb	1.6×10^{10}	0.59×10^{10}	0.434
Fe	21.5×10^{10}	8.39×10^{10}	0.291

variation of viscosity with time (cf. Chapter 2). Studies of viscoelastic behavior in inorganic glasses are relatively few in number in contrast to the case of organic polymers where they constitute a current investigative method.[347]

If a stress is applied instantaneously at $t = 0$ and maintained constant for the duration of the test, a deformation curve for the glass (creep curve) is obtained with a behavior as shown in Fig. 9.10 in which there is:

– an instantaneous deformation ($0a$);

– a deformation with decreasing rate (ab);

– a viscous deformation at a nearly constant rate (bc).

At the elimination of the stress at $t = t_1$, there is:

– an instantaneous deformation (cd) = ($a0$);

– a deformation (de) at a decreasing rate which progressively approaches zero.

After the test, a permanent deformation γ_p remains.

Assuming that the viscous deformation occurs at a constant rate, this contribution γ_v as well as γ_i can be separated from the total curve (Fig. 9.10), with the remaining difference being the delayed elastic deformation γ_d which can be inverted and superimposed on the recovery curve observed after unloading. This is a result of the Boltzmann superposition principle; the unloading of the specimen is the same as the application of a supplementary stress $-\sigma_0$ while the initial stress $+\sigma_0$ remains. At the point $t = t_1$, the elastic and viscous components due to $+\sigma_0$ and $-\sigma_0$ cancel. Only the delayed deformation, which occurs in the opposite sense to the deformation caused by $+\sigma_0$, remains.

Creep deformation thus consists of the three components of the rheological model: an instantaneous elastic component, a reversible delayed component and an irreversible viscous component.

In their classic work, De Bast and Gilard[354] studied the viscoelastic behavior of a stabilized soda-lime-silica glass (typical window glass). The specimens were loaded in compression in a testing machine equipped with a specially designed extensometer. The delayed elastic component ϵ_d can be represented by a generalized Kelvin–Voigt model.

For reasons of experimental precision, the authors usually preferred, however, to use in the place of an equation of the type:

$$\epsilon_d = \sum B_i[1 - \exp(-b_i t)]$$

an expression:

$$\epsilon_d = \epsilon_\infty[1 - \exp(-a_2 t)^{b_2}]$$

a_2 and b_2 being constants depending on temperature. Experiments show that $b_2 \sim 0.5$

Figure 9.22 shows the delayed deformation curve at $T = 510.9$ °C and Fig. 9.23 the reduced curve relative to the temperature of 543.8 °C, for which $a_2 = 1$.

Similarly a study of isothermal relaxation was performed, a stress σ_0 was applied at the instant $t = 0$; immediately after being loaded, the sample is unloaded to maintain the deformation at the initial value ϵ_i.

An equation of the general form:

$$\frac{\sigma}{\sigma_0} = \sum A_i \exp(-a_i t)$$

corresponding to the generalized Maxwell model allows a correct representation of the results but the number of terms must be rather large. In experiments using a torsion method, Kurkjian[355] had to use a summation of six terms and, in numerical order, the precision of the terms decreased with the order of the exponential.

De Bast and Gilard again preferred to use an empirical equation with two parameters a_1 and b_1 which are functions of temperature. Under these conditions:

$$\sigma/\sigma_0 = \exp[-(a_1 t)^{b_1}]$$

It is found that $b_1 \sim 0.5$.

Fig. 9.22. Delayed deformation curve for a glass. After Ref. 354.

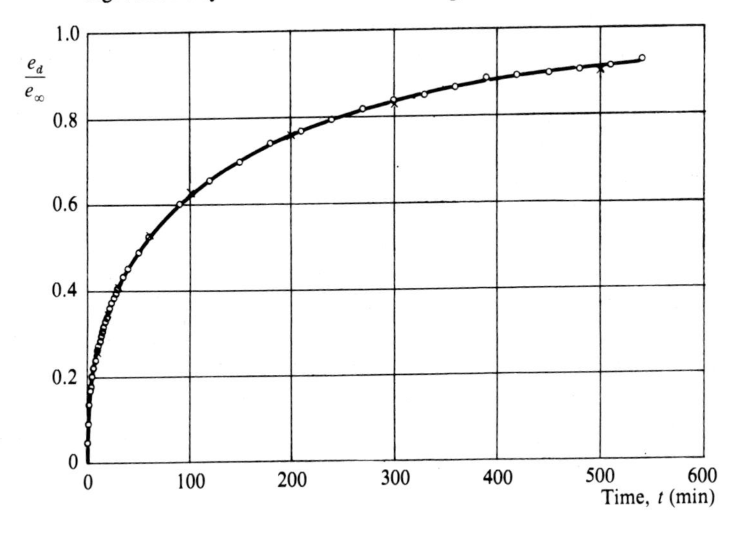

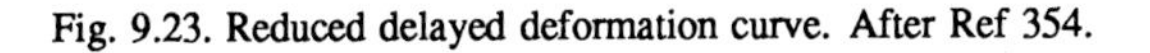
Fig. 9.23. Reduced delayed deformation curve. After Ref 354.

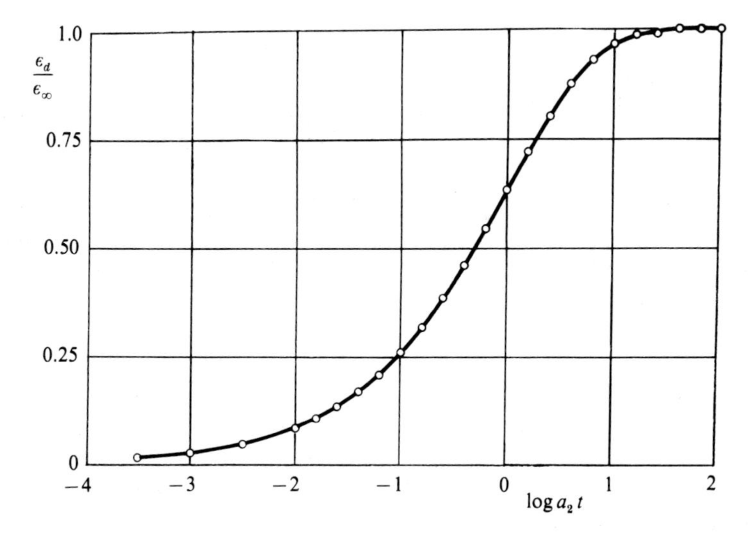

It is then possible to calculate the spectrum of relaxation times (corresponding to the generalized Maxwell model). Figure 9.24 shows the spectrum thus obtained as a function of a reduced variable $z = \tau/\tau_m$ where τ_m is the relaxation time corresponding to the maximum in the spectrum at the given temperature.

Fig. 9.24. Reduced average spectrum and average of the spectra. After Ref. 354.

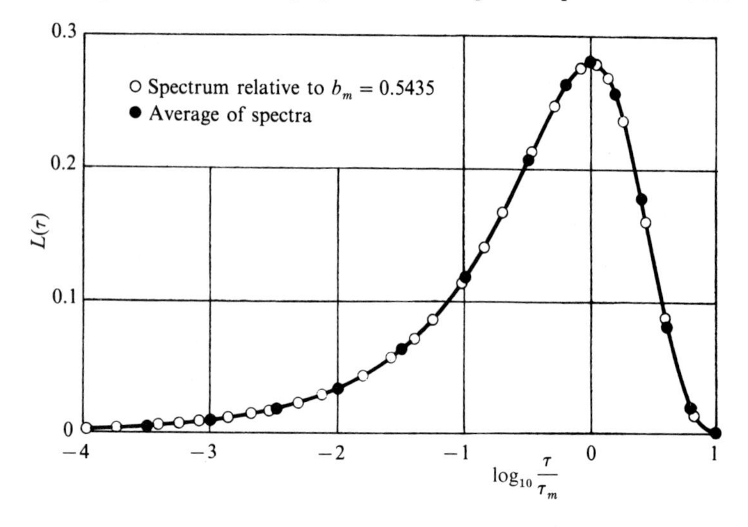

A study of the relaxation modulus in compression of a glass of composition (wt %) 71 % SiO_2, 15 % Na_2O, 8 % CaO, 4.5 % MgO, 1.5 %Al_2O_3, at temperatures up to 150 °C above T_g = 536 °C was made by Larsen, Mills and Sievert.[356] These authors applied the reduction methods explained above. Figure 9.25 shows values of the *moduli* for different temperatures and Fig. 9.26 the curve reduced to 530 °C. The factor a_T obeys the WLF relation with:

$$C_1 = 14.7 \pm 0.5 \quad \text{and} \quad C_2 = 300 \pm 20$$

This glass thus appears as a liquid having a simple thermorheological behavior.

The viscosity calculated using Ferry's relation (cf. Section 9.1.7(c)) deduced from $G(t)$ was compared to values obtained by a fiber elongation method thus confirming the validity of the technique.

De Bast and Gilard also made a dynamic study on window glass using a torsional fiber device coupled to a special generator operating between 5 and 10^{-5} Hz.[354]

A study of the SiO_2–Na_2O system up to 40 mol % Na_2O was made by Mills[357] using a torsion apparatus which applied a sinusoidal stress at a given frequency to a bar, The deformation was measured as a function of time in the frequency range 10^{-3}–1 rad s^{-1}. Figure 9.27 shows reduced curves for the moduli G' and G'' for a glass $2SiO_2{\cdot}Na_2O$ obtained by the method of reduced variables. Analogous results were found for the

Fig. 9.25. Relaxation modulus of a glass as a function of temperature. After Ref. 356.

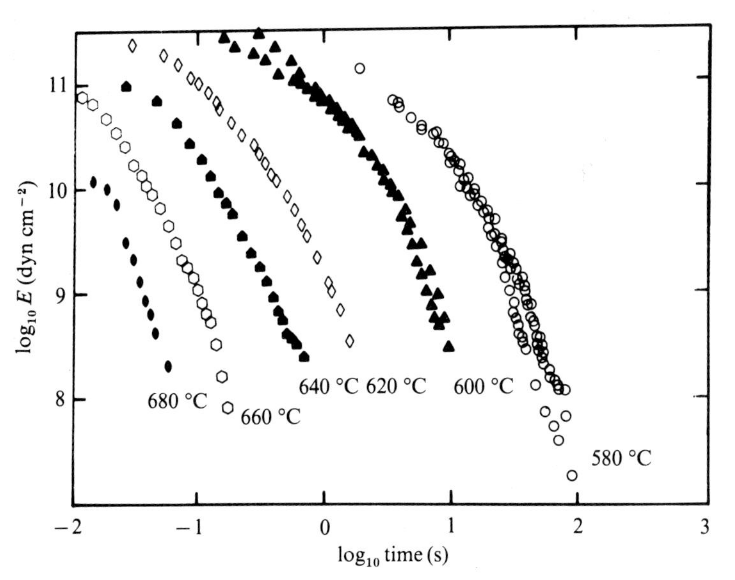

Fig. 9.26. Reduced curve obtained from the results of Fig. 9.25. After Ref. 356.

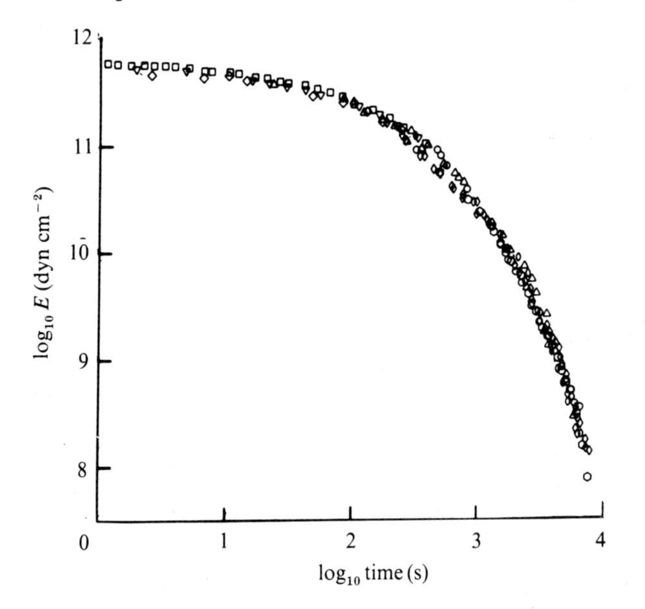

other compositions, $G''(-2)$ and $G''(\infty)$ values decreasing with the Na_2O content.

The full width of the peak at half maximum is independent of the composition. The factor a_T follows the WLF relation.

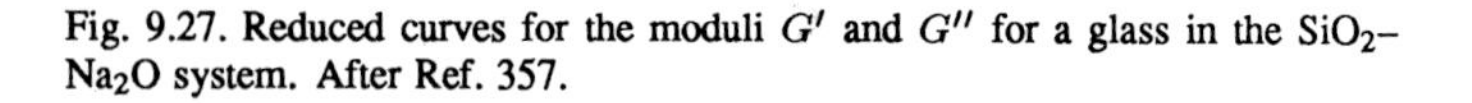

Fig. 9.27. Reduced curves for the moduli G' and G'' for a glass in the SiO_2–Na_2O system. After Ref. 357.

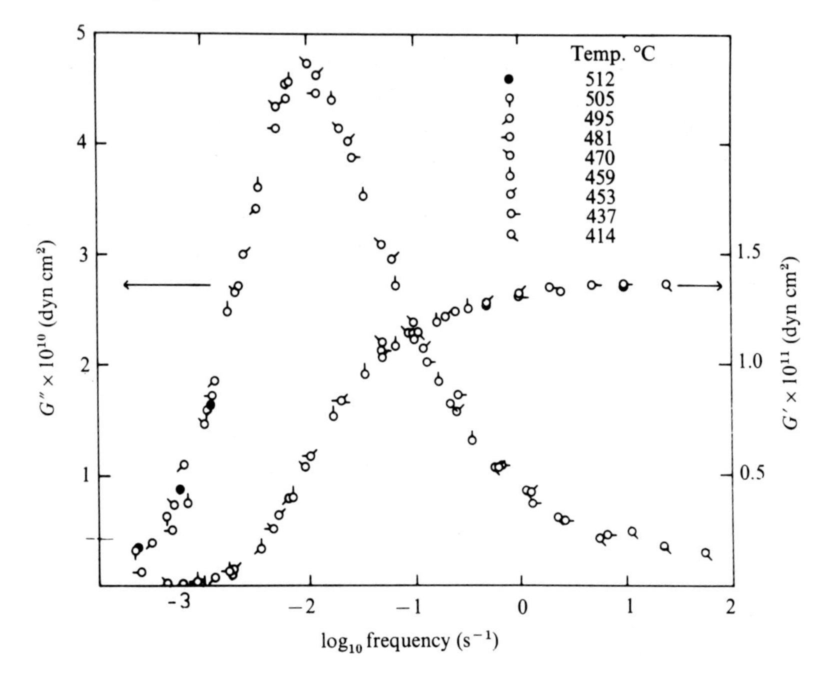

(c) *Internal friction*

Anelasticity can originate from different processes. For binary silicate, borate and phosphate glasses containing only one type of alkali metal ion, the absorption maxima due to mechanical losses or *internal friction* occur below the transition temperature T_g. They are generally measured as a function of temperature at a fixed frequency since it is easier to vary the temperature than the frequency in the techniques normally used. The transition from the temperature variable to the frequency variable is accomplished by assuming that the various relaxation times τ occurring in thermally activated anelastic processes are related to the temperature by an Arrhenius relation:

$$\tau = \tau_0 \exp(E_a/RT)$$

where E_a is an apparent activation energy.

In particular, the absorption maximum T_m corresponds to a frequency ν such that:

$$2\pi\nu\tau_m = \omega\tau_m = 1$$

The temperature T_m increases with frequency and it is possible to determine E_a from a plot of $\log \nu$ versus (1/T).

When there is a distribution of relaxation times this may be due either to a distribution of the pre-exponential terms τ_0, or to a distribution of E or both.

Read and Williams[358] showed that for any distribution of relaxation times (or retardation times), the mean activation energy is given by the relations:

$$\left\langle \frac{1}{E} \right\rangle^{-1} = (G_u - G_R)\frac{R\pi}{2}\left[\int_0^\infty G''\mathrm{d}\left(\frac{1}{T}\right)\right]^{-1}$$

$$\left\langle \frac{1}{E} \right\rangle^{-1} = (J_R - J_u)\frac{R\pi}{2}\left[\int_0^\infty J''\mathrm{d}\left(\frac{1}{T}\right)\right]^{-1}$$

with R being the gas constant. In the case where only the term τ_0 is distributed, the mean reduces to $\langle 1/E \rangle^{-1} = E_a$.

Using the formalism of the equations of Section 9.1.6, a log-normal distribution of relaxation (or retardation) times can be assumed.

Nowick and Berry[359] introduced the distribution function:

$$\phi(z) = \frac{1}{\beta\sqrt{\pi}} \exp\left[-\left(\frac{z}{\beta}\right)^2\right]$$

using the variable $z = \ln(\tau/\tau_m)$ where τ_m is the most probable relaxation time. These authors tabulated the curves $J''/J''(0)$ as a function of $\log \omega\tau$ for different values of the parameter β which is obtained directly from the full width at half maximum of the curve J'' versus $(1/T)$. The parameter β characterizes the variation of the distribution with frequency, i.e. the variation of the peak with T_m.

Nowick and Berry showed that:

$$\beta = \beta_0 + \beta_E/RT_m$$

where β_0 and β_E are respectively the parameters of the log-normal distributions of the entropy ΔS and the activation enthalpy ΔH of the relaxation times written in the form:

$$\tau = \frac{h}{kT} \exp\left(-\frac{\Delta S}{R}\right) \exp\left(\frac{\Delta H}{RT}\right)$$

Figure 9.28 shows an example of an internal friction spectrum for the glass $3\,SiO_2 \cdot Na_2O$.[360] The first peak at about −32 °C corresponds to a migration under stress of the alkali metal ions. The second, near 182 °C, has been associated either with the oscillations of non-bridging oxygen ions[361] or with the presence of hydroxyls (OH) in the structure. When part of the Na is replaced by another alkali metal, e.g. by K, the first peak disappears and a large amplitude intermediate peak near 100 °C appears. This is called the *mixed alkali effect*. It is shown on Fig. 9.28 for the case

Fig. 9.28. Internal friction spectrum Q^{-1} and modulus G' as a function of temperature for a glass $3\,SiO_2 \cdot Na_2O$ and a glass $3\,SiO_2 \cdot 0.4\,Na_2O \cdot 0.6\,K_2O$. After Ref 360.

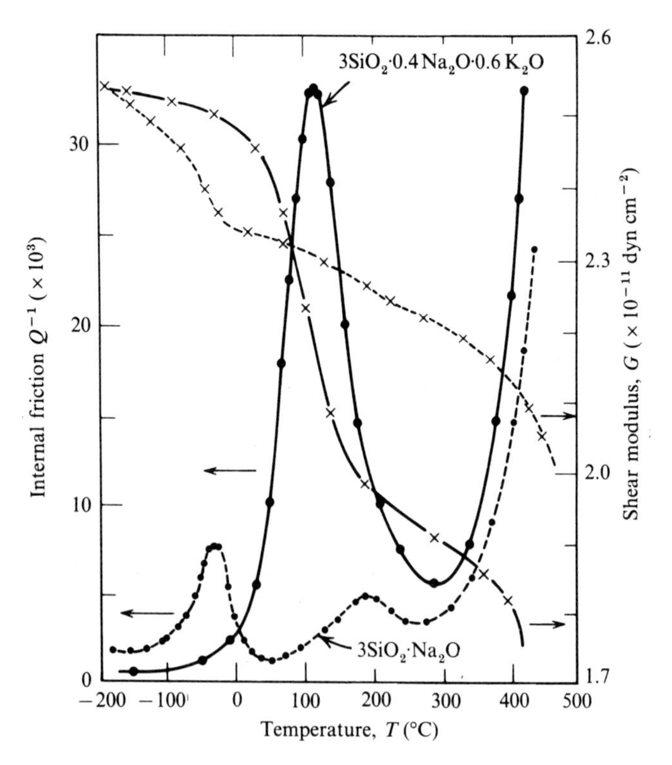

Fig. 9.29. Internal friction spectra of a glass $0.48\,P_2O_5 \cdot 0.05\,Al_2O_3 \cdot 0.47\,Na_2O$ obtained by torsion pendulum, flexural vibration and ultrasonic attenuation methods. After Ref. 362.

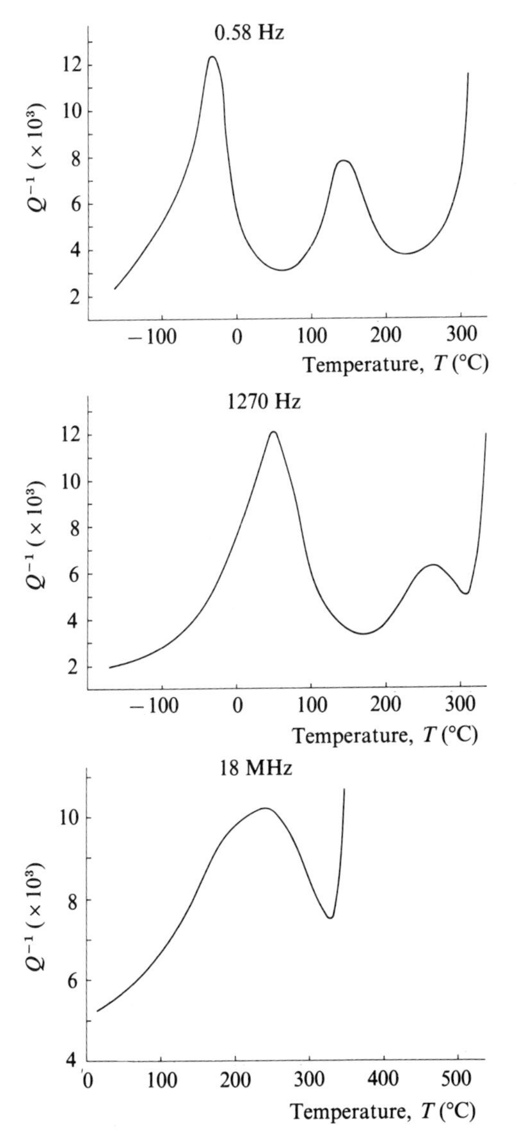

Fig. 9.30. Components D' and D'' of the complex compliance for the first peak of the spectrum of Fig. 9.29 for a frequency of 1270 Hz. After Ref. 362.

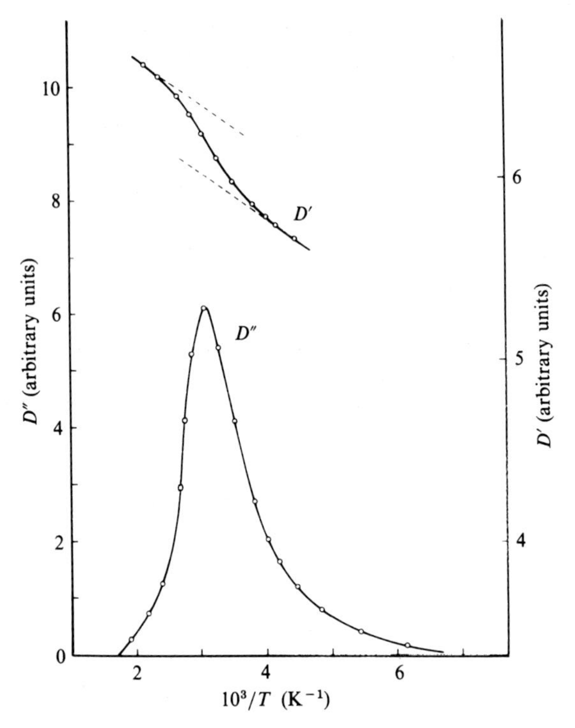

Fig. 9.31. Components M' and M'' of the complex modulus for the glass of Fig. 9.29 for a frequency of 11 MHz. After Ref. 363.

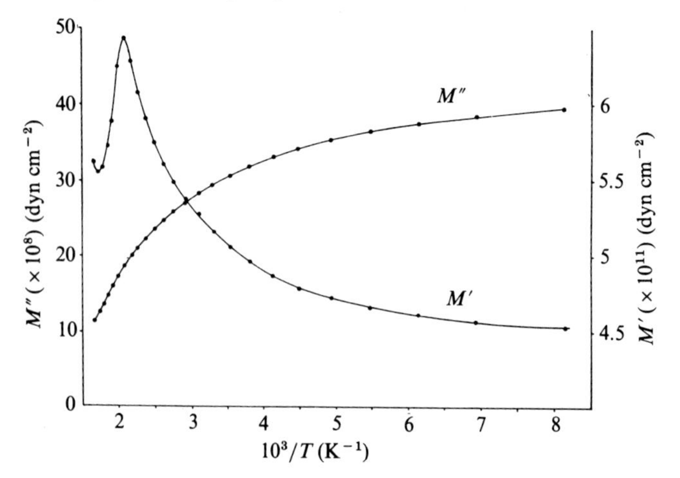

of the glass $3\,SiO_2 \cdot 0.4\,Na_2O \cdot 0.6\,K_2O$. Its origin is not yet understood. Each of the peaks emerges from a continuum representing anelastic effects due to the vitreous transition considered previously and which culminates near T_g. Each peak is moreover associated with a decrease of the modulus G with increasing temperature.

Figure 9.29 shows internal friction spectra for a glass $0.48\,P_2O_5 \cdot 0.05\,Al_2O_3 \cdot 0.47\,Na_2O$ obtained by torsion pendulum (0.58 Hz), forced flexural vibration (1270 Hz) and ultrasonic wave attenuation (18 MHz) methods respectively.

The apparent activation energy calculated from the displacement of the first peak as a function of the frequency is $E_a = 14.8 \pm 1$ kcal/mole.

Figure 9.30 shows the components of the *complex compliance* D^* for 1270 Hz and Fig. 9.31 those of the *complex modulus* M^* for 11 MHz as a function of inverse T. From D^* using the method of Read–Williams, a value of $\langle 1/E \rangle^{-1} = 14.4$ is obtained which is close to that obtained directly.

Generally, it is found that the activation energy is close to that of alkali metal ion diffusion or the ionic electrical conductivity in the same glasses which leads to the association of the observed absorptions with the phenomena of ion migration under stress.

For the preceding glass, the variations of the Nowick and Berry parameter β obtained from measurements at three different frequencies show that $1.6 < \Delta S < 9.6$ cal °C^{-1} and $14 < \Delta H < 15.6$ kcal mole^{-1}; the distribution of relaxation times is thus largely due to that of ΔS.

10 Diffusion phenomena

The phenomena of diffusion, i.e. the transport of atoms or molecules, have great importance in glass science. The formation of a glass from its components by fusion, crystallization or non-crystalline phase separation (unmixing) is controlled by the migration of the participating atomic species. Stress relaxation in the solid state occurs by diffusion. In the case of oxide glasses, the movement of alkali metal ions is responsible for electrical conductivity, dielectric loss or mechanical loss (anelasticity) and all these phenomena are closely related. Ion exchange by diffusion is a phenomenon which occurs in the surface and is used to increase the mechanical strength of glass. It is also applied in the use of glass electrodes which are selective to particular ions. The diffusion of gas through a vitreous network is related to the elimination of gas bubbles in the process of glass fining.

Diffusion is thus of both technological and fundamental interest. For the latter, glass presents a particular attraction as a material capable of being quenched from elevated temperatures without detectable modification of its structure. It possesses, in the case of oxide glasses, extended regions of solubility and flexibility in composition which allow systematic examination over wide limits.

Even though the lack of precise information on the glass network handicaps microscopic interpretations, diffusion may be used as a tool to obtain certain structural information as demonstrated by research.

The following material will concentrate mostly on oxide glasses, and we will essentially consider ionic diffusion where the migrating species are ions participating in the vitreous network in contrast to molecular diffusion which occurs when gaseous species pass through or dissolve in glass. The questions of *permeation* and *solubility* of gas will not be dealt with here; and the interested reader is directed to the article by Doremus.[364]

10.1 Review of the diffusion laws

10.1.1 *Fick's equations*

The analogy between heat conduction in solids and diffusion led Fick to express the diffusion transport law by the relation:

$$J = -D\frac{\partial c}{\partial x} \tag{10.1}$$

where J is the flux in direction x of the diffusing species (i.e. the quantity transported in unit time through a unit cross-section), c the concentration, and D the diffusion coefficient. This equation expresses Fick's first law. To establish the variation of concentration with time, the continuity equation is applied:

$$\frac{\partial c}{\partial t} = \frac{\partial(-J)}{\partial x} = \frac{\partial}{\partial x}\left(D\frac{\partial c}{\partial x}\right) \tag{10.2}$$

In the case where D is a constant independent of concentration:

$$\frac{\partial c}{\partial t} = D\frac{\partial^2 c}{\partial x^2} \tag{10.3}$$

which constitutes Fick's second law. The distribution of the diffusing species is given by the solution to equation (10.2) or (10.3) with specified initial and final conditions. This is a mathematical problem for which numerous solutions are given, particularly in Refs. 365–7.

10.1.2 *Definition of mobility – Nernst–Einstein and Stokes–Einstein relations*

In reality, as shown by Einstein, the driving force for diffusion is due to the chemical potential gradient, and the flux is given by the relation:

$$J = -M\frac{\partial \mu}{\partial x}$$

where μ is the chemical potential and M is the *mobility*.

The *absolute mobility* B_i is defined for an atom i as the transport velocity it attains under a unit applied force. Under the action of a virtual force due to the chemical potential gradient, the velocity v_i is:

$$v_i = -B\frac{1}{N}\frac{\partial \mu_i}{\partial x}$$

and the flux is:

$$J_i = c_i v_i$$

where μ_i is the chemical potential of species i, c_i is its concentration and N Avogadro's number.

Assuming the activity coefficient to be equal to unity:

$$\mathrm{d}\mu_i = RT\ln c_i$$

hence by substitution:

$$J_i = -\frac{RT}{N}\frac{\partial c}{\partial x}B_i$$

Comparing this result with Fick's equation, the diffusion coefficient for species i is:

$$D_i = kT_iB_i$$

This is called the *Nernst–Einstein* relation and will be used in our consideration of electrical conductivity.

Assuming the diffusing species to be a sphere of radius r moving with a velocity v in a medium of viscosity η , the force exerted on it is given by Stokes' law:

$$f = 6\pi r\eta v$$

Using the definition of absolute mobility, the velocity per unit force is:

$$B = \frac{1}{6\pi r\eta}$$

Substituting this in the previous equation:

$$D = \frac{kT}{6\pi rv}$$

known as the *Stokes–Einstein* relation.

This relation is not well followed for glasses, and leads to coefficients D which are often 10^{10} times too small. This shows that the diffusion of alkali metal ions occurs by a mechanism which is faster than viscous flow.

10.1.3 *Different types of diffusion*

Self-diffusion is defined as diffusion in the absence of a chemical potential gradient and is the result of random movement of the atoms forming the solid. A distinction is also made between *volume diffusion* and *surface diffusion.* The *apparent* diffusion coefficient can include several mechanisms.

Diffusion in a chemical potential gradient is called *chemical diffusion.* The corresponding *chemical diffusion coefficient* or *interdiffusion coefficient* is the one generally used.

The *interdiffusion* coefficient $\tilde{D}$ can be obtained from equation (10.2) which is independent of the manner in which D depends on concentration. However, since $\tilde{D}$ is, in general, a function of concentration, only an average value of $\tilde{D}$ between initial and final concentrations in the glass can be obtained in this manner.

It has been shown (see e.g. Ref. 366) that the diffusion coefficient of constituent A in a binary A–B can be expressed in the form:

$$D_A^c = M_{AA} kT \left(1 + \frac{\partial \ln \gamma_A}{\partial \ln N_A}\right)$$

and likewise for B:

$$D_B^c = M_{BB} kT \left(1 + \frac{\partial \ln \gamma_B}{\partial \ln N_B}\right)$$

In these equations M_{AA} and M_{BB} are the mobilities, γ_A and γ_B the activity coefficients and N_A and N_B the atomic fractions of the species A and B: $N_A = c_A/(c_A + c_B)$ where c_A and c_B are the concentrations. D_A^c and D_B^c are called *intrinsic* diffusion coefficients.

In the case of an ideal solution or a sufficiently dilute solution, $\gamma_i = 1$ and $D_i^c = D^*$; D^* is the *self-diffusion coefficient* of the species i and can be obtained using tracer methods.

In the general case:

$$D_i^c = D_i^* \left(1 + \frac{\partial \ln \gamma_i}{\partial N_i}\right)$$

The intrinsic diffusion coefficient is the product of the self-diffusion coefficient multiplied by a term which expresses the departure from ideality. These relations are called *Darken equations*.

The interdiffusion coefficient $\tilde{D}$ is related to the intrinsic coefficients by the relation:

$$\tilde{D} = N_A D_B^c + N_B D_A^c$$

Combining the preceding equations, and taking into account the Gibbs–Duhem equation:

$$\frac{\partial \ln \gamma_A}{\partial \ln N_A} = \frac{\partial \ln \gamma_B}{\partial \ln N_B}$$

$\tilde{D}$ can be expressed as a function of the coefficients D_i^*:

$$\tilde{D} = (N_A D_B^* + N_B D_A^*) \left(1 + \frac{\partial \ln \gamma_A}{\partial \ln N_A}\right)$$

The diffusion coefficient can become negative in certain special conditions. This then produces a diffusion against a concentration gradient. It can be shown that the condition:

$$D_A^c < 0$$

is equivalent to:

$$\frac{\partial^2 G}{\partial N_A^2} < 0$$

which is a relation verified in field II interior to the spinodal (cf. Section 6.3.4).

10.1.4 *Temperature dependence*

In a general way, experiments show that the diffusion coefficient follows the Arrhenius relation:

$$D = D_0 \exp(-Q_D/RT)$$

where Q_D is the apparent activation energy for diffusion, D_0 the frequency factor and R the gas constant. The self-diffusion coefficients and the intrinsic diffusion coefficients obey this relation rather well; it is often extended to the case of chemical diffusion:

$$\tilde{D} = \tilde{D}_0 \exp(-Q_D/RT)$$

In the latter case, it is only an approximation.

10.2 **Diffusion mechanisms**

10.2.1 *Microscopic approach*

Diffusion in glass cannot be considered as transport in a viscous medium because the Stokes–Einstein law is not obeyed. By analogy with crystals, it is often explained through a *network defect* mechanism. In the case of a crystal, the point defects mainly consist of empty lattice sites (*vacancies*) or *interstitial* ions which do not occupy normal sites in the lattice. It remains to be seen if such concepts can be rigorously transposed to the case of a disordered substance. However, point defect models which assume only short-range order seem to be confirmed by spectroscopic results. In the case of oxide glasses, it is probable that vacancies are less mobile than *interstitials* and that the latter very likely play the dominant rôle.

10.2.2 *Diffusion considered as a thermally activated process*

Considering diffusion as the result of random migration of atoms, the diffusion coefficient can be related, in general, to the average of the square of the jump distance $\bar{\lambda^2}$ and the jump frequency Γ by the formula:

$$D = g\bar{\lambda^2}\Gamma$$

where g is a geometric factor close to 1/6.

For a jump to be successful, two conditions must be satisfied: the defect must be formed and the thermal agitation must provide the necessary activation energy to the atom for it to surmount the barrier.

If diffusion is considered as the global result of the passage of atoms from one site to an adjacent site over a potential barrier ΔG_m (Fig. 10.1), the presence of a chemical potential gradient influences the conditions for overcoming the barrier. If $\chi = (1/N)(\partial\mu/\partial x)$ is the chemical potential gradient relative to an atom, the work done during a jump is $\lambda\chi$. The minimum to the left is raised by $u = (1/2)\lambda\chi$ and the one to the right lowered by u.

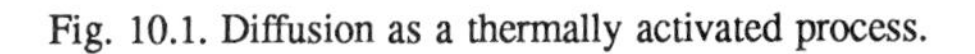

Fig. 10.1. Diffusion as a thermally activated process.

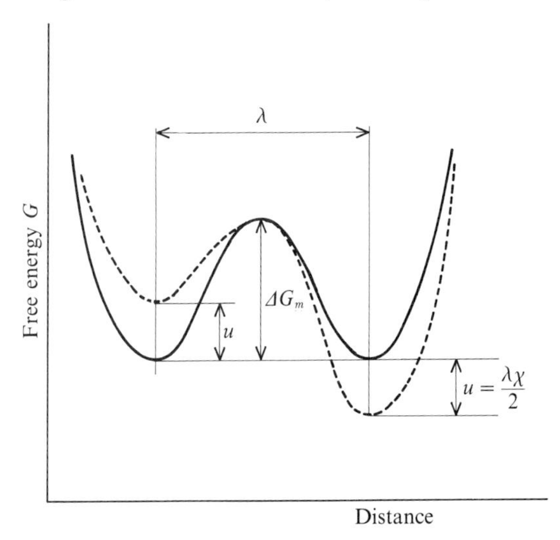

The jump frequency is then:

towards the right:

$$\omega_1 = \nu \exp\left(-\frac{\Delta G_m - u}{kT}\right)$$

and towards the left:

$$\omega_2 = \nu \exp\left(-\frac{\Delta G_m + u}{kT}\right)$$

the transport velocity is:

$$v = \lambda(\omega_1 - \omega_2) = 2\nu\lambda \exp\left(-\frac{\Delta G_m}{kT}\right) \sinh\left(\frac{u}{kT}\right)$$

For diffusion, $\lambda\chi \ll kT$ and $\sinh u$ can be replaced by u:

$$v = 2\nu\lambda \exp\left(-\frac{\Delta G_m}{kT}\right)\left(\frac{\lambda\chi}{2kT}\right)$$

substituting from the definition of χ, the flux is obtained as:

$$J = vc = -\frac{\lambda^2 \nu c}{NkT} \exp\left(-\frac{\Delta G_m}{kT}\right)\left(\frac{\mathrm{d}\mu}{\mathrm{d}x}\right)$$

It follows that the mobility is:

$$B = \frac{\lambda^2 \nu}{kT} \exp\left(-\frac{\Delta G_m}{kT}\right)$$

For an ideal solution:

$$\mathrm{d}\mu = NkT\mathrm{d}(\ln c)$$

$$\frac{\mathrm{d}\mu}{\mathrm{d}x} = \frac{NkT}{c}\frac{\mathrm{d}c}{\mathrm{d}x}$$

$$J = -\lambda^2 \nu \exp\left(-\frac{\Delta G_m}{kT}\right)\frac{\mathrm{d}c}{\mathrm{d}x} = -D\frac{\mathrm{d}c}{\mathrm{d}x}$$

Hence, including the geometric factor g:

$$D = g\lambda^2 \nu \exp\left(-\Delta G_m / kT\right)$$

It thus follows that:

$$\Gamma = \nu \exp\left(-\Delta G / kT\right)$$

In the general case, the activity coefficients must be used.

The *concentration of defects* responsible for diffusion can be introduced. It has been shown that if ΔG_f is the free energy of formation of a *vacancy–interstitial pair*, the concentration of defects is proportional to $\exp(-\Delta G_f/2kT)$. In the general case the diffusion constant will be of the form:

$$D = g\lambda^2\nu \exp\left[\left(\frac{\Delta S_f}{2} + \Delta S_m\right)\Big/k\right] \exp\left[-\left(\frac{\Delta H_f}{2} + \Delta H_m\right)\Big/kT\right]$$

hence

$$D = D_0 \exp\left[-\left(\frac{\Delta H_f}{2} + \Delta H_m\right)\Big/kT\right]$$

Thus the Arrhenius form of the diffusion coefficient is obtained, with:

$$Q_D = \frac{\Delta H_f}{2} + \Delta H_m$$

where ΔH_f and ΔH_m are respectively the enthalpies of formation and migration of the defects.

In the case of glasses, the precision of the results is insufficient to establish with certainty the temperature dependence of D_0. It is not possible, moreover, to separate the terms ΔH_f and ΔH_m. For crystalline solids, these quantities can sometimes be obtained from measurements of electrical conductivity. Doping with selected impurities permits control of the defect concentration in the network and makes it possible to distinguish between the terms which control the probability of a jump and those which relate to the vacancy–interstitial concentration.

10.3 Experimental methods

10.3.1 *Solutions to Fick's equations*

Glasses being essentially solids, the techniques employed to measure the diffusion coefficient are analogous to those used in the case of crystals. Simple geometric situations are sought for which the solution to Fick's equations is known. The diffusion coefficient can be deduced by determining the concentration profile of the diffusing species after a known period of time. By way of illustration, the following solutions to Fick's equations are given.

(a) *Limited source*

$$c(x,t) = c_0\,\delta(x) \quad \text{for} \quad t = 0 \qquad (\delta(x)\text{: Dirac function})$$

$$c(x,t) = 0 \quad \text{for} \quad t > 0 \ \text{and}\ x > 0$$

This corresponds in practice to the application of a thin film containing a surface concentration c_0 of the diffusing substance on a plane face of an unlimited solid (Fig. 10.2(*a*)).

The solution in these conditions is the function:

$$c(x,t) = \frac{c_0}{\sqrt{\pi Dt}} \exp\left(-\frac{x^2}{4Dt}\right)$$

By sectioning the solid in thin slices perpendicular to x and determining $c(x)$ after a certain time t, the diffusion constant is determined from the graph of $\ln c = f(x^2)$ which is linear. If the layer is a sandwich between two specimens, c_0 must be divided by 2.

Fig. 10.2. Typical diffusion experiments and the evolution of the corresponding diffusion profiles: (*a*) limited source, (*b*) constant source, (*c*) interdiffusion. (I_M indicates the position of the Matano surface).

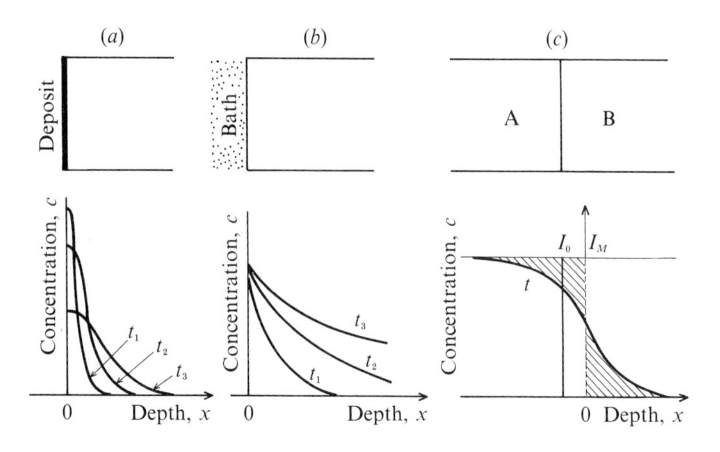

(b) *Constant concentration source*

By maintaining a constant concentration c_0 on the entrance face, the initial conditions become (Fig. 10.2(*b*)):

$$\begin{array}{llll} c(x,t) = c_0 & \text{for} \quad x = 0 & \text{and} & t > 0 \\ c(c,t) = 0 & \text{for} \quad x > 0 & \text{and} & t = 0 \\ \dfrac{\partial c}{\partial x} = 0 & \text{for} \quad x \to \infty & \text{and} & t > 0 \end{array}$$

The solution is given by:

$$c(x,t) = c_0 \operatorname{erfc}\left(\frac{x}{2\sqrt{Dt}}\right)$$

where erfc is the complementary error function:

$$\operatorname{erfc}(z) = \frac{2}{\sqrt{\pi}} \int_z^{\infty} \exp(-y^2)\,\mathrm{d}y$$

This corresponds to the immersion of the solid in a stirred bath containing the diffusing substance (e.g. a fused salt bath containing a radioactive isotope).

The total quantity Q' of the substance diffusing into the solid after a given time t is:

$$Q' = 2c_0\sqrt{\frac{Dt}{\pi}}$$

(c) *Interdiffusion case*

To determine the interdiffusion coefficient $\tilde{D} = \tilde{D}(c)$, the classic Boltzmann–Matano method is used.[366] With the following initial and final conditions:

$$\begin{array}{llll} c(x,t) = c_0 & \text{for} \quad x < 0 & \text{and} & t = 0 \\ c(c,t) = 0 & \text{for} \quad x > 0 & \text{and} & t = 0 \\ \dfrac{\partial c}{\partial x} = 0 & \text{for} \quad x \to \pm\infty & \text{and} & t > 0 \end{array}$$

the solution for $t = Constant$:

$$\tilde{D}(c) = -\frac{1}{2t}\left(\frac{\mathrm{d}x}{\mathrm{d}c}\right)_c \int_c^{c_0} x\,\mathrm{d}c$$

The origin of the abscissa is found from the condition:

$$\int_c^{c_0} x\,\mathrm{d}c = 0$$

which states that the net flux across the plane $x = 0$ (called the Matano surface) is zero (Fig. 10.2(*c*)).

10.3.2 *Use of tracers*

Most of the studies have been made on oxide glasses, but several results are also available for chalcogenide glasses.

In the case of oxide glasses (mostly silicates), measurements of *self-diffusion* have been applied to the evaluation of the mobilities of the most mobile modifying cations and, less often, to network-forming elements Si, O, B; the measurements make use of *tracers*. The easiest and most often used method employs radioactive isotopes emitting γ, β, or α rays which are measured by counting techniques. Table 10.1 lists the isotopes used in diffusion studies in glasses. To satisfy the limiting conditions permitting the application of Fick's equations, the isotope is either deposited as a thin film, e.g. ^{22}Na in ^{22}NaCl or $^{22}Na_2SO_4$ (limited source), or contained in a fused salt bath (constant concentration source). The specimen is held at a given diffusion temperature and then specimen is sectioned in slices parallel to the surface to measure the distribution in depth of the isotope. The successive layers are removed either by abrasion or by chemical dissolution, and the residual activity is measured in the remaining part of the specimen. From these data, the concentration profile $c(x, t)$ is reconstructed.

The self-diffusion of Si is difficult to measure because of the lack of isotopes with sufficient life-times; ^{31}Si has a half-life of only 2.6 h, and ^{30}Si (250 yr) is not commercially available.

The self-diffusion of oxygen has been measured by gas–solid exchange methods with the stable isotope ^{18}O. A specimen of glass in the form of a fiber or sphere with known total surface area is placed in a chamber containing oxygen enriched with ^{18}O isotope and held at a given temperature. Exchange occurs between the atmospheric oxygen and ^{16}O of the network, the ratio ^{18}O/^{16}O in the atmosphere around the specimen is measured with a mass spectrometer.

Another method, which avoids difficulties related to surface adsorption, consists of measuring the ratio ^{18}O/^{16}O in the specimen itself using the technique of *activation analysis*. When the specimen containing ^{18}O is irradiated by protons with energies of 400–500 keV, the following nuclear reaction occurs:

$$^{18}\mathrm{O}(p, \alpha)^{15}\mathrm{N}$$

and measurement of the α particle spectrum allows evaluation of the ^{18}O concentration.[369]

In the case of Li and B, the stable isotopes ^{7}Li and ^{11}B can be used in connection with NMR.

To study the diffusion of atomic species not contained in the glass, all classical chemical analysis methods are used as well as X-ray fluorescence and electron microprobe analysis.

Table 10.1. *Characteristics of isotopes used in studies of diffusion in glasses.*

Stable isotopes	^{2}D	
	^{7}Li	
	^{18}O	
	^{30}Si	
		Period
Radioactive isotopes	^{22}Na	2.6 yr
	^{24}Na	15.06 h
	^{26}Al	7.6 s
	^{31}Si	2.6 h
	^{32}Si	>250 yr
	^{42}K	12.4 h
	^{45}Ca	152 d
	^{85}Sr	65 d
	^{86}Rb	18.7 d
	^{134}Cs	2.2 yr
	^{137}Cs	30 yr
	^{140}Ba	12.8 d
	^{210}Pb	19.4 yr
	^{212}Pb	–

10.4 Examples of applications to glasses

Most of the measurements of diffusion are for oxide glasses; among these, vitreous SiO_2 and glasses based on glass formers SiO_2 and B_2O_3 have been studied the most.

10.4.1 *Self-diffusion of oxygen*

In the case of SiO_2 it is possible to make comparisons with the crystalline form: quartz. Studies show that diffusion is strongly influenced by impurities included in the network (OH^-, alkali metal ions, etc).

The results (Fig. 10.3) show a large scatter probably because of the different origins of the materials used, though this is not indicated by the authors.

For quartz, the results are strongly anisotropic; the values of D for diffusion parallel to the c axis being 50 times higher than those perpendicular to c.

Fig. 10.3. Variation of the self-diffusion coefficient of oxygen as a function of temperature.

(*a*),(*b*),(*c*)	Vitreous SiO_2, Refs. 368a, 368b, 368c.			
(*d*):	42.3 SiO_2,	12.4 Al_2O_3,	45.3 CaO,	Ref. 368d.
(*e*):	71.1 SiO_2,	15.5 Na_2O,	12.8 CaO,	Ref. 368d.
(*f*):	42.2 CaO,	42.1 B_2O_3,	15.7 Al_2O_3,	Ref. 368e.
(*g*):	25.9 CaO,	61.4 B_2O_3,	12.7 Al_2O_3,	Ref. 368e.
(*h*):	31 B_2O_3,	69 PbO,		Ref. 368f.
(*i*):	53.4 SiO_2,	33 PbO,	8.7 K_2O,	Ref. 368f.

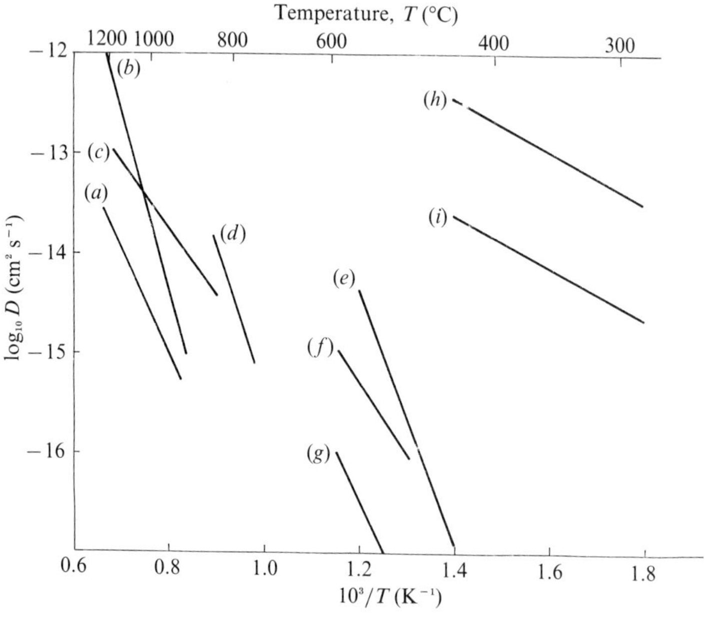

The diffusion mechanism of oxygen in the network is uncertain; the experimental results show considerable scatter for the activation energies (29–71 kcal mole^{-1}). The higher values may indicate a migration of non-bridging oxygens; the lower values may be the consequence of permeation of O_2 molecules.

10.4.2 *Diffusion of alkali metal ions*

It is well known that alkali metal cations, e.g. Li^+ and Na^+, can penetrate into the quartz network and play an important rôle in ionic conductivity of the material. These effects are also found in vitreous SiO_2 where the Na^+ cation is the most mobile, at least at lower temperatures. Diffusion results using the radioactive tracer ^{22}Na are shown on Figure 10.4. It is seen that there is again a strong dependence on the origin of the material as well as on its thermal history.

Fig. 10.4. ^{22}Na tracer diffusion in vitreous SiO_2 (the results depending on the OH content and thermal history of the specimen). After Ref. 370. The values for diffusion of ^{45}Ca and ^{26}Al are also shown. After Ref. 370a.

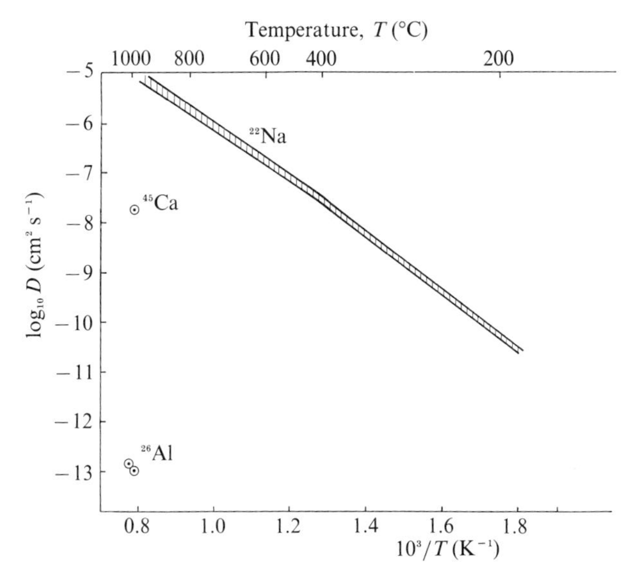

The addition of alkali metal oxide modifiers to SiO_2 causes a depolymerization effect in the network with the appearance of non-bridging oxygens thus increasing the mobility of the alkali metal cations. This effect can be counteracted by alkaline earth cations. Numerous binary and ternary systems have been studied. The measurements mainly apply to the migration of alkali metal and alkaline earth cations.

Figure 10.5 shows the self-diffusion coefficient of Na^+ in Na_2O–CaO–SiO_2 glasses. Measurements of oxygen self-diffusion are more rare; the only known ones are assembled in Fig. 10.3. With respect to self-diffusion of Si, there are very few measurements available.[372]

In a general way, the experimental results reported by different authors show a strong scatter which does not seem to be entirely due to the different methods used, but is perhaps due to structural differences related to impurities and the thermal history, the chemical composition alone being inadequate to characterize the diffusional behavior of these materials.

The preceding results are for glasses with simple compositions. Measurements on industrial glasses with several components, borosilicates of the "PYREX" type or glasses for radioactive nuclear waste storage are difficult to interpret from a structural point of view.

Fig. 10.5. ^{22}Na tracer diffusion in Na_2O–CaO–SiO_2 glass. After Ref. 371.

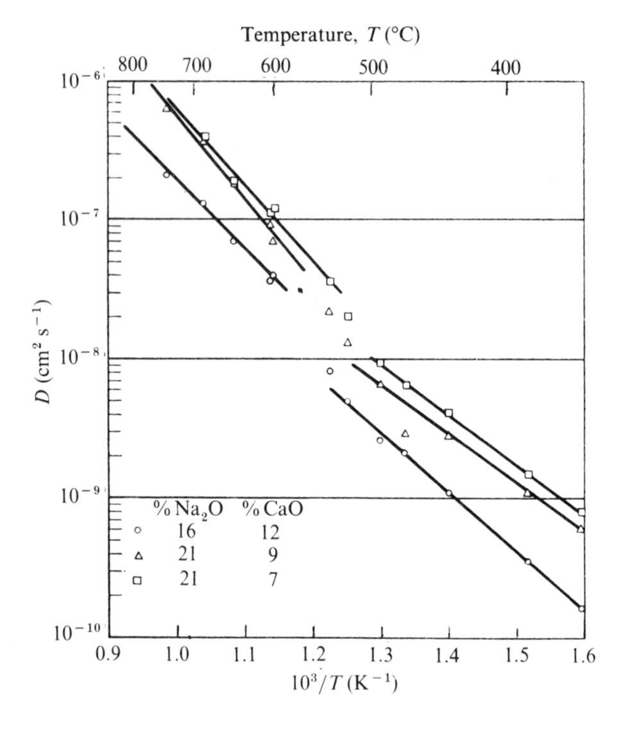

A complete compilation of the available results can be found in the monograph by Frischat.[373] For metallic glasses a compilation of diffusion coefficient data for various species can be found in Refs. 374–6.

11 Electrical properties of glasses

Among the electrical properties of glasses, *conduction* and *dielectric* properties play a pre-eminent rôle. This is a vast subject so here we will limit the following to an overview of its main characteristics. For further details, the reader is referred to specialist monographs.[377–80]

11.1 Electrical conductivity

11.1.1 *Definitions*

The passage of current in a material, or the phenomenon of electrical *conduction*, is characterized by the *conductivity* σ defined by the relation:

$$\vec{p} = \sigma \vec{E}$$

where $\vec{p}$ is the current density and $\vec{E}$ is the applied electric field. In an isotropic material such as glass, σ is a scalar. The inverse of σ is the *resistivity* ρ. The units of ρ and σ are respectively ohm meter (Ω m) and ohm^{-1} meter^{-1} (Ω^{-1} m^{-1}). The conductivity of glass depends primarily on its composition and the temperature. At ambient temperature most oxide glasses are *insulators*, their conductivity being of the order of 10^{-17}–10^{-5} Ω^{-1} m^{-1}.

The variation of σ with temperature is large. In a soda–lime–silica glass, the conductivity, which is of the order of 10^{-9}–10^{-8} Ω^{-1} m^{-1} at 20 °C, can be multiplied by a factor as high as 10^{19} at 1200 °C. The insulator is transformed progressively into a conductor to a point where it is possible to pass a current through the glass and heat it directly by the Joule effect.

The electric current results from particle or *charge carrier* movement through the network. The carriers can be ions (*ionic* conduction), or electrons and/or holes (*electronic* conduction). The two mechanisms can, moreover, be present simultaneously. In a general form, the conductivity σ_i associated with a carrier of species i is of the form:

$$\sigma_i = n_i q_i \mu_i$$

where n_i is the number of carriers per unit volume for which the charge is q_i and the *mobility* is μ_i, related to the *absolute* mobility B_i by:

$$\mu_i = q_i B_i$$

The total conductivity σ_T is equal to the sum of the partial conductivities σ_i, and the *transport number* of the species i is defined as:

$$t_i = \sigma_i/\sigma_T$$

which represents the fraction of the total conductivity attributable to a given type of carrier.

Historically, ionic conductivity in glass was recognized first and has been studied for a long time. Interest in glasses with electronic conduction is more recent but has made significant contributions to our general knowledge of conductivity in glasses.

11.1.2 *Ionic conduction*

It has been known for a long time that oxide glasses containing alkali metal ions (modifiers) act essentially as solid electrolytes. The classic experiments of Warburg (1884) consisted of electrolyzing a tube of glass filled with Na amalgam and placed in a similar amalgam bath at 500 °C, the interior and exterior amalgam serving as electrodes. The experiments showed that the current is carried by Na^+ ions, and the Na amalgam is necessary to compensate for the loss of Na^+ ions at the cathode. These studies confirmed the Faraday laws and showed that the transport number of the alkali metal ions is equal to unity.

Similarly in "pure" SiO_2 containing several ppm Na^+, the conductivity is dominated by these ions. The mobility of ions with valence 2 and higher is generally much less than that of alkali metal ions. In the glasses B_2O_3–CaO–Al_2O_3 ("CABAL" glasses) it has been suggested that the conductivity is by O^{2-} ions.

In the presence of transition metal ions, electronic conductivity can predominate.

(a) *Temperature dependence*

It is found empirically that the resistivity ρ follows the Rasch and Hinrichsen law over an extended temperature range:

$$\log \rho = A + \frac{B}{T} \tag{11.1}$$

where A and B are constants.

Figure 11.1 gives examples for several common glasses. The change of slope in the transformation range (Fig. 2.13(*b*)) is a standard way of determining T_g. Above T_g the response is still linear but at high temperatures the graph takes a curved form. At low temperatures, deviations are observed which are attributed to *surface conductivity* from adsorbed

Fig. 11.1. Resistivity of several common glasses: (*a*) vitreous SiO_2, (*b*) alumino-silicate, (*c*) lead glass, (*d*) Pyrex, (*e*) window glass.

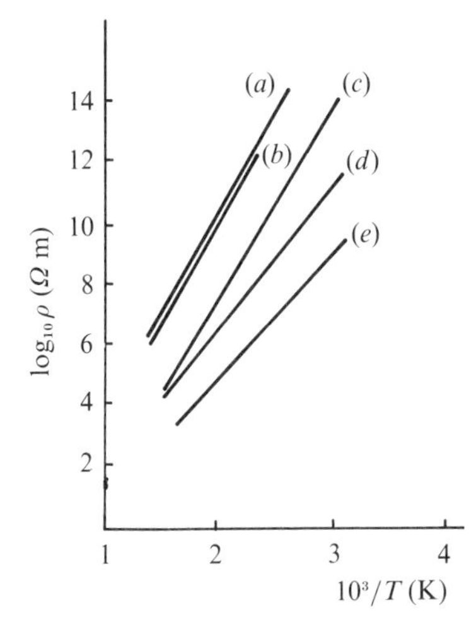

water. This parasitic effect can exceed the volume conductivity and mask the real glass behavior.

In the determination of conductivity, it is essential to use non-polarizing electrodes to avoid a decrease in current due to the lack of charge carriers near the electrode.

(b) *Theoretical calculation of conductivity*

The theoretical calculation of the mobility of an ion in glass uses absolute reaction rate theory, the arguments being analogous to those for diffusion.

The movement of ions in a network subject to thermal agitation is influenced by the applied electric field which modifies the potential distribution.

Consider the one-dimensional case (Fig. 11.2) in which the movement of ions with charge q_i is parallel to axis $0x$. In the absence of an electric field, the ions must surmount the barrier ΔG_m between two potential wells separated by the jump distance a. If ν is the vibration frequency of the ions, the probability of a jump is:

$$\omega = \nu \exp\left(-\Delta G_m/kT\right)$$

This applies to any jump in either direction.

The presence of the electric field E lowers the depth of the well on the right by $u = \frac{1}{2}\, a\, q_i E$ (the work being equal to the force $q_i E$ acting over

Fig. 11.2. Modification of the potential due to an applied electric field E.

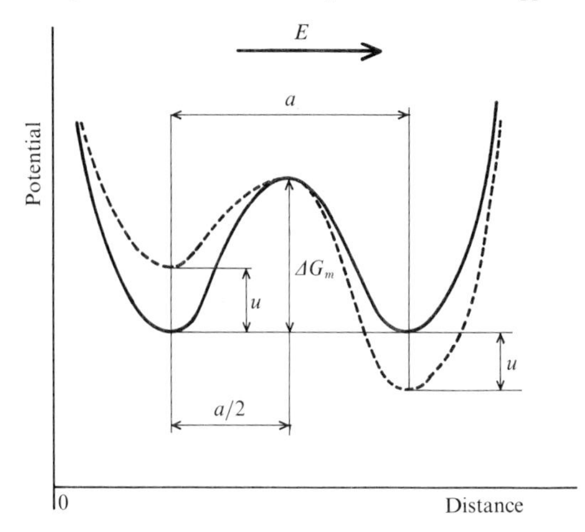

the distance $a/2$) and raises the potential well on the left by the same amount. The jump frequencies become:

$$\omega_1 = \nu \exp\left[-\frac{\Delta G_m - u}{kT}\right] \qquad \text{(towards the right)}$$

$$\omega_2 = \nu \exp\left[-\frac{\Delta G_m + u}{kT}\right] \qquad \text{(towards the left)}$$

The current density is:

$$j_i = n_i q_i a(\omega_1 - \omega_2) = 2n_i q_i a\nu \exp\left(-\frac{\Delta G_m}{kT}\right) \sinh\left(\frac{a q_i E}{2kT}\right)$$

and for sufficiently weak field intensity such that $u \ll kT$:

$$j_i = E\frac{n_i q_i^2 a^2 \nu}{kT} \exp\left(-\frac{\Delta G_m}{kT}\right)$$

For a three-dimensional model, the result must be multiplied by a constant $b \sim 1/4$.

This expression shows that at constant temperature, $j_i/E = \sigma_i$, and the model follows Ohm's law. The number of carriers n_i can be multiplied by the probability of formation of a vacancy–interstitial pair, $\exp\left(-\Delta/G_f/2kT\right)$ determined by the Boltzmann equation.

In these conditions:

$$\sigma_i = \frac{bn_i a^2 q_i^2 \nu}{kT} \exp\left[-\frac{(\Delta G_m + \Delta G_f/2)}{kT}\right]$$

In this expression, valid for a crystal, the apparent activation energy is thus equal to the sum of the jump energy ΔG_m and the defect formation energy, $\Delta G_f/2$.

For a glass, the potential energy is better represented by a curve (Fig. 11.13) where the activation energy ΔG_m and the distance between sites are variable. By comparison the mobility of the ionic species i is deduced:

$$\mu_i = \frac{q_i a^2 \nu}{kT} \exp\left(-\frac{\Delta G_m}{kT}\right)$$

and the absolute mobility $B_i = \mu_i/q_i$.

(c) *Relation between conductivity and diffusion*

If it is assumed that the electrical transport mechanism is the same as diffusional transport, the mobility μ_i is related to the self-diffusion coefficient D_i by the Einstein formula:

$$\mu_i = q_i B_i = \frac{q_i D_i}{kT}$$

where, replacing μ_i by $\sigma_i/n_i q_i$ and introducing the transport number t_i:

$$(D_i)_\sigma = \frac{t_i \sigma_T}{n_i q_i^2} kT$$

which is called the *Nernst–Einstein* relation.

For soda–lime–silica glasses it can be shown that the measured coefficient D_i is smaller than $(D_i)_\sigma$ resulting from the preceding relation:

$$D_i = f(D_i)_\sigma \qquad \text{with} \qquad 0 < f \ll 1$$

Since for these glasses, $t_{Na^+} = 1$, the difference can only be attributed to a correlation effect through the coefficient f; the successive ionic jumps are not statistically independent. The corrected Nernst–Einstein relation is then written:

$$D_i = f_i t_i \frac{\sigma_T kT}{n_i q_i^2}$$

The correlation coefficients vary according to the glasses studied, and different authors report substantially different values.

(d) *The effect of pressure*

Mechanical stresses affect the conductivity. A quenched glass has a higher conductivity than an annealed glass. The effect of hydrostatic pressure P is described by an equation of the form:

$$\sigma = \sigma_0 \exp\left(\frac{V^*P}{RT}\right)$$

where σ_0 and the "activation volume" V^* are independent of P at constant temperature. For a soda–lime–silica glass, $V^* \sim 3.7\,\text{cm}^3\,\text{mole}^{-1}$.

(e) *Compositional dependence*

The conductivity is related to composition in a complex way. The pre-exponential factor depends on the concentration of carriers, the jump frequency ν, and the jump length a. The quantities ν and a determine the mobility of the ions or the bond force and again depend on parameters such as the strength of the ionic field of the carriers, the polarizability of the neighbors, steric obstructions, and the rigidity of the glass structure. These factors not only act in a cooperative manner but also influence the apparent activation energy. Thus, for activation energies in SiO_2–X_2O glasses (where X = Li, Na, K) at low concentrations of X, $E_K > E_{Na} > E_{Li}$, i.e. they follow the same order as the ionic radii, while at higher concentrations they follow the order of the ionic forces: $E_{Li} > E_{Na} > E_K$. The highest activation energy then corresponds to the Li^+ ion which has the highest ionic field.

Figure 11.3 shows the classic example from Fulda[381] of the effect of the partial substitution of SiO_2 by other oxides in a glass of composition $82\,SiO_2 \cdot 18\,Na_2O$ (wt %). The largest increase of resistivity is observed for CaO, BaO and B_2O_3 which consolidate the network and thus reduce the mobility of the Na^+ ion. In contrast, the replacement of SiO_2 by Al_2O_3 increases the mobility of Na^+ by suppressing non-bridging oxygens and forming Al–O–Al bridges which do not restrain the alkali metal cations as much.

Technical glasses with high electrical resistance contain large proportions of B_2O_3 or PbO and only small amounts of alkaline oxides.

It has been shown[382] that the ratio of the conductivies of glasses is related to the ratio of the thermodynamic activities of the alkaline oxide:

$$\sigma_1/\sigma_2 = (a_1/a_2)^{1/2}$$

where a_i is the activity of the alkaline oxide X_2O in the glass i and σ_i is the conductivity.

Fig. 11.3. Effect of the partial replacement of SiO_2 by other oxides on the electrical conductivity of the glass $82\,SiO_2 \cdot 18\,Na_2O$. After Ref. 381.

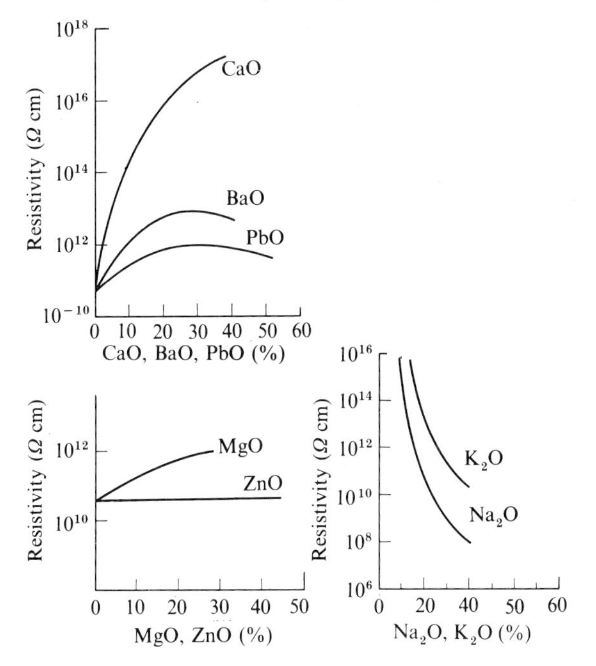

(f) *The "mixed alkali" effect*

When a second alkaline oxide is added in varying proportions, it is observed that the resistivity passes through a pronounced maximum when the two oxides are present in approximately equal proportions (Fig. 11.4). The activation energy and the pre-exponential term A in equation (11.1) pass through a maximum, and although they act with contrary effects, the activation energy dominates.

An analogous effect is observed for diffusion and internal friction (cf. Chapter 9). Semi-quantitative theories have been proposed to explain these effects, but none is satisfactory. More recently Hendrickson and Bray[384] presented a theory based on the interaction energy W of two neighboring oscillating dipoles. W depends on the difference in the resonant frequencies of the two dipoles and is not zero when the dipoles are different (i.e. different cations). This interaction energy is negative, increasing the depth of the potential wells of the participating ions which reduces their mobility.

Fig. 11.4. The "mixed alkali" effect on the resistivity of binary alkali metal–silicate glasses. After Ref. 383.

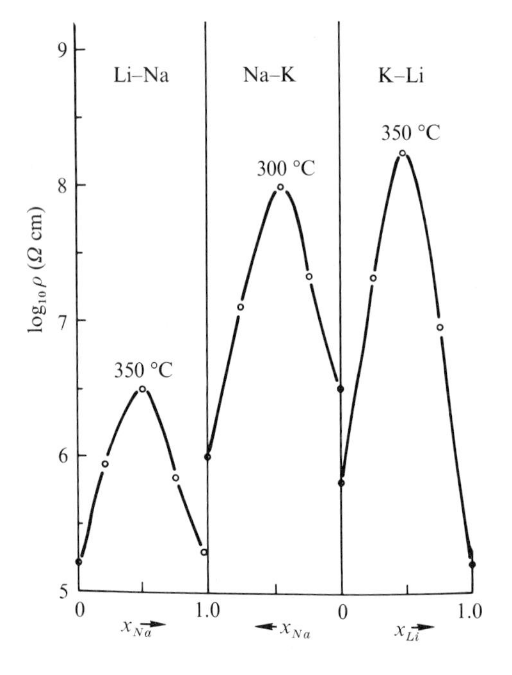

11.1.3 *Electronic conduction*

In certain glasses the conduction is of an electronic origin. If prolonged electrolysis of a glass between metallic electrodes does not lead to a change of electrical conductivity, it may be assumed that the charge carriers are electrons. High conductivity at low temperature with a low activation energy also favors an assumption of an electronic mechanism.

Electronic conduction is present in several classes of glasses.

1. Certain glasses containing transition element ions in several valence states. For example, glasses based on vanadium oxide: ($BaO\,2V_2O_5$), ($0.6V_2O_5\,0.4TeO_2$) and the glasses V_2O_5–P_2O_5 which contain V^{5+} and V^{4+} ions. Different silicate, borate and phosphate glasses containing Fe, Co and Mn are in this category.
2. Chalcogenide glasses: glasses of elements of group VI (S, Se, Te) alone or in combination with elements of group V (P, As, Sb, Bi) often associated with Tl, Ge, etc.
3. Thin films: the classic semi-conductors Ge, Si and InSb as well as Te, SiC, GeSi, etc., deposited from a vapor phase in the form of thin films.
4. Thin films of oxide glasses containing metals of the Pt, Ir, Pd family.

5. Certain vitreous polymers.
6. Metallic glasses.

The properties of electronic conduction in amorphous solids are the subject of numerous studies. For a presentation of the various theories which are in constant evolution, the interested reader can refer to the work of Mott[385] and Refs. 386–8. More recent results are to be found in Refs. 389–96.

(a) *Mechanisms of conduction*

To give an idea of the problems encountered, consider the schematic representation of the distribution of the density of states (Fig. 11.5) where $N(E)$ represents the number of states per unit volume per unit interval of energy as a function of the energy E of the electrons. Different states can be observed according to their electronic character: *extended states* ES, tails of bands T and the forbidden band G. The latter two represent *localized states.*

Fig. 11.5. Schematic diagrams of the density of electronic states for a semiconductor glass: (*a*) an ideal glass, (*b*) glass with localized states, (*c*) Cohen–Fritzsche and Ovshinsky model. After Ref. 397.

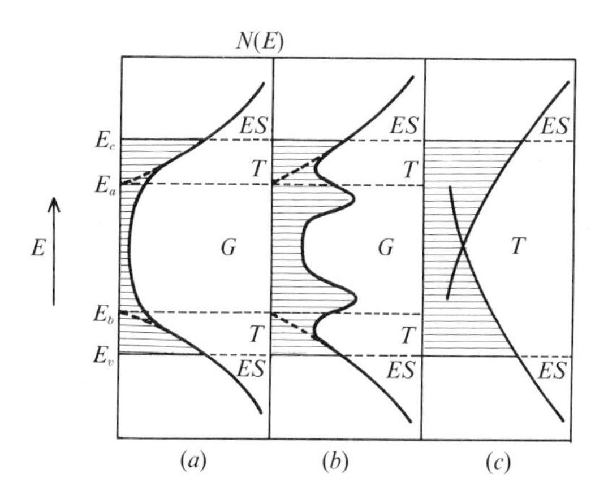

Localized states are those for which the wave functions decrease exponentially with distance. Non-localized or *extended* states refer to those for which the wave functions extend across the entire lattice.

There is currently general recognition of the following electronic conduction mechanisms in glasses.

1. *Band conduction.* Conduction in the extended states by electrons possessing energies just above E_c (or by holes just below E_v).

The mean free path of the electrons and the coherence length are then of the same order of size as the interatomic distances. Following Cohen's theory[398] the conduction is essentially comparable to a diffusion process, the electron jumping from one site to another with a frequency $\nu \sim 10^{15}\,\mathrm{s}^{-1}$ without the necessity of thermal activation. In these conditions, the mobility μ of the electron (or of a hole) is:

$$\mu = \frac{1}{6}\frac{ea^2}{kT}\nu_e$$

where a is the average interatomic distance.

The mobility estimated from this equation is $\sim 10\ \mathrm{cm}^2\ \mathrm{V}^{-1}\ \mathrm{s}^{-1}$. According to Cohen, the limit for diffusive mobility in the extended states is $\sim 10^{-2}\ \mathrm{cm}^2\ \mathrm{V}^{-1}\ \mathrm{s}^{-1}$. Below E_c localization occurs and this sudden transition is considered to be fundamental in the electronic conduction theory for glasses.

2. *Phonon-assisted conduction.* Conduction by jumps ("hopping") across the localized states (T and G) requires the aid of phonons.

In this thermally activated mechanism, the mobility $\mu(E)$ is:

$$\mu(E) = \frac{eR^2(E)}{kT}\exp(-2\alpha R)\,\nu\exp\left(-\frac{W}{kT}\right)$$

where R is the average jump distance which depends on the distribution $N(E)$; the term $\exp(-2\alpha R)$ describes the overlap of the wave functions of neighboring sites, α being the parameter which controls this decrease, and $\nu\exp(-W/kT)$ represents the probability for the jump per second of a localized electron to a site situated at W above the initial site. This equation predicts mobilities (at ambient temperature) which are less than or equal to $10^{-2}\ \mathrm{cm}^2\ \mathrm{V}^{-1}\ \mathrm{s}^{-1}$.

Near to E_c or E_v, the mobility is thus reduced by about 10^3 which defines a "mobility gap," a concept which became important in the case of non-crystalline solids.

3. *Conduction by tunneling near the Fermi level.* At low temperatures, a "variable range hopping" mechanism exists in which it becomes energetically favorable for a carrier to jump towards a site which is more remote but which is energetically ($\lesssim kT$) closer to its initial site. The conduction is then in the neighborhood of the Fermi level E_F, and Mott showed that the conductivity σ in this case is:

$$\sigma = \sigma_0 \exp\left(-\frac{T_0}{T}\right)^{1/4}$$

with

$$T_0 =\sim 18\,[\alpha/kN(E_F)]$$

The diagram in Fig. 11.5(*a*) is relative to an "ideal" glass. In a real glass (Fig. 11.5(*b*)) it is necessary to add the localized states caused by structural defects in the disordered network.

A different model was proposed by Cohen, Fritzsche and Ovshinsky[399] in which the valence and conduction bands have tails extending across the forbidden band (Fig. 11.5(*c*)). It seems at the present time that the model of Fig. 11.5(*b*) is more appropriate, even for more complex glasses. The question which presents itself is that of the origin and nature of the localized states, their discrete levels having to be associated with specific structural defects.

(b) *Switching phenomena*

"Switching" or "memory" effects have been shown in chalcogenide glasses such as As–Te–(I, Br, Se); As–Tl–Se; Tl–As–(Se, Te); As–Te–Si–Ge, in non-cyrstalline Si and Ge as well as in vanadium phosphates, sodium titanates, and certain organic semi-conductors – such effects probably occur in all the semi-conducting glasses.

Switching devices based on chalcogenide glasses can be classified into two categories: *monostable* (threshold devices) and *bistable* (memory devices). The character depends essentially on the composition of the glass. The behavior of a monostable device is shown in Fig. 11.6(*a*). The glass has two very different conductivity states. The current–voltage characteristics first show a high resistance (state (1)) then, above a threshold voltage V_{TH}, the system suddenly becomes conducting with the voltage falling to a much lower value (state (2)). This state maintains itself indefinitely provided the voltage does not fall too low; below I_H the system reverts to state (1). A glass showing this type of behavior is $As_{30}\,Te_4\,Si_{12}\,Ge_{10}$ ("STAG" glass) or $As_{36}\,Te_{39}\,Si_{17}\,Ge_7\,P_1$. The changes do not depend on the polarity and can be repeated indefinitely.

The behavior of a memory device is shown in Fig. 11.6(*b*). The system passes from state (1) to state (2) at V_{TH} and stays indefinitely in state (2) even if the current is reduced to zero. The passage from (1) to (2) is obtained by a millisecond pulse of several milliamperes. To return from state (2) to state (1), it is necessary to apply a pulse which is much more intense but of shorter duration (1–10 μs). A glass of this type is $Ge_{15}\,Te_{81}\,As_4$. The change from state (1) to state (2) is attributed to the formation of crystalline filaments by local devitrification of the glass under the thermal action of a sufficiently intense current; the inverse change

Fig. 11.6. Switching phenomena in semiconducting glasses: (*a*) monostable device, (*b*) bistable device. After Ref. 400

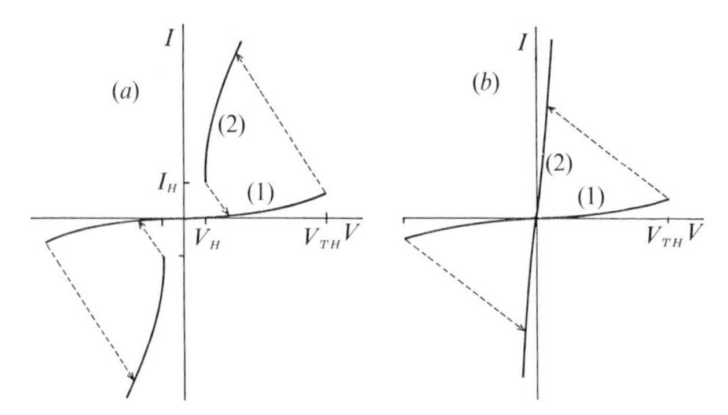

from (2) to (1) necessitates the remelting of the crystalline part followed by quenching to freeze the system in the vitreous state.

The work of Pearson *et al.*[400] and Kolomiets[401] mentioned these phenomena and Ovshinsky[402] drew attention to the practical applications in microelectronics (computers, displays, etc.) resulting in a large body of work which is reviewed in Refs. 387 and 388.

Initial hopes for the practical application of these devices ("OVONICS") have been dampened by their lack of reliability due to aging. From a theoretical point of view, all these studies posed numerous problems, and the mechanism of switching is not yet totally clear.

11.2 Dielectric properties

11.2.1 *Definitions*

The application of an external electric field $\vec{E}$ produces a flux $\vec{D}$ or electric displacement related to $\vec{E}$ by:

$$\vec{D} = \epsilon \vec{E}$$

which defines the electric permittivity ϵ of the substance. In the case of an isotropic substance, such as a glass, ϵ is a scalar.

The preceding relation can be compared with the relation established in Chapter 9:

$$\gamma = J\sigma_t$$

where the compliance J was defined as the ratio of the deformation γ to the stress σ_t. There exists an almost complete correspondence between dielectric properties and viscoelastic mechanical properties of glasses.

If a fixed potential is applied to a capacitor which has glass as the dielectric, the charge on the plates is $Q = AD$ (A being the plate area). Measurement of the charge as a function of time gives curves which are similar to those representing flow under constant shear stress (Fig. 11.7). The part of the current ($\mathrm{d}Q/\mathrm{d}t$) corresponding to the delayed elasticity is called the "absorption current" or "anomalous current." When the capacitor is short circuited, a current flows in the opposite direction and this is called the "discharge current." These currents, due to *delayed polarization* are present in all dielectrics, but ordinarily are difficult to demonstrate in liquids and crystals because the relaxation times are generally very short. In the case of common glasses, the time taken to return to equilibrium can be several minutes or more.

Fig. 11.7. Behavior of a dielectric: (*a*) ideal, (*b*) real.

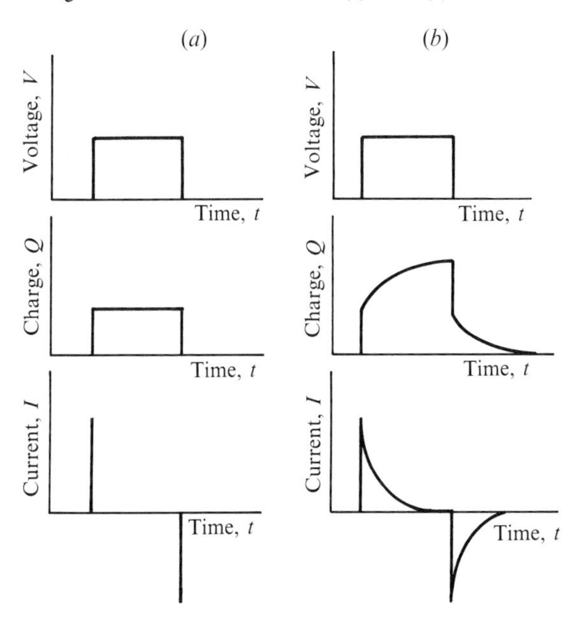

As for delayed elasticity, permittivity as a function of time can be defined as $\epsilon(t) = D/E$. In particular, the application of a constant field E_0 at $t = 0$ produces a displacement $D(t)$ such that:

$$\epsilon(t) = D(t)/E_0 = \epsilon_u + (\epsilon_R - \epsilon_u)F(t)$$

The permittivity changes progressively from the instantaneous (non-relaxed) value ϵ_u to the relaxed value ϵ_R, the relaxation being governed by a function $F(t)$. In general, $F(t)$ is not of the form $1 - \exp\left(-t/\tau\right)$; it is necessary to define a *distribution of relaxation times* (in viscoelasticity usually called *retardation times*).

Likewise, an alternating field with angular frequency ω:

$$E = E_0 \exp \mathrm{i}\omega t$$

produces a displacement:

$$D = D_0 \exp\left[\mathrm{i}(\omega t - \delta_E)\right]$$

which is out of phase by an angle δ_E relative to the field.

In complete analogy with mechanical phenomena, *complex permittivity* ϵ^* is defined by the relation:

$$D^* = \epsilon^* E^* = (\epsilon' - \mathrm{i}\epsilon'')E^*$$

where ϵ' and ϵ'' are real and imaginary components of the permittivity ϵ^*:

$$\epsilon' = |\epsilon| \cos \delta_E \qquad \text{and} \qquad \epsilon'' = |\epsilon| \sin \delta_E$$

ϵ'' is the dielectric *loss factor* and δ_E the *loss angle* defined by:

$$\tan \delta_E = \epsilon''/\epsilon'$$

Taking ϵ_0 as the permittivity of the vacuum, the *relative permittivity* κ^* (or complex dielectric constant) is defined as:

$$\kappa^* = \epsilon^*/\epsilon_0 = \kappa' - \mathrm{i}\kappa''$$

A capacitor formed from two conductor plates of area A separated in vacuum by a distance d has a capacitance :

$$C_0 = \frac{A}{d}\epsilon_0$$

Replacing the vacuum by a dielectric, the capacitance becomes:

$$C = \frac{A}{d}\epsilon$$

the ratio $C/C_0 = \epsilon/\epsilon_0 = \kappa'$

The current density j across a capacitor charged under a potential $V = Ed$ is:

$$j = \frac{\mathrm{d}D}{\mathrm{d}t} = \epsilon^* \frac{\mathrm{d}E}{\mathrm{d}t} = \mathrm{i}\omega\epsilon^* E = (\mathrm{i}\omega\epsilon' + \omega\epsilon'')E$$

The quantity $\chi = \omega\epsilon''$ in phase with the field E is thus equivalent to a *dielectric conductivity*. It represents all the dissipative effects resulting in losses by steady conduction (ohmic) as well as losses depending on the frequency (dielectric dispersion).

Certain authors use a complex electric modulus $\mathcal{M}^* = 1/\epsilon^*$ by analogy with the moduli defined in elasticity:

$$\mathcal{M}^* = \mathcal{M}' + \mathrm{i}\mathcal{M}''$$

where:

$$\mathcal{M}' = \epsilon'/(\epsilon'^2 + \epsilon''^2)$$

$$\mathcal{M}'' = \epsilon''/(\epsilon'^2 + \epsilon''^2)$$

This type of representation masks the effect of the electrode polarization, the high values of ϵ' appearing in the denominator, which is sometimes an advantage in some applications.

A theory formally identical to that for viscoelastic phenomena, based on the principle of linear superposition called the Curie–Hopkinson principle (the equivalent in viscoelasticity is the Boltzmann principle), relates $\epsilon(t)$ to the components $\epsilon'(\omega)$ and $\epsilon''(\omega)$ of the complex permittivity.

By introducing a normalized distribution $\psi(\tau)$ of the relaxation times (the corresponding term in viscoelasticity would be *retardation times*) of the variable $\ln\tau$:

$$\int_{-\infty}^{+\infty} \psi(\tau)\,\mathrm{d}\ln\tau = 1$$

the following expressions are obtained:

$$\epsilon(t) = \epsilon_\infty + (\epsilon_S - \epsilon_\infty) \int_{-\infty}^{+\infty} \psi(\tau)\left[1 - \exp\left(-t/\tau\right)\right] \mathrm{d}\ln\tau$$

$$\epsilon'(\omega) = \epsilon_\infty + (\epsilon_S - \epsilon_\infty) \int_{-\infty}^{+\infty} \frac{\psi(\tau)\,\mathrm{d}\ln\tau}{1 + \omega^2\tau^2}$$

$$\epsilon''(\omega) = (\epsilon_S - \epsilon_\infty) \int_{-\infty}^{+\infty} \frac{\omega\tau\psi(\tau)\,\mathrm{d}\ln\tau}{1+\omega^2\tau^2}$$

In these expressions, ϵ_∞, the permittivity at infinite frequency corresponds to ϵ_u while ϵ_S at zero frequency corresponds to ϵ_R. In the expression for $\epsilon''(\omega)$, the term corresponding to ohmic conductivity χ/ω is sometimes separated.

In a manner analogous to that discussed in Chapter 9 for viscoelastic functions, it is possible to pass from $\epsilon(t)$ to $\epsilon'(\omega)$ and $\epsilon''(\omega)$ by Fourier transformations. In addition, the components $\epsilon'(\omega)$ and $\epsilon''(\omega)$ are not independent; they are related by the Hilbert transformations (the Kramers–Kronig relation of optics).

To obtain the curves in reduced coordinates, the following values are used:

$$\frac{\epsilon' - \epsilon_\infty}{\epsilon_S - \epsilon_\infty} \quad \text{and} \quad \frac{\epsilon''}{\epsilon_S - \epsilon_\infty}$$

as a function of ω/ω_{max} where ω_{max} represents the frequency of the maximum. The temperature–frequency correspondence principle permits the derivation of a reduced curve (master curve) as for viscoelasticity.

11.2.2 *Polarization mechanisms*

The relation between permittivity and polarization is furnished by the classic Clausius–Mossotti equation:

$$\frac{\epsilon^* - 1}{\epsilon^* + 2} = \frac{1}{3\epsilon_0} \sum N_i \alpha_i$$

where N_i is the number of species of polarizability α_i per unit volume (for a derivation, see e.g. Ref. 403).

There are four primary polarization mechanisms in glasses: each one corresponds to a short-range charge displacement which contributes to the total polarization:

—electronic polarization,
—atomic polarization,
—orientation polarization,
—interfacial polarization.

Electronic polarization is due to the deformation of the electron cloud relative to the nucleus. This mechanism is activated at very high frequencies (10^{15} Hz) in the UV. The index of refraction of glass thus depends on the electronic polarization.

Atomic polarization is produced at IR frequencies (10^{12}–10^{13} Hz). It corresponds to the displacement of positive and negative charges. In the case of glasses, the range of this absorption can be quite extensive.

Orientation polarization or dipole orientation is related to the perturbation of the thermal agitation movement of ionic dipoles or molecules when they are subjected to an electric field.

Two cases are distinguished:

(a) The case where a permanent dipole is returned to an equilibrium position by a restoring force; e.g. oscillation of a Si–O–Si bond under the action of an oscillating field of frequency $\sim 10^{11}$–10^{12} Hz. (These mechanisms have been called *deformation polarization* by Stevels.[404])

(b) The case where the dipoles oscillate between two equivalent equilibrium positions. This is particularly true for interstitial ions (modifiers) in the glass network. The oscillations are always present, but the application of a field makes jumps in the direction of the field more probable. The jump distance being appreciable, the frequencies of these migration losses are between 10^3 and 10^6 Hz. Mobile cations contributing to this process are the same ones responsible for ionic conductivity, but this does not imply long-range migrations.

Interfacial polarization or *space charge* occurs when the charge carriers encounter a physical barrier (surface, etc.) which blocks their movement and produces a charge accumulation and local polarization. Barriers separated by distances of the order of a centimeter correspond to frequencies $\sim 10^{-3}$ Hz. However, internal inhomogeneities (unmixing, etc.) can give rise to polarizations for frequencies near 10^3 Hz which are difficult to separate from orientation polarization processes.

Figure 11.8 summarizes the effects discussed above.[404]

The effects due to electronic or atomic polarization are related to optical properties; only losses by orientation and interfacial polarization will be examined here.

11.2.3 *Experimental methods*

Two classes of experimental methods are used:

1. *Charging or discharging currents* are measured as a function of time (Fig. 11.9) and the permittivity is calculated as a function of the frequency.

2. The dielectric characteristics are measured by *dynamic* methods as a function of the frequency.

In methods using a steady or low frequency current, the choice of electrodes is critical; it is necessary to use electrodes capable of furnishing or accepting the charge carrying ions (amalgam or fused salt electrodes for example) to avoid measuring the properties of the electrodes rather

Fig. 11.8. Various types of dielectric losses in a glass: (*a*) conduction, (*b*) dipole relaxation, (*c*) deformation, (*d*) vibration. After Ref. 404.

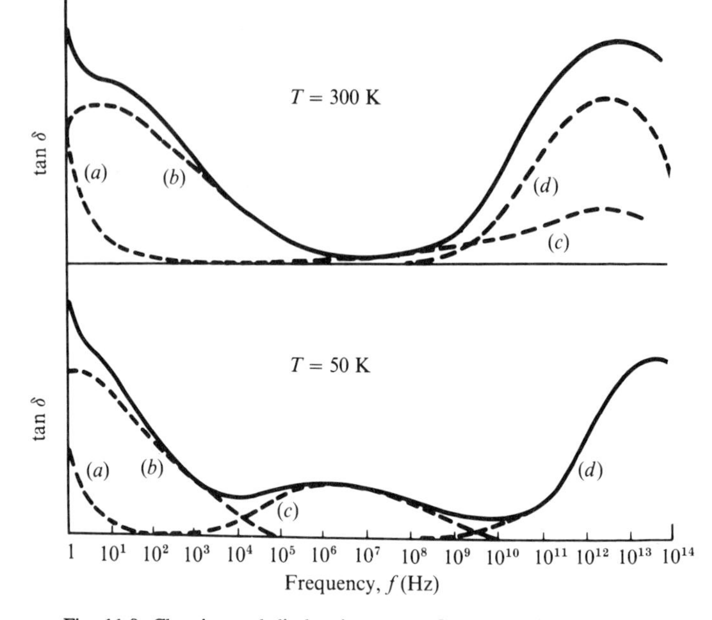

Fig. 11.9. Charging and discharging curves for a soda–lime–silica glass. After Ref. 405.

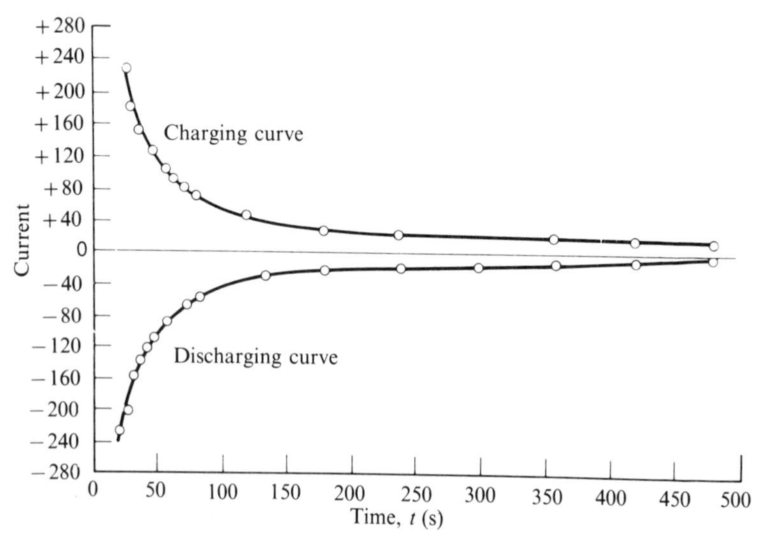

than those of the dielectric. These *electrode polarization* effects are related to phenomena at the interface and are explained by space charge mechanisms. They can, however, be used for the study of glass surfaces.

Figure 11.10 gives ϵ' as a function of the frequency for several surface conditions. Figure 11.11 shows an example of the determination of ϵ' and ϵ'' as a function of frequency and temperature and reduced loss curves for several glasses are given in Fig. 11.12. The experiment shows that the frequency f_m corresponding to maximum of the peak has the same activation energy as the steady current conductivity, which implies that the same charged species are responsible for both phenomena.

It is found that:[408]

$$f_m = \frac{\sigma}{2\pi\epsilon_0\epsilon_\infty}$$

or:[409]

$$f_m = \frac{\sigma}{2\pi\epsilon_0\Delta\epsilon}$$

with $\Delta\epsilon$ representing the amplitude of the dielectric dispersion: $\Delta\epsilon = \epsilon_s - \epsilon_\infty$.

Fig. 11.10. Effects of electrode polarization on the permittivity ϵ' for different surface conditions. After. Ref. 406.

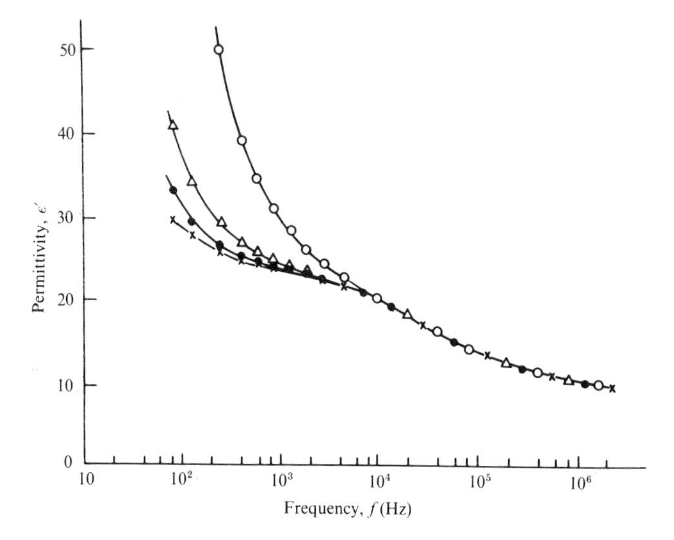

Fig. 11.11. Determination of the components ϵ' and ϵ'' as functions of the frequency and temperature for a typical soda–lime–silica glass. After Ref. 407.

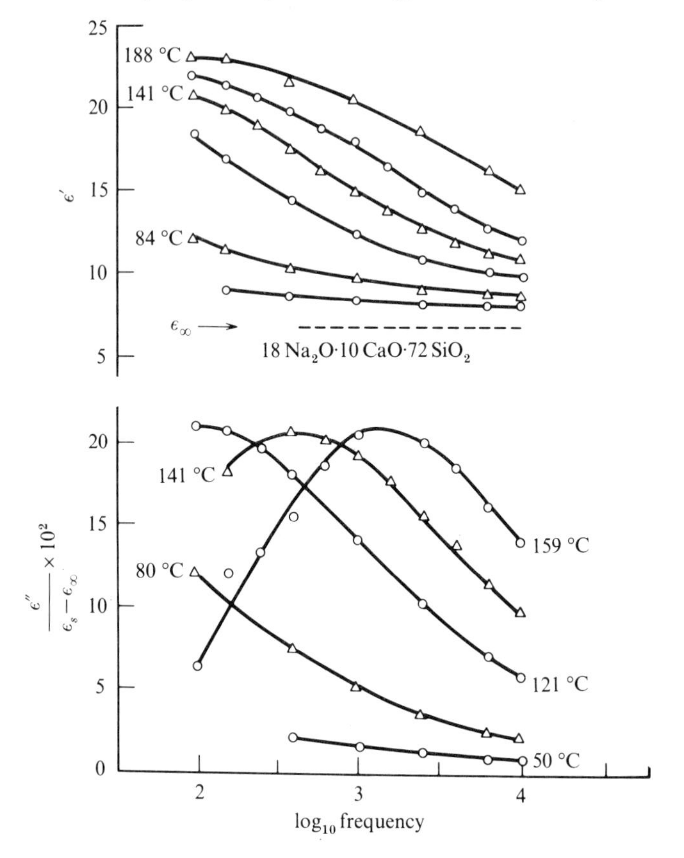

It is apparent that the peaks are wide and asymmetric which is incompatible with a unique relaxation time and necessitates the introduction of a distribution of relaxation times.

11.2.4 *Theoretical models*

Several models have been proposed to account for the experimental results.

1. Stevels[410] and Taylor[411] present a model where the potential barriers are of variable heights (Fig. 11.13). For steady current conduction, the largest barrier must be surmounted, while in an alternating field, the migration is related to shorter distances and to overcoming lower barriers. Such a mechanism has been considered compatible with a disordered net-

Fig. 11.12. Reduced dielectric loss curves for different soda–lime–silica glasses. After Ref. 407.

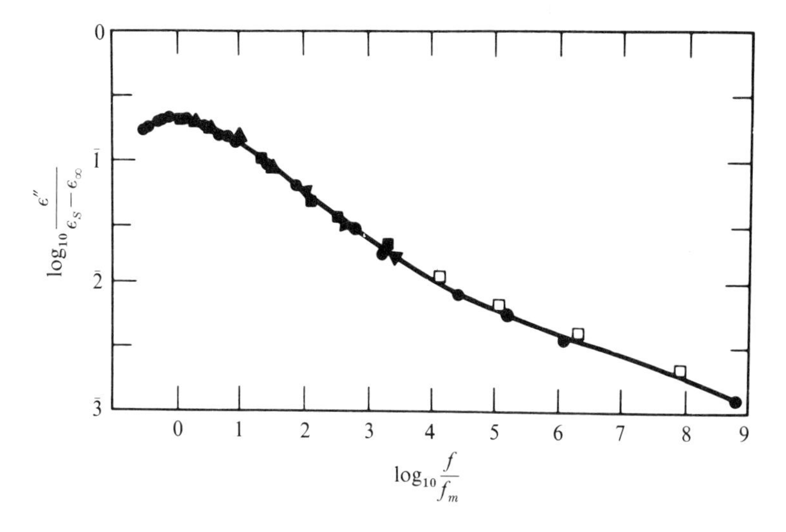

Fig. 11.13. The random potential model. After Ref. 410.

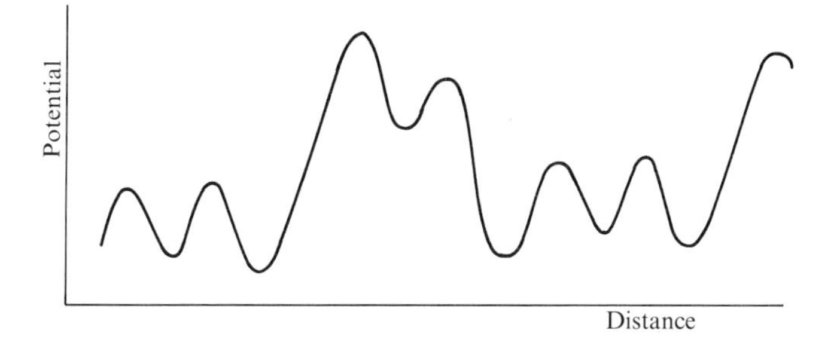

work. However, it predicts different activation energies for the two types of conduction which is contrary to experimental evidence.

2. Charles[412] proposed a defect migration model where the alkali metal ion in the network remains near a non-bridging oxygen. He assumed that close to such an atom a certain number of equivalent stable positions exist. For example (Fig. 11.14), considering three non-bridging oxygens, the passage of a Na^+ ion from position 1 to 1′ followed by the movement of another Na^+ ion from 2 to 2′ is simultaneously equivalent to a migration (conduction) and a change of orientation in the polarization. This model leads to the preceding relations between f_m and ϵ. In contrast, the oscillation of the same ion between two positions around an oxygen leads to polarization without conduction.

Fig. 11.14. Model of the migration of a defect with polarization. Large circles are O, small circles are Na. After Ref. 412.

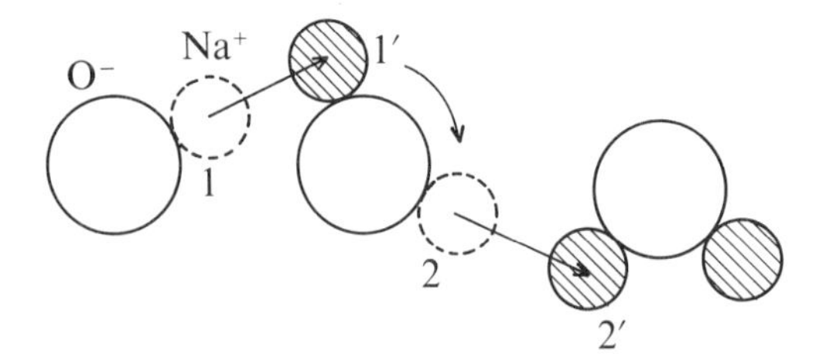

3. The classical theory of *Maxwell–Wagner–Sillars*[413] for heterogeneous dielectrics permits the interpretation of the results in terms of a distribution of particles having spherical or ellipsoidal form. To confirm this hypothesis, it is necessary to have independent information on the structure of the glass. Excluding the possibility of electrode polarization, the dielectric response can be attributed to either a superficial layer or the presence of regions of unmixing.

The position of the maximum of ϵ'' depends on the form of the particles and the conductivity of the dispersed phase. Such effects have been observed in glasses undergoing phase separation.[414]

4. The *Debye–Falkenhagen* theory, founded on the classic *Debye–Hückel* theory of electrolytes, analyses the formation of an ionic "atmosphere" of opposite sign to the ion considered and the deformation of the atmosphere under the influence of an electric field. It leads to a very extended maximum in the dielectric loss and to a proportionality relation between f_m and $\sigma/\epsilon_0\epsilon_\infty$.[406]

It appears that this interaction can be a sufficiently general mechanism to explain dielectric relaxation in the most diverse glasses.

12 Optical properties of glasses

The optical properties of glasses are based on the interaction of the material with the energy of electromagnetic waves.

Spectroscopic methods have already been reviewed relative to the study of glass structure (cf. Chapter 5), and we have seen that they cover an extended range from γ rays to Hertzian frequencies. These methods are not specific to glasses. Indeed, we have seen that their application to non-crystalline substances suffers from difficulties of interpretation due to a lack of appropriate theoretical support.

In this chapter, we examine glass as an optical *material* and point out several recent specific applications. The interest in glasses in optics is related to certain intrinsic characteristics: *isotropy, a high degree of homogeneity*, the possibility of *extended and continuous* variation of their properties by variations in composition (absence of stoichiometric constraints), the possibility of obtaining pieces without dimensional limitations (as opposed to single crystals). In the majority of cases, only short-range order influences optical phenomena — interesting properties of crystals are obtained but without their inconvenience.

Optical glasses are used to make prisms, lenses, filters, mirror supports, etc. Recent applications have used glass in the construction of lasers and light guides (photonic light conductors). The properties of birefringence under stress are utilized in the control of tempering states.

12.1 Review of definitions

The propagation of an electromagnetic wave in a material produces a displacement of electric charges. For a sinusoidal wave, the change of speed and intensity are contained in the *complex index of refraction* n^* which is related to the complex relative permittivity $\kappa^* = \kappa' + \mathrm{i}\kappa''$ by the relation:

$$n^{*2} = \kappa^*$$

Taking $n^* = n - \mathrm{i}k$ where n is the *refractive index* and k the *absorption index*, the following relations are obtained:

$$n^2 - k^2 = \kappa' \qquad 2nk = \kappa''$$

The quantity $\chi = \kappa - 1$ is the electric susceptibility.

12.1.1 *The refractive index and dispersion*

The refractive index is equal to the ratio of the speed of light in vacuum v_0 to that in the material v_m; it depends on the wavelength and normally decreases with increasing wavelength. This variation is called *dispersion* and can be defined by the relation:

$$D = \frac{\mathrm{d}n}{\mathrm{d}\lambda}$$

The variations of n and k are related (Fig. 12.1), the index undergoing a variation in the opposite sense in the region of strong absorption (*anomalous* dispersion).

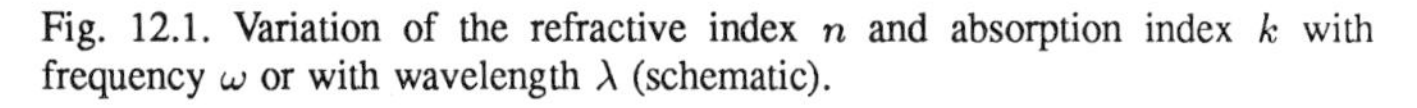

Fig. 12.1. Variation of the refractive index n and absorption index k with frequency ω or with wavelength λ (schematic).

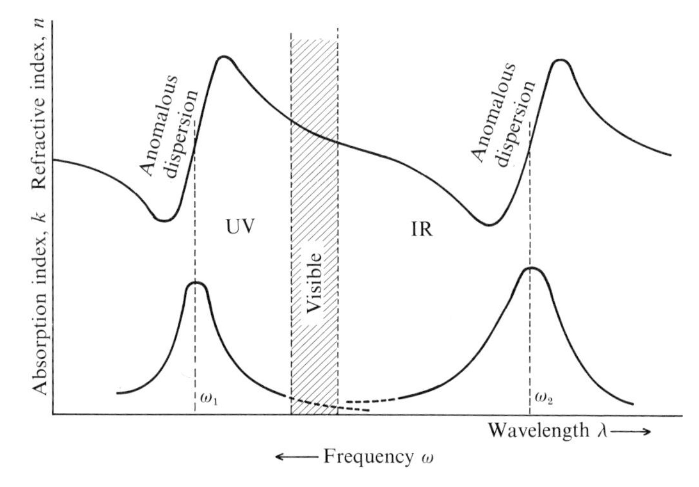

In classical theory these regions correspond to natural resonance frequencies of the vibrators excited by the incident electromagnetic wave. It is shown that for an angular frequency ω:

$$(n - \mathrm{i}k)^2 - 1 = \frac{e^2}{m\epsilon_0} \sum_{j=1}^{l} \frac{N f_j}{\omega_j^2 - \omega^2 + \mathrm{i}g_j\omega}$$

for l oscillators of strength f_j and natural frequency ω_j; g_j is the damping constant of the oscillator, m the mass of an electron, e the electron charge and ϵ_0 the permittivity of a vacuum. (Quantum mechanics leads to a result of the same form but with different definitions for f_j and g_j.)

Separating the real and imaginary parts:

$$n^2 - k^2 = 1 + \frac{e^2}{m\epsilon_0} \sum_1^l \frac{N f_j (\omega_j^2 - \omega^2)}{(\omega_j^2 - \omega^2)^2 + \omega^2 g_j^2}$$

$$2nk = \frac{e^2}{m\epsilon_0} \sum_1^l \frac{N f_j \omega g_j}{(\omega_j^2 - \omega^2)^2 + \omega^2 g_j^2}$$

Far from the fundamental frequencies $k \sim 0$ and, when $g_j \sim 0$, the first equation is equivalent to the Sellmeier relation:

$$n^2 - 1 = \sum_1^l \frac{A_j \lambda^2}{\lambda^2 - \lambda_j^2}$$

where λ is the wavelength and A_j and λ_j are constants. This formula is frequently used in the interpolation of current dispersion curves.

12.1.2 *Non-linear refractive index*

The dependence of the refractive indices on the intensity of the light is governed by the third-order electric susceptibility. The refractive index is related to the time average $\langle E^2 \rangle$ of the electric field of the beam by the formula:

$$n = n_o + n_2 \langle E^2 \rangle$$

where n_o is the usual linear index and n_2 the coefficient of the non-linear refractive index.

12.1.3 *Absorption coefficient*

From the absorption index k, a function of the wavelength λ, the *absorption coefficient* $\alpha = 4\pi k/\lambda$ is defined. For a homogeneous material, the relative fraction of the light intensity absorbed in traversing a thickness $\mathrm{d}x$ depends on α:

$$\frac{\mathrm{d}I}{I} = -\alpha\, \mathrm{d}x$$

The attenuation of the light traversing a material of thickness x is thus given by the formula:

$$I = I_0 \exp(-\alpha x)$$

where I is the transmitted intensity, I_0, the incident intensity and α the linear absorption coefficient. Instead of *transmission*: $(I/I_0) \times 100$, the *optical density* (OD) is often used where:

$$\mathrm{OD} = \log_{10}(I_0/I)$$

(For example 1 % transmission corresponds to an OD = 2.)

The technology of optical conductors has popularized the use of the *decibel*, which is related to the OD by:

$$1\,\mathrm{dB} = \frac{1}{10}(\mathrm{OD})$$

The losses in optical fibers are given in dB km^{-1}. Note that if the units (OD) or dB are used, it is necessary to specify the thickness traversed x:

$$\alpha = \ln\left(\frac{I_0}{I}\right)\Big/x = 2.303\,(\mathrm{OD})/x = 23.03\,\mathrm{dB}/x$$

The absorption can be due to a particular ion (chromophore), the absorption coefficient then being proportional to the concentration c of the ion. In these conditions, $\alpha = \epsilon c$ where ϵ is the *extinction coefficient* and:

$$I = I_0 \exp(-\epsilon c x)$$

which is the Beer–Lambert law.

Reflection:

The intensity fraction R reflected under normal incidence is given by the Fresnel formula:

$$R = \frac{(n-1)^2 + k^2}{(n+1)^2 + k^2}$$

which far from the absorption bands ($k \sim 0$), reduces to:

$$R = (n-1)^2/(n+1)^2$$

Review of units

The refractive index and the absorption index are given as functions of the wavelength or the energy of the incident photons. The wavelengths are generally indicated in micrometers (μm), millimicrons (mμm), Angstroms (Å) or nanometers (nm).

$$1\,\mathrm{nm} = 1\,\mathrm{m}\mu\mathrm{m} = 10\,\text{Å} = 10^{-3}\,\mu\mathrm{m}$$

The energy E and λ are related by: $E(\mathrm{eV}) = 1239.8/\lambda\,(\mathrm{nm})$. The frequency $\nu\,(\mathrm{cm}^{-1}) = 1/\lambda\,(\mathrm{cm})$.

12.1.4 *Induced anisotropy — Birefringence — Photoelastic effect*

Under certain conditions glasses can become anisotropic. The most frequent cause is the application of mechanical stress which induces birefringence. The propagation velocity and thus the index of refraction, depend on the orientation of the plane of polarization. Under the action of an axial stress σ_z (Fig. 12.2), the glass behaves like a uniaxial substance. The propagation velocity of light travelling parallel to σ_z is independent of the orientation of the plane of polarization. For rays propagating perpendicularly to σ_z, the velocity varies according to whether the plane of polarization is perpendicular to σ_z ("ordinary" ray) or parallel to σ_z "extraordinary" ray). The glass then has two refractive indices corresponding to n_o and n_e and the *birefringence*, which can be positive or negative, is defined by:

$$\Delta n = n_e - n_o$$

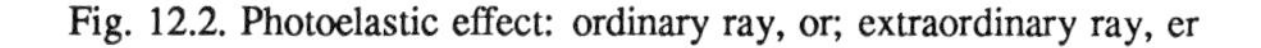

Fig. 12.2. Photoelastic effect: ordinary ray, or; extraordinary ray, er

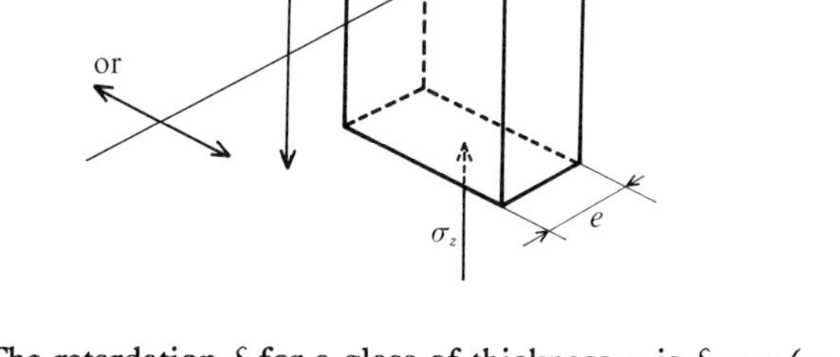

The retardation δ for a glass of thickness x is $\delta = x\,(n_e - n_o)$: usually it is evaluated by the ratio δ/x in nanometers per centimeter. The relative retardation is:

$$r = \frac{n_e - n_o}{n}$$

where n is the index of stress-free glass. The relative retardation is a measure of the *deformation.* The relation:

$$n = C\,(p - q)$$

where p and q are principal stresses, defines the photoelastic constant C which is measured in square meters per Newton in the international system. A spectral unit called a *brewster* (B) was proposed by Filon:

$$1\,(\mathrm{B}) = 10^{-12}\,(\mathrm{m}^2\ \mathrm{N}^{-1})$$

For an ordinary glass, $C = 2.6$ B.

In other cases the birefringence can arise from freezing-in elastic deformation[415,416] or from a micro-structure in the glass caused by phase separation or the presence of oriented particles.[417]

12.2 Optical glasses

12.2.1 *Transmission properties*

(a) *Oxide glasses*

Current use of oxide glasses is based on their good transmission in the "optical" part of the spectrum (UV + visible + near IR). This "optical window", corresponding to the spectral sensitivity of the human eye, is bounded at the UV end by electronic transitions from the valence band to the conduction band and at the IR end by the natural vibration frequencies of the constituent ions in the network (Fig. 12.1). Absorption in the visible region results from the superposition of the "tails" of the electronic and vibrational transitions to which must be added contributions from impurities such as transition element ions and color centers.

Traces of H_2O in oxide glasses can be a troublesome impurity in certain applications. Studies of the concentration of H_2O as a function of the partial pressure show that H_2O is often present in the form of hydroxyls OH^-. Its presence is apparent in the near IR, near 2.8 μm (3570 cm^{-1}). In the case where the OH^- can interact with non-bridging oxygen in the network, additional absorption bands are present near 3.5 and 4.5 μm (Fig. 5.7). These various absorption bands are used to determine the OH^- content of a glass.

The bands due to H_2O play an important rôle in the choice of glasses for optical fibers (Section 12.3). A combination band found near 0.945 μm is in an important region for applications using laser sources. The problem of "water in glass" is more fully discussed in Refs. 418 and 419.

(b) *Chalcogenide glasses*

The replacement of O by S, Se, or Te as network- forming anions displaces the IR absorption limit towards longer wavelengths. These glasses are opaque in the visible and generally begin to transmit from 1 to 1.5 μm for selenides and from 2.0 μm for tellurides. The most used glasses for IR optical windows are: $Ge_{28}Sb_{12}Se_{60}$ (1–14 μm) and $Ge_{33}As_{12}Se_{55}$ (0.8–16 μm).

The transmission is very sensitive to O impurity which must be excluded. These glasses are therefore usually prepared in sealed tubes under vacuum or protective atmosphere.

Figure 12.3 shows the transmission curves for several materials based on chalcogenides.

Fig. 12.3. Transmission of chalcogenide glasses in the IR (schematic). Dashed line, reflection maxima.

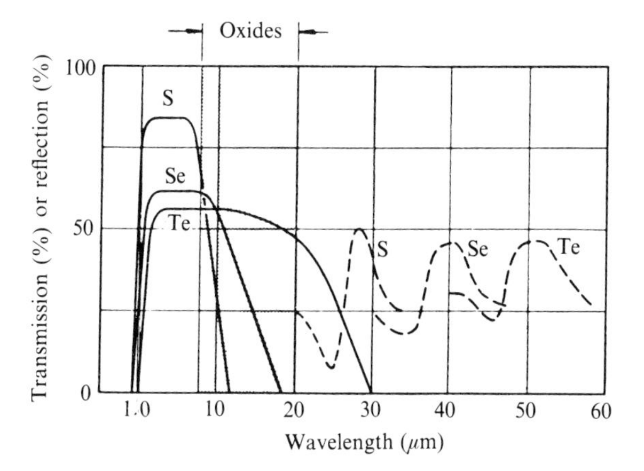

(c) *Halide glasses*

Halide glasses based on ZrF_4 and HfF_4 have excellent transmission properties in the IR region. The best compositions are found in the ZrF_4–BaF_2–MF_3 systems where M is Al, La, or Gd. The glass-forming limits of some of the systems are shown in Fig. 8.1 and their optical transmission compared to that of vitreous SiO_2 in Fig. 8.2.

12.2.2 *Classification of optical glasses*

Optical glasses for the formation or transmission of images are used in the construction of many different instruments. For these applications, they must have good homogeneity and a constant and well-defined index. The index variation must not exceed 10^{-4} in ordinary applications, 10^{-5} in ordinary optics and 10^{-6} in glasses destined for certain scientific applications or for astronomy. This distinguishes optical glasses from ordinary industrial glasses where larger index variations are tolerated.

The older fabrication processes start with fusion in a refractory crucible or pot (cf. Chapter 17). The crucible is then cooled slowly to ambient temperature to avoid thermal stresses, then broken and the most homogeneous pieces of glass are sorted out. Currently the fusion is done in a Pt lined furnace and the fined glass poured into a mold.

Moreover, in optical applications, it is not sufficient that the index n be constant, it must also be known with high precision. Following the dispersion relations, n varies with the wavelength: Fig. 12.4 shows an example of this for several common optical glasses.

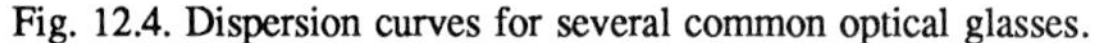

Fig. 12.4. Dispersion curves for several common optical glasses.

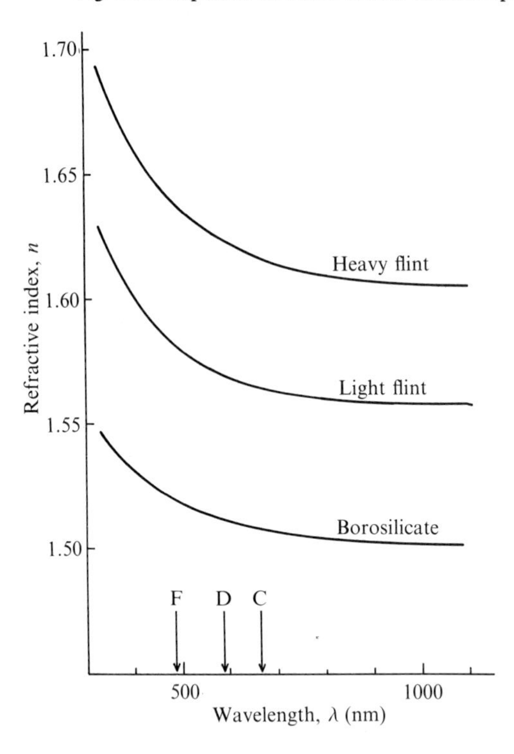

In practice, an optical glass is identified by a value n_D which is the refractive index for an intermediate wavelength 589.6 nm (yellow D line of sodium) and the dispersion is characterized by Abbe's number ν or *constringence*.

$$\nu = \frac{n_D - 1}{n_F - n_C}$$

where n_F and n_C are respectively the refractive indices for reference lines of hydrogen: F (486.1 nm, blue) and C (656.3 nm, red). Other lines may be used, as in the German standard (DIN 58 925).

A very large number of optical glasses exist: it is possible to obtain series of glasses each with a different index n_D and for each n_D, a whole range of dispersions ν. Figure 12.5 shows different types of optical glasses furnished by industry and characterized by the parameters n_D and ν.

Fig. 12.5. Regions of known optical glasses classified by their refractive index n_D and Abbe's number ν. The lines of constant non-linear index n_2 are indicated. After Ref. 316.

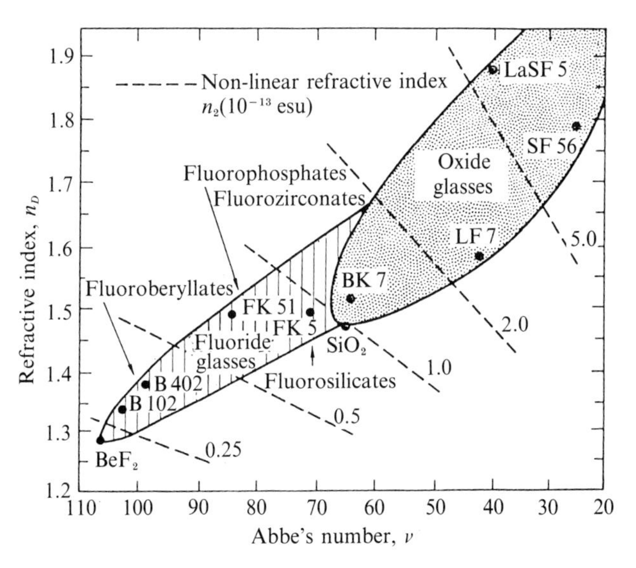

Glasses with low n_D and $\nu > 55$ are called *crown glasses*, while those with high n_D and $\nu < 50$ are called *flints*. Numerous intermediate varieties have been produced and they find application in the *chromatic correction* of objectives and other optical systems.

Manufacturers' catalogs use a numerical code where the type of glass is followed by three decimals of the value $n_D - 1$ and then three digits of ν (rounded). For example, a barium-crown 5 for which $n_D = 1.556710$ and $\nu = 58.65$ would be coded as 557 586.

It is necessary to note that the common industrial glasses (cf. Chapter 8) are nearly all of the "crown" type. Only the PYREX glasses tend toward the borosilicate-crown field and the "crystal" glasses towards the flints.

Considerable effort has been expended to extend the fields of Fig. 12.5. There has been some success by the substitution or addition of other glass-forming oxides to SiO_2 e.g. B_2O_3, P_2O_5, GeO_2, etc. and the introduction of heavy modifying oxides such as BaO, La_2O_5, PbO, etc. The demand for optical glasses with diverse characteristics has stimulated glass research in the past (e.g. the work of Abbe) and this effort in synthesis continues to the present time. The preceding graph is constantly being extended with the addition of new types, for example fluoroberyllate glasses, fluorophosphates, fluoroborates, fluorosilicates, etc.

The importance of the non-linear refractive index in the technology of high-intensity lasers has led to research on glasses having a very small n_2. It has been shown that glasses with low refractive index and low dispersion such as fluoroberyllates also have the smallest n_2.[316] Fluoroberyllate glass systems, based on BeF_2 as the glass-former, were progressively extended towards fluorophosphates which are now used in the USA as laser hosts in gigantic installations testing the possibility of nuclear fusion by inertial confinement.

The methods of measuring the optical constants of glasses are well presented in Ref. 420, and a survey of optical properties of glass and optical materials is to be found in Ref. 421.

12.2.3 *Colored glasses*

If a glass selectively absorbs or scatters light in the visible part of the spectrum, the light will not be transmitted equally at all frequencies and the glass will appear colored. The impression of *color* is a subjective sensation, depending partly on the spectral sensitivity of the eye and partly on the nature of the incident light.

In contrast, a *spectral transmission* curve is precisely defined and physically measurable. It can give quantitative measurements of glass color. Numerous examples will be found in the book by Weyl.[422]

(a) *Colors due to transition metal ions*

In common glasses, absorption in the visible is principally due to ions of the transition elements characterized by incomplete 3d levels, particularly V, Cr, Mn, Fe, Co, Ni, Cu and, to a lesser degree, rare earths having incomplete 4f levels, or to color centers.

The ligand-field theory (Chapter 5) gives a quantitative interpretation of the phenomena and permits the study of the ion sites responsible for the absorption.

Figure 12.6 shows the transmission curves of several glasses containing transition metal ions. The colors depend on the oxidation state and the coordination number of the ion responsible (cf. Table 5.3). For example Co^{2+} in a silicate glass is in tetrahedral coordination relative to O and produces a *blue* color while in a metaphosphate (coordination 6) or a borosilicate, the color is pink.

Fig. 12.6. Transmission curves for glasses colored by transition metal ions. Below, the spectral sensitivity curve of the eye.

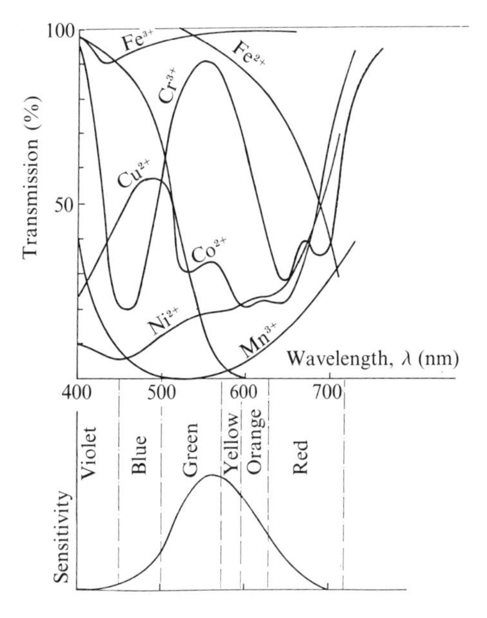

The valence of the central cation controls the spectral absorption. The change $M^{2+} \rightarrow M^{3+}$ decreases the ionic radius of the cation which can often lead to a change in coordination.

Cations of a given element with different valencies may coexist, the proportions depending on the O_2 partial pressure according to the following equilibrium:

$$\frac{4}{n} M^{x+} = O_2 \rightleftharpoons \frac{4}{n} M^{(x+n)+} + 2O^{2-}$$

In common glasses the case of Fe is very important. Fe exists as an impurity, either as Fe^{2+} (green color) or as Fe^{3+} (pale yellow color). Because Fe^{3+} is a less active colorant, the Fe^{2+} is changed to the trivalent state by the addition of oxidizers (nitrates, As_2O_3) to obtain a practically colorless glass. To compensate for the residual yellowish tint, selenium and cobalt oxide which absorb respectively in the blue and red, are added. This flattens the absorption curve of the glass which then appears totally colorless.

The spectra also depend on the nature of the anions which surround the central ion. Tetrahedrally coordinated Fe^{3+} with a S substituting for an O produces the "amber" color widely used in the bottling industry because of its strong absorption in the UV (photoprotection).

Another combination of absorbing ions or *chromophore* is the Cd–S pair. Cd^{2+} and S^{2-} ions alone are colorless; together, they produce a strong yellow tint. Increasing the temperature modifies the absorption and at a few hundred degrees the glass color shifts to orange-red.

Rare earth (Y, La, Gd, Yb, and Lu) ions produce absorption effects because of the separation of the levels in their 4f shells. These ions do not have bands in the visible part of the spectrum; in contrast, Nd gives a strong red-violet coloration (Fig. 12.7). (Glasses doped with Nd^{3+} are the basis of laser techniques.) Likewise, the Pr^{2+} ion gives a green color and the Er^{3+} ion a pink color. For a discussion, see Ref. 423.

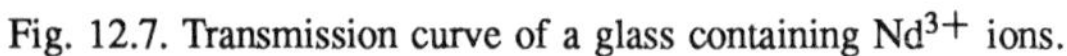

Fig. 12.7. Transmission curve of a glass containing Nd^{3+} ions.

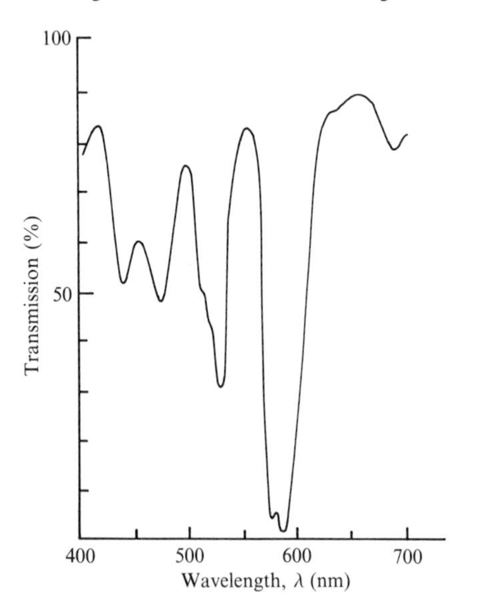

(b) *Color centers – Photochromism*

The prolonged exposure of glass to solar UV radiation produces a coloring due to valence changes of certain ions or combinations of ions; this is the *solarization* phenomenon. If the glass contains Mn and Fe as impurities the following can occur:

$$Mn^{2+} + h\nu \rightarrow Mn^{3+} + e^-$$

where $h\nu$ is a UV photon and e^- an ejected electron which will be trapped somewhere in the structure, for example on a Fe^{3+} site where:

$$Fe^{3+} + e^- \rightarrow Fe^{2+}$$

The solar color center is thus stabilized and the glass takes on a violet tint due to the presence of Mn^{3+}. (This has been observed in old glass which has been subjected to prolonged solar exposure.)

Silicate glasses which are strongly reduced and contain Eu^{2+} and Ti^{4+} ions subjected to photon action develop color centers which progressively disappear when the light source is removed. Such glasses are called *photochromic glasses*. The reaction in this case is:

$$Eu^{2+} + Ti^{4+} \rightleftharpoons Ti^{2+} + Eu^{3+}$$

and the center responsible for the color is the Ti^{3+} ion. Other mechanisms link photochromism with crystalline particles which from a dispersed phase in the glass.

Colors can result from the absorption of light by interaction with electrons which are not associated with a specific ion but are trapped by network defects (cf. Chapter 5). The variety of trapping sites in a glass means that the transitions are rather uniformly distributed across the spectrum which produces a uniform gray absorption rather than a well-differentiated color.

The bombardment of glass by energetic particles or radiation (X-rays or γ rays) produces a modification of the transmission which can be removed by a reheat treatment (thermal bleaching). The atomic mobility becomes sufficient to restore the non-perturbed structure. A progressive modification of the transmission under radiation can be troublesome in certain technical applications (nuclear engineering, space technology) and is avoided if possible. In contrast, it is deliberately sought in other cases – as in the case of variable transmission glass for spectacles and *glass dosimeters* for radiation monitoring.

(c) *Colors due to particles dispersed in the glass*

Glass constitutes a medium in which it is possible to produce various precipitation reactions by thermal treatments or by the action of light (photosensitive reactions).

Certain metallic ions such as those of Cu, Au, Ag, Pt dissolved in a glass can be reduced to the metallic state by incorporating reducing agents in the glass such as Sb_2O_3 or SnO. The famous *gold ruby glass* is made by first dissolving in the glass small amounts of Au (0.01–0.02 wt%). The glass containing Au^{3+} ions is then heated to a temperature above the annealing point and the ions are reduced:

$$Au^{3+} + 3e^- \rightarrow Au^0$$

the necessary electrons being furnished by the reaction:

$$Sn^{2+} \rightarrow Sn^{4+} + 2e^-$$

or

$$Sb^{3+} \rightarrow Sb^{5+} + 2e^-$$

Thermal treatment then produces an agglomeration of the Au atoms; first as colloids, then as small crystallites. The glass, initially colorless, takes on a strong ruby hue in the course of this treatment, the process is called "striking." The color results from the interaction of light with the metallic particles; the action is not purely a scattering phenomenon but is accompanied by an *absorption* by the Au sol. Mie,[424] using Maxwell's equations, provided the theory for this effect. Doremus[425] showed that the action was a plasma resonance effect, the electrons in the particles oscillating collectively at a characteristic frequency if the dimensions are of the order of 200 Å.

Similar absorption effects are produced in glasses containing Ag in solid solution. The dissolved Ag^+ ions (colorless) can be reduced to the metallic state Ag^0, the atoms then being fluorescent. Upon agglomeration of the Ag atoms to a colloidal state, the fluorescence disappears and a yellow color appears which can be explained by Mie's theory. The corresponding absorption (near 396 nm) has been used in studies of H diffusion in glasses doped with Ag which serves as a tracer.[426]

The reduction of Cu^+, Au^+, Au^{3+} can be accomplished by photosensitive reactions by incorporating in the glass small quantities (0.05 %) of photoreducing elements such as cerium oxide (CeO_2). Under the action of UV radiation at ambient temperature there is an emission of an electron:

$$Ce^{3+} + h\nu \rightarrow Ce^{4+} + e^-$$

which serves for example to reduce the Cu^+ ions:

$$Cu^+ + e^- \rightarrow Cu^0$$

The Cu atoms act as nucleation centers, and color is developed in the irradiated portions by a thermal treatment. Such *photosensitive glasses* containing Cu, Ag, or Au have been produced; they allow the creation of true photographs by exposure to UV light (sensitization) followed by "development" by thermal treatment at a temperature near 500–600 °C. (The application of these processes to make nucleation centers will be considered in the section on glass-ceramics (cf. Chapter 16.)

It is possible to precipitate, in suitable glasses, small crystals of Ag halides and thus obtain transparent glasses having *photochromic* properties. Typical glasses are alkali metal boroaluminosilicates containing AgCl, AgBr or AgI in the form of small crystals from 80–150 Å in diameter precipitated by a thermal treatment at 400–800 °C. The average distance between particles is about 1000 Å. Glass composition, the nature of the halogen, particle size and thermal treatment influence the sensitivity and the kinetics of the processes of darkening and return to normal in these systems. The addition of Cu increases the sensitivity to light.

The system functions as a reversible photographic plate: the absorption of a photon causing a dissociation into Ag^0 and halogen. Ag^0 absorbs light and colors the glass gray. In contrast to usual photographic films, the pair can recombine when the light is removed causing bleaching.

Such systems are reversible and do not show signs of fatigue up to 300,000 cycles of darkening and bleaching unlike photochromic organic substances which progressively degrade (age). These glasses find application as sunglasses. A review article on this subject has been published by Araujo.[427]

12.2.4 *Luminescent glasses*

The colors considered above result from *absorption* in the visible spectrum. Colors due to *fluorescence* result from electron transitions with *emission* of a photon in the visible. An atom raised to an excited state by the absorption of a photon returns to the ground state with the emission of light, either immediately (fluorescence) or with a delay (phosphorescence).

The fluorescence centers in glass can be either metallic atoms (e.g. Ag) or crystalline phases (e.g. CdS), or ions, the most important of which are the rare earths used in coherent light amplification or lasers.

12.3 Recent optical applications

12.3.1 *Laser glasses*

A solid state laser is a luminescent material in which the light emitted by fluorescence in one of the centers in turn stimulates other centers to cause emission of light in phase with that of the first center and in a single direction.

To obtain this *stimulated emission*, it is necessary to establish a population inversion, i.e. create a situation where the species in the excited state are more numerous than those in the ground state. Limiting ourselves to the case of photon excitation (optical pumping), it can be shown that the ion which is susceptible to excitation must have at least *three* energy levels (Fig. 12.8(*a*)).[428] The excitation (pumping) raises the atoms to level 3 (or 3′) from which they can return either to the ground state (level 1) with an emission of a photon or pass by way of a non-radiative transition to an intermediate level 2. The fluorescent level 2 is essential in the lasing mechanism. When the atoms return from level 2 to the ground state 1, they emit light of the same wavelength as that stimulating the transition which in turn can stimulate emission from other centers. (In the absence of level 2, only equal populations in level 3 and 1 can be obtained.) Some systems are four-level (Fig. 12.8(*b*)), where the lasing action is produced between levels 2 and 1 above the ground state 0. The excitation is generally provided by an external lamp which emits light absorbed by the participating ions. The active solid is placed between two mirrors M_1, M_2, with respective reflectivities $R_1 = 100\,\%$ and $R_2 < 100\,\%$. The repeated internal reflection of the emitted light between the mirrors causes stimulated emission by an avalanche effect. If N_1 and N_2 are the populations per unit volume in the higher and lower states, for an inverted population with $N = N_2 - N_1 > 0$ and a gain coefficient per

Fig. 12.8. Simplified diagram of the energy levels for a laser system: (*a*) three levels; (*b*) four levels. E, pumping energy; Q, Q', Q'', non-radiative transitions; A, radiative transition; B, laser emission. After Ref. 428.

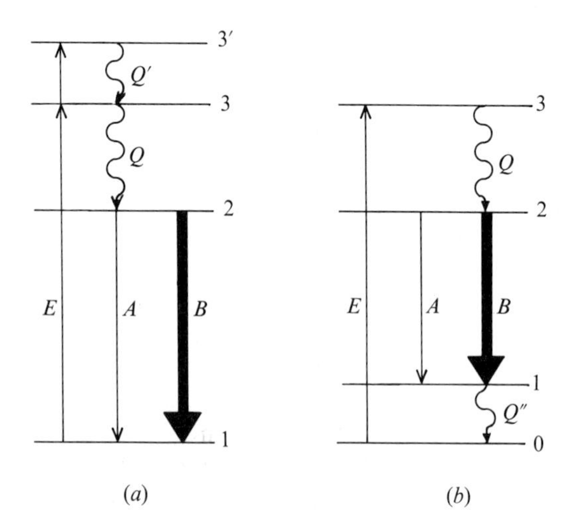

ion β, then light amplification occurs if:

$$R_1 R_2 \exp\left(\frac{\beta N - \alpha}{2L}\right) > 1$$

where α is the classical absorption coefficient and L the length of the rod. β depends on the refractive index n, the wavelength λ, the width $\Delta\lambda$ of the fluorescence line, the Einstein coefficient A and the velocity of light c:

$$\beta = \frac{1}{8\pi c} \frac{\lambda^4}{n^2} \frac{A}{\Delta\lambda}$$

Originally, solid lasers were essentially ruby (Al_2O_3 doped with Cr^{3+} ions) or YAG (Yttrium–Aluminum–Garnet, $Y_3Al_5O_{12}$ containing Nd^{3+} ions). Ruby lasers emit near 690 nm and YAG near 1060 nm.

Glasses can be host materials for lasers and are widely used for this purpose. Table 12.1 shows the rare earth ions used in various types of glasses. The most important is the Nd^{3+} ion which is a four-level system operating with good efficiency at ambient temperatures. Erbium, Er^{3+}, a three-level system, is also interesting, the emission at 1.54 μm being less dangerous to eyesight. These glasses are doped with ytterbium, Yb^{3+} which sensitizes the Er fluorescence by transferring the absorbed energy to it. Although Nd^{3+} can be used in a variety of host glasses, the durability and ease of fabrication favor alkali metal and alkaline earth silicates.

Compared to crystals, glasses are less limited in size and can be made in large isotropic and homogeneous volumes with good optical quality in which the temperature coefficient of the refractive index can be adjusted by variation of the composition. Their thermal conductivities are probably not as good, which imposes limitations on continuous or high flash frequency use. The main difference between glass lasers and crystals is that the glass structure implies a variable environment for the active ions (more than five main types of sites for certain glasses). From the shape of the spectra, it was determined that Yb^{3+} is in coordination 6 in a distorted octahedral site. As a consequence, fluorescent lines are broadened: for example if the width of the Nd^{3+} emission in YAG is 10 Å, it will be about 300 Å in a glass. A continuous mode glass laser is more difficult to make than a crystal laser containing the same ion.

In contrast, the broadened fluorescent lines of the glass lasers can be used to advantage in the "Q-switched" mode. In this mode of operation, the totally reflecting mirror M_1 is replaced by a weak reflector during the pumping. After population inversion, the reflectivity of M_1 is quickly increased producing a much more intense light pulse than in the continuous

Table 12.1. *Laser ions in glasses. After Ref. 428.*

Ion	Host glass (oxides)	$\lambda(\mu m)$
Nd^{3+}	K–Ba–Si	1.06
	La–Ba–Th–B	1.37
	Na–Ca–Si	0.92
	Li–Ca–Al–Si	1.06
Yb^{3+}	Li–Mg–Al–Si	1.015
	K–Ba–Si	1.06
Ho^{3+}	Li–Mg–Al–Si	2.1
Er^{3+}	Yb–Na–K–Ba–Si	1.543
	Li–Mg–Al–Si	1.55
	Yb–Al–Zn–P	1.536
Tm^{3+}	Li–Mg–Al–Si	1.85
	Yb–Li–Mg–Al–Si	2.105

mode operation. In certain glasses it is possible to use transient saturable color centers which lose their absorbing property at a certain level of intensity thus allowing Q-mode operation.

A technologically important problem is the *life-time* of glass lasers. The high luminous intensity and Q-mode operation produce irreversible damage. This is often due to inclusions (principally Pt particles from the fusion crucibles) and to surface contamination. It is also an intrinsic effect of auto-focusing of the beam resulting from non-linear variations of the refractive index due to electrostriction or the Kerr effect. The resulting local energy concentration leads to local vaporization of the glass in a very small volume.

The applications of glass lasers range from industry, where energies of several hundred joules can be produced, the determination of distances (telemetry), e.g. the height of clouds (Er glasses), to controlled thermonuclear fusion. For this latter application, either a CO_2 gas laser or a Nd glass laser is envisaged. In the SHIVA and NOVA experiments at Lawrence Livermore Laboratories, a set of laser beams is focused on hollow glass micro-spheres containing a mixture of deuterium and tritium to induce fusion by inertial confinement. The gigantic lasers use Nd^{3+} doped fluorophosphate glasses. NOVA is the largest optical device in the world, containing 2000 l of laser glass, 1000 l of fused SiO_2, 10,000 l of crown glass and 200 m^2 of optical quality surfaces.

A compilation of different optical characteristics and compositions of laser glasses is available in Ref. 429

12.3.2 *Optical fibers*

At an interface between two substances with refractive indices n_i and n_e such that $n_i > n_e$, a light ray propagating in substance n_i undergoes total internal reflection if its angle of incidence α is greater than the limiting reflection angle defined by $\sin \alpha_c = n_e/n_i$ (Fig. 12.9(*a*)).

This property allows a light ray to be channeled in the interior of a substance placed in a medium of lower refractive index so that it serves as an *optical conductor*. A well-known example is that of the "light fountain."

Consider a straight glass fiber of index n_i with circular cross section and indefinite length. The lateral wall is in contact with a substance of index n_e and the entry face is normal to the axis and in contact with a substance of index n_0. A light ray entering by this face and parallel to the axis propagates without obstruction. If it is inclined at an angle β, it can continue in the interior of the fiber if it undergoes successive total reflections. An elementary calculation shows it is necessary for this angle β to be less than a critical angle β_c such that (Fig. 12.9(*b*)):

$$\sin \beta_c = \frac{1}{n_0} \sqrt{n_i^2 - n_e^2}$$

the quantity:

$$n_0 \sin \beta_c = \sqrt{n_i^2 - n_e^2}$$

is called the *numerical aperture* of the system. The numerical aperture increases with the index n_i and the index difference $\Delta n = n_i - n_e$. As an example, for $n_i = 1.75$ and $\Delta n = 0.25$, the value is 0.90.

From these considerations the total number of reflections in a given length can be determined. Calculations show that the transmission of a system decreases when the length is increased, the diameter decreased or the angle β is increased. If the fiber is not straight, the total internal reflection will still channel the propagation, provided that the curvature is not too great. It has been established that the losses are still acceptable if the radius of curvature is more than 20 times the diameter of the fiber core.

Fig. 12.9. Operating principle of an optical conductor: (*a*) total reflection; (*b*) reflection condition; (*c*) multiple reflections in a curved "conductor"; (*d*) cross-section of an optical fiber.

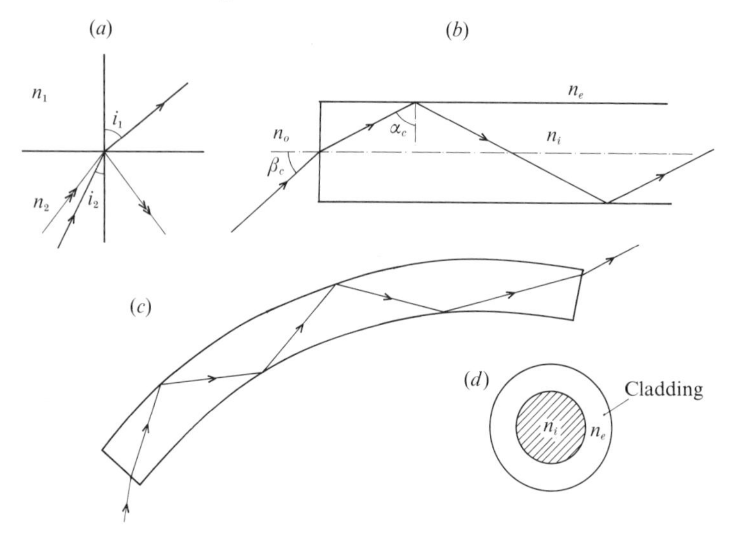

The medium of index n_e could be equal to n_0, e.g. the fiber being placed in air. However, in practice, to avoid losses between two contiguous fibers or by contamination of the surface (effectively increasing n_e), the fiber is generally encased in a sheath of index n_e, which is generally also a glass. In these conditions, the propagation is maintained by an internal interface free from external influences. (Fig. 12.9(*d*)).

Optical fibers grouped in a bundle which can transmit an image from one point to another constitute a *fiber optic lens*. The resolution of such a device is determined by the distance between the axes of two adjacent fibers in the bundle. See Chapter 17 for a review of fiber fabrication processes.

In all these applications, the choice of glass is of primary importance. In addition to suitable optical properties, the glasses must be chosen so that their thermal expansion coefficients match and their viscosities (fixed points) are comparable in order to satisfy the forming (drawing) requirements.

12.2.3 *Photon conductors*

(a) *Ultrapure glasses*

The essential problem here is the attenuation of the light transmitted by the fiber. A standard optical fiber of very good quality absorbs about 20 % of the light per meter traversed (attenuation of 1 dB m^{-1}, i.e. 1000 dB km^{-1}). This is acceptable for short-distance applications but prohibitive for information transmission over long optical paths (optical fiber telephone lines).

In such a system, the carrier wave is light emitted by a laser in the visible or near IR which is then modulated. Because of the very high frequencies (10^{15} Hz), compared to currently used electromagnetic frequencies (10^6–10^9 Hz) or the proposed (10^9–10^{12} Hz), the available bandwidth is greatly extended. A single optical telephone line could theoretically carry 500 million simultaneous communications.

Until recently, this technology was not possible because of the problem of obtaining glasses with extremely high transparency. For practical reasons the signal intensity at the exit end must be at least 1 % of the entrance signal to allow amplification and transfer to the following section. Assuming a distance of 10 km between relays, an attenuation less than 2 dB km^{-1} is required for the optical conductor.

The need for glasses with ultralow absorption for optical fibers has led to new production methods of extremely pure SiO_2 and SiO_2-based glasses by vapor phase oxidation of $SiCl_4$ (cf. Chapter 17). Glasses containing less than 1 ppm of transition metal ions and attenuations as low as 0.2 dB km^{-1}, close to the theoretical limit, were obtained. The transmission window depends on several attenuation mechanisms (Fig. 12.10). A search for even better candidates is leading towards fluoride glasses for which the theoretical losses are much smaller, of the order of 10^{-2} dB km^{-1} (Fig. 12.11). Much work is still needed, however, to overcome the purification problems as well as the difficulties with fiber forming due to rapid viscosity variations with temperature ("short" glasses).

Renewed interest in chalcogenide glasses is also motivated by their possible use in fiber optics functioning in the infrared range.

(b) *Single mode and multimode fibers*

In ordinary optical fibers the goal of a high numerical aperture is achieved by increasing the refractive index difference $n_i - n_e$. In the case of telecommunication fibers, the need to avoid signal distortions introduced by the different propagation velocities of different *modes* leads to the elimination of all rays which are not parallel or nearly parallel to the fiber axis. *Waveguide theory* indicates that for a single mode of wavelength λ to propagate in the center of the fiber, it is necessary that:

$$\frac{2a\pi}{\lambda}\sqrt{n_i^2 - n_e^2} < 2.405$$

Fig. 12.10. Intrinsic transmission window in SiO_2-based glasses. The different attenuation mechanisms are indicated. After Ref. 419.

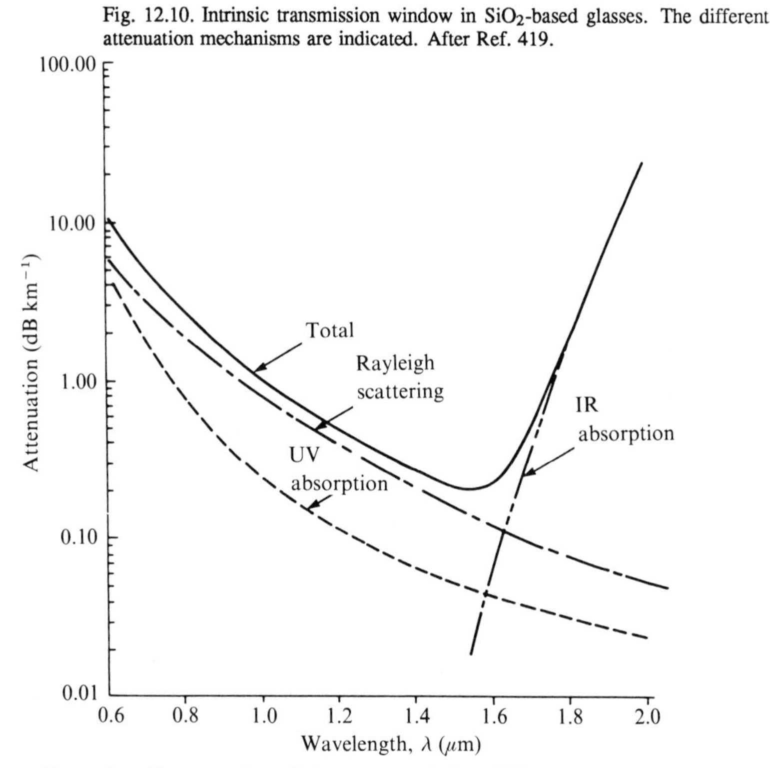

Thus the diameter $2a$ of the core and the difference $n_i - n_e$ must be small. In practice the fibers are classified as:

1. *Multimode* fibers for which $2a \gg \lambda$, e.g. 10–100 μm; such fibers only have a usable bandwidth of $\sim$ 20 MHz km^{-1}. These fibers will be restricted to short distance applications (Fig. 12.12(*a*)).

2. *Single mode fibers* for which $2a \sim \lambda$, e.g. $2a \sim 1\,\mu$m, the bandwidth then being of the order of 25 GHz km^{-1} (bandwidths greater than 1000 GHz km^{-1} are possible). Application of the Maxwell equations shows that the wave also propagates in the sheath as a damped transverse wave. The sheath must be sufficiently thick to reduce the impact of this effect. In general the diameter is much larger, 50–100 μm to ensure the mechanical strength of the conductor assembly. Only this type of fiber is suitable for long-distance transmission (Fig. 12.12(*b*)). Fabrication techniques for such fibers are described in Chapter 17.

The problems of ultrapure glasses for telecommunication fibers can now be considered solved; mechanical properties and installation in the field (junctions, coupling etc.) still pose technological problems.

Fig. 12.11. Comparison of intrinsic transmission windows for various types of glasses.

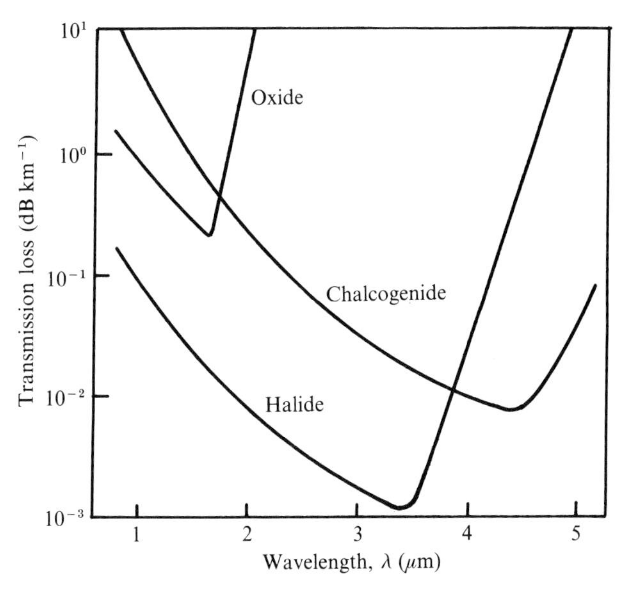

Fig. 12.12. Cross-sections and refractive index profiles of different types of optical fibers: (*a*) multimode, (*b*) single mode; (*c*) unclad; with continuous index variation (*R*, radial coordinate).

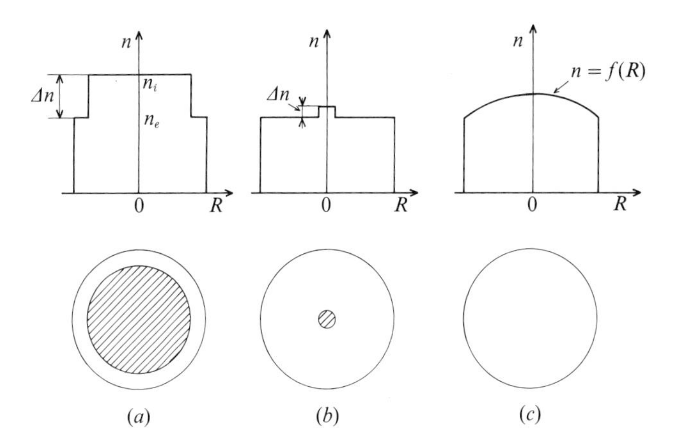

(c) *Unclad fibers*

It is possible to obtain the propagation of light along a fiber by creating a refractive index gradient (Fig. 12.12(*c*)). This can be accomplished through ion exchange by immersing the fiber in a fused salt bath (cf. Section 14.4.2 chemical tempering). In principle, different modes can travel with the same velocity in this type of conductor. In practice application is limited by index inhomogeneities and absorption problems arising from the need to use multicomponent glasses to obtain the ion exchange. A review of optical fiber technology can be found in Ref. 419.

13 Thermal properties of glasses

Thermal properties are directly related to temperature changes. They are essentially: the specific heat, the thermal expansion coefficient and the thermal conductivity. They play an important rôle in practical applications. The specific heat measures the quantity of thermal energy necessary to change the temperature of the body. Combined with the thermal conductivity, it determines the rate of temperature change which the glass can sustain during fabrication and use. Different thermal expansions produce thermal stresses which strongly influence the mechanical properties of glass articles.

13.1 Specific heat

13.1.1 *Temperature dependence*

The specific heat at constant pressure C_p and at constant volume c_v are interrelated by the formula:

$$C_p - c_v = \alpha^2 VT/\beta$$

where α is the volume expansion coefficient, β the compressibility and V the molar volume at temperature T.

The energy necessary to raise the temperature of a body from absolute zero consists of: the vibrational energy of the atoms around their equilibrium positions, the rotational energy of atomic groups possessing rotational degrees of freedom, the energy for transitions between electronic energy levels and the energy for the creation of defects, magnetic orientations, structural transformations, etc. In all these changes, there is an increase in internal energy and configurational entropy. In the classical theory, each degree of freedom is associated with an average kinetic energy $\frac{1}{2}kT$ and an equal average potential energy. For an atom having three degrees of freedom, the total energy is thus $3kT$, or $3NkT$ per gram atom (g a), hence:

$$c_v = \left(\frac{\mathrm{d}E}{\mathrm{d}t}\right)_v = 3Nk = 5.96 \text{ cal (g a)}^{-1}\ {}^\circ\mathrm{C}^{-1}$$

This value corresponds well to experimental results at high temperature. At low temperatures it is necessary to take into account the distribution of network vibration frequencies γ. Introducing, from Debye, a characteristic temperature $\theta_D = h\nu_{max}/k$, related to the maximum frequency ν_{max}, one can write:

$$c_v = 2Nkf(\theta_D/T)$$

where $f(\theta_D/T)$ is the Debye function:

At high temperatures $f \rightarrow 1$, and the heat capacity, c_v is found again to be independent of temperature. At low temperatures, c_v varies as $(T/\theta_D)^3$. The heat capacities of most of the glasses attain 70–90 % of their limiting theoretical value near the lower end of the glass transition region. The heat capacity undergoes a noticeable increase during the change to the liquid state in the transition interval. This corresponds, as seen in Chapter 2, to an increase in the configurational entropy of the liquid state, the times necessary for molecular rearrangements becoming small relative to the time of the experiment. Below the transformation range, the thermal properties of glass become similar to those of crystals. This is not the case at very low temperatures where glasses have an unusual behavior.

13.1.2 *Phonons in glasses*

The problem of vibrational excitations in crystals has been the subject of many studies. For a perfectly harmonic crystal, the vibrations take the form of plane waves (phonons), characterized by a frequency ω, a polarization vector $\vec{\alpha}$ and a wave vector $\vec{Q}$. The anharmonic effects from vibrations and the crystal defects lead to dispersion relations and the crystalline imperfections themselves become more or less localized excitation sites. In the case of a disordered substance, the situation is less clear because the simplifications introduced by the periodicity of the crystalline network are not possible. It is currently recognized[431] that the normal modes are no longer plane waves. The vibrational displacements may still be formally described in these terms, but each wave is characterized by a band of frequencies corresponding to the modes making up the wave. In other words, the phonons will be *attenuated*. This attenuation, which shows as a broadening of the $\omega(Q)$ relations, increases with Q and ω when the phonon wavelength becomes of the same order as the interatomic distance. The "phonons" in glass must be more localized than those in crystals and there must exist in glass a more localized excitation spectrum related to specific characteristics of the disordered structure. At very low frequencies and very long wavelengths, glasses like crystals behave as an elastic continuum. Precise $\omega(Q)$ relations exist and thermal waves, longitudinal as well as transverse, can be shown directly (e.g. results obtained for frequencies of 10^{10} Hz by the Brillouin effect).

The presence of localized excitations further complicates the situation because the distribution of the two types of excitations is not clear. In certain cases such excitations can predominate to the point of rendering a description in terms of phonons completely invalid. At still higher frequencies, the modes can be considered as deriving from the *optical modes* of the crystal. It is not certain that the description in terms of phonons would be appropriate here and in any case there does not currently exist an experimental means to study such high-frequency phonons in glasses.

13.1.3 *Properties at very low temperatures*

At low frequencies ($< 10\,\mathrm{cm}^{-1}$) the heat capacity of crystals is proportional to T^3, conforming with the Debye theory and corresponding to a density of states $g(\omega)$ proportional to ω^2. It was thought that glasses followed the same law until measurements at very low temperatures showed a totally different behavior; the density of states in a glass is much higher and varies more rapidly than in a crystal.

Figure 13.1 shows the heat capacity $C(T)$ of vitreous SiO_2 divided by T^3 as well as that of allotropic crystalline forms (cristobalite and quartz) as a function of T. A significant maximum is seen which is found not only for oxide glasses but also for chalcogenide glasses and organic polymers. At first it might be thought that this is a purely structural effect, but, given the large variety of amorphous solids which show this anomaly, a more general explanation must be sought. It seems that the maximum is due to very low frequency modes which can be produced in the network formed of groups with low coordination (maximum 4),

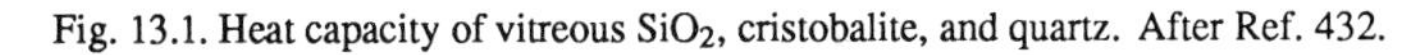

Fig. 13.1. Heat capacity of vitreous SiO_2, cristobalite, and quartz. After Ref. 432.

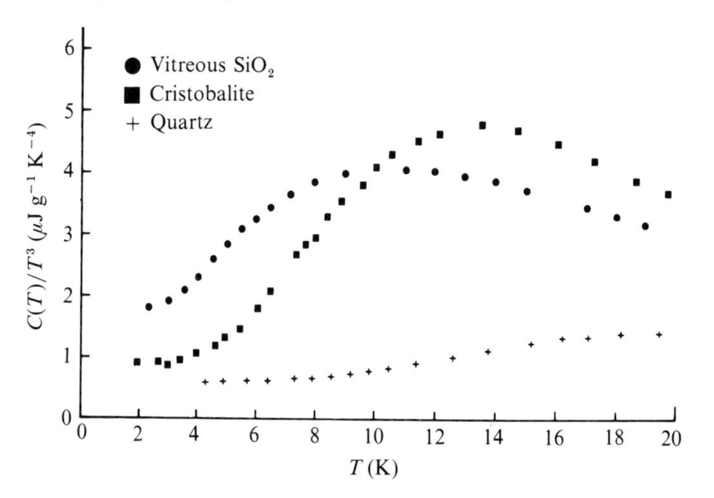

and having great flexibility, i.e. having weak bending constants.[433] These modes imply the participation of more distant neighbors. The simple type of Raman effect analysis using the interaction between the nearest neighbors is inadequate here. It is necessary to consider the groups formed of at least five SiO_4 tetrahedra.[432] However, even though the calculations include a large number of atoms,[284] they do not provide sufficient detail at very low frequencies. The only disordered material for which a direct comparison has been attempted is amorphous Ge.[434] At temperatures less than 2 K, the specific heat of nearly all amorphous solids (Fig. 13.2) varies as $C_1T + C_3T^3$, although at even lower temperatures, this description no longer applies. The constant C_1 is between 1 and 5 μJ g^{-1} K^{-1} except for metallic glasses where there is also an electronic contribution. In a general way these results show that the density of vibrational states is higher in an amorphous solid than in the corresponding crystal and varies more slowly with energy below 10^{-4} eV.

Fig. 13.2. Heat capacity of non-crystalline solids below 1 K. PS, polystyrene; PC, polycarbonate (Lexan); PMMA, polymethylmethacrylate. After Ref. 435.

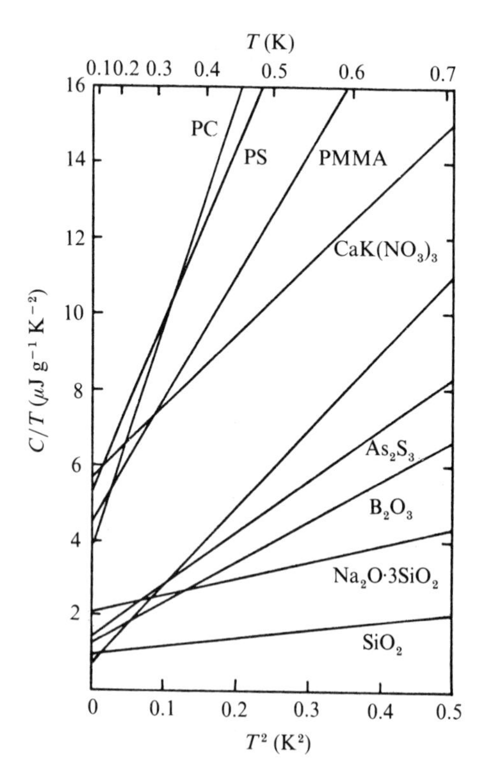

13.2 Thermal conductivity

13.2.1 *Definitions*

The fundamental equation:

$$\frac{1}{A}\frac{\mathrm{d}Q_T}{\mathrm{d}t} = -K\frac{\partial T}{\partial x}$$

expresses the fact that the heat flux $(1/A)\,\mathrm{d}Q_T/\mathrm{d}t$ flowing perpendicularly across a surface A is proportional to the temperature gradient $\partial T/\partial x$ in the direction of flow. The proportionality constant K, a material property, is the *thermal conductivity.*

For an ideal gas, kinetic theory provides the relation:

$$K = \frac{1}{3}cvl$$

where c is the heat capacity per unit volume, v the average velocity and l the mean free path.

Heat conduction in a dielectric solid can be considered as due either to the propagation of anharmonic elastic waves, or to interaction between the thermal energy quanta: the phonons. The interactions between the waves (otherwise called scattering) depends on the phonon frequency. By analogy with the preceding formula, the thermal conductivity will be represented by the relation:

$$K = \frac{1}{3}\int c(\omega)vl(\omega)\,\mathrm{d}\omega$$

where $c(\omega)$ is the contribution to the specific heat at frequency ω and $l(\omega)$ is the attenuation length. The process giving rise to the finite thermal conductivity is the phonon–phonon interaction. Moreover, various imperfections in the network introduce anharmonicities and scatter the phonons which reduces the length $l(\omega)$.

13.2.2 *Photon conductivity*

In addition to the vibrational energy, part of the energy content is due to high-frequency electromagnetic radiation. For a black body, the energy density is:

$$E_R = \frac{4\sigma n^2}{v_0}T^4$$

in which v_0 is the speed of light in vacuum, n the index of refraction and σ the Stefan–Boltzmann constant ($1.37 \times 10^{-12}\,\mathrm{cal\,cm^{-2}\,s^{-1}\,K^{-4}}$). The corresponding heat capacity per cubic centimeter is obtained from:

$$c_R = \frac{\partial E_R}{\partial T} = \frac{16\sigma n^3}{v_0}T^3$$

Denoting the speed of the radiation by $v = v_0/n$, and the mean free path of the radiant energy by l_R, the photon thermal conductivity is:

$$K_R = \frac{16}{3}\sigma n^2 T^3 l_R$$

This conductivity becomes important at high temperatures. Since c_R and l_R depend strongly on the wavelength, this variation must be taken into account, using the form already indicated for the phonon conductivity. The thermal transfer by photons depends in a critical way on l_R. When $l_R \sim 0$ (opaque material) or when l_R is large relative to the dimensions of the specimen, the interaction is negligible. It is only when l_R attains values which are small relative to the dimensions of the specimen that the photon conduction becomes appreciable. This is particularly true for glasses melted at high temperatures, and the photon conduction mode plays an important rôle in the technology of glass-melting furnaces.

13.2.3 *Temperature dependence*

(a) *Behavior at high temperatures*

Figure 13.3 shows the temperature dependence of thermal conductivity for several glasses. Very few good, systematic studies of this subject have been made except on vitreous SiO_2.

Figure 13.4 shows the effect of fast neutron irradiation on the thermal conductivity of quartz. A strong decrease in conductivity as the radiation dose is increased is apparent. This progressively introduces structural disorder. The curves tend towards a limit corresponding to the value for vitreous SiO_2.

The mean free path $\bar{l}$ of the phonons tends towards a value which is independent of the temperature and is of the order of interatomic distances in the disordered structure. This limitation of $\bar{l}$ by the structure causes the thermal conductivities of different glasses to be quite similar.

A value for $\bar{l}$ may be estimated by the equation:

$$K = \frac{1}{3} c_v v \bar{l}$$

where c_v is the heat capacity and K the measured thermal conductivity. Taking v equal to the sonic velocity, it is found that the values of $\bar{l}$ for different glasses lie between 3 and 5 Å, about the same size as the interatomic distances.

The temperature dependence of the thermal conductivity follows the variations in the heat capacity. If the part due to photon conductivity is subtracted, the conductivity due to photons becomes nearly constant

Fig. 13.3. Thermal conductivity of several glasses. After Ref. 436.

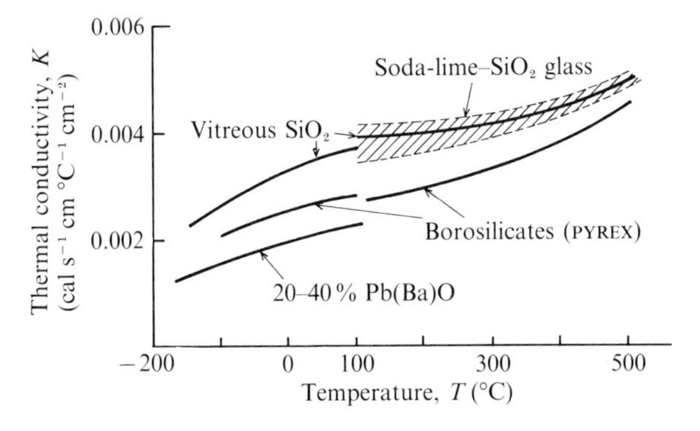

Fig. 13.4. Effect of neutron irradiation on the thermal conductivity of SiO_2: (*a*) quartz, (*b*) the same after 0.03 irradiation units, (*c*) 1 unit, (*d*) 2.4 units, (*e*) 19 units (1 unit = $1.8 \times 10^{18}\,cm^{-2}$), (*f,g*) vitreous SiO_2. After the results of Berman and Cohen cited in Ref. 437.

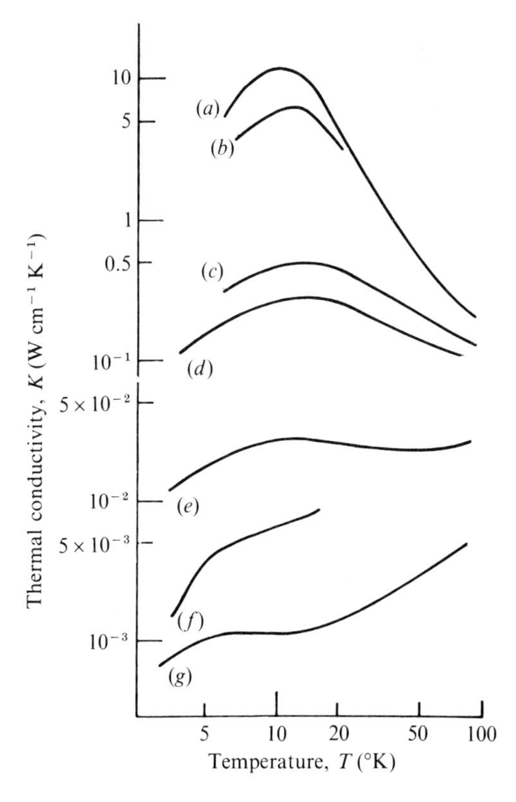

above several hundred degrees. Measurement of the photon mean free path as a function of temperature shows that l_R increases with T. The photon conduction does not obey the T^3 law, but shows a more rapid increase, the temperature exponent being between 3.5 and 5.

(b) *Behavior at very low temperatures*

At very low temperatures the thermal conductivity of vitreous solids has a characteristic appearance. Between 2 K and 10 K the curves have a plateau. Figure 13.5 shows the results for several glasses and amorphous solids; it is generally the same for all non-crystalline solids and has not yet been explained satisfactorily. The transport of heat in this temperature domain is by acoustic phonons and the temperature dependence reflects a scattering process for these phonons. One suggestion is scattering by density fluctuations. The temperature dependence predicted by the theory is correct, but the amplitude of the necessary fluctuations is too large. Note that the plateau of K is near the temperature at which a maximum occurs in the heat capacity curve C/T^3. This suggests that the thermal conductivity above 2 K can be directly related to the vibration spectrum and thus depends only indirectly on the structure.

At lower temperatures, less than 2 K, the thermal conductivity of all non-crystalline solids varies approximately as T^n with $n \sim 1.8$. This is equally true for the phonon contribution to the thermal conductivity in metallic glasses. Two main approaches have been used to explain the preceding facts; in one it is necessary to consider only the *static* disorder of the glass, while in the other, a resonant scattering of phonons by *localized excitations* is considered in addition.

Phonon scattering has also been studied by measuring the attenuation of ultrasound in the glass. Below 1 K, if the acoustic intensity is weak, the attenuation increases as the temperature decreases. However, the attenuation decreases at higher intensities. The values for the mean free path deduced from the attenuation agree with those from the thermal conductivity and this constitutes supplemental proof that acoustic phonons exist in glasses.[432] The saturation of the attenuation observed at high acoustic intensities is an important element limiting the number of different possible models.

The model presently most favored is that of Anderson, Halperin and Varma.[438] The model is based on a universal characteristic of the vitreous state: the state of non-equilibrium. It is assumed for this purpose that the atoms (or groups of atoms) in a non-crystalline solid have two (or several) equilibrium positions corresponding to the minima of double asymmetric potential wells. The movement of such atoms can be described approximately as an oscillation around one of the two minima. The potential

Fig. 13.5. Thermal conductivity of non-crystalline solids. PS, polystyrene; PMMA, polymethylmethacrylate. After Ref. 435.

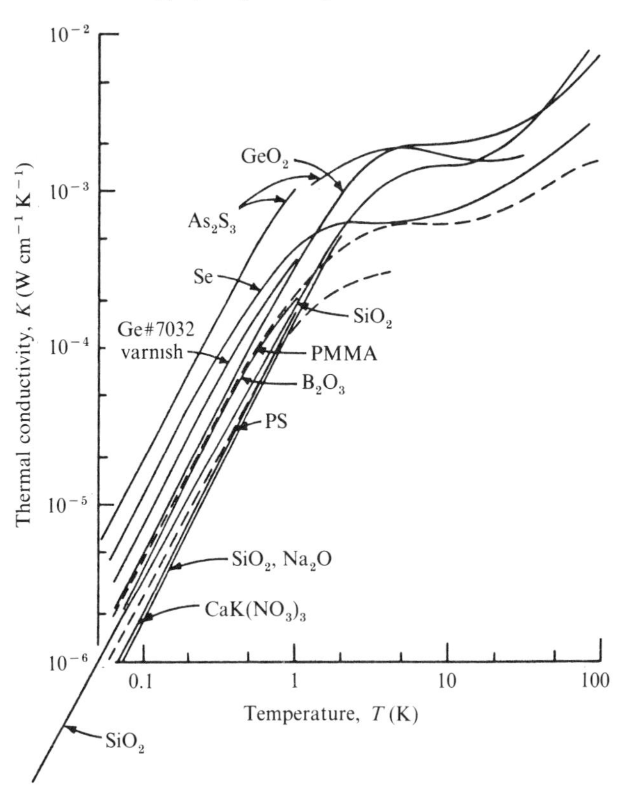

barriers are sometimes sufficiently low to allow transitions by tunneling with the emission (or absorption) of a resonant phonon. The model gives a linear term for the heat capacity as a function of temperature.

A similar model was developed by Phillips.[439] Although the phenomenological description is satisfactory, the explanation in terms of atomic structure is much less so. In particular, the nature of the entities responsible for this effect is not clear. The OH^- groups were studied from this point of view; vitreous SiO_2 containing 1200 ppm OH has a heat capacity 30 % greater than SiO_2 containing less than 1.5 ppm. (Fig. 13.6).[440]

The thermal conductivity decreases only 10 % in the presence of OH. Another microscopic model postulates the existence of two preferential orientations for the Si–O–Si bond.[441] It is improbable that a single microscopic model applicable to all glasses can be defined to explain the observed behavior at very low temperatures.

Fig. 13.6. Heat capacity of vitreous SiO_2 with different OH contents. After Ref. 440.

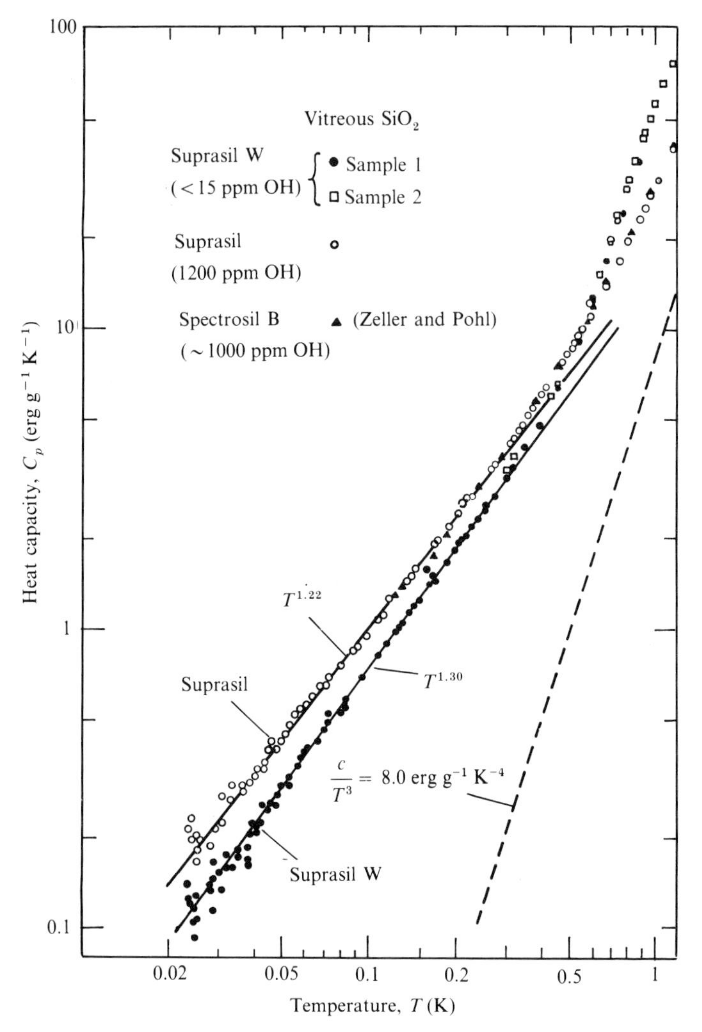

13.3 Thermal expansion

13.3.1 *Microscopic interpretation*

As with crystals, the thermal expansion of glass is due to anharmonicity of the atomic vibrations. The increase of the volume with temperature is mainly determined by an increase in the amplitude of atomic vibrations around a mean position. Since the repulsion term changes more rapidly with interatomic distance than the attraction term, the potential energy curve is asymmetric (Fig. 13.7). As the energy of the network increases, the mean separation between the atoms also increases. The potential curve depends to a first approximation on the nature of the atoms and the short-range order which are similar in the glass and the parent crystal. The resulting expansion coefficient will be nearly the same for the two forms. The depth of the minimum in the potential energy curve depends on the bond energy E (measured e.g. by the latent heat of sublimation). For a crystal, the fusion point T_f is given approximately by $kT_f \sim \frac{1}{2}E$, the high fusion points corresponding to deep minima and more symmetric curves – thus high melting point solids will have lower expansion (Fig. 13.7(*b*)). The glass → liquid transition is accompanied by an increase in the expansion coefficient, a property already discussed relative to the definition of the glass transition temperature (cf. Chapter 2).

Fig. 13.7. Potential energy and expansion of a solid with (*a*) low, (*b*) high melting point.

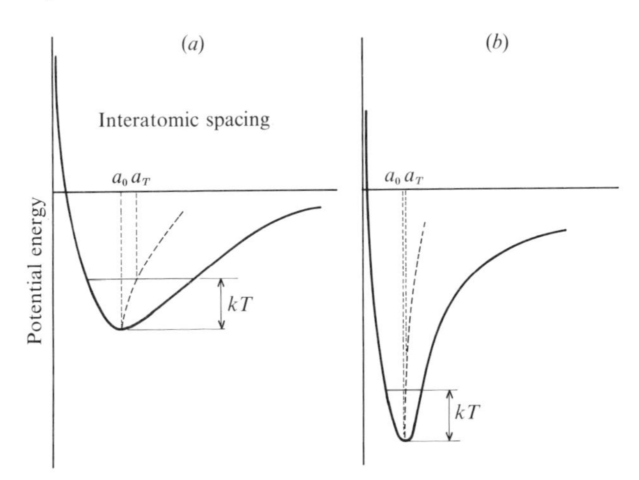

13.3.2 *Examples of practical interest*

The linear coefficient of expansion, $\alpha = (l/l_0)\,\Delta l/\Delta T$, and the cubic coefficient are a function of temperature, and the *average* values are used over a defined interval of temperature. As an example, Table 13.1 gives the average expansion coefficient over the indicated temperature intervals for an ordinary glass. Glasses for which $\alpha < 6 \times 10^{-6}\,\mathrm{K}^{-1}$ are called *hard* glasses and those with $\alpha > 6 \times 10^{-6}\,\mathrm{K}^{-1}$ *soft* glasses.

SiO_2 has a very low expansion coefficient; $\alpha = 0.5 \times 10^{-6}\,\mathrm{K}^{-1}$. Glasses with coefficients which are near zero and even negative have been obtained in the SiO_2–TiO_2 system ($\alpha = -0.3 \times 10^{-6}\,\mathrm{K}^{-1}$ for 11 wt% TiO_2).

Table 13.1. *Average values of the linear thermal expansion coefficient* (K^{-1}) *for the glass* 75 SiO_2, 15 Na_2O, 10 CaO.

$\alpha_{20/100}$	$\alpha_{20/200}$	$\alpha_{20/300}$	$\alpha_{20/400}$	$\alpha_{20/500}$
8.9×10^{-6}	9.1×10^{-6}	9.35×10^{-6}	9.6×10^{-6}	9.85×10^{-6}

Additive formulae permitting the calculation of α from composition can be found in Refs. 3 and 8.

From a technological point of view, thermal expansion is a most important thermal characteristic; in particular it controls the possibility of bonding by welding.

Glass can be welded to a variety of metals, the bonding layer being a thin adherent oxide on the metal. In glass-to-metal sealing, which is frequent in electronics, the expansion coefficients of the two materials must be very close over an extended temperature range. A difference between the expansion coefficients as the temperature falls below the softening temperature can lead to fractures from differential stresses (cf. Section 14.4). It is thus necessary to *match* the metal and the glass, and a number of solder glasses have been developed to match the expansions of various materials. W and Mo, which have very low values of α, can be sealed to borosilicate glasses. Alloys of Ni, Fe, Co, which are easier to work than W and Mo, have been produced. For common lighting applications, wires of Ni–Fe are coated with Cu and sealed to a lead silicate based glass with high electrical resistivity, which in turn is easily sealed to the body of the soda–lime–silica glass envelope. This method of *graded joints* is common; by using a series of intermediate glasses, the difference in expansion coefficients is distributed over multiple interfaces and thus remains within a tolerable limit at each point.

The same matching problem is found in glazing ceramic ware, an operation which consists of applying a thin layer of glass on the surface to make an impermeable product. Failure to match the glaze expansion to the support material results in a regular cracking called crazing. Sometimes it is possible to put the surface layers into compression to increase the strength of the objects. The quality of the match between two glasses can be verified quickly by pulling a fiber from the two glasses placed in contact. On cooling, the fiber behaves like a bi-metallic strip and its curvature allows the evaluation of the difference between expansion coefficients of the two glasses.

14 Mechanical properties – Fracture

14.1 Mechanical strength of glass

14.1.1 *Brittle fracture*

The brittleness of glass is well known; it is a property which is easy to recognize but difficult to define in a quantitative manner. In a tensile test (Fig. 14.1(*a*)), a metallic specimen undergoes first a reversible deformation ϵ proportional to the stress σ (Hooke's law), then an irreversible plastic deformation above the elastic limit σ_e up to failure which occurs at a rupture stress σ_r.

Fig. 14.1. Rupture mode in a tensile test: (*a*) ductile material, (*b*) brittle material.

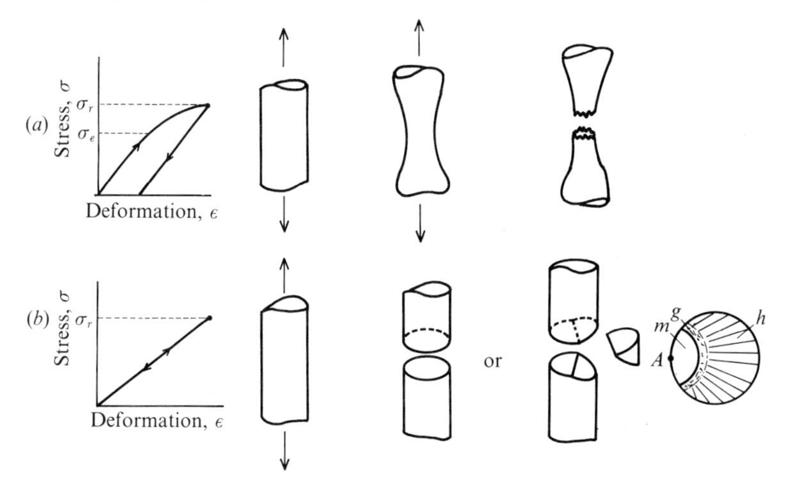

The rupture is preceded by a necking phase where the cross-section of the specimen decreases. This behavior is characteristic of a *ductile material*. The fracture surface is generally granular.

Glass has a completely different behavior (Fig. 14.1(*b*)); under the action of a tensile stress, it deforms elastically up to failure without permanent deformation. The fracture begins in a direction approximately perpendicular to the maximum tensile stress. If, for example, a cylindrical rod is subjected to a tensile test, the fracture initiates at point A at the surface and then continues in the same plane across the entire section or possibly branches to produce a crescent shaped wedge. This mechanism of sudden failure without a preceding plastic phase is characteristic of a *brittle* material of which glass is a typical example. The fracture is a *conchoidal* type. Detailed examination of the fracture surface (Fig. 14.1(*b*)) shows successive features from point A in the direction of propagation; a smooth surface (m), "the mirror," followed by a "mist zone" (g), then a "hackle" zone (h) in which the surface irregularities radiate from point A. The end of the mirror region corresponds to the point where the fracture begins to change direction before the multiple branching which leads to the formation of visible lines in the portion (h).

14.1.2 *Theoretical cohesive strength of glass*

The theoretical cohesive strength σ_{th} of a material depends on the bond forces which exist between the atoms of the structure – it can be an indication of the stress level necessary to initiate fracture. This allows, in principle, a comparison of the potential strength of different materials, and for a given material, it provides an indication of a maximum possible fracture stress.

Usually, fracture is considered as a process of separating a solid into two parts, each being bounded by a new surface called the *fracture surface*. The attractive forces acting across the interface vary during the separation process – their maximum giving the *maximum cohesive strength* of the material. Representing this behavior by a characteristic potential function U, e.g. by a Morse potential in the case of covalent bonds:

$$U = U_0 \left\{\exp\left[-2a(x - x_0)\right] - 2\exp\left[-a(x - x_0)\right]\right\}$$

where U represents the potential energy per unit area of the fracture surfaces separated by a distance x. The stress on the surfaces $\sigma = -\partial U/\partial x$ passes through a maximum at $\partial^2 U/\partial x^2 = 0$ which defines the theoretical cohesive strength σ_{th} (Fig. 14.2).

The constants U_0, x_0 and a which enter into the definition of the potential function can be obtained from measurable properties of the material. The equilibrium distance x_0 in the absence of stress is determined from the distance between neighboring atoms (e.g. Si–O for SiO_2) and results from the condition:

$$\left(\frac{\partial U}{\partial x}\right)_{x=x_0} = 0$$

The parameter U_0 of the potential function depends on the surface energy γ of the two new surfaces created by the fracture:

$$2\gamma = \int_{x_0}^{\infty} \frac{\partial U}{\partial x} \mathrm{d}x = -U_0$$

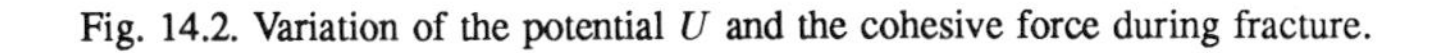
Fig. 14.2. Variation of the potential U and the cohesive force during fracture.

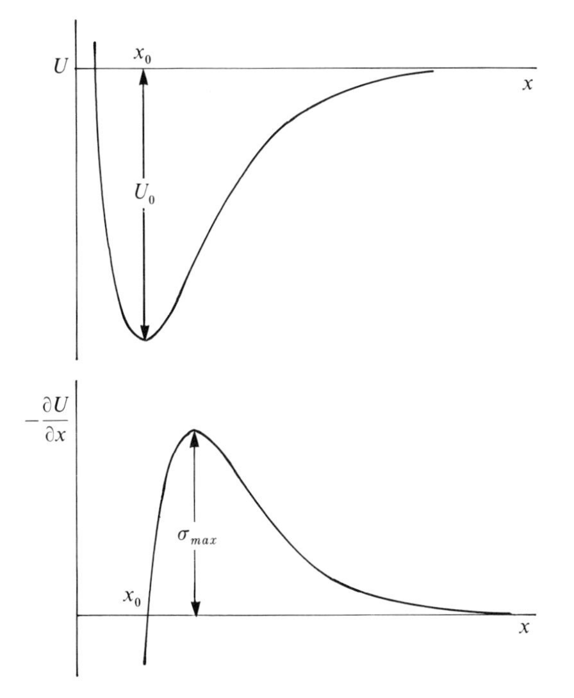

and Hooke's law relates the curvature of U to Young's modulus E of the solid:

$$d\sigma = E\left(\frac{dx}{x_0}\right)$$

hence:

$$E = x_0\left(\frac{\partial^2 U}{\partial x^2}\right)_{x=x_0}$$

Under these conditions, the theoretical fracture stress σ_{th} or the maximum cohesive strength, corresponding to the maximum of $\partial U/\partial x$ is:

$$\sigma_{th} = \sqrt{\frac{E\gamma}{4x_0}}$$

Using estimates of γ and x_0 for SiO_2 this calculation gives values between:

$$1.8 \times 10^{10} \quad \text{and} \quad 2.2 \times 10^{10}\ \mathrm{N\ m^{-2}}$$

14.1.3 *Experimental determination of failure stress*

Because tensile specimens of brittle materials are difficult to align and grip in mechanical testing machines, bending tests on bars with circular or rectangular cross sections are usually used.

In a *three-point bending* test (Fig. 14.3(*a*)) maximum stress σ_{max} occurs on the surface at a point directly opposite the central load point and is:

$$\sigma_{max} = M/D$$

where $M = Pl/4$ is the maximum bending moment and $D = (1/6)a^2b$, the moment of inertia of the section. For a circular section of radius r, $D = (1/4)\pi r^3$.

The fracture can occur at a point which is not at the maximum M thus introducing some uncertainty. To minimize this, the *four-point bending* test is preferred (Fig. 14.3(*b*)). M is then constant between the central loading points: $M = Pd/2$. For a load P leading to failure, the stress σ_{max} is taken as a measure of the failure stress or the experimental mechanical strength of the glass.

Fig. 14.3. Bending test: (*a*) three-point, (*b*) four-point. The lower diagrams indicate the corresponding bending moments.

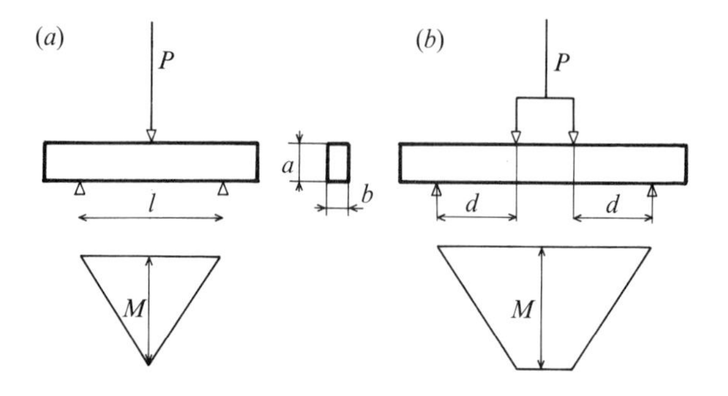

The tests show *large scatter* in the results which is related to the *state of the specimen surface*. Specimens which are *flame-polished* (cf. Section 15.1) and protected from all mechanical contact before the test show a higher average strength but also a greater scatter in the results (Fig. 14.4).

The highest strengths only approach the theoretical values in exceptional cases; e.g.[442]

1.35×10^{10} N m^{-2} for bulk SiO_2 glass, flame-polished and measured at -196 °C.

or

1.47×10^{10} N m^{-2} for a vitreous SiO_2 fiber measured at -196 °C.

Fig. 14.4. Histograms of the fracture strength of test specimens: (*a*) ordinary as-received; (*b*) without apparent defects; (*c*) flame-polished.

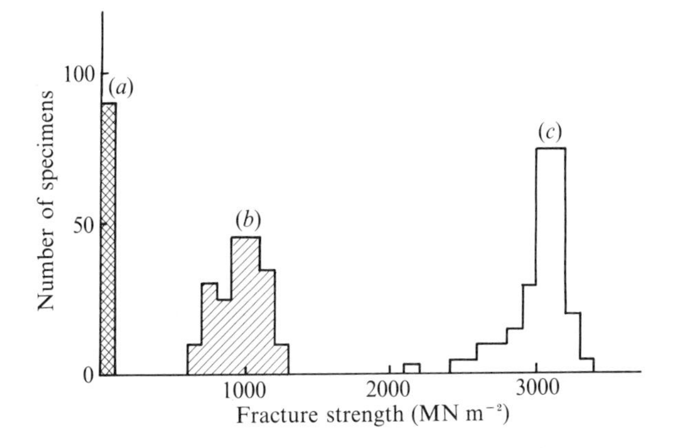

The values usually measured for SiO_2 are approximately 1×10^8 N m^{-2}, i.e. about 100 times weaker than the calculated theoretical values. This is even more pronounced in the industrial soda–lime–silica glasses which can commonly have strengths as low as 3×10^7 N m^{-2} which are 1000 times lower than the theoretical maximum values.

14.1.4 *Griffith's micro-crack theory*

(a) *Flaws – initiation criteria*

The first explanations of the low values observed for the mechanical strength of glass were given by Griffith.[443] This author suggested that all specimens of ordinary glass contain micro-cracks or surface *flaws* resulting from handling which act as stress *concentrators*. Griffith's explanation is based on the work of Inglis[444] who calculated the stress distribution in a thin plate containing an elliptical hole and subjected to a uniform tensile stress σ_L (Fig. 14.5). If the length of the major semi-axis of the ellipse is designated by c and the radius of curvature at C by ϱ, the largest tensile stress σ_{yy} occurs at the tip of the major axis and is:

$$\sigma_{yy} = 2\sigma_L \sqrt{\frac{c}{\varrho}}$$

The term $2\sqrt{c/\varrho}$ acts as a stress concentration factor. Assuming that $\sigma_{yy} = \sigma_{th}$ at the moment of rupture, the measured fracture strength σ_r will be:

$$\sigma_r = \sigma_{th} \bigg/ 2\sqrt{\frac{c}{\varrho}}$$

Fig. 14.5. Plate containing an elliptical hole and subjected to a uniform tension σ_L.

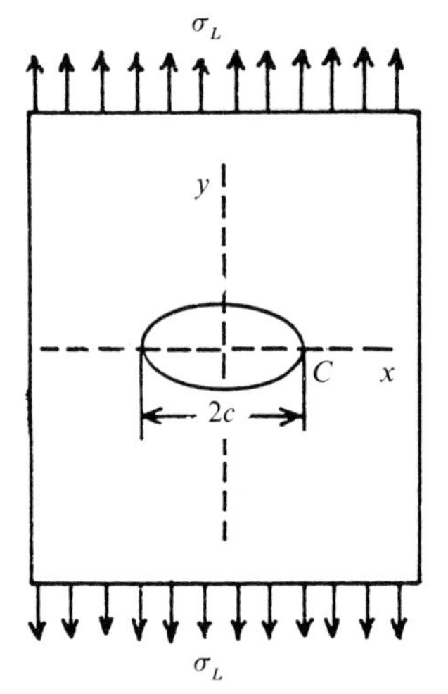

The stress concentration factor depends mainly on the *form* of the cracks and can take values significantly higher than unity for sharp cracks. At the limit, a crack with tip dimensions of the order of interatomic distances, e.g.

$$\varrho = 2\,\text{Å} \qquad c = 1\,\mu m \qquad 2\sqrt{\frac{c}{\varrho}} \sim 140$$

accounts for the normally observed strength values which are low relative to the theoretical strength. Moreover, this hypothesis explains the large scatter in measured results, the specimens differing in the number and size of micro-cracks which they contain.

In spite of this fundamental advance, the explanation of the failure mechanism remained obscure. In particular, if the Inglis analysis remains valid down to atomic dimensions, why do larger cracks usually propagate more easily than small cracks, and what is the significance of the radius of curvature in a real crack?

Griffith restated the question in terms of a reversible thermodynamic process.

(b) *Energy balance – criterion of propagation*

Consider a system consisting of an elastic body with a crack of length $2c$ and subjected to loads at its edges and seek the configuration which minimizes the total free energy of the system; i.e. the configuration corresponding to the equilibrium state for the crack.

Given that the work of the exterior forces W increases the elastic energy U_E and supplies the energy U_S of newly created surfaces:

$$\mathrm{d}W = \mathrm{d}U_E + \mathrm{d}U_S$$

The total energy of the system is:

$$U = (-W + U_E) + U_S$$

where the term in parentheses corresponds to the mechanical energy of the system. This term decreases as the crack extends while the surface term U_S increases (creation of new surfaces against cohesive forces). The first term favors the extension while the second opposes it. At equilibrium:

$$\frac{\mathrm{d}U}{\mathrm{d}c} = 0$$

from which comes the energy equilibrium concept of Griffith. A crack would have a tendency to extend if $\mathrm{d}U/\mathrm{d}c < 0$ or to close up reversibly if $\mathrm{d}U/\mathrm{d}c > 0$. This constitutes the Griffith propagation criterion.

It is shown by linear elastic theory that for all bodies under *constant* applied stress during the formation of the crack:

$$W = 2U_E$$

Taking the solutions corresponding to the model of an elliptical crack (Fig. 14.5) and integrating the strain energy density over distances which are large relative to the dimensions of the crack (the thickness of the plate is taken as unity) one obtains:

$$U_E = \pi c^2 \sigma_L^2 / E \qquad \text{for a thin plate (plane stress)}$$

or:

$$U_E = \pi(1 - \mu^2) c^2 \sigma_L^2 / E \qquad \text{for a thick plate (plane strain)}$$

where σ_L is the tensile stress, E is Young's modulus and μ Poisson's ratio. Since $U_S = 2\,(2c\gamma)$ where γ is the surface energy per unit area, the balance for the case of plane stress is:

$$U = -U_E + U_S = -\pi c^2 \sigma_L^2 / E + 4c\gamma$$

Application of the Griffith criterion $(\mathrm{d}U/\mathrm{d}c = 0)$ leads to the critical stress for crack propagation:

$$\sigma_f = \sqrt{\frac{2E\gamma}{\pi c}}$$

In the case of plane strain, E is replaced by $E/(1 - \mu^2)$. Since $\mathrm{d}^2U/\mathrm{d}c^2 < 0$, the system is unstable and when the applied stress exceeds the value of σ_f the crack propagates with no limit.

To verify his theory Griffith performed a series of experiments in which *macroscopic* cracks of known length, $2c$, were systematically introduced into the walls of the glass envelopes of light bulbs. The bursting strength, σ_f was then measured by increasing the interior pressure and the general form of the relation was verified as:

$$\sigma_f\sqrt{c} = \text{Constant} = 2.63 \times 10^5\ \mathrm{N\ m^{-3/2}}$$

Combining this relation with the definition of σ_r and taking $\varrho \sim 5 \times 10^{-10}$ m, the resulting theoretical cohesive strength of glass was found to be:

$$\sigma_{\mathrm{th}} \simeq 2.3 \times 10^{10}\ \mathrm{N\ m^{-2}}$$

However, using the value of Young's modulus, the corresponding strain would be 0.3–0.4. It is difficult to accept that Hooke's law would still be valid in these conditions, the bond force–separation relation certainly becoming non-linear before the failure.

With his hypothesis concerning flaws as fracture initiators and his concept of energetic equilibrium determining the propagation, Griffith established a solid basis of a general theory for the fracture of glass and other brittle solids.

Numerous discussions have been devoted to the respective merits of the two criteria proposed by Griffith. Taking into account the heat energy exchanges, Doremus[445] established from thermodynamic principles that Griffith's energy balance is a necessary but not sufficient condition for propagation. On the other hand, he also showed that the existence of radii of curvature less than about 10 interatomic distances is improbable, the system increasing ϱ automatically by fluctuations up to a thermodynamically acceptable value.

(c) *Statistical aspect of mechanical strength*

The fact that fracture is initiated by Griffith flaws implies that the fracture strength of glass is essentially random and depends on the probability that a flaw which is capable of initiating fracture under a given stress would be present in this particular position.

This explains the scatter of results observed for glass and implies that the strength should be related somehow to the *volume* or *surface* of the piece considered. The measured fracture strength values thus vary with the test method; in particular, higher apparent strength is found in a bending test where the effective volume subjected to the critical stress is much less than in a uniform tensile test.

Many statistical flaw theories have been developed with the goal of predicting the theoretical strength by taking various distributions of flaw lengths for a given sample lot. They are of limited validity because the most serious flaws will inevitably be located in the tails of the statistical distributions and thus easily escape detection.

The problem remained in this state until the sixties when it was reconsidered, mainly in the work of Irwin[446] on brittle fracture, motivated by the necessity of having more reliable criteria of mechanical strengths in post-war advanced technology. This led progressively to a general theory of fracture mechanics.

14.1.5 *Mechanics of brittle fracture*

The work of Irwin was oriented in two distinct directions constituting generalizations of the two Griffith criteria.

(a) *Stress intensity factor*

Taking up the first criterion of Griffith, the brittleness of glass can be considered as due to the fact that flaws tend to redistribute the internal stress fields to cause a high stress concentration at the tips of the cracks. Fracture occurs when the stress fields attain a critical value. To determine the fracture strength, it is thus necessary to study the stress distribution around the cracks. Irwin[446] and Williams[447] determined the most general stress field near the tip of an infinitely thin crack in an infinite continuous material.

Three modes of fracture can be distinguished (Fig. 14.6), and the most general case is obtained by a linear superposition of these three fundamental modes. Mode I, called the *opening mode*, is the most applicable to glass, mode II (sliding) and mode III (tearing) are less important since glass generally fails in tension.

In a cylindrical coordinate system, with the origin of (r, θ) located at the crack tip (Fig. 14.7), the components of the stress tensor σ_{ij} are of

Fig. 14.6. Modes of fracture.

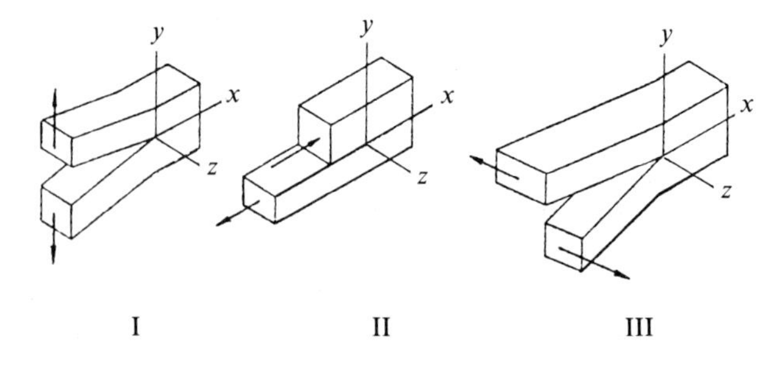

Fig. 14.7. Representation of the stress field near a crack tip. After Ref. 448.

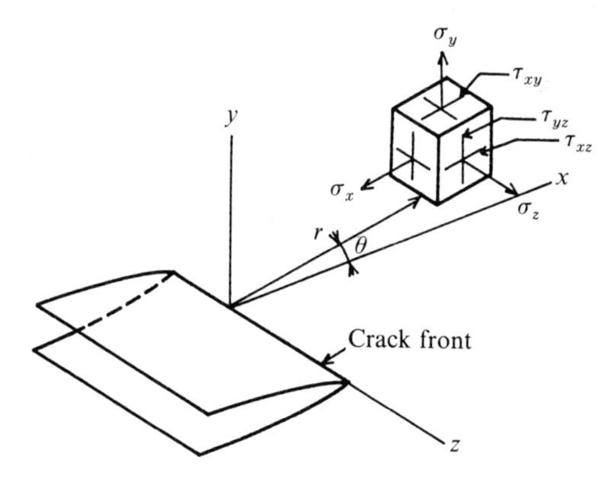

the form:

$$\sigma_{ij} = \frac{K}{\sqrt{2\pi r}} f_{ij}(\theta)$$

the corresponding displacements u_i being:

$$u_i = \sqrt{\frac{K}{2E}}\sqrt{\frac{r}{2\pi}}\, f_i(\theta)$$

in which $f_{ij}(\theta)$, $f_i(\theta)$ are calculated functions (cf. e.g. Ref. 449) and K is a constant. In all cases, σ_{ij} appears as the product of three distinct factors. The radial dependence is always proportional to $r^{-1/2}$, for any loading mode, crack geometry or load level. The angular variation, given by $f_{ij}(\theta)$ depends only on the type of stress and the mode of load application., but not the level of the loads or the crack geometry. These functions thus determine the *distribution* of the local field. The factor K, called the *stress intensity factor*, depends only on crack geometry and the manner of the exterior load application. K determines the *intensity* of the local field. The most significant result of this analysis is the separation of the factors which determine the *geometry* of the field and those which determine the intensity.

According to the preceding formulae, the terms σ_{ij} become infinite at the tip of the crack for $r \to 0$. This obviously does not correspond to reality, but is a mathematical singularity due to the assumption of an infinitely thin crack in a continuum material. The singularity disappears for a material with discrete atomic structure. In a real material the perturbation in the immediate neighborhood of the crack tip, related to the discrete character of the material, would have negligible influence on the more distant stress field (Saint Venant's principle).

The radial and angular distributions will be identical for a given mode while the stress levels depend only on K. In mode I, for example, the stress intensity factor K_I will be of the form: $K_I = \alpha P$ where P is the applied load and α is a geometric factor calculated for a given configuration. Figures 14.8–14.10 give examples of the most suitable techniques used in tests with the corresponding calculation of K_I indicated. Numerous other configurations have been used.

The essential fact of this analysis is the possibility of establishing a correspondence between the maximum tensile stress near the crack and the value of the stress intensity factor. At failure, the material can thus be characterized by a critical value K_C of K, e.g. K_{IC} for mode I. This factor has been given the name *fracture toughness*.

Fig. 14.8. Three-point notched beam bending test (NBT). Calculation of K_I. After Ref. 448.

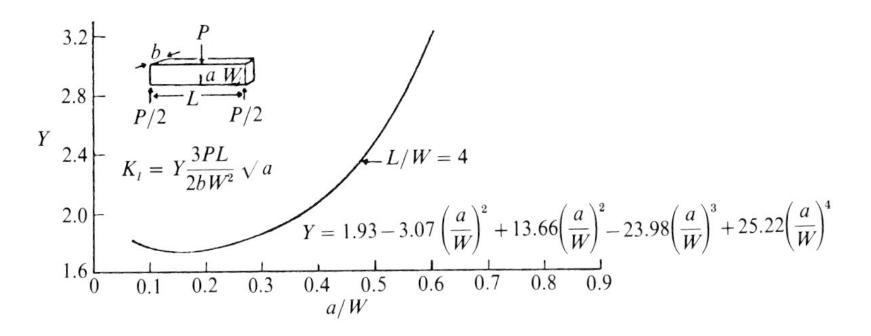

Fig. 14.9. Double cantilever beam. Calculation of K_I. After Ref. 448.

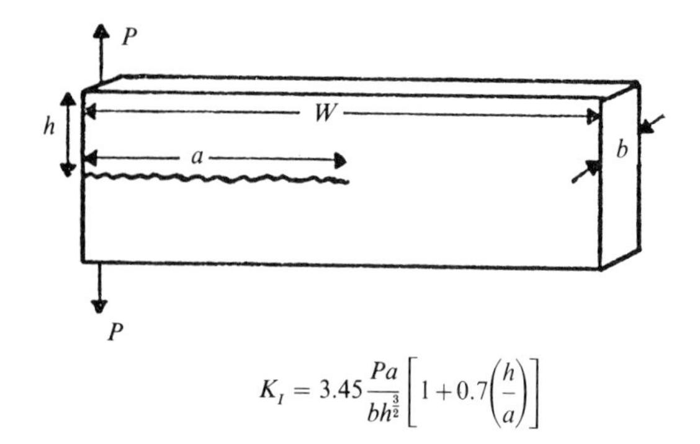

Fig. 14.10. Thin plate in double torsion. Calculation of K_I. After Ref. 448.

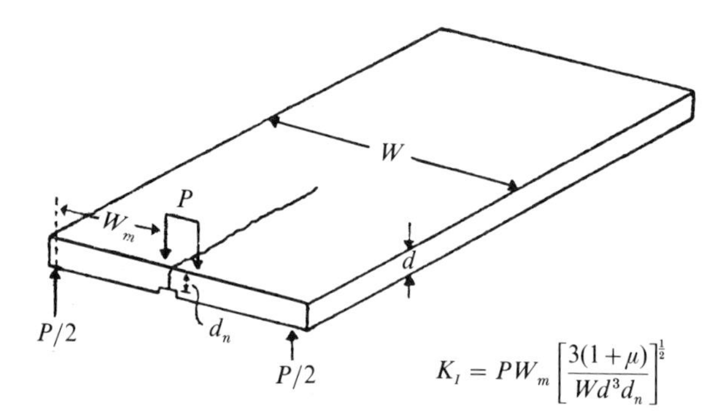

K_{IC} is thus an intrinsic material constant which depends on physical processes at the crack tip during fracture, but is independent of the geometry, form of the specimen or type of test. K_{IC} permits the difficulty of crack propagation in a material to be quantified without regard to the flaw distributions, one of the difficulties in preceding characterizations.

(b) *Strain energy release rate*

This second approach developed by Irwin[446] aims at incorporating the preceding solutions from the theory of elasticity into the Griffith energy equilibrium criterion.

Consider the system represented in Figure 14.11 which contains a crack of length $2c$ with its walls free from tensile forces. With fictive constraints "pinching" the ends of the crack to prevent their advance under the application of an external force F, the system undergoes an elongation u

Fig. 14.11. Elastic solid containing a crack.

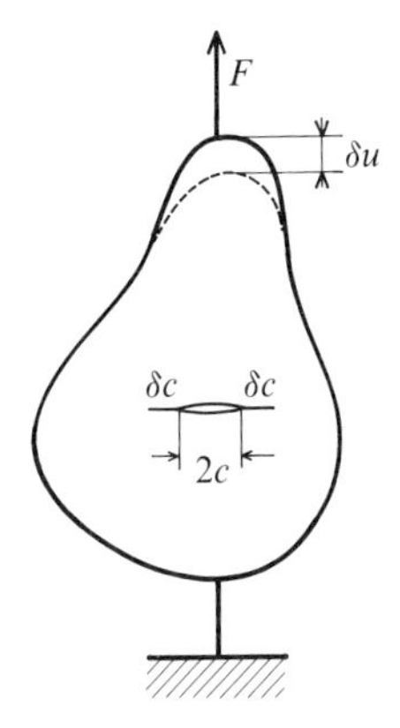

given by Hooke's law:

$$u = \lambda F$$

where λ is the (constant) elastic compliance of the system.

Increasing the force F progressively from 0 to F, the stored elastic energy is:

$$U_E = \int_0^u F(u)\,\mathrm{d}u = \frac{1}{2}\,Fu = \frac{1}{2}\,F^2\lambda = \frac{1}{2}\,\frac{u^2}{\lambda}$$

If, with the specimen under tension, the constraints on the crack tips are now relaxed to permit the growth of the crack, the specimen elongates by δu; the compliance λ is then modified:

$$\delta u = \lambda\,\delta F + F\,\delta\lambda$$

but, since $\delta u > 0$ and $\delta F < 0$, it is seen that $\delta\lambda > 0$, and the compliance can only increase or remain constant.

To evaluate the variations of the mechanical energy $-W + U_E$, two limiting configurations are considered:

1. Fixed load: the force F remains constant during crack extension:

$$\delta W \simeq F\,\delta u = F^2\,\delta\lambda$$
$$\delta U_E \simeq \frac{1}{2}\,F^2\,\delta\lambda$$
$$\delta(-W + U_E) \simeq -\frac{1}{2}\,F^2\,\delta\lambda$$

2. Fixed displacement: the displacement u remains constant during crack extension:

$$\delta W = 0$$

$$\delta U_E = \frac{u^2}{2}\,\delta\left(\frac{1}{\lambda}\right) = -\frac{1}{2}\frac{u^2}{\lambda^2}\,\delta\lambda = -\frac{1}{2}\,F^2\,\delta\lambda$$

from which:

$$\delta(-W + U_E) \simeq -\frac{1}{2}\,F^2\,\delta\lambda$$

giving the same value as in the preceding case. This is expressed by saying that the mechanical energy liberated during crack growth is independent of the loading mode.

It can be shown that this result is completely general thus permitting the definition of a generalized force for crack extension (per unit thickness).

$$G = \frac{\mathrm{d}}{\mathrm{d}c}\,(-W + U_E)$$

Since G is independent of the loading mode, only the case of fixed displacement u need be considered; i.e. one can define G by the relation:

$$G = -\left(\frac{\partial U}{\partial c}\right)_u$$

G is called the *strain energy release rate* (the rate being defined relative to the length c of the crack and not relative to time).

(c) *Relation between the parameters G and K*

The relation between G and K can be determined by evaluating the work of closure of the crack. Using the general expressions for σ_{ij} and u_i it is found by integration (see e.g. Ref. 449) for mode I (Irwin's formula) that:

$$G_I = \frac{K_I^2}{E} \qquad \text{for plane stress}$$

and

$$G_I = \frac{K_I^2}{E}\,(1 - \mu^2) \qquad \text{for plane strain}$$

The rates of energy release for the different modes are additive and the influence of the modes can thus be taken separately into account.

(d) *Generalization of the Griffith concept*

To introduce the variation of surface energy, a reversible variation (opening–closing) of the crack is considered (see Fig. 14.11). For a unit thickness:

$$\delta U_S \simeq 2\gamma \, \mathrm{d}c$$

(the factor 2 coming from the *two* surfaces created during fracture).

A *generalized surface tension force* can then be defined:

$$-\frac{\mathrm{d}U_S}{\mathrm{d}c} = -2\gamma$$

which opposes the crack extension.

Expressing the thermodynamic balance of the process:

$$\delta U \sim \delta(-W + U_E) + \delta U_S$$

or

$$\delta U \sim -G\,\delta c + 2\gamma\,\delta c$$

The Griffith propagation criterion $\delta U < 0$ is thus equivalent to $G > 2\gamma$. At equilibrium, when $\delta U = 0$, $G = G_C = 2\gamma$, and the process is called *stable* crack growth.

For example, for mode I fracture (plane strain case):

$$G_{IC} = \frac{K_{IC}^2}{E}(1 - \mu^2) = 2\gamma$$

which constitutes a generalization of the Griffith energy balance.

The preceding analysis is applicable to the case of an ideal brittle solid, the material being linearly elastic and obeying Hooke's law. Difficulties arise when processes acting at the crack tip are represented in this way leading to the introduction of the concept of a certain zone near the crack tip in which non-linear, energy dissipating plastic deformation phenomena can occur.

Irwin and Orowan generalized the Griffith concept by including dissipative components in the surface energy term.[446] If it is assumed that these effects of cohesive forces are localized in a weak zone around the crack tip, the term for the variation of the mechanical energy remains unchanged and one can write:

$$\delta U \sim \delta(-W + U_E) + \delta U_S$$

whence:

$$\delta U \sim -G\,\delta c + \frac{dU_S}{dc}\,\delta c$$

and, defining a quantity Γ such that:

$$-2\Gamma = -\frac{dU_S}{dc}$$

at equilibrium: $\delta U = 0$, $G = G_c$ and $G_c = 2\Gamma$

The term Γ is called the *fracture surface energy* or *work of fracture*; it is a measure of the intrinsic fracture resistance of the material; Γ is related to the *fracture toughness* by the Irwin formulae. For an *ideal* brittle material, Γ reduces to the surface energy γ.

Different models of the plastic zone have been developed. In the Dugdale[450]–Barenblatt[451] model, the plastic zone extends over a distance D_c in front of the fracture (Fig. 14.12). The extent of D_c can be evaluated from the fracture toughness K_{IC} if the elastic limit of the material is known:

$$D_c = \frac{\pi}{8}\left(\frac{K_{IC}}{\sigma_e}\right)^2$$

The *crack opening displacement* is:

$$2u_c = \frac{8\sigma_e D_c}{\pi E}$$

Fig. 14.12. Dugdale–Barenblatt model including the plastic zones at the crack tips.

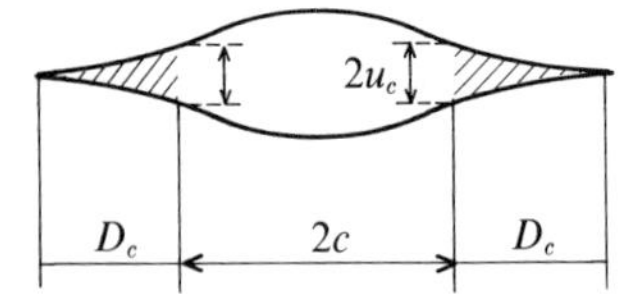

(e) *Dynamic effects*

The preceding analysis relates to a static or quasi-static system in which the cracks propagate slowly in a fracture test under controlled conditions. The kinetic energy components are insignificant relative to the elastic energy components. However, the case exists where the fracture propagates in a catastrophic manner; a certain critical crack length having been exceeded, the system acquires kinetic energy through the separation of the sides of the crack. The system then moves rapidly to a limiting rate which depends on the speed of elastic wave propagation in the material, transmitting information ahead of the crack tip. A dynamic situation can also be created when the material is subjected to rapidly varying load (shock effect).

Mott[452] extended the static concept of Griffith in adding a kinetic term U_K to the energy:

$$U = (-W + U_E) + U_S + U_K$$

In order for the system to remain in thermodynamic equilibrium when the crack extends δc, it is necessary that:

$$\frac{\mathrm{d}U}{\mathrm{d}c} = 0 \qquad \text{i.e.} \qquad G - 2\Gamma = \frac{\mathrm{d}U_K}{\mathrm{d}c}$$

The kinetic term serves to dissipate the excess energy resulting from the non-equilibrium. To calculate U_K effectively it is necessary to consider the velocity of all the elements near the crack and to carry out integration over the entire volume during the deformation – an extremely complex problem.

Mott examined the simple case of the propagation of an internal crack in a system under uniform tensile stress (fixed load). Assuming that the stress σ is increased progressively up to a level where the Griffith equilibrium is exceeded, then held constant:

$$U = -\frac{\pi c^2 \sigma^2}{E} + 4c\Gamma + U_K$$

At equilibrium

$$c = c_0, \qquad 4\gamma = \frac{2\pi\sigma^2 c_0}{E}, \qquad U_K = 0$$

These conditions allow the evaluation of the constant U and the elimination of Γ from which:

$$U_K = \frac{\pi\sigma^2 c^2}{E}\left(1 - \frac{c_0}{c}\right)^2$$

The kinetic energy of a volume element $\mathrm{d}x\,\mathrm{d}y$ of density ϱ is:

$$\frac{1}{2}\varrho\,(\dot{u}_x^2 + \dot{v}_y^2)\,\mathrm{d}x\,\mathrm{d}y$$

where

$$\dot{u}_x = \left(\frac{\partial u}{\partial c}\right) v_c \qquad \dot{v}_y = \left(\frac{\partial v}{\partial c}\right) v_c$$

with:

$$v_c = \frac{\mathrm{d}c}{\mathrm{d}t}$$

Mott used an argument based on a similarity concept: the deformation field (u, v) is proportional to the crack length as are $\partial u/\partial c$ and $\partial v/\partial c$ which, moreover, must be proportional to σ/E giving the following result:

$$U_K = \frac{1}{2} k \varrho v_c^2 \frac{c^2 \sigma^2}{E^2}$$

where k is a proportionality constant.

The velocity of the fracture v_c is:

$$v_c = \left(\frac{2\pi E}{k\varrho}\right)^{1/2} \left(1 - \frac{c_0}{c}\right)$$

The velocity v_c tends asymptotically towards a terminal velocity v_m:

$$v_m = \alpha v_e$$

where $v_e = \sqrt{E/\varrho}$ is the velocity of elastic waves in the substance and α is a numerical constant of the order of 0.3–0.6. For soda–lime–silica glass $v_m \sim 1500$ m s^{-1} or about one third of the velocity of longitudinal elastic waves (velocity of sound in glass). The terminal velocity does not seem to be affected by the fracture stress.

14.1.6 *Methods of measuring fracture toughness*

In the case of glasses, it is mode I which predominates, modes II and III occurring in materials with some plasticity. The following discussions are limited to mode I.

Three experimental methods are used to obtain K_{IC} (or G_{IC}).

(a) *The analytical method*

Theories of brittle fracture introduce, as we have seen, the idea of a *critical threshold*, not for the stress (theoretically infinite at the tip of the crack) but for the factor K_I; at failure, $K_I = K_{IC}$.

K_{IC} is thus a physical characteristic of the material and its determination is an experimental problem. To evaluate the mechanical resistance of a glass, the measurement of K_{IC}, the critical value of the coefficient K_I, is carried out in a simple geometric configuration where K_I can be calculated *a priori* by analysis as an elastic problem.

In a general way, K_I is a function of the length of the crack, the geometry of the system and the mode of loading. For fixed material and environmental conditions, K_{IC} must be independent of the type of test used. The configurations used in practice are essentially:

–three-point notched beam (NBT) (Fig. 14.8)
–double cantilever beam (DCB) (Fig. 14.9)
–thin plate in double torsion (Fig. 14.10)

The expressions for K_I are shown on the above figures. Ref. 449 gives values of K_I for many other configurations.

The advantage of the notched beam specimen is its simplicity, but the preparation of the notch requires some precautions; it is generally necessary to pre-crack the specimen by thermal shock to obtain a final notch tip.

The thin double torsion plate is a specimen configuration with constant K_I (independent of the crack length). It is useful in situations where crack length measurement is difficult (high temperature or on radioactive materials); it requires, however, a specimen with a larger volume.

(b) *The compliance method*

It was shown that for a fixed load F:

$$\delta(-W + U_E) = -\frac{1}{2}F^2\,\delta\lambda$$

hence;

$$G = \frac{1}{2}F^2\left(\frac{\partial\lambda}{\partial c}\right)_F$$

The method consists of preparing a series of specimens with different crack lengths c. By measuring the displacement under a load F, the value $\lambda = u/F$ is obtained which is a function of c, and then $(\partial\lambda/\partial c)_F$. This method is applicable to specimens of an arbitrary form, but necessitates very precise measurement of u.

(c) *Work of fracture method*

In this method the work to propagate a crack across the entire section of the specimen is determined. The test is done in three-point bending on a notched specimen. The method is based on the fact that when the crack extends, if the compliance λ decreases the propagation can be produced in a stable and continuous manner as the load decreases.

It is essential to establish *stable* fracture propagation conditions. By tracing the load–deformation curves for different notch lengths, the different points for fracture initiation are obtained which together constitute Berry's curve[453] (Fig. 14.13). Thus the stability condition for a given notch depth can be predicted. If at an imposed displacement $\delta = \delta_i$, the system is on the lower part of the curve (point B), extension of the crack leads to a reduction of the compliance and the load will be insufficient to continue propagation; this corresponds to stable crack growth. In contrast, on the upper part of the curve (point A), any advance of the crack leads outside the curve and gives instability. For a given initial length, the crack will stop (semi-stable fracture) or not (catastrophic fracture) according to the kinetic energy acquired at the time of crossing the equilibrium curve.

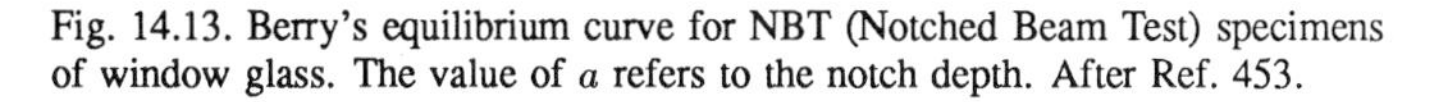

Fig. 14.13. Berry's equilibrium curve for NBT (Notched Beam Test) specimens of window glass. The value of a refers to the notch depth. After Ref. 453.

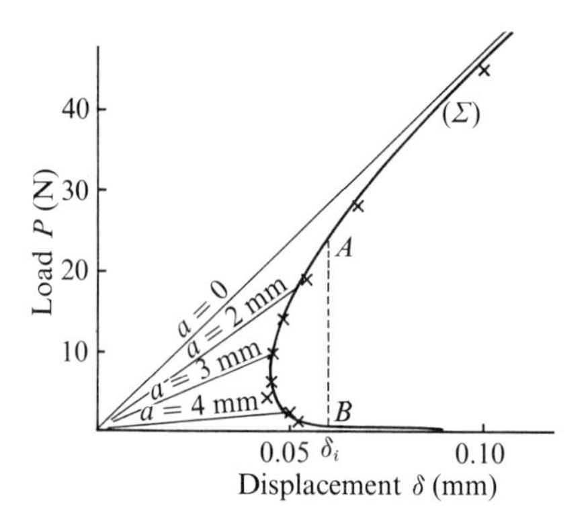

Figure 14.14 shows examples of different types of propagation stability. In this type of test the specimen is slowly deformed at a constant rate up to fracture and the load–displacement curve is recorded (Fig. 14.14(c)). The area under the curve represents the work done during the test. Since at the end of the test there is no stored elastic energy in the machine and no loss of kinetic energy, the elementary work which produced the fracture is:

$$\mathrm{d}W = \gamma \,\mathrm{d}A + \mathrm{d}\Phi$$

where A is the area of the fracture surface and $\mathrm{d}\Phi$ a correction factor which takes into account parasitic dissipated energies (plastic deformation at the load points, heating, etc.).

Fig. 14.14. Different types of stability in the propagation of a crack: (*a*) catastrophic fracture, (*b*) semi-stable fracture, (*c*) stable fracture.

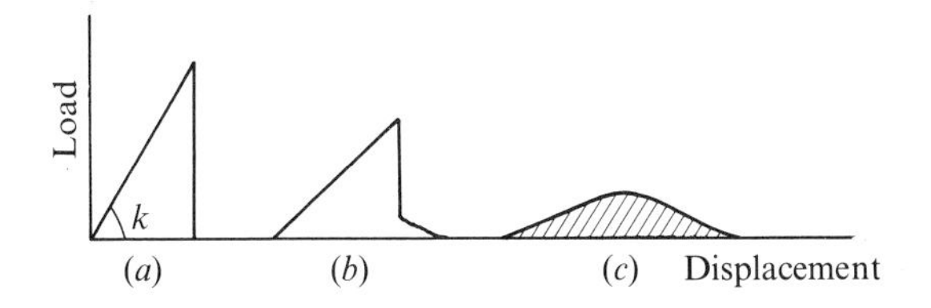

The total work done is:

$$W = \int_0^{d_F} P \, \mathrm{d}\delta$$

where d_F is the displacement at the moment of failure.

The fracture surface energy is:

$$\gamma = \frac{W}{2A} - \Phi$$

The contribution of Φ is eliminated by extrapolating the curve $\gamma = f(c)$ to a notch depth c equal to the total specimen thickness (zero fracture surface).

The three methods are based on different fracture criteria. The analytic method implies that the movement of the crack begins when the stress intensity at the crack tip attains a certain value – thus reflecting the physical processes at this point but insensitive to the dimensions of the crack. The compliance method is based on the Griffith energy balance; the fracture occurs when the energy supplied is equal to the required surface energy, the process being extrapolated up to equilibrium. Both methods measure the conditions of failure *initiation*. In contrast, the work of fracture method integrates all parts of the failure and gives values which relate rather to the *propagation* of the crack.

14.1.7 *Experimental results*

Table 14.1 gives fracture surface energy values for common glasses obtained by the double-cantilever method. These values of γ lead to a cohesive strength of the order of 2×10^{10} N m^{-2}.

In spite of the importance of the subject, only a few measurements of K_{IC} exist for glasses as a function of chemical composition.[454–6]

Table 14.1. *Data for the fracture of some glasses. After Ref. 442.*

Glass	Fracture surface energy γ (N m^{-1})			Fracture toughness (MN m$^{-3/2}$)		
	77 K	196 K	300 K	77 K	196 K	300 K
SiO_2	4.56	4.83	4.37	0.811	0.839	0.794
96% SiO_2	4.17	4.60	3.96	0.741	0.779	0.722
aluminosilicate	5.21		4.65	0.963		0.910
borosilicate	4.70		4.63	0.774		0.768
soda–lime–silica	4.55	4.48	3.87	0.820	0.812	0.754
alkali silicate	4.11		3.52	0.734		0.680

Figs. 14.15–14.18 show several results obtained for silicates, germanates, borates and borosilicates.

For the GeO_2–Na_2O system, K_{IC} increases with the percentage Na_2O while the system B_2O_3–Na_2O shows the opposite effect. These variations of K_{IC} with composition seem contrary to what might be expected from a depolymerization of the network (a reduction in K_{IC}) in the first case and a change of the B coordination from 3 to 4 in the second (which must lead to a reinforcement of the network). An explanation of these results has been proposed which assumes a micro-heterogeneous structure.[455]

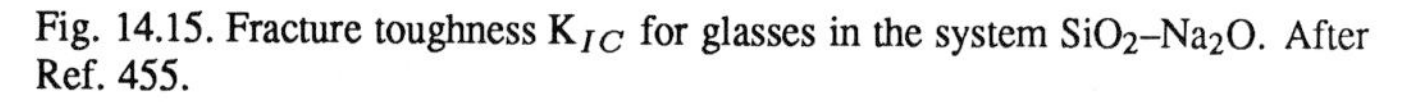

Fig. 14.15. Fracture toughness K_{IC} for glasses in the system SiO_2–Na_2O. After Ref. 455.

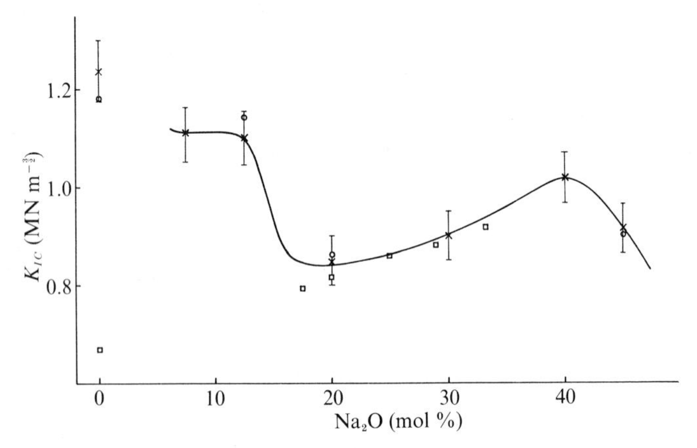

Fig. 14.16. Fracture toughness K_{IC} of glasses in the system GeO_2–Na_2O. After Ref. 455.

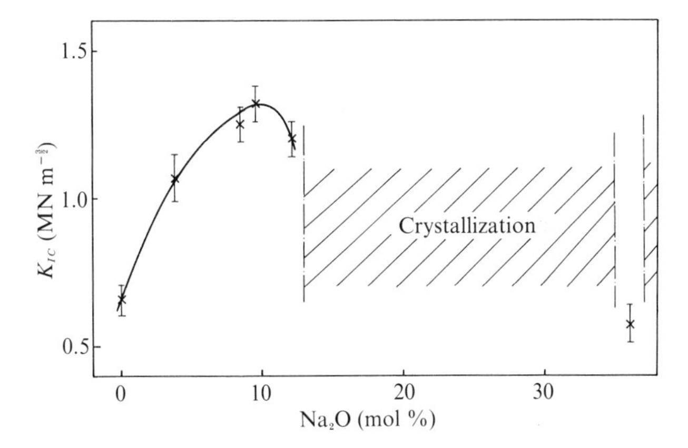

14.1.8 *Possibility of occurrence of plastic phenomena*

Although glass seems to behave as an ideal elastic solid up to the failure point, several authors have not excluded the possibility of localized plasticity at the crack tip.[457]

A few observations show the possibility of flow under very high stresses. Observation of crossed scratches on a glass surface shows a backflow of the material on the edges and the filling of a groove by the following one (Fig. 14.19). During a micro-hardness test on glass, it is seen that the imprint of the diamond after unloading shows a "cushion" form (Fig. 14.20). Effects which are to a greater or lesser degree irreversible are apparently produced under concentrated stresses, perhaps due to local elevation of the temperature. Systematic studies of the densification of glasses under high pressure have shown that irreversible effects occur.[458] Based on these observations, Marsh[457] developed a theory on the mechanical strength of glass taking into account micro-plasticity effects.

Application of the Dugdale model permits an estimation of the extent of the hypothetical plastic zones at the crack tip. Taking Marsh's values[457] for σ_e, the parameters of Fig. 14.12 were calculated. For vitreous SiO_2 $D_c = 6$ Å and $2u_c = 4$ Å while for soda–lime–silica glass, $D_c \sim 26$ Å and $2u_c \sim 9$ Å which shows the extremely small size of these hypothetical zones.

14.2 Environmental effects – Static fatigue

14.2.1 *Delayed fracture*

In the case of silicate glasses, it is observed that the measured

Fig. 14.17. Alkali borate glasses $(100 - x)B_2O_3, xR_2O$. Variations: (*a*) of the critical stress intensity factor K_{IC} and (*b*) the fracture surface energy γ with the alkali oxide content. After Ref. 454.

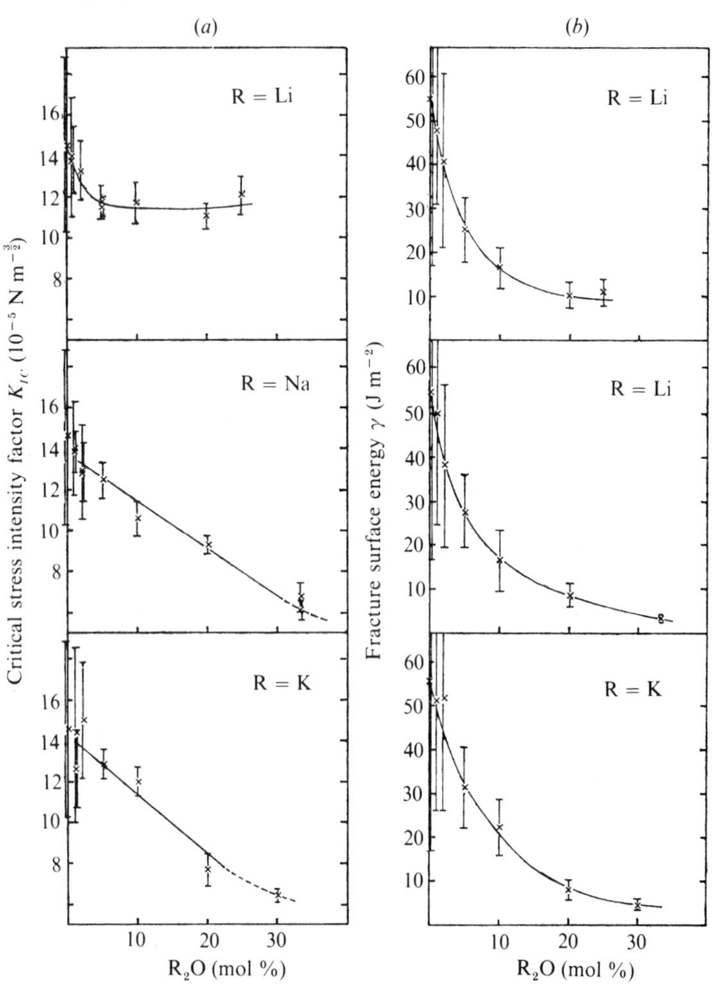

fracture strength depends on the duration of the load application or the rate at which the load is applied.

This phenomenon, called *fatigue*, can be described by functions of the type:

$$\log t = \frac{A}{\sigma} + B \qquad \text{or} \qquad \log t = a + b \log \sigma$$

Fig. 14.18. Fracture toughness K_{IC} of glasses in the ternary system $20\,Na_2O \cdot (80 - x)\,B_2O_3 \cdot x\,SiO_2$. After Ref. 455.

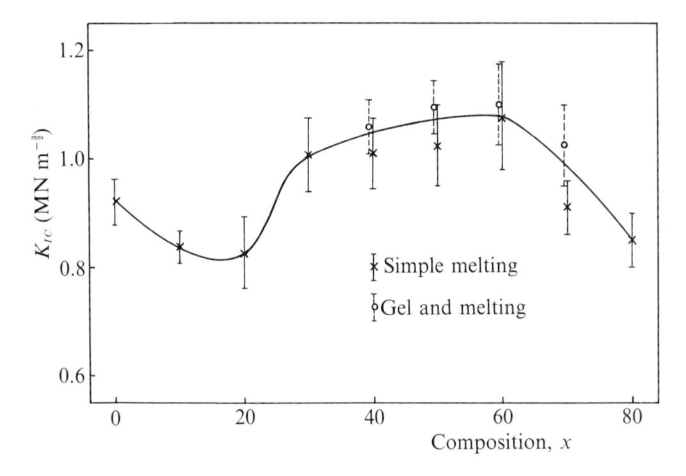

Fig. 14.19. Crossed scratches in the surface of glass. Replica electron microscope photomicrograph. (By courtesy of Saint-Gobain Co.)

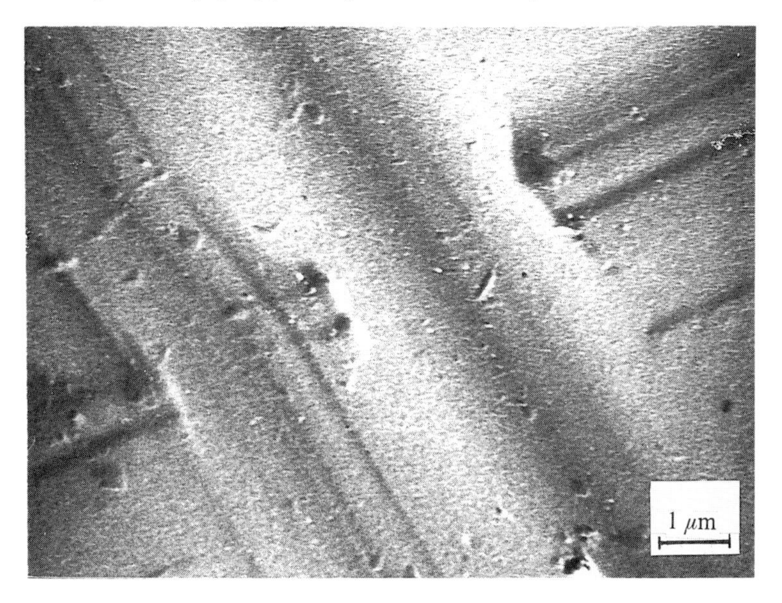

where t is the "life-time," σ the failure stress and A, B, a, b are experimentally determined constants. The result is that if a glass can support a

Fig. 14.20. Indent after a diamond micro-hardness test on glass: (*a*) direct examination; (*b*) in an interference microscope.

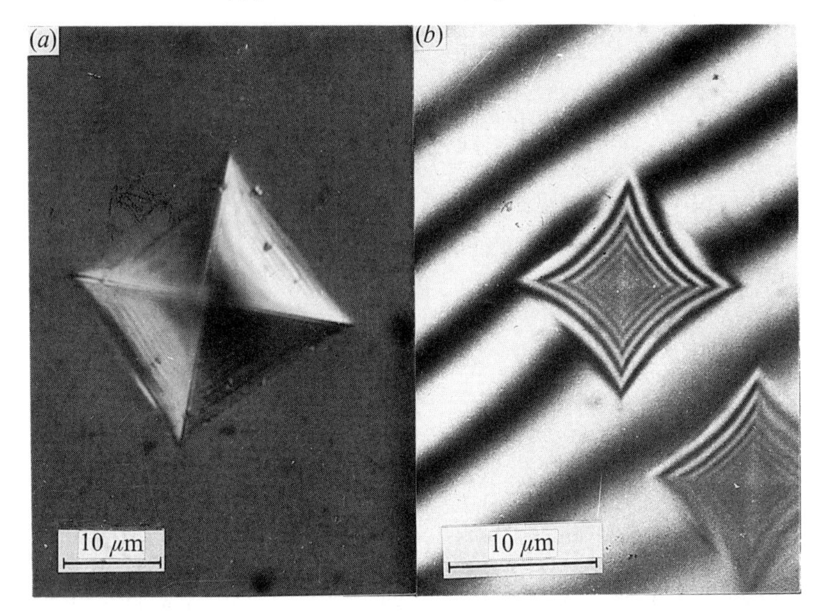

certain stress for a short time, it will fail under a lower stress provided it is applied for a sufficiently long time.

Experiments show that fatigue or *delayed fracture* is more pronounced when the atmosphere contains water vapor and is reduced when the glass is in *vacuum* or in a perfectly dry atmosphere. The chemical nature of this process is also evident from its temperature dependence. At low temperatures when the rate of attack by water vapor is reduced, the results are close to those obtained in vacuum.

It is found that the condition of the surface plays a rôle, with different abrasions producing different fatigue curves. However, using a reduced coordinate system plotting the ratio of the observed σ to σ_N at 77 K (the latter observed in liquid N_2) as a function of the ratio of the life-time t to $t_{0.5}$ (for which the strength falls to half that at 77 K) produces a unique reduced curve for each surface condition (Fig. 14.21). This is called the *universal fatigue curve* for a given glass.

The same behavior is found with a periodically varying stress; the dynamic fatigue curve is independent of the frequency and coincides with the results of static tests.

Fig. 14.21. Static fatigue of glass. Comparison of results obtained from fracture propagation with the universal fatigue curve. After Ref. 459.

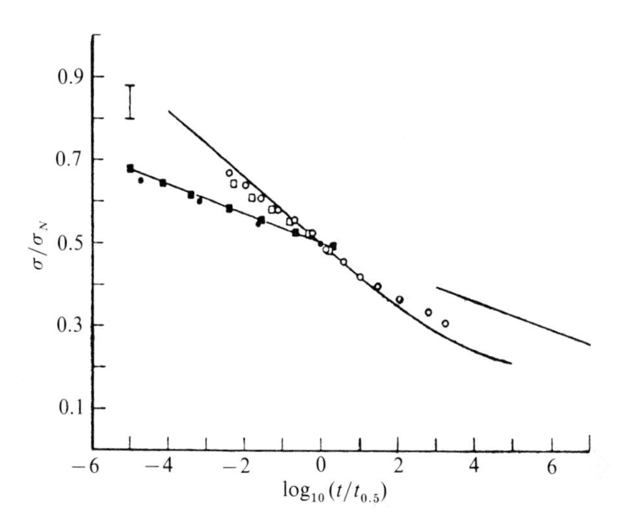

14.2.2 *Theory of Charles and Hillig*

The time dependence of the mechanical strength is explained by the concept of *stress corrosion*.[460] Silicate glasses age by adsorption of ambient humidity and a hydroxyl reaction:

$$\mathrm{H{-}O{-}H} + {-}\overset{|}{\underset{|}{\mathrm{Si}}}{-}\mathrm{O}{-}\overset{|}{\underset{|}{\mathrm{Si}}}{-} \longrightarrow {-}\overset{|}{\underset{|}{\mathrm{Si}}}{-}\ \mathrm{OH} \quad \mathrm{HO}{-}\overset{|}{\underset{|}{\mathrm{Si}}}{-}$$

converting a thin surface layer of the glass into a gel.

In the absence of applied stress, the corrosion processes occur uniformly over the entire accessible surface of the glass including the flaws or microcracks (Fig. 14.22). This results in an increase in the radius of curvature at the tip of the flaw and a small increase in the mechanical strength (Fig. 14.22(*b*)).

In contrast, if a tensile stress is applied which is just insufficient to cause immediate fracture, the rate of reaction progresses much faster at the crack tips where the structure is extended i.e. the chemical potential of the glass is higher at this point.

The local acceleration of the process causes the cracks to become sharper until the Griffith criterion is fulfilled and fracture is initiated (Fig. 14.22(*c*)). An intermediate stress, called *static fatigue limit*, must then exist for which the stress corrosion effect is just compensated by the aging effect (which rounds the crack tip), the strength then remaining constant.

Fig. 14.22. Explanation of static fatigue of glass: (*a*) two-dimensional model of a crack introduced by water in SiO_2 glass. The filled circles represent atoms reacting with the environment (H_2O); (*b*) uniform attack on the crack in the absence of stress; (*c*) preferential attack at the tip of the crack under tensile stress.

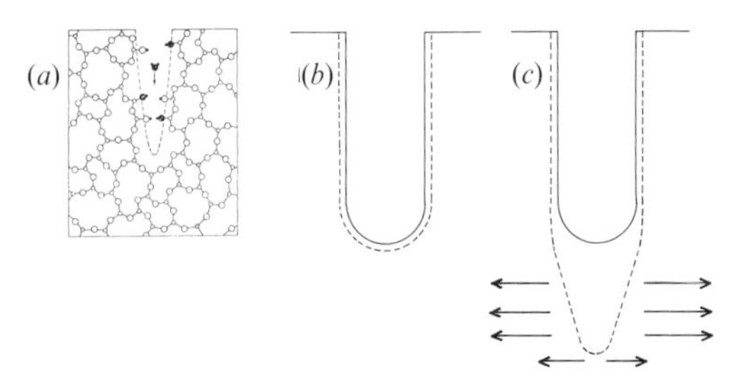

14.2.3 *Propagation of sub-critical cracks*

Static fatigue in glass depends on the rate of crack propagation. Apparently stationary cracks (sub-critical) can grow at a very low rate resulting in often unexpected failure of the object.

Wiederhorn[461] used the double cantilever method (Fig. 14.9) to measure the rate of propagation v of a crack as a function of the stress intensity factor K_I, temperature and environment. Figure 14.23 shows a group of typical results for a soda–lime–silica glass.

For each water vapor concentration (indicated in % on the figure), the curve shows three distinct regions. For values of $K_I \lesssim 6 \times 10^5$ N m$^{-3/2}$ (region I), the velocity depends exponentially on K_I and on the percentage of water vapor. In region II (plateau), the velocity is independent of K_I, but depends on the ambient atmosphere. Finally, in region III the velocity again shows a dependence on K_I but is no longer dependent on the environment.

Regions I and II can be explained by the theory of Charles and Hillig; in region I, the velocity is limited by the chemical reaction while in region II the velocity is limited by the transport of water vapor from the environment to the crack tip. Region III has not yet received an adequate explanation.

Study of the temperature dependence gives a relation of the type:

$$v = v_0 \exp\left[\left(-E^* + bK_I\right)/RT\right]$$

in agreement with the Charles and Hillig theory. Often a relation of the following type is used in practice:

$$v = AK_I^n$$

where A and n are constants.

Fig. 14.23. Dependence of the crack velocity v on the stress intensity factor K_I for a Na_2O–CaO–SiO_2 glass. After Ref. 461.

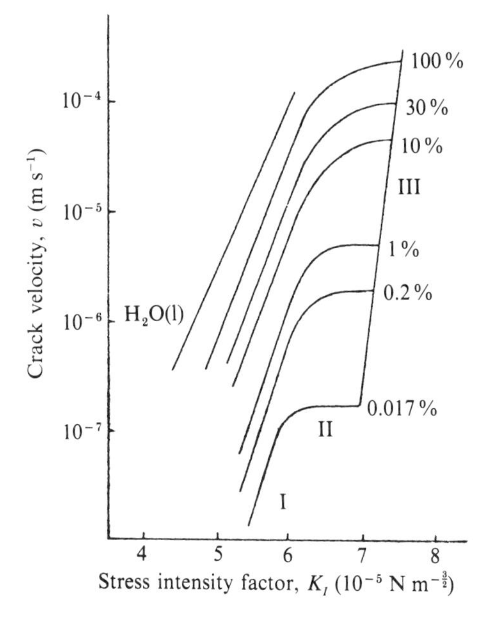

14.2.4 *Estimation of the life-time of brittle components*

The importance of these measurements rests in the possibility of applying the results to the problem of prediction of the minimum time to failure (service life-time) of glass objects.[462] The failure times can be evaluated using methods combining the concepts of fracture dynamics and statistical estimations.

The time to failure for a specimen under tensile stress can be evaluated as follows. The stress intensity factor K_I for a flaw of dimension a under an applied stress σ_a is of the form:

$$K_I = \sigma_a Y \sqrt{a}$$

where Y is a geometric constant depending on the form and location of the flaw (for a surface flaw, $Y \sim \sqrt{\pi}$). Assuming σ_a constant, the derivative relative to temperature gives:

$$\frac{dK_I}{dt} = \left(\frac{\sigma_a^2 Y^2}{2K_I}\right) v$$

where $v = (\mathrm{d}a/\mathrm{d}t)$, the propagation velocity of the crack. By integration, the time t to failure (life-time) is:

$$t = \frac{2}{\sigma_a^2 Y^2} \int_{K_{Ii}}^{K_{IC}} \left(\frac{K_I}{v}\right) \mathrm{d}K_I$$

where K_{IC} is the critical stress intensity factor and K_{Ii} the initial stress intensity factor of the most dangerous flaw.

The relation between K_I and velocity v can be evaluated in the form:

$$v = AK_I^n$$

where a and n are experimentally determined constants. Since $n \sim 20$ and $K_{IC} > K_{Ii}$:

$$t \simeq 2\,(K_{Ii})^{2-n}/A\sigma_a^2 Y^2 (n-2)$$

Two approaches can be used to estimate K_{Ii}.

(a) *Statistical estimate*

K_{Ii} is expressed in terms of the probability of failure and a global estimate for the group of specimens is used. In the statistical theory developed by Weibull,[463] the cumulative failure probability P, i.e. the fraction of specimens which will break at a given stress level σ_{IC}, has the form:

$$P = 1 - \exp\left[-\left(\frac{\sigma_{IC} - \sigma_l}{\sigma_0}\right)^m\right]$$

where σ_0, σ_l and m are empirical constants.

σ_{IC} is the strength measured in conditions where there is no sub-critical crack growth before fracture; it is determined in inert atmosphere with rapid loading.

It is generally assumed that the minimum observed $\sigma_l = 0$. The parameters σ_0 and m are determined from the graph $\log\log\left(1/(1-P)\right)$ versus $\log\sigma_{IC}$. The slope of this linear plot is m and the intercept at the origin gives $m\,\log\sigma_0$.

Designating the initial dimension of the most severe flaw in the specimen by a_i, at failure:

$$K_{IC} = \sigma_{IC} Y \sqrt{a_i}$$

while under a "service" stress σ_a:

$$K_{Ii} = \sigma_a Y \sqrt{a_i}$$

thus:

$$K_{Ii} = K_{IC}\,(\sigma_a/\sigma_{IC})$$

from which, applying the Weibull relation (with $\sigma_l = 0$):

$$K_{Ii} = K_{IC}\,(\sigma_a/\sigma_0)\left(\log\frac{1}{1-P}\right)^{-1/m} \tag{14.1}$$

The life-time[462] is thus of the form:

$$t = \sigma_a^{-n} f(P)$$

where $f(P)$ is a function of the cumulative failure probability P. A graph is obtained which gives the probability P of a life-time t for an imposed service stress σ_a.

Figure 14.24 shows the application of this principle in the case of ultralow expansion SiO_2–TiO_2 glass used for windows in the spacecraft Skylab.

(b) *Proof testing*

In this test, *all* components are subjected to a mechanical strength test under conditions identical to the service conditions but at a systematically higher stress level.

This test eliminates those components with a higher probability of failure, and components which survive (pass the proof) can be used at a lower stress level with a higher degree of confidence than that furnished by statistical methods.

For a proof test with a stress σ_P:

$$K_P = \sigma_P Y \sqrt{a_i}$$

and, if the specimen survives, $K_{IC} > K_P$. Thus, in service, if $K_{Ii} = \sigma_a Y \sqrt{a_i}$:

$$K_{Ii} < K_{IC}\,(\sigma_a/\sigma_P)$$

Substituting this upper limit for K_{Ii} in the integral of equation (14.1), a "pessimistic" estimate is obtained for the (minimum) life-time t_{min} which, as can be easily verified, depends on the ratio σ_P/σ_a:

$$t_{min} = \sigma_a^{-2} f(\sigma_P/\sigma_a)$$

Fig. 14.24. Estimation of the life-time of window glass in the spacecraft Skylab as a function of the service stress σ_a. The lines of equal probability P from 10^{-1} to 10^{-5} correspond to estimates following the Weibull method. The straight lines marked 1.5 to 3.5 correspond to the proof test ratios σ_P/σ_a. After Ref. 464.

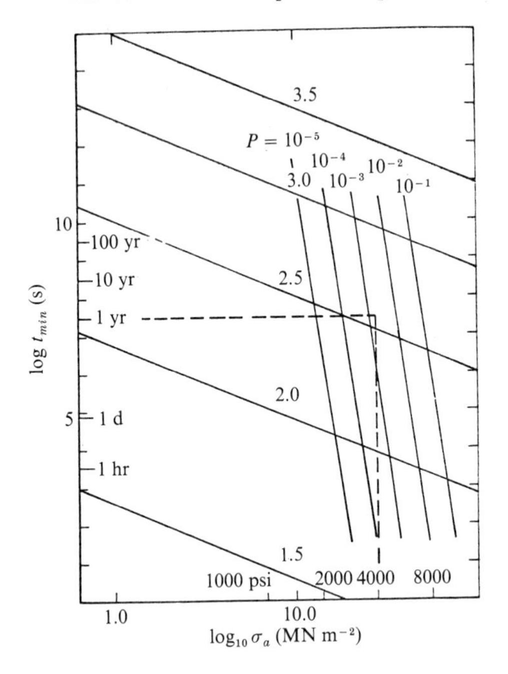

Figure 14.24 shows the corresponding graph. For a minimum life-time of one year under 28 MN m^{-2}, a proof test ratio of 2.5 is necessary – the specimens which pass this test (about 77 MN m^{-2}) will have a zero probability of failure within one year. As a comparison, without the proof test, the failure probability would be 2×10^{-3}.

The production of large quantities of optical fibers in recent years has made available *convenient sample material*, protected by coating immediately after drawing. This has enabled more accurate measurements of strength and a better understanding of fatigue effects.

These and other recent results are to be found in Ref. 472.

14.3 Study of fracture

14.3.1 *Measurement of the rate of propagation*

The rate of fracture propagation can be quite variable, from values near zero for quasi-stationary (sub-critical) cracks up to velocities approaching 1500 m s^{-1} for explosive failures.

Given the high values of the limiting velocity, studies of propagation use high-speed photographic methods capable of recording images at rates of the order of 10^6 frames per second.

In the Schardin method,[465] the principle of which is shown in Fig. 14.25, the object undergoing fracture is successively illuminated by a series of concentrated sparks powered by a stabilized generator. A photographic chamber with separate objectives corresponding to the spark sources records sequentially a series of images. The objectives are coupled to the light sources by an intermediate concave mirror M. Figure 14.26 shows a series of time-lapse photographs of the propagation of a multiple fracture.

14.3.2 *Fractography*

The fracture surface displays certain zones or features which can provide information on how the propagation occurred and, in some cases, information on the cause of the failure. The "post-mortem" study of the pieces of fractured glass is the object of *fractography.*

(a) *Mirror*

This is the smooth portion of the fracture surface which corresponds to the region of crack before branching occurs (Fig. 14.27). It is found that the mirror radius r is inversely proportional to the square of the fracture stress σ_F:

$$r = c/\sigma_F^2 \qquad \text{with} \qquad c \sim 4 \times 10^{11}\ \mathrm{N^2\ m^{-3}}$$

Fig. 14.25. Principle of the camera for ultrarapid time-lapse photography. (e_1, e_2,... spark sources, M mirror with 6 m radius of curvature, A object, O_1, O_2... objective lenses, A_1, A_2,... successive images.

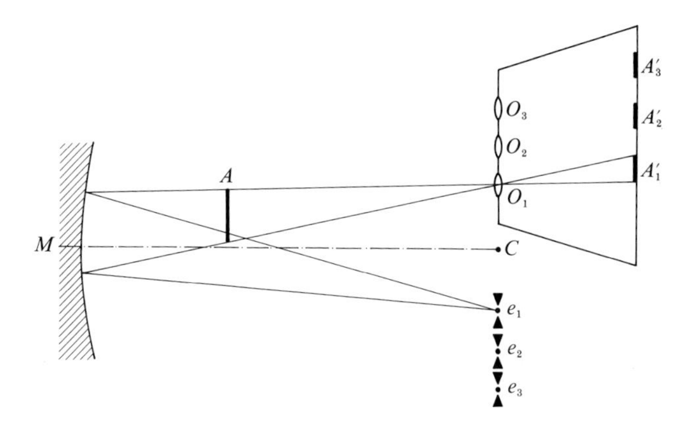

Fig. 14.26. Series of 30 sequential photographic images taken at a rate of 200,000 s^{-1} showing the progression of a fracture of a glass plate after impact (read line by line, left to right). The dimension of the object is 30 cm. (By courtesy of Saint-Gobain Co.)

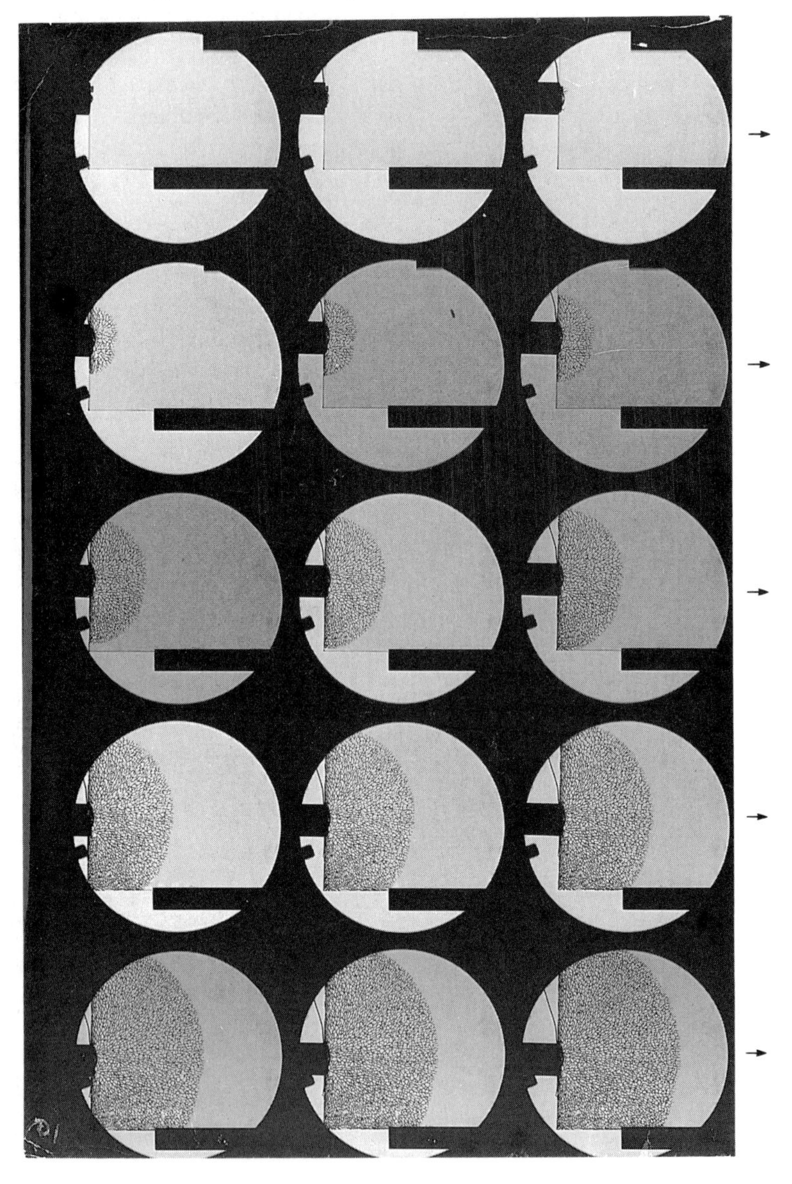

Fig. 14.26. (continued)

Fig. 14.27. Fracture surfaces of two glass rods which failed at different stresses. (By courtesy of Saint-Gobain Co.)

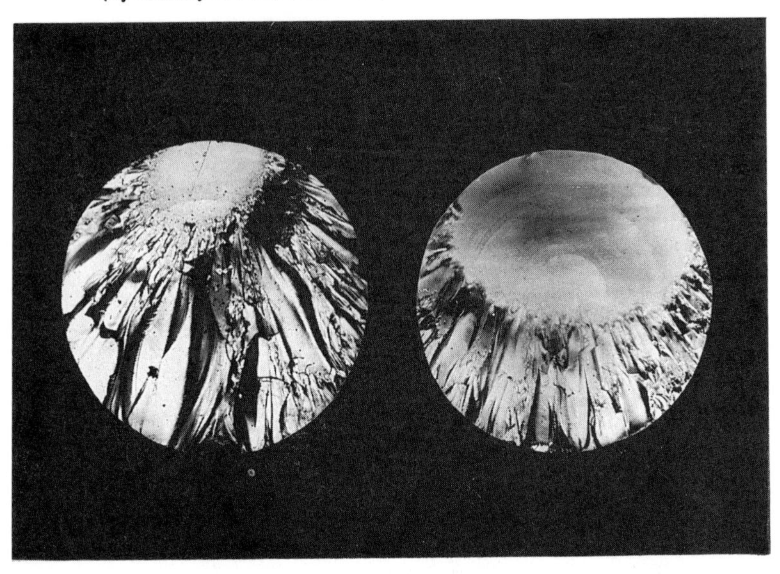

Thus the probability of branching increases with the stress in the material.

(b) *Wallner lines*

These are regular lines which sometimes appear on the "mirror" surface (Fig. 14.28). They come from the interaction of the crack front with elastic waves produced by collision of the crack front with surface defects. This interaction has the effect of modifying the plane of propagation of the cracks and produces a mark.

Figure 14.29(a) shows successive positions of a crack front originating at point A and propagating with a velocity v_c separated by equal time intervals Δt. The front encounters a defect at D which generates an elastic wave propagating with (much higher) velocity V_e. The geometric locus of the two interfering wavefronts corresponds to the Wallner line originating at D. Examination of the patterns formed by the Wallner lines allows the measurement of the crack front velocity.

In Figure 14.29(b), a and b are two points on the Wallner line corresponding to instants t and $t + \mathrm{d}t$:

$$cb = v_c \,\mathrm{d}t = ab \sin\theta$$

$$ad = V_e \,\mathrm{d}t = ab \cos\alpha$$

Fig. 14.28. Wallner lines on a fracture surface of a glass rod. (By courtesy of Saint-Gobain Co.)

hence:

$$v_c = V_e \frac{\sin \theta}{\cos \alpha}$$

Knowing V_e and the geometry of the line, v_c can be obtained. This is applicable for sufficiently high propagation velocities (at least equal to $\frac{1}{4} V_{limit}$, otherwise the angle θ is too small for satisfactory precision).

For two Wallner lines crossing at M with an angle β (Fig. 14.29(c)):

$$v_c = V_e \frac{\sin \theta_1}{\cos \alpha_1} = V_e \frac{\sin \theta_2}{\cos \alpha_2} \qquad \theta_1 = \pi - \beta - \theta_2$$

from which velocity can be found knowing only the angles θ_1 and θ_2.

Kerkhof[466] improved this method by modulating the fracture during propagation by means of ultrasound. Using transverse waves propagating in a direction parallel to the applied stress, undulations are produced in the surface which mark the fracture front positions at equal time intervals. Frequencies of about 10^6 Hz are necessary for rates near the propagation velocity limits.

Numerous examples of fracture diagnoses by fractographic methods can be found in the work of Frechette.[467]

Fig. 14.29. (*a*) Generation of a Wallner line; (*b*) determination of the local fracture propagation velocity; (*c*) case of two Wallner lines crossing at M (see text).

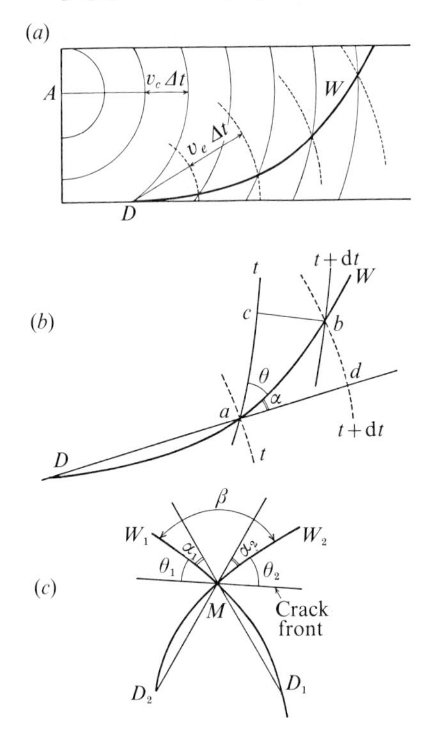

14.4 Processes for strengthening glass

Thermal expansion can create mechanical stresses. If an elastic body is homogeneous and isotropic, free expansion does not produce stress. If, however, the body is prevented from expanding, a stress is developed which can be evaluated by assuming that the specimen first expands freely and is then forced to assume its constrained length.

The necessary compressive stress σ is proportional to the modulus of elasticity E of the material and the original thermal deformation which is equal to the product of the linear thermal expansion coefficient α and the temperature interval ΔT:

$$\frac{\Delta l}{l} = \alpha\,\Delta T = \frac{\sigma}{E}$$

In the case of heating, σ will be a *compressive* stress as the body expands against the constraints. For the case of cooling a body constrained at its ends, σ will be a *tensile* stress.

If two elastic solids A and B with respective expansion coefficients $\alpha_1 > \alpha_2$ (Fig. 14.30(*a*)) are cooled, solid A will shorten more than solid B. If they are joined (welded) at elevated temperature, on cooling, solid A is prevented from contracting freely by solid B; it will be in tension, and to maintain equilibrium, B will be in compression.

Fig. 14.30. Mechanism of the origin to thermal stresses. (*a*) Two solids A, B with expansion coefficients $\alpha_1 > \alpha_2$. (*b*) Two identical solids taken to two different temperatures T_1 and T_2 ($T_1 > T_2$; see text).

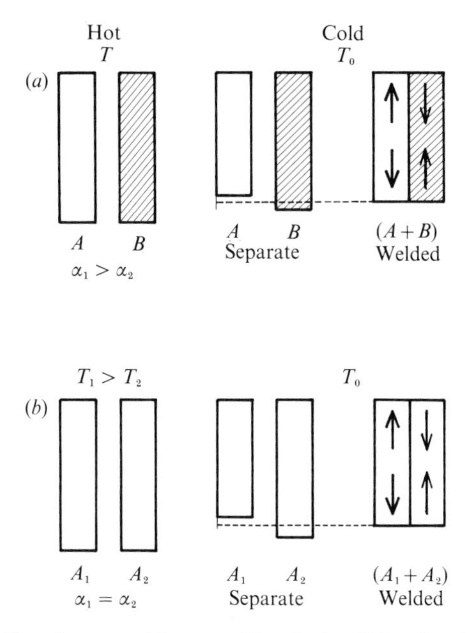

Likewise, consider two identical solids A_1, A_2 taken at different temperatures T_1, T_2 (Fig. 14.30(*b*)). On cooling, if they are free, A_1 contracts more than A_2. If they had been joined while at high temperature and then cooled, A_1 would be placed in tension and A_2 would be put in compression.

These considerations find application in the processes for strengthening (tempering) glass.

Because the origin of the fracture of glass is related to the activation of a surface flaw by a tensile stress, these flaws can be neutralized by putting the surface layer of the glass into a state of *pre-stressed compression.*

Two methods are possible to attain this goal; either the glass is subjected to controlled cooling (thermal tempering) or the nature of the surface layer is modified (chemical tempering).

14.4.1 *Thermal tempering*

The principle of this method consists of heating the finished glass object to a temperature near the softening point, removing it from the furnace and cooling the surface rapidly (generally by jets of compressed air).

In these conditions the surface layers (the "skin") rapidly become rigid while the internal layers (the "core") remain viscous. The initial thermal contraction of the skin causes the interior, still near the softening point, to flow and relieve the stresses. A temperature gradient is established through the thickness of the glass, and the interior temperature soon reaches a point sufficiently low that viscous relaxation can no longer occur. When the object is finally cooled to ambient temperature, the warmer interior must contract more than the "skin." This differential contraction, corresponding to the mechanism shown in Fig. 14.30(*b*), leads to the appearance of compressive stresses in layers near the surface equilibrated by tensile stresses in the interior.

The stress distribution as a function of depth is approximately parabolic. The maximum compressive stress in the surface is about double the maximum tensile stress at the center. The plane where the residual stress is zero is located at a depth of about 1/5 the thickness (Fig. 14.31(*a*)).

In this process, the equilibrium stress system at ambient temperature results from the disappearance of a temperature gradient introduced at high temperature, where the stresses cannot be maintained because of viscous relaxation of the glass.

The magnitude of the stresses developed depends on: (a) the expansion coefficient α, and (b) the possibility of maintaining a temperature gradient within the object, to a point where the internal layers can no longer flow. The gradient thus depends on the thickness of the object and the rate of heat removal.

For soda–lime–silica glass, a surface stress close to $100\,\mathrm{MN\ m^{-2}}$ can be currently attained; Figure 14.31(*b*) shows the effect of the pre-stress for a piece in flexure. The usual mechanical strength being of the order of $20\,\mathrm{MN\ m^{-2}}$, a substantial strength gain is obtained by this process.

The degree of temper (maximum pre-stress) can be evaluated by the classical methods of photoelasticity using the birefringence resulting from the mechanical stresses (cf. Chapter 12). See the work of Le Boiteux[468] for detailed information on these methods.

The degree of temper depends on the initial temperature T_0 and the rate at which the heat is removed from the surface of the glass. For a given heat transfer coefficient, the degree of temper first increases with T_0 and then attains a constant limiting value.

A given piece of tempered glass is in mechanical equilibrium and

Fig. 14.31. (*a*) Parabolic distribution of stress layers in thermally tempered sheet glass; (+) compressive stress, (−) tensile stress, σ_c surface compression stress, σ_{ex} maximum tensile stress in the interior. (*b*) Top: distribution of tempering stresses. Middle: flexure stresses. Bottom: superposition of the preceding distributions showing the effect of the initial pre-stress.

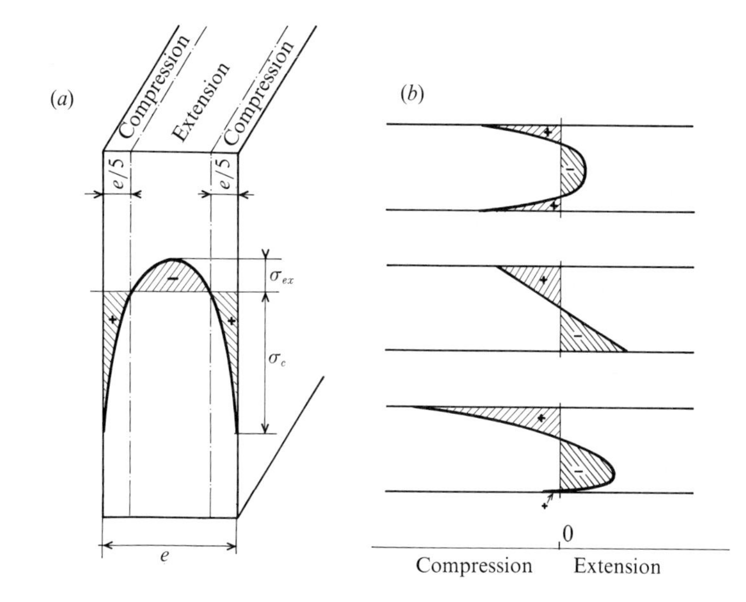

consists of a system of tensile stresses equilibrated by a system of compressive stresses. Any upset of this equilibrium leads to the fracture and destruction of the entire volume. In particular, if a surface flaw reaches the interior tensile zone, a fracture is immediately initiated and then bifurcates. Each branch then bifurcates in turn resulting in total fracture of the part into small pieces; this spontaneous fragmentation phenomenon often occurs in automobile glass. Once tempered, glass cannot be cut; it must therefore be shaped into its final form before tempering.

Multiple fracture in tempered glass ("dicing") produces fragments with edges which are less likely to cut and is the reason for its use in vehicle windows (safety glass). The degree of fragmentation depends on the elastic strain energy density stored during the quenching – a higher degree of temper thus leads to a smaller particle size fragmentation.

Thermal tempering implies that a temperature gradient must be established during freezing. Thus it is not practical to quench thin objects nor glasses in which the expansion coefficient α is too small (e.g. vitreous SiO_2).

14.4.2 *Chemical tempering*

A second method of obtaining pre-stress consists of modifying the expansion coefficient α of the surface layers to utilize the mechanism explained in Figure 14.30(a). This is done by modifying the *chemical* nature of the glass, and several possibilities exist for soda–lime–silica glasses.

(a) *Schott process*

A surface layer of glass having a lower expansion coefficient can be joined to the object at a temperature high enough to obtain welding. On cooling, the surface layers contract less and are put into compression (Fig. 14.30(a)).

This is the Schott process which has been used since 1891. It can produce surface stresses of the order of 240–280 MN m^{-2}.

(b) *Alkali metal removal*

A glass containing a smaller percentage of Na_2O modifier has a smaller expansion coefficient α. By reducing the amount of Na_2O in the surface of a glass, a surface layer of reduced *alkali metal* content is produced which, on cooling, will be placed in compression by the preceding mechanism. In practice, this is done by treating the glass objects in an atmosphere containing sulfur dioxide and water vapor.

Schematically, the dealkalization reaction is written:

$$\begin{array}{c} | \\ -\mathrm{Si}-\mathrm{O}^- \\ | \end{array} \begin{array}{c} \mathrm{Na}^+ \\ \\ \mathrm{Na}^+ \end{array} \begin{array}{c} | \\ {}^-\mathrm{O}-\mathrm{Si}- \\ | \end{array} + SO_2 + H_2O + \tfrac{1}{2}O_2 \rightarrow$$

$$\begin{array}{c} | \\ -\mathrm{Si}-\mathrm{OH} \\ | \end{array} \quad \begin{array}{c} | \\ \mathrm{OH}-\mathrm{Si}- \\ | \end{array} + SO_4Na_2$$

The extracted Na^+ ions form a soluble sulfate powder on the surface and are replaced by protons which enter into the glass structure in the form of hydroxyls. This treatment is mainly used in bottle manufacture with the dealkalization occurring during annealing (cf. Chapter 17).

(c) *Ion exchange*

Modern processes of chemical tempering are based on ion exchange by diffusion. The glass object to be chemically tempered is immersed in an appropriate bath of fused salts causing a surface exchange. The alkali metal cations of the glass diffuse from the surface into the bath and are replaced by cations from the fused salt moving in the opposite direction into the glass.

Two variants are possible:

1. The cation in the glass is replaced by a cation with a smaller radius. For example, Na^+ ($r = 0.98$ Å) is replaced in the surface by Li^+ (r = 0.78 Å) from a fused salt bath containing Li^+. This operation must be done at a temperature *above T_g to allow flow*, i.e. complete relaxation of the stresses at the temperature of treatment. Glass containing Li^+ has a lower expansion coefficient than glass containing Na^+, and on cooling, the surface layers are put into compression as in the Schott process.

2. The cation in the glass is replaced by a cation with a *larger* radius. For example, Na^+ is replaced by K^+ ($r = 1.33$ Å) by exchange in a bath of molten KNO_3. The larger K^+ ion enters into the network and causes an expansion thus causing compressive stress in the surface layer.

The surface stress associated with a relative volume variation ($\Delta V/V$) can be approximated by:

$$\sigma = \frac{E}{3(1-2\mu)} \frac{\Delta V}{V}$$

where E is Young's modulus and μ is Poisson's ratio. In this case, it is necessary to *avoid flow* which causes stress relaxation and would cancel the desired effect. It is thus necessary to operate at temperatures *below* T_g. The temperature selected is the result of a compromise: the diffusion coefficient must not be too low (treatment time would be prohibitively long), and the stresses must not relax during the exchange.

To accelerate the exchange, the composition of the glass can be modified. The addition of Al_2O_3 significantly increases the exchange diffusion coefficients and allows the achievement of sufficient exchange depth in reasonable times.[469]

Chemical tempering leads to a stress profile very different from that of thermal tempering. The surface compressive stresses are concentrated in a very thin surface layer (several tenths of millimeters) and are equilibrated by a "plateau" of very low internal tensile stresses (Fig. 14.32).

Very high surface stresses can thus be attained, commonly 350–700 MN m^{-2}. Chemical tempering is applicable to any shape and to very thin objects where thermal tempering is impossible. Its weakness is principally the thinness of the compression layer which can easily be penetrated by a surface flaw. Moreover, the low internal tensile stresses do not ensure the formation of small and non-cutting fragments in the case of fracture; breakage produces large, sharp pieces similar to those obtained from ordinary non-tempered glass.

Fig. 14.32. Stress profile through the thickness of a chemically tempered glass.

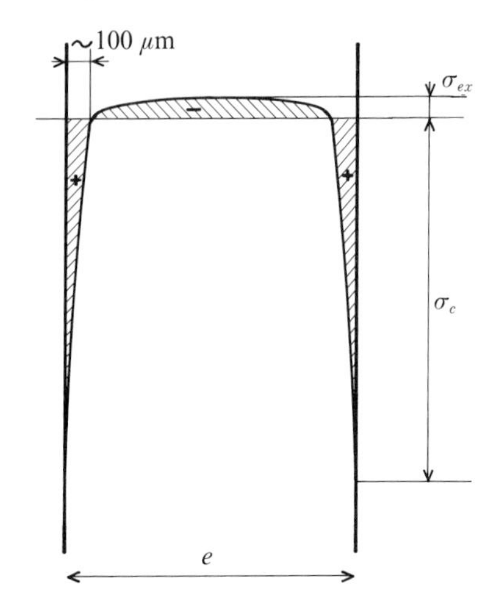

The low rate of exchange and the need for regeneration of the salt baths which become progressively richer in extracted Na^+, which further reduces the exchange rate, are important technological problems. The exchange can be accelerated by operating in an electric field.

Other chemical tempering processes combine ion exchange with controlled devitrification of the surface layers and will be mentioned in the chapter on glass-ceramics (Chapter 16).

Measurement of the high surface stresses in such thin layers is not possible by ordinary photoelastic methods. Special procedures based on the study of the polarization of surface waves have been utilized.[470]

14.5 Elimination of residual stresses – Annealing

The quenching process introduces controlled residual stresses, in contrast to most ordinary cases where it is desirable to avoid the development of residual stresses. Forming operations (blowing, pressing, etc. cf. Chapter 17) introduce stresses of various distributions which are related to variations in properties and cause the breakage of objects on cooling. In particular, for cutting, the glass must not contain any appreciable residual stresses.

Optical glasses must be free of internal stresses which produce birefringence. The level of residual stress tolerated in glasses for ordinary applications is ~ 2.6 MN m^{-2} and must be less than 0.35 MN m^{-2} in high quality glasses. The thermal treatment which eliminates residual stresses is *annealing*.

One method consists of reheating the glass to a uniform temperature near the transformation temperature (in the *annealing range*) for a sufficient time to allow the elimination by relaxation of the stresses initially present. The object is then slowly cooled in order to prevent the generation of new permanent stresses during cooling. Reheating a finished object is avoided as far as possible for economic reasons, and an appropriate cooling profile after forming is sought. During annealing, the glass simultaneously undergoes stabilization which complicates the study of this phenomenon.

Maxwell's model (cf. Chapter 9) leads to the relation for relaxation:

$$\frac{d\sigma}{dt} = -\frac{1}{\tau}\sigma$$

the rate of relaxation being proportional to the stress remaining. If the relaxation time $\tau = \eta/G$ is independent of time:

$$\sigma = \sigma_0 \exp\left(-\frac{t}{\tau}\right)$$

Experiments show that this type of equation does not account for the observations which are better represented by an empirical equation due to Adams and Williamson:[471]

$$\frac{1}{\sigma} - \frac{1}{\sigma_0} = At$$

where A is a constant depending on the annealing temperature:

$$A = A_0 \exp\left[-C(T - T_0)\right]$$

A_0, C and T_0 being constants. This is explained by the variations of viscosity with time during stabilization. It is as if the relaxation time τ were a function of time, $\tau(t)$ which introduces a nonlinear effect into the Maxwell equation which is then written:

$$\frac{d\sigma}{dt} = -\frac{\sigma}{\tau(t)}$$

In practice, the rate of cooling depends on the dimensions of the object. Typically, for a coarse optical anneal the rate must be less than 5 °C min^{-1}; for a fine optical anneal, less than 0.7 °C min^{-1}. It is not necessary to keep the same slow cooling rate during the entire thermal treatment. It is preferable, for reasons of economy, to cool the glass rapidly to the temperature where the stress is relieved in a few seconds (above the annealing point) then adopt the calculated cooling rate to a lower limit down to the "strain point." The cooling rate can then be increased without the introduction of residual stresses (Fig. 14.33).

Fig. 14.33. Programs of thermal annealing treatment of a mirror glass: (*a*) continuous cooling, (*b*) accelerated cooling, (*c*) economic cycle.

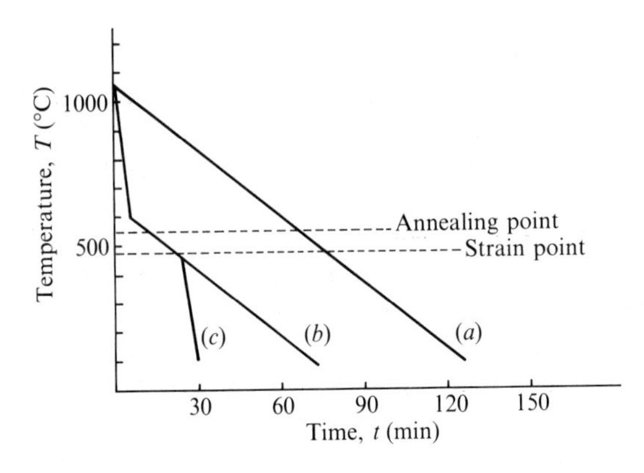

15 The surface of glasses

15.1 Importance of the surface

The surface plays a determining rôle in the mechanical strength of glass. The phenomena of chemical and mechanical surface degradation cause a spectacular reduction in the normal mechanical properties. The removal of surface layers by HF attack causes a (temporary) increase in the mechanical performance.

The nature of the surface differs according to the method of fabrication of industrial material (cf. Chapter 17). The *flame-polished* surface obtained in a drawing process is closer to ideal conditions than the *mechanical polish* of standard plate glass. In the case of "float" glass which has become the preponderant high-quality flat glass, the conditions of formation of the two surfaces are different, one being in contact with a bath of molten Sn while the other is formed in the (reducing) atmosphere of the furnace.

The processes of chemical strengthening: dealkalization by SO_2 or chemical tempering by ion exchange, modify the chemical composition of the surface layers. On the other hand, processes to improve the surface, e.g. the vacuum deposition of thin metallic films to modify the spectral transmission characteristics (treatments for anti-reflection, anti-UV, anti-moisture etc...) depend on the state of the surface.

The fabrication of polymer–glass composites, the production of layered windscreen safety glass and the sealing of glass to metal are related to interfacial properties. Finally, in the medical field, "bio-glass," used in bone prosthesis, is based on natural bonding established between the glass and the living tissue.

15.2 Characterization of surface layers

15.2.1 *Classical techniques*

For many reasons, the study of glass surfaces has always been considered as fundamental but technically difficult to achieve. The surface layer of glass consists of a zone which extends to a depth of 10 μm and more. The physico-chemical properties in this zone can undergo rapid variations, and the lack of appropriate analytical techniques has hampered the studies of surfaces.

For a long time, analysis of the chemical composition of surface zones was done using classical chemical techniques. The extraction of alkali metal ions from glasses subjected to the action of water, acid or alkaline solutions (elution methods) has been used to determine the global concentration of alkali in surface zones. The analysis was done by titration or by measuring the conductivity of the extracting solutions.[473,474] IR spectroscopy and adsorption isotherms of water have also been used with glass which has been powdered to increase its specific surface area.

15.2.2 *Modern techniques*

A total change occurred in the seventies when recently perfected surface spectrometric analysis techniques were systematically applied to the study of surface layers of glass.[475,476] All of these methods study the emission of photons, electrons or ions from surface regions of the material under the action of a particle flux bombarding the specimen. The emitted flux which is detected and analyzed contains information on the chemical composition of the layers which, depending on the technique, are from $10–10^5$ Å thick. Table 15.1 shows the various techniques used to study glass according to the modes of excitation and emission

These techniques have been applied to a large variety of glasses, particularly soda–lime–silica glasses as well as special glasses such as "bioglass." The first studies of a qualitative character were on the effects of polishing, prolonged heating and water attack. These types of investigations were quickly followed by quantitative measurements and then by determinations of elemental concentration profiles in the surface layers. Particular mention must be made of systematic studies of float glass[477] which involve the distribution of Sn ions in the face in contact with the metallic bath.

The choice of method depends first on the *depth of analysis*. Several methods may be combined to explore the concentration profile or to obtain better resolution in the surface layers. It is also possible to remove successive layers by ion milling *in situ* in the spectrometer and to study the modifications of the corresponding spectra. The following sections will review the principal techniques employed and the most significant results obtained.

It must be noted, however, that methods which use charged particles present some problems when applied to usual dielectric glasses because of charge accumulation, and the interpretations can sometimes be difficult.

15.3 **Review of the principal modern methods**

15.3.1 *Auger spectroscopy*

When a beam of electrons (of several keV energy) bombards a specimen, secondary electrons (Auger electrons) are emitted by nonradiative relaxation of the atoms ionized by the bombardment. Analysis

Table 15.1. *Techniques to study glass surfaces.*

Excitation ⇒ Emission ⇓	Photons	Electrons	Ions
photons	X-Ray Fluorescence (XRF) IR Reflection Spectrometry (IRRS) Ellipsometry	Electron Microprobe (EMP)	Surface Composition by Analysis of Neutral Ion Impact Radiation (SCANIIR) Charged Particle X-ray Spectrometry (CPXS) Charged Particle Activation Analysis (CPAA)
electrons	Electron Spectrometry for Chemical Analysis (ESCA)	Auger Electron Spectrometry (AES) Scanning Electron Microscopy (SEM) Transmission Electron Microscopy (TEM)	
ions	Laser microprobe		Secondary Ion Mass Spectrometry (SIMS) Ion Microprobe Mass Analysis (IMMA) Ion Scattering Spectrometry (ISS) Rutherford backscattering

of their energy spectrum permits, in principle, the characterization of the chemical elements in surface layers less than 100 Å thick. Figure 15.1 shows the Auger spectrum of the surface of a soda–lime–silica glass cooled by liquid N_2.

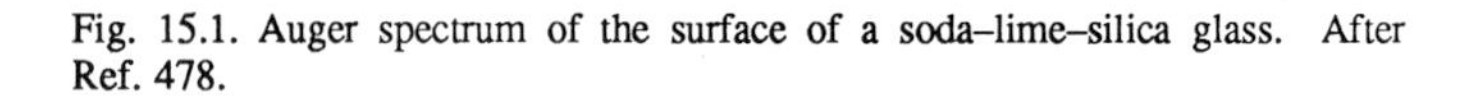

Fig. 15.1. Auger spectrum of the surface of a soda–lime–silica glass. After Ref. 478.

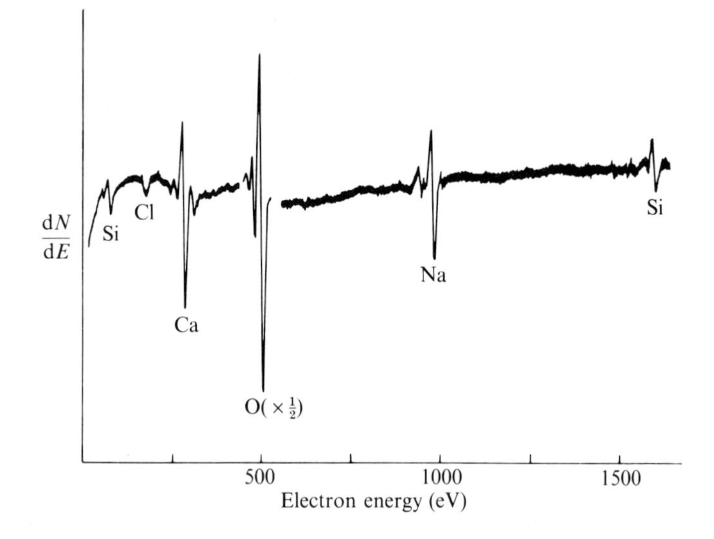

The polarization of a specimen and local heating under the impact of the primary beam can modify the local concentration of mobile elements (Na^+).[477] Chemical effects such as the reduction of SiO_2 have also been observed.[479] It is preferable to resort to other modes of excitation as in the case of photoelectron spectroscopy.

15.3.2 *ESCA spectroscopy*

A specimen irradiated by X-photons of sufficient energy $h\nu$ emits electrons with kinetic energy E_c given by the Einstein formula:

$$E_c = h\nu - E_B$$

where E_B is the bond energy in the atom from which the electron originated. The values of E_B for all the elements have been measured and tabulated, and so measurement of E_c permits the determination of the elements present in the surface layers (i.e. less than 100 Å deep). The X-rays used for the excitation are usually obtained from anodes of Mg

(Kα_{12} = 1254 eV) or Al (Kα_{12} = 1487 eV). The kinetic energy E_c is measured using an electrostatic analyzer. The interpretation of the spectra requires an evaluation of the mean free path of the electrons which is difficult. A frequently employed method consists of depositing a thin layer ($\sim$ 10 Å) on a substrate and comparing the signal intensity emitted by the film and the substrate. It is also possible to calculate these values.[480,481] For vitreous SiO_2 Colombin[477] found a mean free path of 29 Å for photoelectrons of 1148 eV and 36 Å for 1381 eV while calculation gives 18.2 and 21 Å respectively. (Flitsch[482] on the other hand found 25–27 Å.)

Figure 15.2 shows a typical example of an ESCA spectrum from a float glass. For glasses subjected to accelerated aging, experimental results show a rapid decrease of the Na concentration as a function of time in a surface region $\sim$ 100 Å thick.

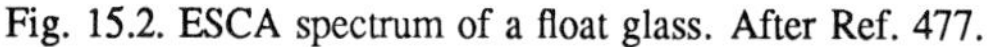
Fig. 15.2. ESCA spectrum of a float glass. After Ref. 477.

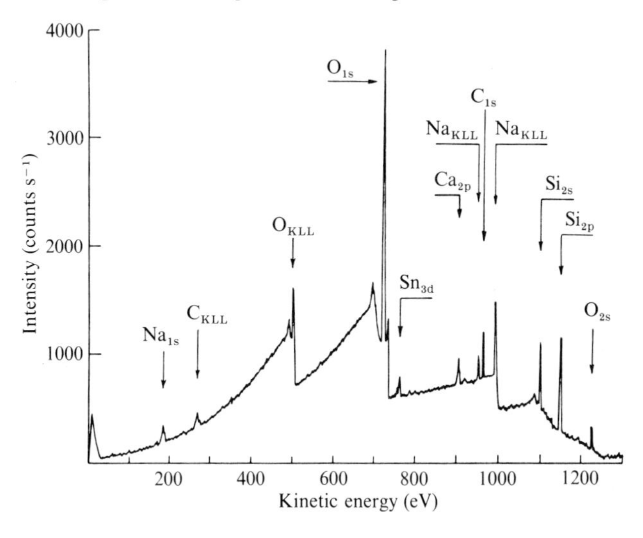

15.3.3 *X-ray fluorescence*

X-ray spectroscopy has become a standard method of quantitative chemical analysis.[483–5] The specimen is excited by an X-ray flux; the X-radiation emitted, called fluorescent radiation, comprises a frequency continuum on which characteristic emission lines are superimposed. The fluorescence yield is determined by whether the excited atom relaxes by the Auger effect with the emission of an electron or by X-ray fluorescence. The intensity of characteristic X-ray fluorescence lines is related to the concentration of the emitting element in depths to 10^5 Å (10 μm).

In the case of glasses the preferred excitation is polychromatic radiation from a Cr anode, and the spectral analysis uses a PET (pentaerithritol) crystal for elements with Z equal or greater than 13 (Al) and a RbAP (rubidium acid phtallate) crystal for analysis of Kα from Na and Mg.

In these analyses it is important to take into account the secondary fluorescence (induced by the primary X-ray fluorescence) and the effects of variations in the concentration with depth. It is necessary to remove successive layers by mechanical abrasion or chemical etching to determine concentration profiles. The thickness of the layers removed can be determined from the specimen weight change or by optical interferometry.

Figure 15.3 shows the concentration profiles of the elements Si, Na, Ca, Mg and S in the upper face and for Sn in the lower face of float glass.[477] A significant reduction in Na and an increase in Si are seen near the surface.

Fig. 15.3. Concentration profiles of Si, Na, Ca, Mg and S (in wt.% of the element) for the upper surface of float glass and the profile for Sn for the lower surface. After Ref. 477.

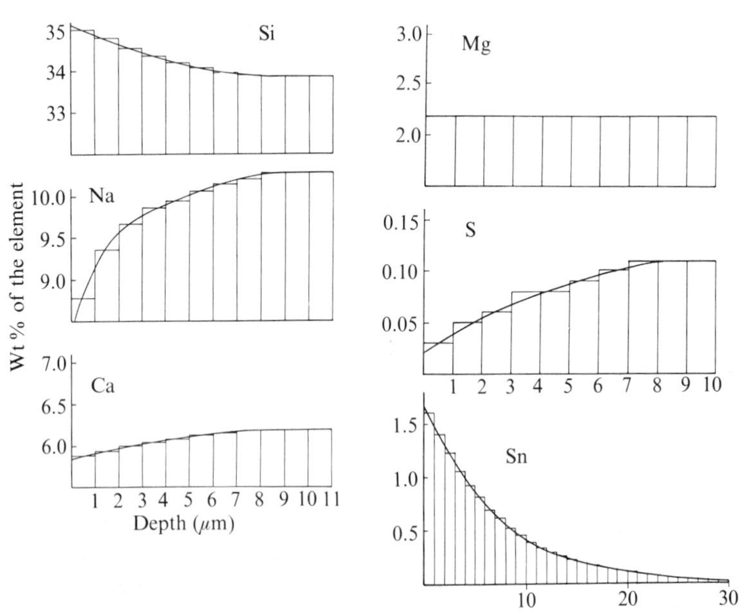

15.3.4 *Elastic scattering of charged particles – Rutherford backscattering*

By bombarding a target with particles having an energy between 0.4 and 4 MeV, scattering is observed at significant angles relative to the incident beam in both the forward and reverse directions. The energy of the backscattered particles is characteristic of the atoms present in the target and this permits their identification. The distribution of the elements as a function of depth can be deduced from the energy loss of the particles during their trajectory from the scattering site towards the surface. The depth resolution for particles of 1 MeV is from 100–300 Å to a depth of 5000 Å.[486]

For elastic scattering it can be shown from the principles of energy conservation and momentum transfer that the kinetic energy E_d of the incident particle after collision is related to its initial kinetic energy E_0 by the relation:

$$E_d = kE_0$$

where the *kinetic factor* k is calculated as a function of the masses of the particles and the scattering angle θ. Experimental measurement of E_d of the incident particle after scattering allows the identification of the element responsible for the scattering. E_d corresponds to the scattering particles located on the surface itself; for deeper particles, losses reduce the kinetic energy which can decrease to zero.

The spectrum from a glass (Fig. 15.4) consists of a series of "plateaus" separated by discontinuities or "fronts" where the inflection corresponding to E_d is used to identify the scattering element. Figure 15.5 shows the concentration profile of Sn between 300 Å and 3500 Å obtained in this way for a float glass. It agrees well with the profile determined to a greater depth by X-ray fluorescence.

15.3.5 *Ion bombardment induced photon emission (SCANIIR)*

When a solid material is bombarded by a neutral or charged low-energy particle beam ($\sim$ 5 keV), particles, atoms, ions or molecules are ejected from the surface. This is the well-known phenomenon of "sputtering." A certain proportion of the ejected particles are in an excited state and relax by radiative transitions corresponding to UV, visible or IR frequencies. The emitted spectral lines permit the identification of the chemical elements ejected. Since the penetration depth of the particles is very limited, the information relates to the first few monolayers. The erosion of the specimen in the course of sputtering (of the order of 0.5 Å s^{-1}) means that sequentially deeper and deeper layers are exposed. The observation of the intensity of emitted lines as a function of time thus permits the direct determination of concentration profiles.

Fig. 15.4. Float glass, lower face; general backscattering spectrum for α particles (1.7 MeV) scattered at 135° relative to the incident beam. After Ref. 477.

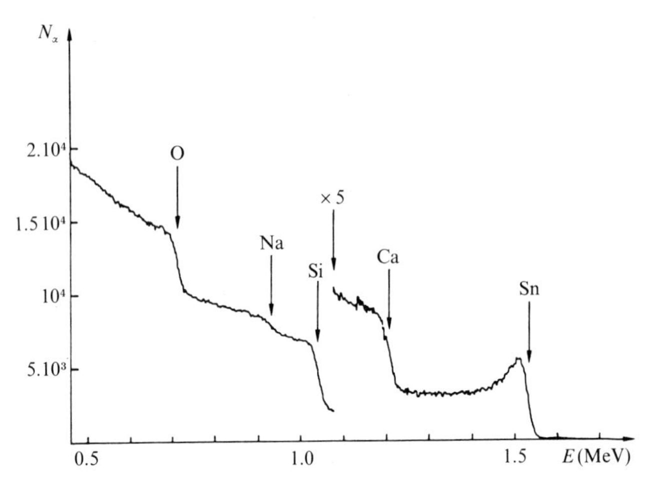

Fig. 15.5. Float glass, lower face: concentration profile of Sn between the surface and 0.3 μm. After Ref. 477.

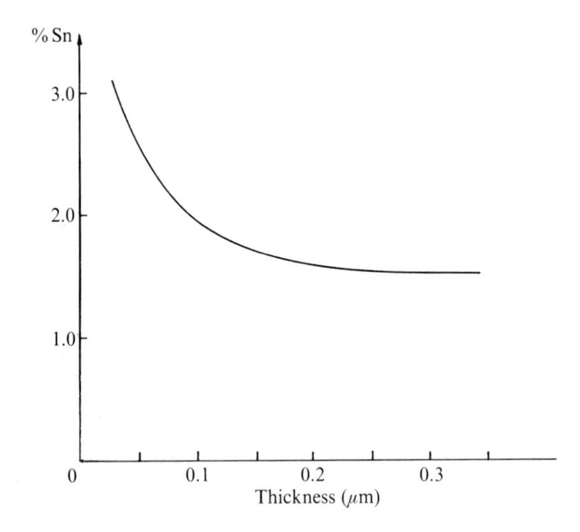

The penetration depth of Ar^+ ions (5 keV) being several tens of angströms, Bach[487] estimated that the resolution of the technique is between 30 and 50 Å. The emission spectra are relatively simple since the optical lines correspond to weakly excited states.

Figure 15.6 shows the general appearance of a spectrum for float glass. This technique has been applied[477] to the determination of the distribution of Sn in the lower face of float glass in layers between 0 and 100 Å. This required increasing the sensitivity of the method because Sn was concentrated on the surface and the signal decreased rapidly as the erosion progressed into the first 100 Å.

Fig. 15.6. Float glass, upper face. General emission spectrum induced by the SCANIIR technique. After Ref. 477.

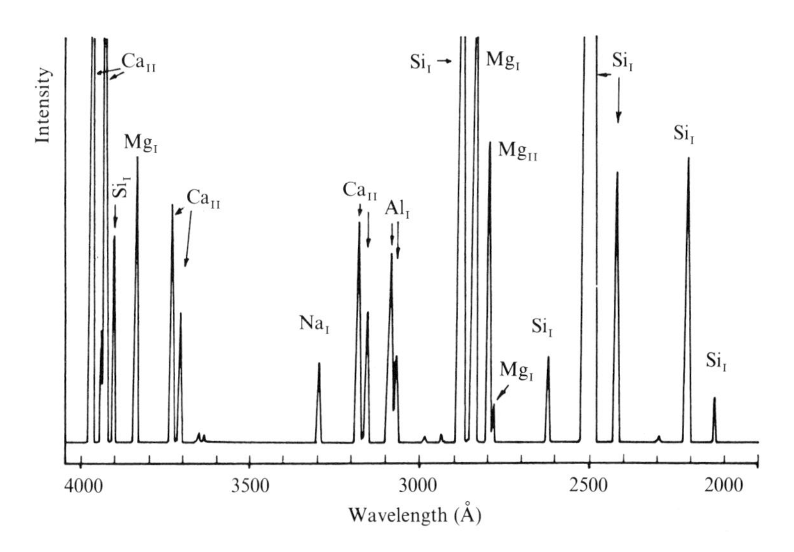

15.3.6 *Ellipsometry*

This method studies the change of polarization of polarized light reflected from a surface. By measuring the new polarization state of the reflected beam, the optical constants n and k of the complex refractive index $n^* = n - ik$ can be calculated. If a thin film is present, its thickness can also be determined. The principles of the method were established by Drude (1840), but it is only since 1950 that ellipsometry has gained interest as a working tool for the study of thin films.[488] Experimentally the measurement of the polarization ellipticity of the reflected light is simple – the practical difficulties are in the numerical calculation of the optical constants. The method is non-destructive, but does not provide a precise measurement of the chemical composition of the surface layer. At most, by repeating the measurements after the removal of successive layers, the thickness of the surface zone can be measured. Thus, for example, in the case of an aged drawn glass, it is found that upon removal of 10^4 Å (1 μm) the optical characteristics become that of the original material.

Other difficulties come from the fact that corrosion films are often inhomogeneous and there does not yet exist an applicable ellipsometry theory for *porous* materials. At the present time, the application of this method to glass surfaces remains mostly qualitative.

15.4 Formation of surface layers

15.4.1 *Importance of the specimen's history*

The surface of a glass is the result of interaction with the environment. The composition of the glass and the melting and forming methods impose the initial conditions from which the surface layers evolve in a complex dynamic process. This dependence on the history of the specimen makes it difficult to draw conclusions about the physical chemistry of the surfaces. It is necessary to resort to simple systematic experiments in which a given glass is subjected to treatments so that the history is known with precision. The methods of study described previously can then serve to determine the state of the surface and to follow its evolution with time.

The rate of layer formation can be very rapid. Uncontaminated specimens have been obtained by drawing glass micro-fibers directly in the column of an electron microscope and it has been shown[489] that contact with ambient air for several seconds is sufficient to form a surface layer obscuring the structure initially observed. Exposure for several minutes to air containing traces of water causes a significant swelling of the fiber as shown by the diameter increase in Fig. 15.7.

To have a pristine surface in electron spectroscopy experiments, it is necessary to remove material by ion milling. The formation of new surfaces by fracture under ultrahigh vacuum can also be considered.

15.4.2 *Formation of surface layers in contact with gas*

When a solid is placed in contact with a gas, an excess concentration of the gas is produced at the interface. This process is called *adsorption* and must be distinguished from absorption which implies penetration into the volume of the solid.

Adsorption is a spontaneous process which is accompanied by a decrease in the free energy ΔG of the system ($\Delta G < 0$). At the same time the mobility of the adsorbed phases decreases ($\Delta S < 0$), it follows that $\Delta H < 0$, i.e. the process is exothermic.

The value of ΔH distinguishes between physical adsorption ($-\Delta H \sim$ 0.5–10 kcal mole^{-1}) due to weak van der Waals type interactions and

Fig. 15.7. Ultrathin fibers of a glass ($72\,SiO_2 \cdot 16\,Na_2O \cdot 12\,CaO$) drawn *in situ* in the electron microscope in an atmosphere of purified Ar. Top: before contact with the atmosphere. Middle: after 2 min contact with the atmosphere. Bottom: after 15 min contact. (The bar represents 1 μm.) After Ref. 489.

chemisorption ($-\Delta H \sim$ 5–150 kcal mole^{-1} or more) which implies the formation of chemical bonds. Whereas physical adsorption is non-specific and the adsorbate can be removed by increasing the temperature, chemisorption is generally very specific and only occurs on atomically clean surfaces.

At ambient temperature, a SiO_2 glass surface consists of OH groups and a layer of physically adsorbed H_2O molecules (Fig. 15.8(*a*)). When the temperature is increased, the physically adsorbed water and some of the OH groups leave. The OH groups remaining consist of two types: those sufficiently near to form hydrogen bonds and those remaining isolated (Fig. 15.8(*b*)). The different types of groupings can be identified by IR reflection spectroscopy; the corresponding bands are indicated on Fig. 15.8. The process is reversible up to about 400 °C; beyond this, new adsorption of water becomes more difficult because the surface has undergone the modifications shown in Fig. 15.8(*c*). After heating above 850 °C the surface no longer adsorbs water and a special treatment would be necessary to reform the hydroxylated surface.

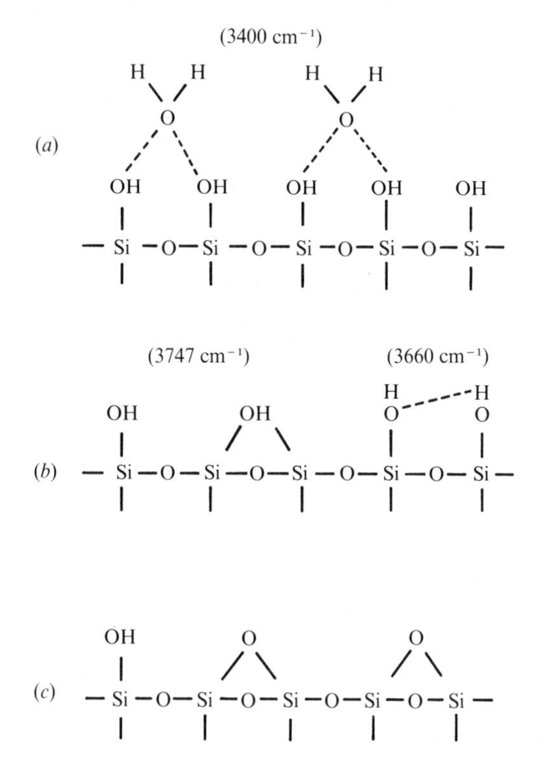

Fig. 15.8. Surface of SiO_2 glass (schematic). (*a*) surface OH^- and H_2O adsorbed, (*b*) modification by moderate temperature increase, (*c*) modification after heating near 800 °C. The corresponding IR bands are indicated.

15.4.3 *Formation of surface layers in contact with solutions*

For silicate glasses, Hench[490] distinguishes five principal surface types (Fig. 15.9). Type I corresponds to glasses which are only covered with a thin hydrated layer ($< 50\,\text{Å}$). It does not produce significant composition change (alkali removal or dissolution). Below this exterior layer, the composition of the surface is nearly the same as that of the bulk glass. This is the case for vitreous SiO_2 exposed to solutions of neutral pH.

Type II surfaces have a protective layer of SiO_2 due to the selective elimination of alkali metal ions. This corresponds to the case of low-alkali glasses subjected to an alkali metal removal treatment (cf. Chapter 12). The chemical durability of such glasses is good, especially in contact with solutions of pH < 9.

Type III corresponds to the formation of a double protective layer. The addition of Al_2O_3 or P_2O_5 to the glass composition results in the formation of layers of SiO_2–Al_2O_3 or P_2O_5–CaO covering a layer rich in SiO_2. Such films can form by alkali removal or by structural modification in contact with a solution.

In type IV, the SiO_2 content at the surface is insufficient to protect the glass against dissolution. Glasses rich in alkali metals have this type of surface and the chemical durability is poor.

Type V corresponds to a uniform dissolution with equivalent losses in alkali metal and SiO_2 . The composition in the surface remains the same as the internal composition. Glasses in contact with a medium of pH > 9–10 show such behavior.

15.4.4 *Formation mechanisms*

It has been established[490] that when glasses containing alkali metal ions are placed in a medium containing H_2O they undergo two reaction stages. The first stage corresponds to *alkali removal* from the surface and the replacement of alkali metal ions by protons H^+:

$$\begin{array}{c} | \\ -\mathrm{Si}-\mathrm{O}^- \\ | \end{array} \begin{array}{c} \mathrm{Na}^+ \\ \\ \mathrm{Na}^+ \end{array} \begin{array}{c} | \\ {}^-\mathrm{O}-\mathrm{Si}- \\ | \end{array} + 2\mathrm{H_2O} \rightarrow \begin{array}{c} | \\ -\mathrm{Si}-\mathrm{OH} \\ | \end{array} \quad \begin{array}{c} | \\ \mathrm{HO}-\mathrm{Si}- \\ | \end{array} + 2\mathrm{NaOH}$$

This reaction dominates for pH < 9. The second stage corresponds to the dissolution of the SiO_2 network by rupture of the Si–O–Si bridges:

$$\begin{array}{c} | \qquad\quad | \\ -\mathrm{Si}-\mathrm{O}-\mathrm{Si}- \\ | \qquad\quad | \end{array} + \mathrm{H_2O} \rightarrow \begin{array}{c} | \\ -\mathrm{Si}-\mathrm{OH} \\ | \end{array} \quad \begin{array}{c} | \\ \mathrm{HO}-\mathrm{Si}- \\ | \end{array}$$

It is predominant for pH > 10.

Fig. 15.9. Different types of glass surfaces: I inert glass, II formation of a protective layer, III double protective layer, IV non-protective layer; V soluble glass. After Ref. 490.

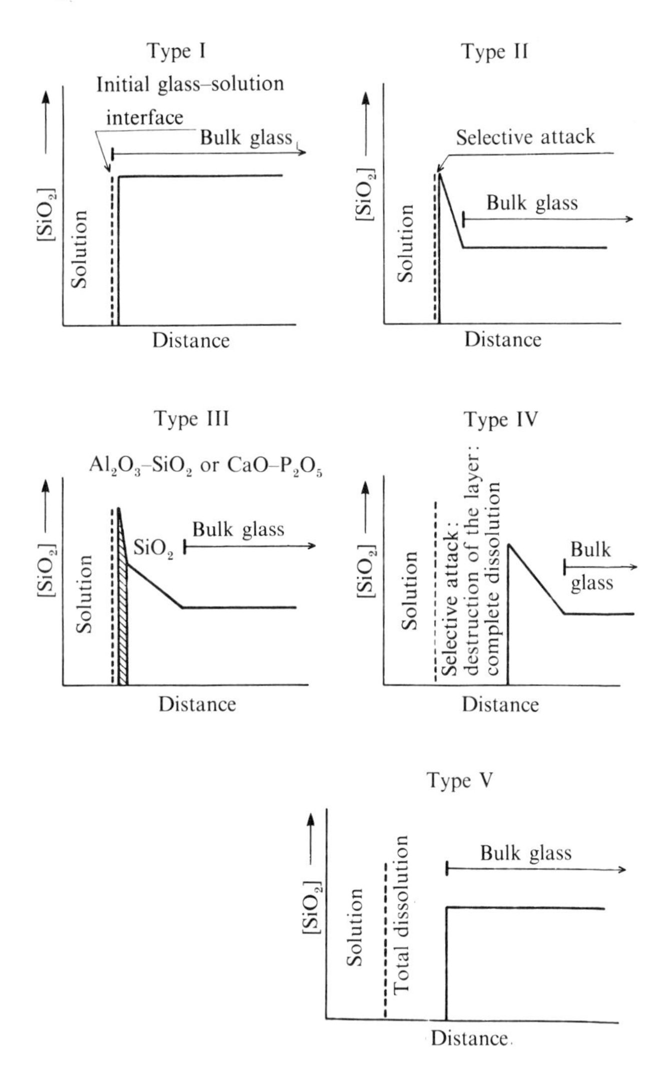

However, it has been established that the two stages occur simultaneously over an extended pH range. The type of surface formed (I–V) depends on the relative rates of the two processes during the specimen's history.

Stage 1 corresponds to diffusion-controlled kinetics (dependence on $t^{1/2}$) while for stage 2 the dissolution shows interface-controlled kinetics (linear time dependence). The change from one stage to the other seems to depend on the time necessary for the pH to attain a value sufficient to begin the depolymerization of the protective layer.

Because the corrosion depends on the surface/volume ratio, it is possible to accelerate the tests by using powders of known particle size or by increasing the temperature.

Glass of composition $45\,SiO_2{\cdot}24.5\,CaO{\cdot}24.5\,Na_2O{\cdot}6\,P_2O_5$ ("bio-glass") has been studied in detail[491] because it has the property of bonding to living bone and is of interest for prosthetic repairs. All the methods of surface analysis have been applied to elucidate this mechanism. It seems that a stable $Ca_3(PO_4)_2$ film forms on contact with the living tissue (type III surface) which incorporates the organic constituents during growth.

Figure 15.10 shows the composition profile obtained by Auger spectroscopy of the surface of an implant subjected to ion milling (Ar^+). The phosphate film crystallizes in the form of microcrystalline *apatite* which has the effect of linking the surface to the growing bone. The presence of the $Ca_3(PO_4)_2$ layer and a layer rich in SiO_2 protects the interior glass from outside attack.

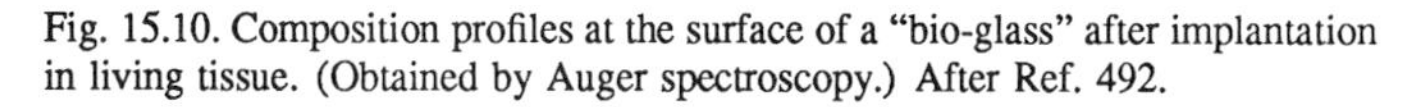

Fig. 15.10. Composition profiles at the surface of a "bio-glass" after implantation in living tissue. (Obtained by Auger spectroscopy.) After Ref. 492.

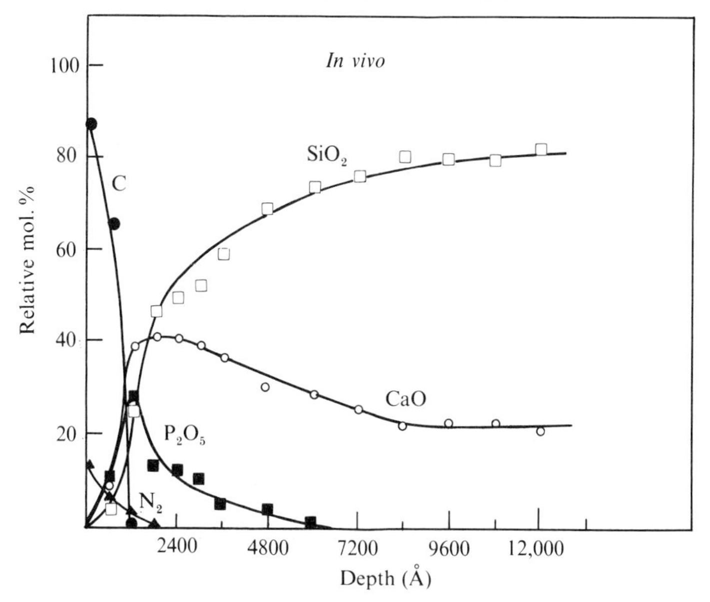

16 Glass-ceramics

In glass technology, care is taken to avoid accidental crystallization (devitrification). The usual compositions are selected so that risks of this are reduced. However, there exists a class of materials called *glass-ceramics* which are obtained by controlled crystallization of glass. The appropriate glasses are subjected to carefully programmed thermal treatments which cause the nucleation and growth of crystalline phases. The process is designed to convert the initial vitreous phase into a polycrystalline material. A certain portion of the initial glass generally remains at the end of the treatment. According to the nature of the starting glass and the thermal treatment, a large variety of materials can be obtained from this process.

16.1 History of the process

For a long time it was known that maintaining a glass at an appropriate temperature for a sufficient time would cause crystallization. Réaumur attempted to produce polycrystalline materials from glass. He showed that bottles heated red hot in a mixture of sand and gypsum transformed into an opaque material resembling porcelain. However, materials produced in this way lacked mechanical strength because of the large size of the crystals which formed on the surface and grew towards the interior.

This "Réaumur porcelain" remained an object of curiosity without application up to the early sixties when the systematic work of Stookey at the Corning Glass Co. (USA) resulted in the development of glass-ceramics in their modern form.[493]

The basic idea was to replace the normal devitrification process, which starts from a limited number of nucleation centers, by a devitrification process catalyzed by a large number of centers distributed uniformly in the volume of the glass.

The first step was the discovery of *photosensitive* glasses (cf. Chapter 5) containing small quantities of metals (Cu, Ag, Au) which could be precipitated in the form of very small crystallites by thermal treatment. Stookey was able to show that this precipitation is aided by UV irradiation prior to thermal treatment; a process which could produce photographic images in the volume of the glass.

Thermal treatment at a temperature above that necessary to "develop" the latent image leads to the opacification of the product by precipitation of a crystalline phase from the glass adjacent to the metallic crystals which act as primary nucleation centers.

It was quickly recognized that the process could lead to a total conversion of the vitreous matrix to a polycrystalline material and that the properties of the final product would be very different from those of the initial glass. In particular, the mechanical and electrical (insulation) properties were remarkably improved. The large number of nucleation centers and their uniform distribution resulted in a polycrystalline micro-structure by devitrification which ensured the rigidity and strength of the final product.

Research has since led to the discovery of other means of producing controlled nucleation which are not necessarily based on the properties of photosensitive glasses.

16.2 Controlled crystallization of glass

16.2.1 *Nucleation catalysts*

(a) *Metals*

A nucleation catalyst must be present in the glass in the form of particles of colloidal dimensions. Metal oxides which decompose on heating or which can be reduced to the metallic state during the glass melting can be used. Decreasing solubility with decreasing temperature leads to the precipitation of extremely small metallic particles. The atoms assemble first in aggregates, then attain colloidal dimensions which can give micro-crystals.

Techniques similar to those used to produce ruby glass use Cu, Ag, Au, or Pt with SnO_2 as the reducing agent (cf. Chapter 5). The concentrations of Cu used here (0.05 % in Li_2O–Al_2O_3–SiO and Li_2O–ZnO–SiO_2 glasses) are much less than that necessary for a ruby glass (0.2–1 %). The necessity of using a reducing agent decreases in going from Cu to Pt, thermal treatment being sufficient for the latter.

(b) *Halogens and other compounds*

The addition of fluorides, cryolite, Na_2AlF_6, or sodium fluorosilicate, Na_2SiF_6, to glass (2–4 %) causes opacification, a property used for a long time to make *opal* glasses. By using smaller proportions, it is possible to obtain, on cooling, a transparent glass which will only become opaque during a thermal treatment. The micro-crystalline fluorides can then serve as nucleation centers.

Sulfides can also precipitate in glasses on cooling and form colloidal aggregates. The sulfides and selenides of Cd have also been used as catalysts.

(c) *Oxides*

A number of oxides induce phase separation (unmixing) leading to the formation of an extremely fine dispersion. *Titanium oxide* TiO_2 is used the most and is added in amounts from 2 to 20 wt% in a large number of compositions. In such concentrations, TiO_2 can no longer be considered as a "catalyst" added in very low concentrations, but more as an integral component of the glass.

Other oxides: ZrO_2, P_2O_5, V_2O_5, Cr_2O_3, MoO_3, WO_3 have also been used. Their use is often limited by their low solubility in glasses; e.g. it is necessary to add P_2O_5 to enhance ZrO_2 solubility.

16.2.2 *Mechanisms of controlled crystallization*

The catalysis mechanism which leads to a subsequent volume crystallization is not yet entirely clear. Several possibilities are put forth:

1. the formation of a significant interface during phase separation can serve as a preferential site for the nucleation of a crystalline phase;
2. phase separation causes a shift in the concentration of constituents towards compositions closer to phases susceptible to crystallization;
3. separation of metastable phases constitutes a precursor state for the precipitation of a more stable phase.[494]

In certain cases, microscopic examination shows the formation of crystalline precipitates located entirely within one of the phases resulting from unmixing (Fig. 16.1), while in other cases the growth of crystalline phases without apparent relation to the unmixing surface is observed (Fig. 16.2).

16.2.3 *Glass-ceramic process*

The process consists of melting a glass of given composition incorporating the appropriate nucleants and forming the object into its final shape using common glassworking techniques e.g. pressing, casting, drawing, etc. (cf. Chapter 17). After annealing to eliminate stresses, the object undergoes a thermal treatment shown schematically in Fig. 16.3.

In the first stage, the object is reheated at a rate of 2–5 °C min^{-1} from ambient (or the anneal temperature) up to the nucleation temperature T_1 and maintained for a given time. The optimum nucleation temperature, generally corresponding to a viscosity in the range 10^{11}–10^{12} dPa s, is determined experimentally.

At the end of the nucleation stage, the temperature is further increased at a rate less than 5 °C min^{-1} to the optimum growth temperature T_2. This temperature is selected for the maximum development of the crystalline phase without deformation of the material by viscous flow. It is generally 25–50 °C below the re-solution temperature of the crystalline phase. After holding at this temperature for the required time (determined experimentally), the crystallized object can be rapidly cooled (10 °C min^{-1}) to ambient temperature without the need for further annealing treatment.

Fig. 16.1. The glass $77\,B_2O_3 \cdot 18\,PbO \cdot 5\,Al_2O_3$(wt) treated 1 h at 520 °C. Growth of a crystalline phase in one of the phases resulting from unmixing. (By courtesy of Saint-Gobain Co.)

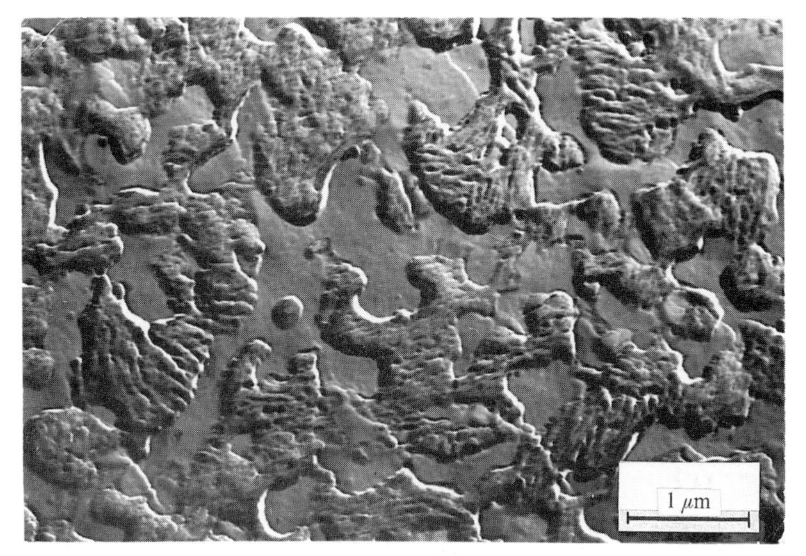

The perfection of a glass-ceramic process consists essentially of studying the details of the devitrification conditions to ascertain the nucleation and growth temperatures. Because of their fineness, the precipitated phases must be examined in the electron microscope. X-ray diffraction is used to determine their nature. Determination of the level of crystallization poses difficult problems – the methods of quantitative X-ray diffraction, IR spectroscopy, or solution calorimetry can be used.

16.2.4 *Modification of properties as a result of thermal treatment*

As a result of a carefully controlled thermal treatment, the initial glass is converted into a polycrystalline material in which the final properties depend directly on the nature of the precipitated phases, the final degree of crystallinity attained, the size of the crystallites, etc. The material is generally opaque although *translucent* and even *transparent* glass-ceramics have been produced in certain cases (the crystallinity only being revealed by X-ray diffraction). The nature of the precipitated phases depends not only on the base glass, but to some degree on the nucleation and thermal treatment. Non-equilibrium phases are often observed as well as transformations involving several successive crystalline phases.

Fig. 16.2. Examples of the growth of crystalline phases independent of domains of unmixing: (*a*) transmission electron micrograph of a glass $86\,B_2O_3 \cdot 9\,PbO \cdot 7\,Al_2O_3$ treated 15 min at 630 °C, (*b*) replica of a glass $73\,B_2O_3 \cdot 12\,PbO \cdot 15\,Al_2O_3$ treated 18 h at 800 °C. (By courtesy of Saint-Gobain Co.)

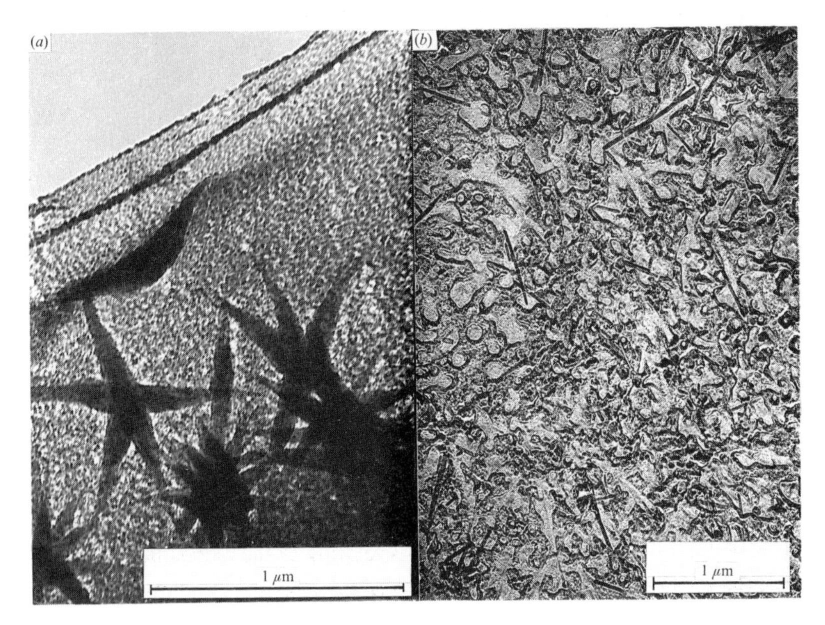

Fig. 16.3. Thermal treatment in a glass-ceramic process.

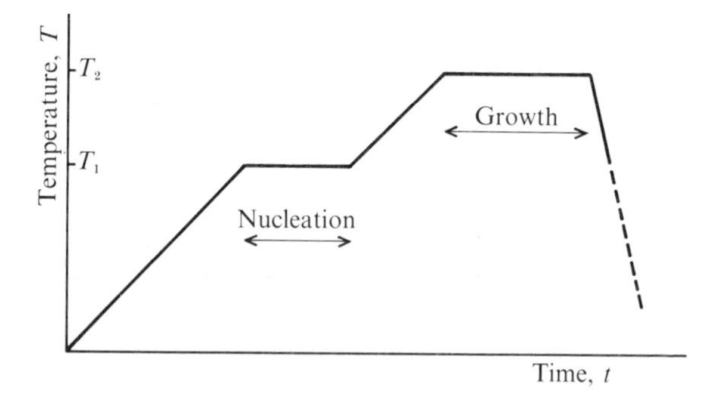

In principle, any glass can be converted to a glass-ceramic by finding a suitable nucleating agent and the appropriate thermal treatment. In practice, a number of technical factors (among which is the production of the base glass) limit the choice.

The glass-ceramic process has the advantage of using the same rapid forming techniques employed in glassworking permitting the economical production of objects with complex shapes or thin walls which are difficult or impossible with classical ceramic techniques.

Extreme fineness of the grains and the total absence of porosity are characteristic of glass-ceramics and distinguish them from traditional ceramics.

The size of the crystals in a glass-ceramic generally does not exceed 1 μm (200–300 Å are possible) while in the usual ceramics, the dimensions of the grains are of the order of 10–20 μm for Al_2O_3 and 40 μm for porcelain. The distribution of crystallites is isotropic although in certain special cases (e.g. extrusion) preferred orientations can be introduced.

The mechanical properties are excellent. The strength of glass-ceramics can be high, the best are based on cordierite (350 MN m^{-2} versus 70 MN m^{-2} for the starting glass). This is explained in part by the action of the micro-crystallites which limit the propagation of Griffith flaws.

The strength depends strongly on the degree of crystallinity (volume % of the crystalline phase). In the theory of Myiata and Jinno,[495] the system is treated as a composite material distinguishing the stages of *initiation* and *propagation* of cracks in the vitreous matrix. When cracks tend to go around the crystalline grains (intergranular cracks), a decrease in strength is seen for low crystalline concentrations followed by a strong increase for higher crystalline levels. The problems relating to the mechanical strength behavior of these materials are far from being solved at the present time.

16.3 Examples of some systems

A large number of systems has been studied and numerous patents cover particular compositions. We will mention a few characteristic cases. Many other examples of glass-ceramics may be found in McMillan's monograph,[496] and new compositions are regularly proposed.

1. SiO_2–Li_2O

This system is interesting because the solubility of crystalline lithium disilicate ($2\,SiO_2{\cdot}Li_2O$) in dilute HF is about ten times greater than that of glass. Using a photosensitive nucleating agent and masking, soluble crystallized patterns can be obtained. This *chemical process* is used in particular for the fabrication of perforated plates for color television tubes and for complex micro-circuits.

2. SiO_2–Al_2O_3–Li_2O

The family of glasses derived from this system has been used the most, the nucleating agents being TiO_2, ZrO_2 or their mixtures. The precipitated crystalline phases: β-eucryptite $Li_2O \cdot Al_2O_3 \cdot 2\,SiO_2$ and β-spodumene $Li_2O \cdot Al_2O_3 \cdot 4\,SiO_2$ result in materials having very low thermal expansion coefficients. Their use includes cookware, supports for distortion free mirrors for astronomy etc. Figures 16.4 and 16.5 show examples of the micro-structure.

3. SiO_2–Al_2O_3–MgO

The glass-ceramics derived from this system have been thoroughly investigated. The most effective nucleating agents are TiO_2, ZrO_2 and SnO_2, the precipitated crystalline phases being cordierite; $2\,MgO \cdot 2\,Al_2O_3 \cdot 5\,SiO_2$ and solid solutions derived from β-quartz (high form). These materials are characterized by high strength, good insulating properties and moderate expansion coefficients. They are mainly used to make radomes and insulators.

Fig. 16.4. Transmission electron micrograph showing the structure of a Corning glass-ceramic. The β-spodumene solid solution contains small particles of rutile and spinel. (By courtesy of Corning Glass Co.)

4. SiO_2–Al_2O_3–Na_2O

Catalysed by TiO_2 alone, or in combination with other divalent oxides, the glass-ceramics based on this system contain nepheline (variable composition solid solution derived from tridymite).

Fig. 16.5. Replica electron micrograph of a transparent, low-expansion glass-ceramic containing a quartz solid solution. (By courtesy of Corning Glass Co.)

16.4 Special applications

The conversion of blast-furnace slag into glass-ceramics has been undertaken in some countries to utilize industrial waste. The products thus obtained are called "Sitalls" in USSR and "Slag-Cerams" in Great Britain. The procedure consists of adding missing oxides and appropriate nucleating agents.[497]

It is possible to increase the strength of a glass significantly by controlled devitrification of a thin surface layer which introduces compressive stresses in the surface. The surface crystallization of a glass in the system SiO_2–Al_2O_3–Li_2O creates strengths as high as 600 MN m^{-2}; the crystals are so small that they are undetectable and do not alter optical transmission properties. Another process variation starts with a base glass of the system SiO_2–Al_2O_3–Na_2O with TiO_2 as the nucleating agent. Ion exchange in a bath of Li salts (5–10 min at 900 °C) exchanges Li^+ for the Na^+ in the surface and the layer thus obtained is converted to a glass-ceramic with the precipitation of β-eucryptite and quartz. The strength reaches 350–700 MN m^{-2}.

A family of glass-ceramics has been developed where mica microcrystals are nucleated from glasses whose composition is near to fluorophlogopite $KMg_3 AlSi_3O_{10}F_2$. The sub-division of the vitreous matrix by crystallites forming a "house of cards" structure allows *machining* of these ceramics with conventional steel tools: they can be drilled, cut or turned on a lathe. Figure 16.6 shows the characteristic microstructure of these materials.

Fig. 16.6. Machinable glass-ceramic showing mica flakes randomly distributed in a vitreous phase ("house of cards" structure). Scanning electron micrograph of a fracture surface. (By courtesy of Corning Glass Co.)

Another original application of glass-ceramics is their use as "biomaterials" in bone prosthesis.[498] $Ca_3(PO_4)_2$ glasses can be made into glass-ceramics to form a material resembling the mineral part of bone, apatite, which can serve as an artificial skeleton part in restorative surgery.

The porous bone-like texture is obtained by first producing a foam-glass by decomposition of a carbonate in the molten glass mass. This material simultaneously undergoes a controlled devitrification, transforming it into a porous micro-crystalline product. The dimensions of the interconnections between the pores must be sufficient to allow the ingrowth of living bone tissue which thus ensures a permanent joint with the surface of the prosthesis (Fig. 16.7).

Fig. 16.7. Texture of a porous glass-ceramic for medical applications. After Ref. 498.

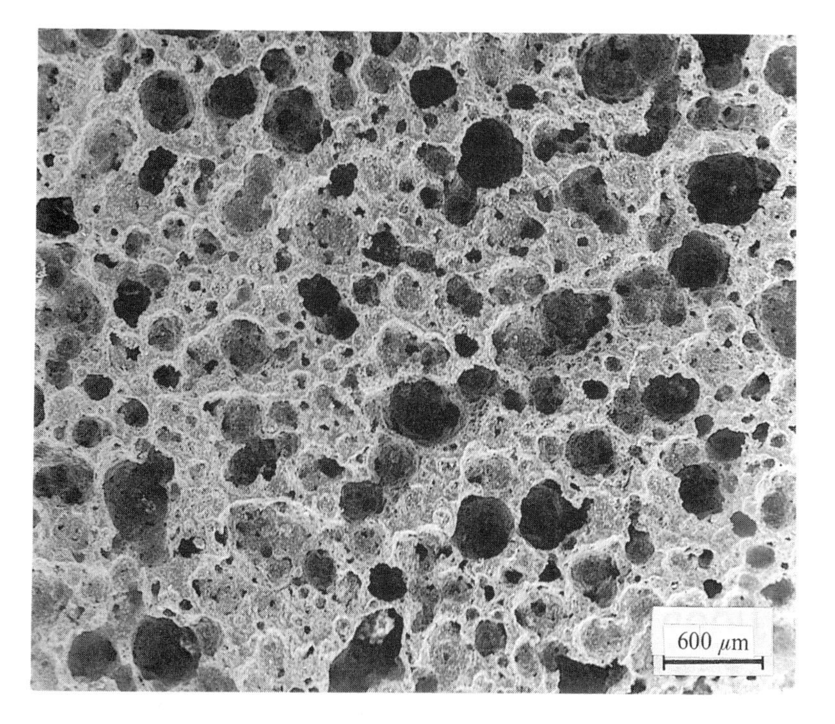

17 Elements of glass technology

This chapter is only intended to give a brief overview of the techniques used for the industrial production of glass. Readers desiring a deeper understanding of this subject should consult specialized works, e.g. those of Tooley[18] and Shand.[20]

17.1 Production of industrial glasses

Glass production starts with a mixture of primary materials consisting for the most part of natural minerals (sand, limestone) and in lesser quantity, industrial chemicals (Na_2CO_3) taken in definite proportions. This vitrifiable heterogeneous mixture called "batch" by glass workers is melted in an appropriate furnace. Except for the volatile parts which escape in the process, all constituents of the initial mixture (including undesirable impurities) fuse together forming an homogeneous liquid called *melt* which vitrifies on cooling, i.e. solidifies as *glass*.

The glass melting process is almost always immediately followed by *forming* and controlled cooling of a definite object or a continuous sheet (flat glass) which is cut into pieces and may undergo other thermal, mechanical or chemical treatments.

Except for certain modifications due to these specific treatments, the properties of the glass are essentially fixed from the very beginning by the "batch" composition which determines the quality of the glass and conditions for the glass-forming process.

17.1.1 *"The batch"*

Nearly all industrial glasses are oxide glasses. The *ability to vitrify* is a term designating the tendency of a liquid solution to vitrify on cooling; it depends on the proportion of glass-forming oxides and modifiers in the composition. Most chemical elements exist in nature as oxides (rocks) or are provided by industry in oxide form or as compounds able to transform into oxides during fusion. The choice of constituents and their proportions is nearly unlimited; physical chemistry considerations determine the composition choice for desired glass properties while economics determine the constituents able to provide oxides in the right proportions in the "batch."

(a) *Glass-forming oxides*

The major industrial glass-forming oxide is SiO_2 or silica, found abundantly in nature in the form of sand (quartz). All sand deposits cannot be used for glass because of impurities. Sands with more than 99 % SiO_2 and less than 0.2 % troublesome impurities are normally used. The purest sands or "Fontainebleau sands" contain only 0.01–0.02 % Fe_2O_3: they are reserved for optical and crystal glasses.

Anhydrous boric oxide B_2O_3, sometimes used as the only glass-former in enamels and very low-melting glasses, is often associated with SiO_2 (borosilicate glasses). It is an expensive former and is introduced either as pure $B(OH)_3$ or more commonly as $2\,B_2O_3 \cdot Na_2O$, either anhydrous or hydrated. Ordinary glasses contain less than 1 % B_2O_3.

Alumina, Al_2O_3 (acting as a glass-former with SiO_2) normally occurs in volcanic rocks (complex aluminosilicates). Certain feldspars (sodium and potassium aluminosilicates) are pure enough to be used in glass making. Al_2O_3 is generally introduced in the form of refined aluminum hydrate $Al(OH)_3$ from bauxite, an impure hydrated alumina. Phosphorous oxide, P_2O_5 may be used in the composition of special glasses.

(b) *Modifier oxides*

The oxides of sodium, Na_2O, and potassium, K_2O, alkali metal oxides improperly called "soda ash" and "potash" in glass works are the "fluxes" necessary to lower by several hundred degrees the temperature for processing and vitrification of SiO_2. They are introduced as industrially manufactured carbonates, sulfates and nitrates. The compositions are essentially based on sodium carbonate, Na_2CO_3, with part of the Na_2O being introduced in the form of sulfate which decomposes at higher temperature liberating bubbles of SO_3 which facilitates fining. Nitrate is added for its oxidizing properties.

K_2O, which is more expensive, is used in lead glasses (crystal) and some colored glasses. In ordinary glass, it is provided by feldspars along with Al_2O_3 and Na_2O.

Calcia, CaO, is an important component which improves the chemical stability of ordinary glass. It is added in the form of $CaCO_3$ as chalk, limestone and marble, or associated with magnesia, MgO, in dolomite.

Barya, BaO, is added in the form of sulfate or carbonate. Other oxides used in optical glasses are manufactured by the chemical industry.

(c) *Secondary constituents*

Several constituents, the total not exceeding 1 %, are added to modify the color or processing conditions of the glass. As_2O_3 and Sb_2O_3, used to facilitate fining, MnO_2 or Se to compensate the residual color from traces of iron oxide. Coloring oxides can be added to obtain special shades.

Table 17.1 indicates the proportions expressed in wt % oxide for the composition of ordinary industrial glasses and some special glasses.

17.1.2 *Production*

There are three steps in the process of transforming a vitrifiable mixture into molten glass which can be made into a finished form:

(a) *Melting*

Depending on the composition, the temperature is progressively raised to 1300–1400 °C (for ordinary glass). In the course of heating, several complex reactions occur: dehydration, dissociation of carbonates and sulfates with release of CO_2, SO_2, SO_3, local formation of compounds (silicates) by solid state reactions, general sintering of the mass, melting of certain components and finally dissolution of the most refractory constituents at temperatures much lower than their actual melting points.

(b) *Fining and homogenization*

Molten glass is not homogeneous; moreover, it contains numerous gas bubbles from dissociation of compounds, reaction with refractories and furnace atmosphere, etc. which are trapped in the high-viscosity mixture.

The process of *fining* eliminates these inclusions. It mainly consists of raising the temperature of molten glass to a practical maximum (1450–1550 °C for ordinary glass) to reduce viscosity. The bubble ascension rate is thus increased (according to Stokes' law) and homogeneity is enhanced. A mechanical agitation (stirring) or injection of air or steam from below (bubbling) can also be used, and finally, fining agents Na_2SO_4 or As_2O_3 can be added. This very complex chemical fining consists of liberating large quantities of gas at the end of the melting process, which precipitate on the fine bubbles already present, increase their volume and thus take them to the surface faster. Thus, at temperatures greater than 1200 °C:

$$Na_2SO_4 \rightarrow Na_2O + SO_2\nearrow + \frac{1}{2}O_2\nearrow$$

Arsenic oxide, As_2O_3, has a more complex action. In a vitrifiable mixture, disproportionation into $As + As_2O_5$ occurs first and As is reoxidized by air (or nitrates) into As_2O_3. As_2O_5 gives arsenates which decompose above ($\sim$ 1 %) liberating O_2:

$$\text{Arsenates} \rightarrow As_2O_3 + O_2\nearrow$$

The finest oxygen bubbles dissolve when the temperature decreases and the inverse reaction forms arsenates.

Table 17.1. *Composition (in wt %) of some industrial glasses.*

	SiO_2	B_2O_3	Al_2O_3	P_2O_5	Na_2O	K_2O	CaO	BaO	MgO	PbO	ZnO	Fe_2O_3
plate glass, window glass	72.5		1.5		13	0.3	9.3		3			0.1
container glass	73		1		15		10					0.05
electric light bulbs	73		1		16	0.6	5.2		3.6			
borosilicate (Pyrex)	80.6	12.6	2.2		4.2		0.1		0.05			0.05
aluminosilicate (fibers)	54.6	8.0	14.8		←0.6→		17.4		4.5			
"crystal" glass	55.5					11.0				33		
optical glass (heavy flint)	28				1	1				70		
sodium lamp envelopes		36	27					27	10			
radiation shielding glass	29							9		62		
HF resistant glass			18	72							10	

Sodium sulfate has only limited solubility ($\sim$ 1 %). When there is an excess, it floats to the surface of the bath where it can be eliminated by the addition of carbon thus preventing corrosion of the refractories.

(c) *Conditioning*

At the end of fining, the viscosity of the glass is too low for working. Depending on the forming process employed, its viscosity is increased by cooling to the temperature range 1200–1000 °C (for ordinary glass).

The three production stages: melting, fining-homogenization and conditioning depend on viscosity. The techniques described above make possible the elimination of visible inclusions (bubbles) and the attainment of a high degree of homogeneity (deviation of the refractive index less than $\pm 5 \times 10^{-4}$ for ordinary glass, and of the order of 10^{-5}–10^{-6} for optical glass).

17.2 Glass melting furnaces

17.2.1 *Intermittent processes*

(a) *Pot furnaces*

When only small quantities of glass are needed (e.g. special optical glass or glass for hand-working) it is often made in individual crucibles or "pots" placed in gas or oil heated furnaces with a recovery system to pre-heat the combustion air. To reduce production costs, furnaces containing about ten pots of about 1000 l are used, all subjected to the same production cycle. Figure 17.1 shows a melting chamber for a single pot.

The pots, made of clay, are heated slowly to 900 °C in a special pre-heat furnace then taken out and placed, without cooling, in the actual glass furnace where the final firing occurs near 1450 °C.

The batch is put into the pot in several stages. The addition of 20–25 % broken glass (cullet) of the same composition aids the melting process.

For a 1000 l pot, the production time is, for instance: melting, 6–8 h; fining, 4–6 h; conditioning, 4–6 h. After production, the pot is either left in the furnace and emptied little by little by the glass blowers, or removed from the furnace, emptied mechanically, and immediately put back in the furnace.

The pot-life can be about 20 melts or 10–12 weeks. To prevent cracking, pot temperature should not go below 850 °C .

(b) *Pt crucible furnaces*

For special optical glasses with a strong tendency to crystallize or aggressive melts, Pt crucibles heated by the Joule effect or induction

Fig. 17.1. Melting chamber for a single pot: (1) arch of SiO_2 bricks; (2) opening for charging the glass batch; (3) glass batch in the course of fusion; (4) molten glass; (5) fire-clay glass pot; (6) gas or oil burner (the combustion air is usually pre-heated by the exhaust gases); (7) the exhaust gases are led to a recuperating heat exchanger before being released through a chimney; (8) drain port for the removal of the glass in case of rupture of the pot. After Ref 499.

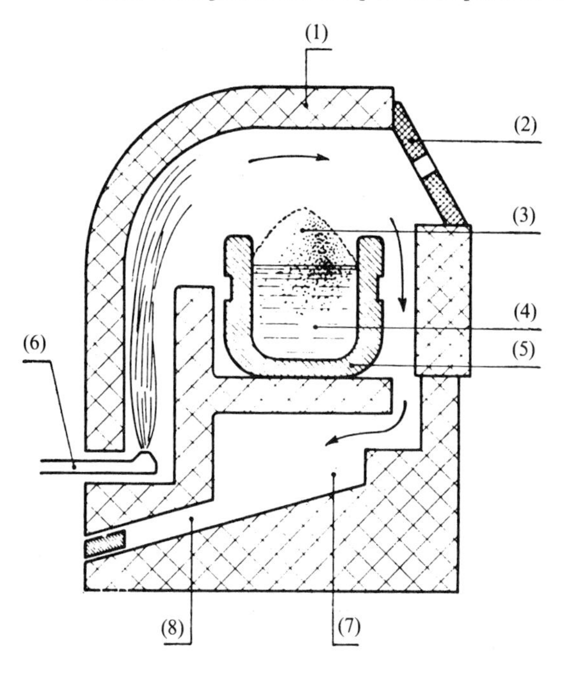

are used (Fig. 17.2) with a pouring orifice normally blocked by solidified glass. At the time of pouring, this passage is heated and glass pours into the mold.

17.2.2 *Continuous furnaces (glass tanks)*

Large scale industrial glass production is done only in continuous furnaces called tanks.

(a) *Furnace for flat glass*

Figure 17.3 represents such a furnace. The essential part is the *tank* built of refractory blocks (Zac, Corhardt, Monofrax) resistant to corrosion by the glass and set dry without cement. Tightness of the joints is provided by the glass itself which solidifies in the cooler zones. The depth of the tank is variable: 1–1.5 m for clear glass, 0.6–0.8 m for colored glass (transmitting heat less effectively). Heat is provided by

Fig. 17.2. Pt crucible for melting optical glass (schematic): (1) crucible; (2) electrical leads; (3) glass; (4) stirrer; (5) casting conduit; (6) heater.

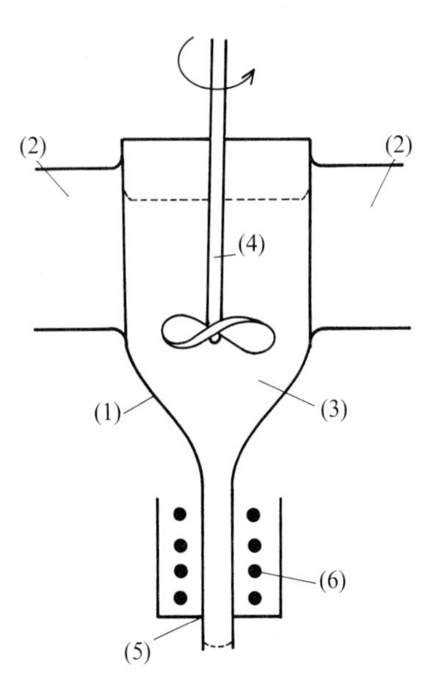

coal gas, natural gas or more often heavy oil with heat recuperation in brick-filled chambers (*checkers*) heated by the burnt gas. Furnaces have two lines of burners and recuperators arranged symmetrically relative to the longitudinal axes and functioning alternately. About every quarter of an hour, the process is *inverted*: the burnt gas is diverted toward cooler recuperators while combustion air is passed over the checkers which had been heated during the preceding cycle. The flames impinge directly on the glass surface in the tank. The burners are individually regulated, which allows control of the furnace temperature profile. The raw materials (batch) are loaded mechanically at one end, and the finished glass is fed to forming machines at the other end by a flow channel or a drawing reservoir. The production cycle is continuous, with successive processes in different parts of the tank. The furnace atmosphere in the melting and the fining zones (where the maximum surface temperature is found) is partially isolated from the cooling zone by a low arch to allow the "refining" of the glass in the latter part of the process. The slow and continuous longitudinal flow of the glass, caused by the "pull" of the machines, is accompanied by a complex system of thermal convection currents which ensures the mixing necessary for fusion and fining and exerts a chemical and thermal homogenizing effect on the glass.

Fig. 17.3. Glass tank for plate or window glass: (1) bottom, alumina-based refractories; (2) melting zone; (3) cooling zone; (4) zone of maximum surface temperature; (5) mechanical batch feeder; (6) glass batch in the course of fusion; (7) silica brick arch; (8) throat arch; (9) burners; (10) recuperator ports; (11) recuperators; (12) skimming recesses; (13) pulling trough (fabrication of window glass); (14) forehearth (fabrication of cast glass by pressing). After Ref. 499.

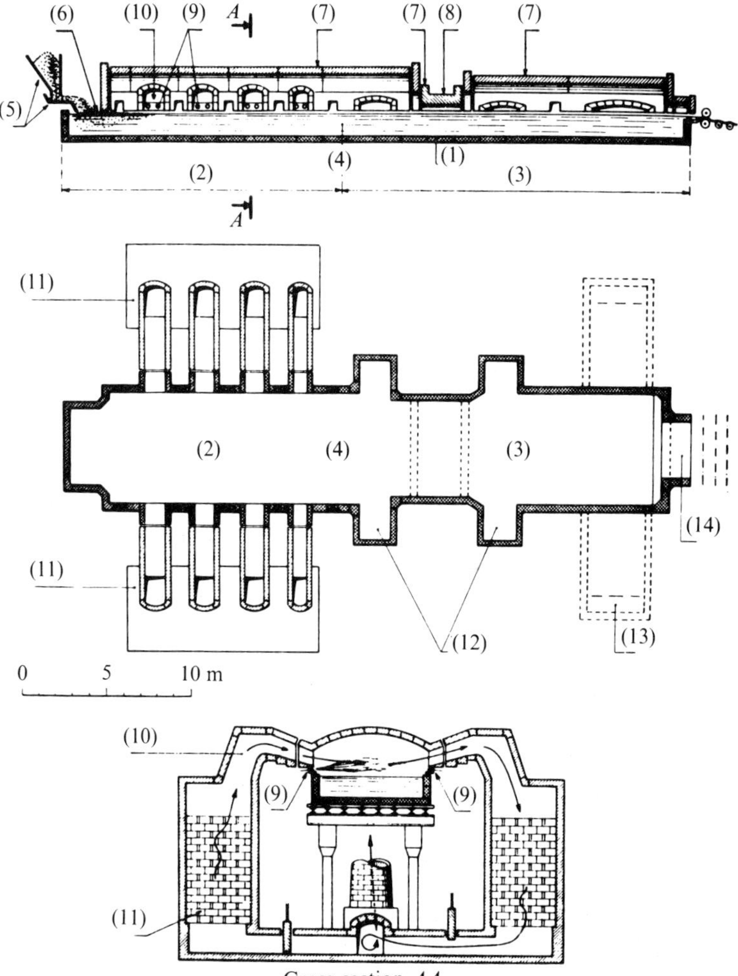

Cross-section *AA*

Tank melting allows a considerable production cost saving through fuel economy, the current trend being towards larger and larger units. Large tank furnaces for flat glass contain about 1000 tons of glass with a tank surface of 300 m^2. The production of these units is about 2 tons per square metre per day with a life-time of 6–8 years. Oil consumption is of the order of 0.2 kg per kilogram of product.

(b) *Furnace for hollow ware*

Furnaces of the preceding type produce very high-quality glass at the cost of some energy waste, since a significant part of the cooler glass is recycled several times towards hotter zones of the furnaces by convection currents.

For products which can be of lower quality, such as "hollow ware" (bottles), the refining zone is separated from the rest of the tank by a "throat" acting as a siphon (Fig. 17.4). The glass, cooled in the refining chamber, is led towards the forming machine by a series of distribution canals or feeders (forehearths) fanning out from the furnace. This allows control of the glass viscosity and supply to multiple working stations. Some glass tanks can have additional electric resistance heating, Mo or graphite electrodes carrying the current.

Total electric melting is done in some countries (Switzerland, Sweden); it requires about 1 kW h per kilogram of glass. Research on this method was stimulated in the 1970s by an oil shortage and the prospect of resorting to electricity generated using nuclear energy.

17.3 Forming processes

There are three main types of manufactured products:

–flat glass (mirror glass, window glass),
–hollow glass (bottles, goblets, lamp bulbs),
–glass fibers,

to which can be added a fourth category grouping glasses for various applications (optical glass, tubes, rods, marbles, etc.).

The main forming processes vary according to the category.

17.3.1 *Manufacture of flat glass*

(a) *Rolling*

The glass flows from the furnace onto the spillway (Fig. 17.5) and passes between cooled metal rollers which solidify it into a continuous ribbon (3–15 mm thick, up to 3.60 m wide) moving at 0.5–5 m min^{-1} and directed towards an annealing lehr to relax internal stresses. The raw sheet is not perfectly flat nor transparent. After cooling, it must undergo *grinding* and *polishing* resulting in a transparent sheet called *plate glass* with two perfectly plane and parallel faces .

Fig. 17.4. Container glass furnace: (1) melting and fining section; (2) automatic batch feeder; (3) burners; (4) recuperator port; (5) throat; (6) loose stack of silica bricks separating the two sections; (7) conditioning section; (8) feeder: (9) forehearth; (10) automatic gob distributor; (11) gob in formation. After Ref. 499.

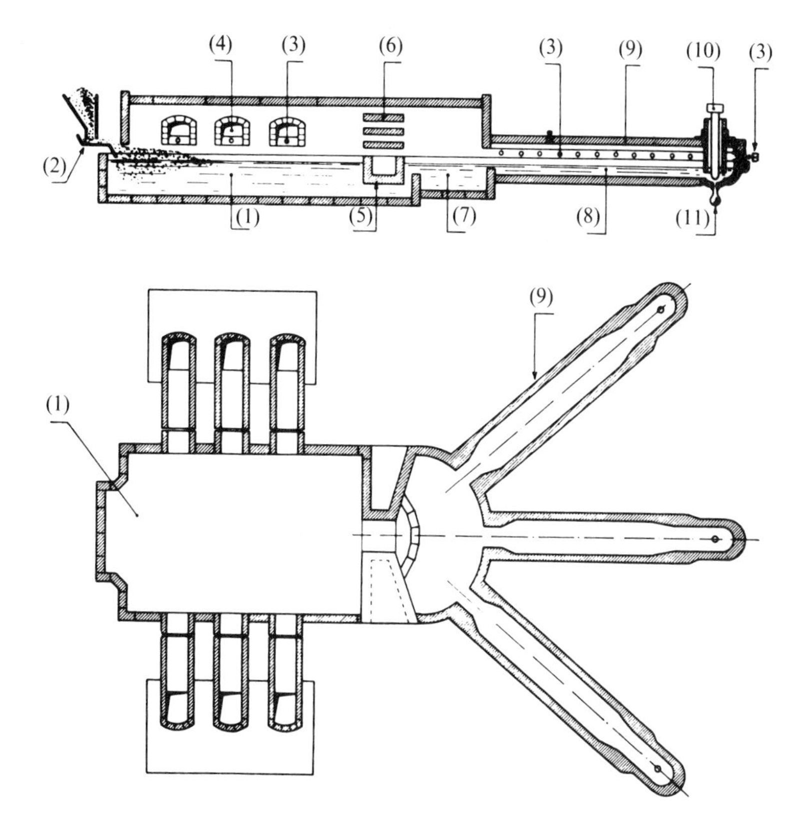

The surface of the raw glass is flattened by grinding using cast iron wheels with sand abrasive and water lubricant. As the grinding progresses, the abrasive used is finer and finer finally producing a very fine satiny surface. The operation is completed by polishing with felt pads and a suspension of iron oxide ("rouge"). The work can be entirely mechanized with continuous simultaneous grinding and polishing on both faces.

Currently, this method of making plate glass is being replaced by the float glass method which produces a glass quality approaching that of plate glass at a lower price by virtue of elimination of the mechanical grinding and polishing operations.

Fig. 17.5. Continuous casting of plate glass. After Ref. 439.

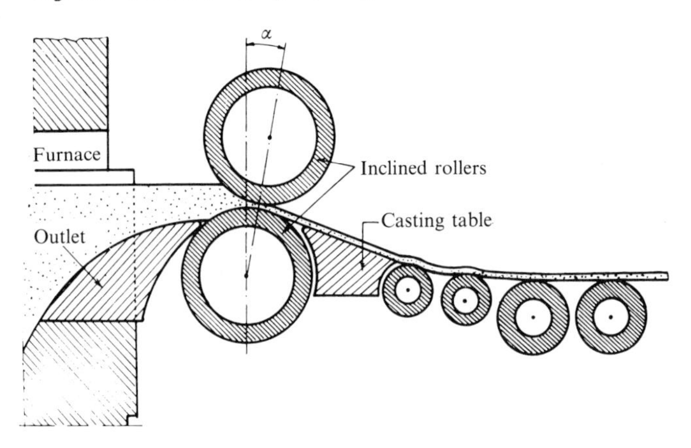

(b) *Drawing*

When a solid metal plate is dipped into molten glass and slowly withdrawn, it drags up some of the liquid which rises while thickening by cooling up to the time where rupture occurs in the emerging part (Fig. 17.6(*b*)). The progressive increase of the viscosity in the "bulb" or meniscus permits drawing but the surface tension tends to reduce the width and leads to rupture of the sheet. To counter this action and allow continuous drawing without rupture, the borders of the sheet are cooled and solidified, and form a rigid framework stabilizing the width of the sheet.

This principle is used to manufacture *window glass* i.e. glass sheets which are sufficiently flat and free of distortions for ordinary applications.

Several variations have been designed:

Fourcault Process (1904) : (Fig. 17.7(*a*)

A refractory block with a slot, called a "*débiteuse*" (Fig. 17.6(*c*)), is submerged in the surface of the glass bath in the working chamber of the furnace. The glass passing through this slot forms the meniscus, the pulling being started by placing a metal bar in contact with the slot; the sheet is pulled vertically supported by guide rollers, while coolers acting by radiation (without direct contact with the sheet) continuously solidify the edges. The sheet passes through a vertical annealing lehr before being cut.

The thickness of the drawn glass depends on the meniscus which is determined by the width of the slot in the guide channel (5–8 cm), and, most of all, by the drawing rate. This is of the order of 70 m h^{-1} for glass called "single strength" (1.9 mm) and is reduced to 25 m h^{-1} for "double

Fig. 17.6. Principle of drawing: (*a*) drawing a circular rod from the surface of molten glass; (*b*) impossibility of pulling a planar sheet: formation of a neck; (*c*) débiteuse and cooling allow the formation of a sheet of constant width.

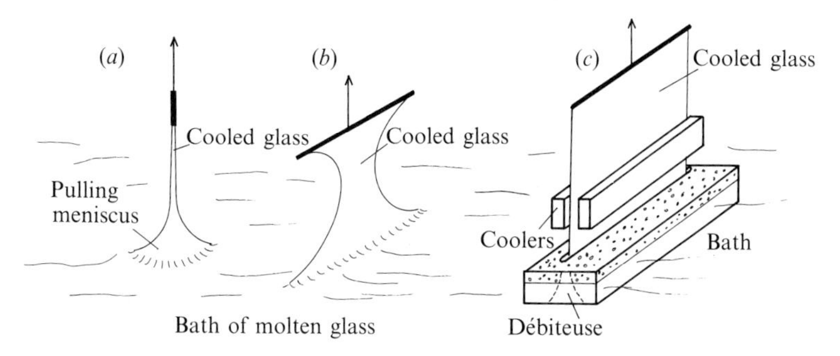

strength" (6 mm). The width can reach 2.70 m. A tank furnace can feed several Fourcault machines simultaneously.

Pittsburgh process (1925) (Fig. 17.7(*b*)).

In this process, the débiteuse is replaced by a completely submerged refractory piece (draw-bar), which lowers the temperature of the glass below the meniscus. The absence of the débiteuse allows the drawing rate to be increased (100 m h^{-1} for single-strength glass) and eliminates drawing defects but, on the other hand, requires better control of the glass temperature and homogeneity.

Libbey-Owens process (1917) (Fig. 17.7(*c*)).

This process from Colburn also operates without a débiteuse. The drawn glass sheet is bent at right angles, about 1 m above the surface of the bath, on a polished chrome-nickel alloy roll. This avoids the excessive height necessary for the equipment of the two preceding processes. A very shallow bath is required for this process and generally only two machines can be installed on one furnace. The drawing speed of single-strength glass is 140 m h^{-1} and up to 3.60 m width.

(c) *"Float process" (1959)*

This process was developed by the British firm Pilkington and revolutionized the flat glass industry (Fig. 17.8). The original process consisted of first forming a glass sheet by rolling, then placing it in a softened state on the surface of a molten Sn bath. The face in contact with the metal acquires the planarity of the bath surface. The opposite face is made perfectly plane by the action of surface tension, which tends to spread the glass sheet on the bath. In its final form, the process consists

Fig. 17.7. Drawing processes for window glass; (*a*) Fourcault, (*b*) Pittsburg, (*c*) Libbey-Owens. After Ref. 18.

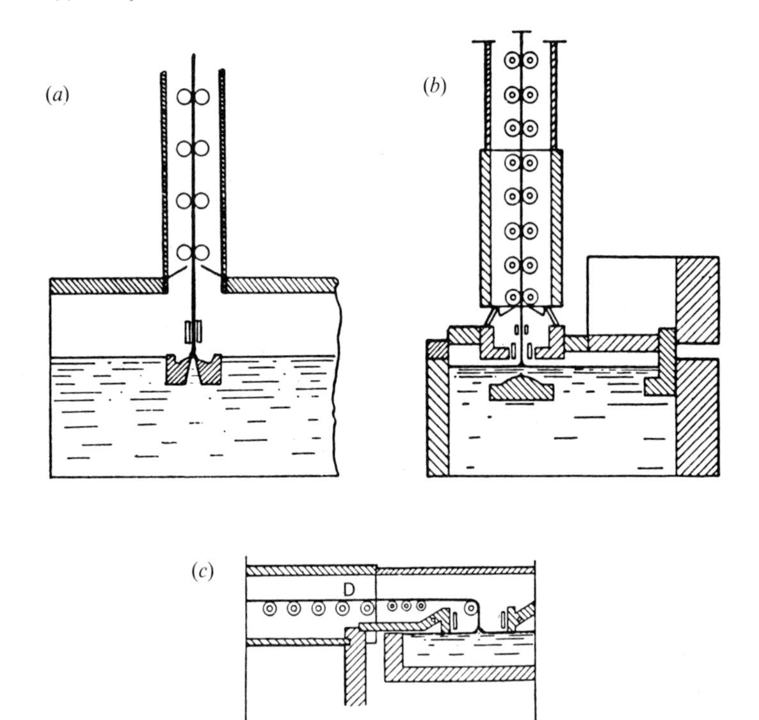

of feeding molten glass directly from the furnace onto a Sn bath. Equilibrium between gravitational forces and surface tension produces sheets of uniform thickness, about 6.5 mm, independent of the ribbon width. To reduce the thickness, traction is applied to the edges by special equipment to keep them from coming together. The glass ribbon leaves the bath and enters an annealing lehr (Fig. 17.8).

The main advantage of this process is immediate high optical quality: the planarity approaches that of plate glass without the need for polishing. Moreover, the output rate is 5–10 times higher than the drawing rate for window glass.

On the other hand, the equipment requires control of the atmosphere above the bath, which must be neutral or slightly reducing to avoid oxidation of the bath and the maintenance of constant surface tension which controls the thickness of the sheet.

The introduction of this process has caused progressive abandonment of mechanically polished plate glass production. Float glass quality may be slightly inferior to that of plate glass but it is clearly superior to window glass.

Fig. 17.8. Float process (schematic).

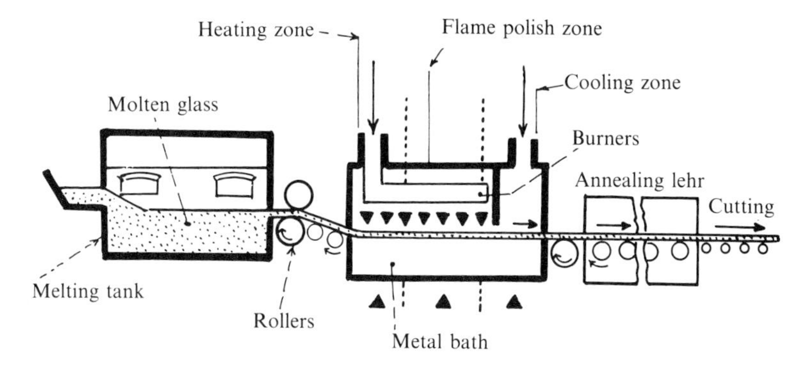

17.3.2 *Fabrication of hollow ware*

(a) *Pressing*

A fixed quantity of molten glass or "parison" is put into a mold and pressed at about 400–450 °C. The molds are made of special steel with chrome-plated surfaces. Manual presses produce up to 300 pieces per hour and automatic presses (with a series of molds) up to 1000 pieces per hour. Dishes, jars, lenses etc. are produced this way.

(b) *Blowing*

The original hand blowing technique, still used in art and science glass working (chemical equipment), is now mechanized to increase productivity. It generally begins with a *blank* in a preparatory mold, which is then transferred into a final mold for the last blow. According to the filling technique of the blank mold, the processes are characterized as: suction and blowing, Owens (1905); blow and blow, Hartford (1925) (Fig. 17.9(*a*)); press and blow (Fig. 17.9(*b*)). The machines have molds on a carrousel for high speed continuous forming up to 70 pieces per minute. They are used for the production of bottles, jars etc.

High-speed production of light bulbs uses a Corning ribbon machine (1926). First, a ribbon of hot glass (1050 °C) is formed by a special roller which imprints a series of circular depressions; it is then carried by two synchronized conveyors: one of them contains blow heads which engage the depressions, the other a series of molds which close on the object during blowing. The production of such machines reaches 5–6 pieces per second, more than 500,000 per day.

Fig. 17.9. Automatic fabrication process for hollow ware: (*a*) formation of the blank by blowing followed by blowing in a mold (blow and blow); (*b*) formation of the blank by pressing followed by blowing (press and blow).

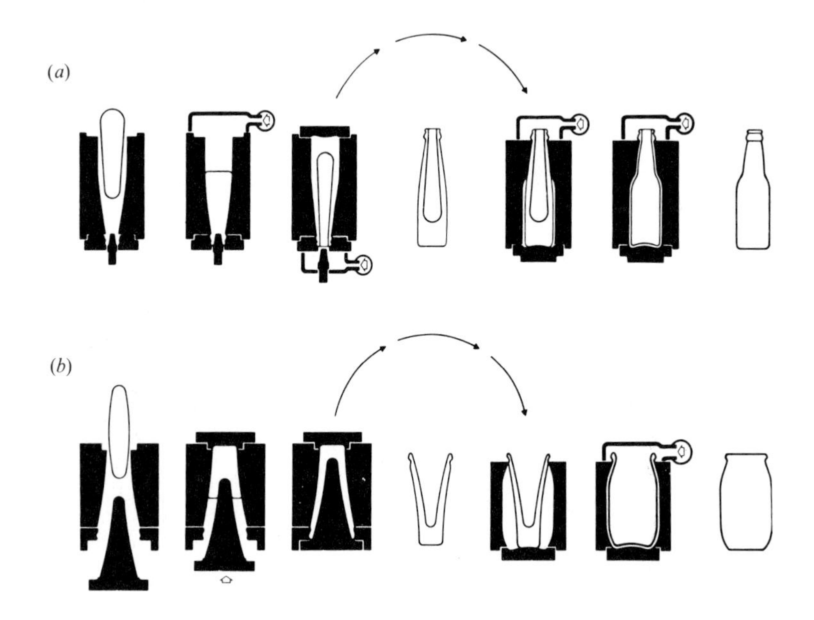

17.3.3 *Production of glass fibers*

Glass can be made into fibers. There are two main categories: textile fibers (continuous filaments) and insulation fibers (short tangled fibers).

There are three methods of production: mechanical drawing, centrifugal drawing, gas drawing: combinations of these can also be used.

(a) *Mechanical drawing*

The Gossler process (1920) consists of an electrically heated furnace with small holes in the bottom from which glass can be drawn into fibers and wound on a rapidly rotating drum (Fig. 17.10(*a*)). This process gives long fibers 14–25 μm diameter, with a drawing rate of 12–20 m s^{-1}, used to make thin or thick mats for heat and sound insulation.

This process has been improved in the United States (Owens-Corning) using a heated Pt bushing with 100–400 holes. The much finer fibers (3–10 μm) are gathered into a thread and rolled onto a drum turning at several thousands revolutions per minute. Before winding, individual fibers are *sized*, i.e. glued together by a plastic coating. The fibers are of a quality suitable for textiles.

Another variant (Schuller 1938) consists of drawing fibers from a row of glass rods heated at their ends by a series of burners (Fig. 17.10(*b*)).

All these processes use a special glass (E-glass) which is low in alkali metals to eliminate water attack. The glass must be totally bubble-free. The bushings are fed with cold glass in the form of marbles and their temperature is well regulated.

Fig. 17.10. Drawing of glass fibers; (*a*) Gossler process, (*b*) Schuller process. After Ref. 499.

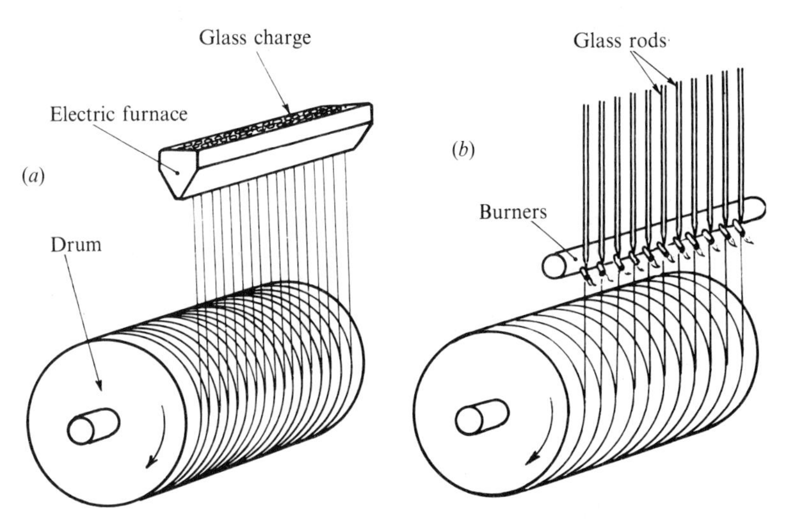

(b) *Centrifugal drawing*

Centrifuge processes produce short bulky fibers (glass wool) which are not appropriate for textiles; they are used in insulation.

In the Hager process (1931), a stream of molten glass falls on a refractory disk turning at high speed (3000–4000 rpm) heated at its edge by flames. The centrifugal force draws fibers of about 25 μm.

To increase output stability, the centrifuge is constructed as a hollow metallic structure with many holes on the periphery and surrounded by an array of burners. This device, based on the same principle as the one used to make "cotton candy" has been under development since 1942 by the Saint-Gobain Co. for the production of insulation fibers (10–12 μm diameter).

(c) *Fluid-assisted drawing*

The method consists of using high-velocity gas jets to blow out small streams of glass, thereby causing a redrawing into very fine fibers.

In the Owens-Corning process, the large diameter primary fibers exiting the bushing are impacted by hot gas jets and flames from a burner in an internal combustion chamber (Fig. 17.11). This high-quality product consists of fibers 0.5–5 μm in diameter (superfine fiber).

Fig. 17.11. Fluid-assisted drawing.

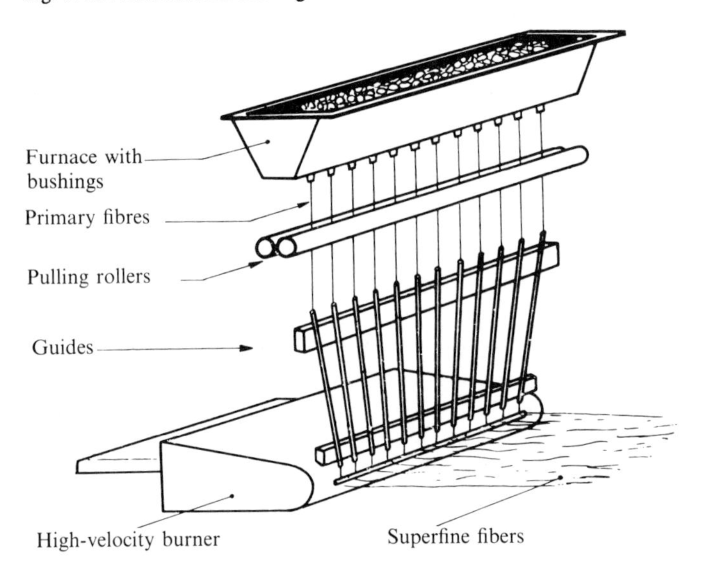

(d) *Mixed process*

In the "TEL" process developed 1954 by the Saint-Gobain Co. (Fig. 17.12), the centrifugal principle is combined with that of fluid drawing. The spinner made of Pt rotates at 3000 rpm. Hot gas issuing from burners redraws the fibers coming out of the spinner. The process is very flexible, allowing the production of fibers 1–6 μm diameter in bulk, suitable for thermal and sound insulation.

Fig. 17.12. TEL process: (1) device to distribute glass to the perforated rim; (2) forming device for the primary fibers; (3) special annular burner; (4) sheet of heated gas from 3; (5) feed layer to the perforated rim; (6) drawing zone; (7) glass; (8) fibers; (Saint-Gobain Co.). After Ref. 499.

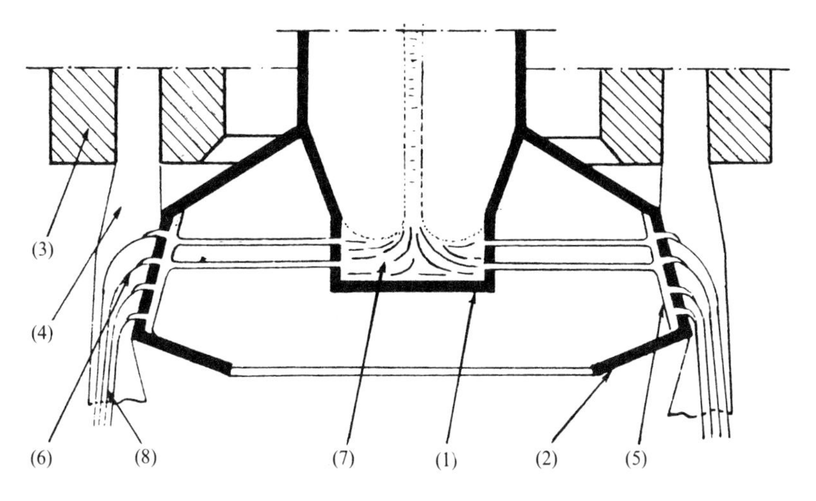

17.4 Production of optical fibers

17.4.1 *Clad fibers and fiber optics*

A clad fiber is produced by placing a high index glass rod inside a glass tube with a lower index and then drawing the assembly in a special furnace (Fig. 17.13(*a*)). Another process consists of drawing a fiber from two concentric crucibles, the interior crucible containing the core glass, the exterior the cladding glass (Fig. 17.13(*b*)). The crucibles are electrically heated and the coaxial fiber drawn from the orifice. Compared with the preceding process, it has the advantage of unlimited fiber length.

To make a *flexible optical imaging fiber*, the fibers are carefully put together in a parallel array, immobilized at the ends with a resin and the faces are polished. Such light pipes are used in the medical field (endoscopy) or, more generally, for the transmission of images of inaccessible objects.

To obtain *rigid* light pipes, used in the form of plates, an array of rods in tubes in a bundle is drawn simultaneously, giving multiple fibers, which are then aligned and redrawn. The operation is repeated to reduce the fibers to the desired core diameter (Fig. 17.14).

These conductors allow the transmission of larger images and are used in photoelectronics, e.g. to transfer an image produced on a fluorescent film to a photoelectric receiver or a photographic film. They allow the transfer of a curved image from an optical device onto a plane. Conical

Fig. 17.13. Drawing of optical fibers: (*a*) from a composite rod, (*b*) from concentric crucibles (cross-sections).

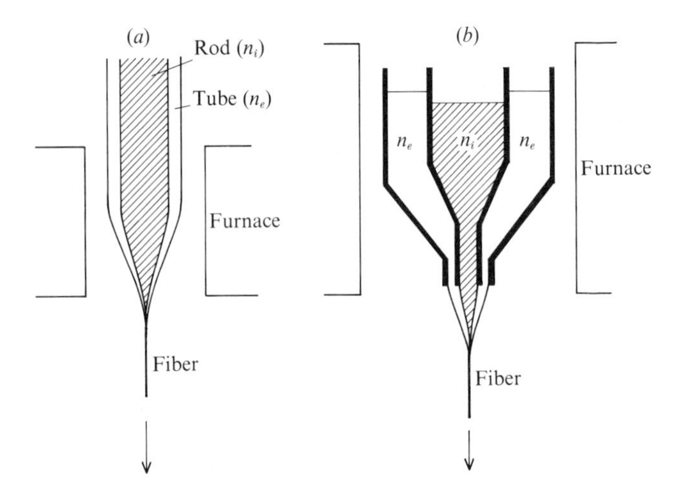

Fig. 17.14. Optical fiber lens: (*a*) cross-section of a multiple fiber array obtained by simultaneous drawing, (*b*) cross-section of an image conductor obtained by sintering the preceding fibers. After Ref. 430.

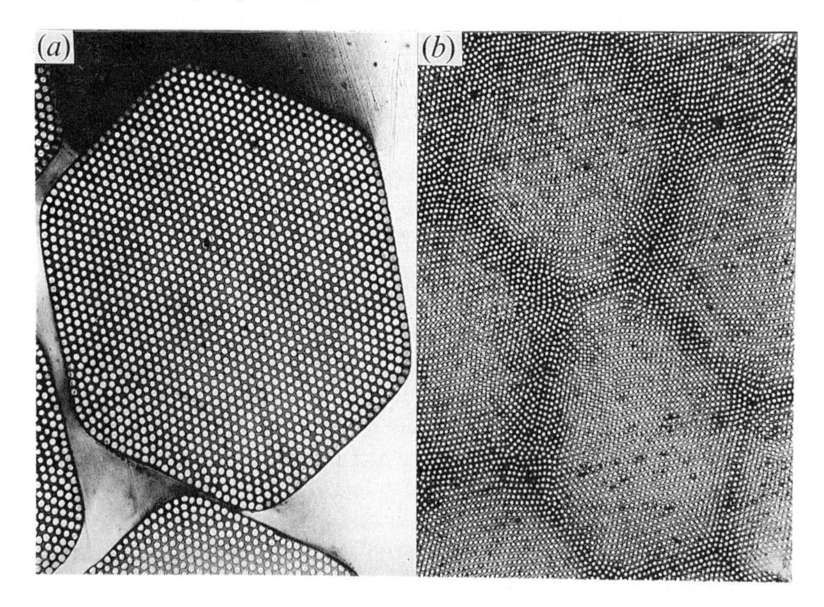

arrays allow changes in geometry (amplification or reduction). Anamorphic devices can also be made, e.g. transferring the image from a circular source to a rectangular window of a spectrograph.

17.4.2 *Optical waveguides*

SiO_2 is practically unrivaled as the constituent material of such fibers; however, it presents a problem in that it is necessary to modify its refractive index slightly to make glasses suitable for the core and the cladding. One solution is to use a core of pure (or nearly pure SiO_2) clad with a SiO_2–B_2O_3 glass which has a slightly lower index than pure SiO_2. Another solution is to raise the index of the core by using, e.g. SiO_2–GeO_2–B_2O_3 glass for the core and SiO_2–B_2O_3 for the cladding. The addition of P_2O_5 minimizes the index dispersion effect.

The creation of *single mode* fibers is difficult; for a fiber with core diameter about 1 μm, the difference in index must be about 10^{-2}.

The problem of producing ultrapure glass of the quality required for optical waveguides has been solved using processes based on the oxidation of halogen compounds in the vapor phase, which gives vitreous deposits or "pre-forms" used for drawing fibers.

The compounds $SiCl_4$, $GeCl_4$, BCl_3 (or BBr_3) and $POCl_3$ are carried by a stream of O_2 to a torch where oxidation reactions occur:

$$SiCl_4 + O_2 \rightarrow SiO_2 + 2\,Cl_2$$

$$GeCl_4 + O_2 \rightarrow GeO_2 + 2\,Cl_2$$

$$2\,BCl_3 + \frac{3}{2}O_2 \rightarrow B_2O_3 + 3\,Cl_2$$

This leads at about 1300–1400 °C, to deposition of oxides in the form of vitreous particles or "soot" which agglomerate into a porous solid with large specific surface area ($\sim$ 20 m^2 g^{-1}). This is then transformed into compact glass by viscous sintering in a hot zone at about 1400–1600 °C (according to composition), in a He atmosphere. This "blank" is then drawn into glass fibers at about 2000 °C.

There are several variants of this process.

(a) *Outside Vapor Phase Oxidation (OVPO)*

Deposition occurs by directing the oxidation products onto a rotating rod where they deposit in semi-sintered form in concentric layers. By varying the composition of the products, the desired index profile is obtained. The rod is then withdrawn and a glass pre-form is obtained by zone sintering of the deposit. The original central hole disappears in this operation.

There are two OVPO processes: horizontal (Fig. 17.15(*a*)) and vertical (Fig. 17.15(*b*)), where the latter can provide a blank of unlimited length.

(b) *Inside Vapor Phase Oxidation (IVPO)*

In this technique, the deposit occurs inside a tube (generally vitreous SiO_2) (Fig. 17.16). The entire piece is then sintered and drawn, which closes the central canal. In another version of this technique, the deposition is carried out in a high-frequency plasma leading directly to a sintered glass deposit.

This process has the advantage of being extremely clean (no OH^-). The tube can be formed by an OVPO process to ensure high-quality cladding glass.

Fibers with attenuation less than $0.2\,dB\,km^{-1}$ for a signal of wavelength 1500 nm have been produced by these processes.[419]

The total world production of the main types of industrial glass is presented in Table 17.2 for the period 1976–86.

Fig. 17.15. Production of optical waveguides by the "OVPO" process: (*a*) horizontal, (*b*) vertical. After Ref. 312.

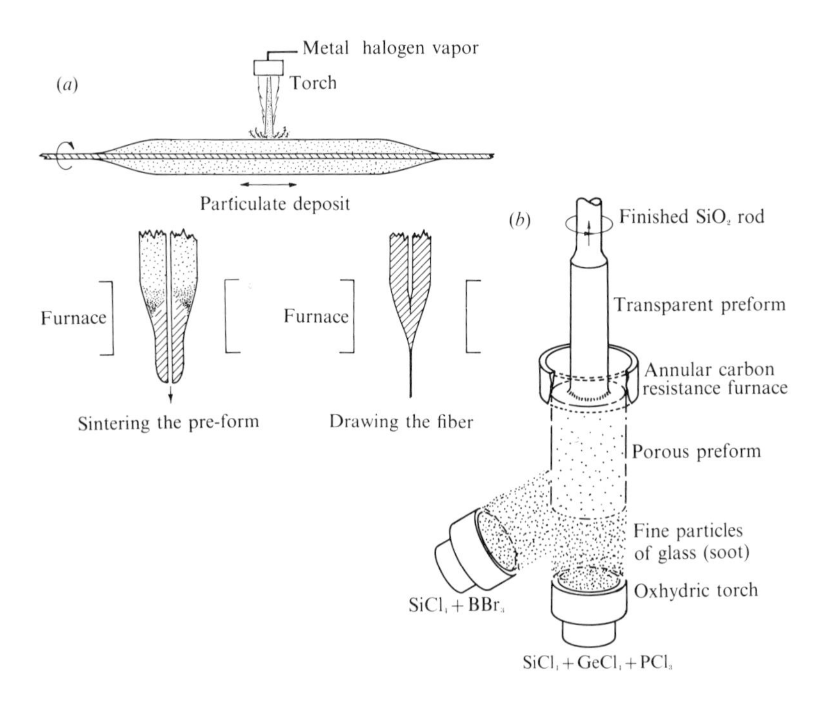

Table 17.2. *World production of glass.*
(From statistics furnished by the Union Scientifique et Continentale du Verre.)

	Total Production (in millions of tons)											1986			
	1976	1977	1978	1979	1980	1981	1982	1983	1984	1985	1986	plate glass	containers	fiber	other
W.Europe (estimated)	17.5	18.5	18.4	17.8	19.0	16.3	17.5	18.0	18.4	17.4	21.0				
EEC	14.7	15.5	15.4	16.6	16.6	15.0	16.3	16.4	17.0	16.4	18.3	4.683	12.436	0.470	0.748
France	3.5	3.7	3.7	3.9	4.1	4.0	4.0	4.2	4.2	4.1	4.0	0.684	3.168	0.146	0.072
W.Germany	4.3	4.5	4.4	4.5	4.7	4.5	4.4	4.4	4.6	4.6	4.7	0.992	3.230	0.193	0.288
UK	2.8	3.0	3.0	3.0	2.9	>2.4	>2.4	>2.4	>2.4	>2.3	>2.3	0.593	1.722		
Italy	2.2	2.3	2.3	2.9	3.0	2.7	2.7	2.8	3.5	3.5	3.5	0.747	2.364	0.080	0.265
Belgium	1.2	1.3	1.2	1.3	1.4	1.2	1.2								
USA			18.1												
Japan			3.5												

Fig. 17.16. Production of optical waveguides by the "IVPO" process. After Ref. 312.

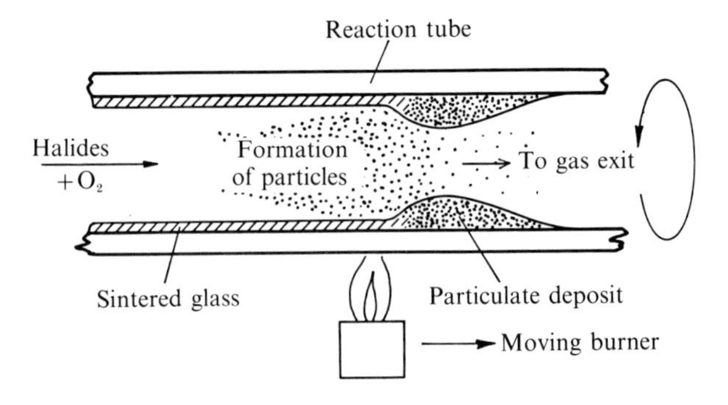

18 Synthesis of glasses from gels

18.1 Introduction

The classical way of obtaining glasses is by quenching a melt. Industrially, silicate glasses are produced by cofusion of SiO_2 with Na_2O and CaO which act as fluxes and then cooling the resulting melt at a sufficient rate to avoid crystallization. This classic technology was the subject of the preceding chapter. We have seen, however, in Chapter 1 that there are other ways to obtain non-crystalline materials: condensation of a vapor on a cold substrate, disordering of a crystalline solid by heavy irradiation or a shock wave, or by various reactions in the liquid phase followed by solvent elimination.

Sol–gel processes belong to this last group of syntheses: they are based on the possibility of forming the disordered network of the glass, not directly at high temperatures from the melt, but at low temperatures from suitable compounds by chemical polymerization in a liquid phase. In this way a *gel* is first formed from which glass may be obtained by successive elimination of the interstitial liquid and collapse of the resulting solid residue by sintering.

This "precursor-based" synthesis of glasses, ceramics and composites is, at the present time, one of the most rapidly progressing fields of materials science and engineering. The "sol–gel" method of preparing glasses is being actively studied in leading laboratories all over the world and, in the 1980s, the number of scientific publications in this field has shown an exponential increase.[500–10]

18.2 General characteristics of the process

18.2.1 *Advantages and disadvantages*

As the sintering operation is carried out at temperatures much *lower* than those required for the melting of glass-forming components, practically in the vicinity of the transition temperature, the process is particularly attractive for the production of those glasses which require high melting temperature (e.g. SiO_2 glass can be made at 1200 °C instead of 2000 °C).

The second important characteristic of the process is that final homogeneity is obtained directly in solution *on a molecular scale*. This can be compared to the difficulties of obtaining homogeneous glasses in the classic way, particularly when one of the components is more volatile or when the resulting melts possess a high viscosity which hinders efficient mixing of the constituents. (In some cases it is then necessary to remelt the original batch several times to reach the necessary compositional uniformity.) This, in turn, increases the likelihood of contamination from crucible walls, particularly at high temperatures, or during repeated crushing procedures.

In the sol–gel route the wet gel may, in principle, be obtained with a degree of purity which depends only on the starting ingredients and the purity of the final glass will depend on the sintering process which is performed at lower temperatures with reduced risk of contamination.

Substantially lower production temperature, excellent homogeneity obtained directly and a high degree of purity are the main advantages usually recognized in the sol–gel process. Is this enough to make the process competitive with the classical glass-melting practice?

The lower production temperature, from which energy saving might be expected, is, however, largely offset by the high cost of the initial ingredients necessary for making the gel. At present organometallic precursors for some of the more exotic cations are not always available. and the initial formulation of the solution leading to a proper gel (without flocculation) can be a difficult task indeed.

The subsequent treatments of the gel, the drying–curing and sintering stages, are also, in practice, more complicated and time-consuming than direct melting and fining in classical glass practice. They are, furthermore, specific to a given composition and the process has to be "tailored" for each new glass requiring a complete preliminary study in each case.

It seems therefore that the sol–gel process can only be competitive in areas of advanced technology (high-tech area) and that neither window nor bottle glass will ever be made industrially in this way.

18.2.2 *Forming processes*

In the usual process, the resulting melt is immediately formed into the desired end products: sheet glass, hollow ware or fibers (Fig. 18.1). In sol–gel technology the various forming operations have to occur before or during the gelling stage: e.g. molding an object, forming a thin coating or spinning a fiber; the drying–curing and sintering stages simply consolidate the original shape produced at low temperature (Fig. 18.2).

It is possible to produce *bulk* pieces of glass if cracking of gels during drying is avoided. This problem of obtaining *monolithic* gels has been studied in recent years – hypercritical solvent evacuation and Drying Chemical Control Additives (DCCA) proved effective in this respect.

The nature of the processes during gelling and subsequent drying, which are of a diffusional nature, favors those configurations in which at least

Fig. 18.1. Classic ways of producing glass. After Ref. 511.

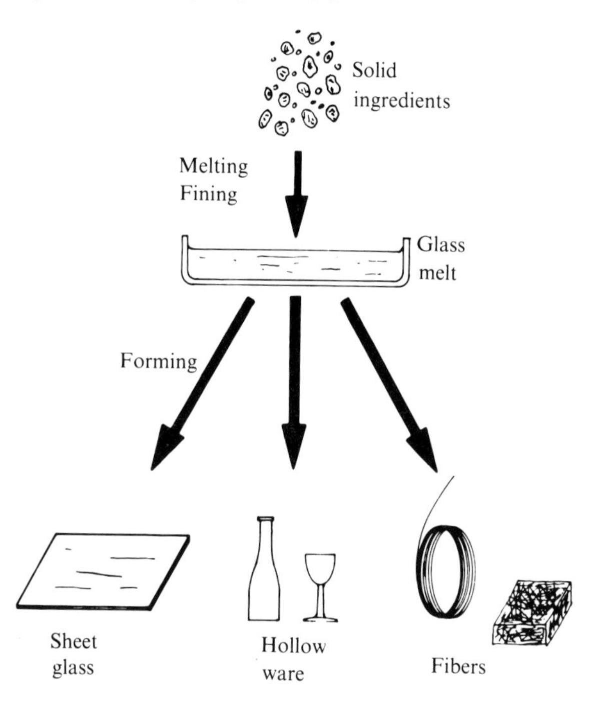

one of the dimensions is small: thin films, fibers and small particles (or shells) are current examples and it is significant that the first established industrial applications of the sol–gel route were precisely in the field of thin coatings.

18.3 Methods of gel formation

The main steps of the process are summarized in Fig. 18.2. The different variants which are developed depend on the way in which the initial gel is obtained. As SiO_2 is the essential ingredient of most of the glasses prepared by sol–gel methods, our description will be centered on SiO_2-based gels. There are three ways of obtaining them:

1. Destabilization of SiO_2 *sols* (e.g. Ludox®, either pure or containing other metal ions added in the form of aqueous solutions of salts.

2. Hydrolysis and polycondensation of *organometallic compounds* (alkoxides) dissolved in *alcohols* in the presence of a limited amount of water.

3. Redispersion of fine dry SiO_2 particles in a suitable medium by mechanical (shearing) action to form a sol which gels spontaneously.

Fig. 18.2. Sol–gel routes for glass formation. After Ref. 511.

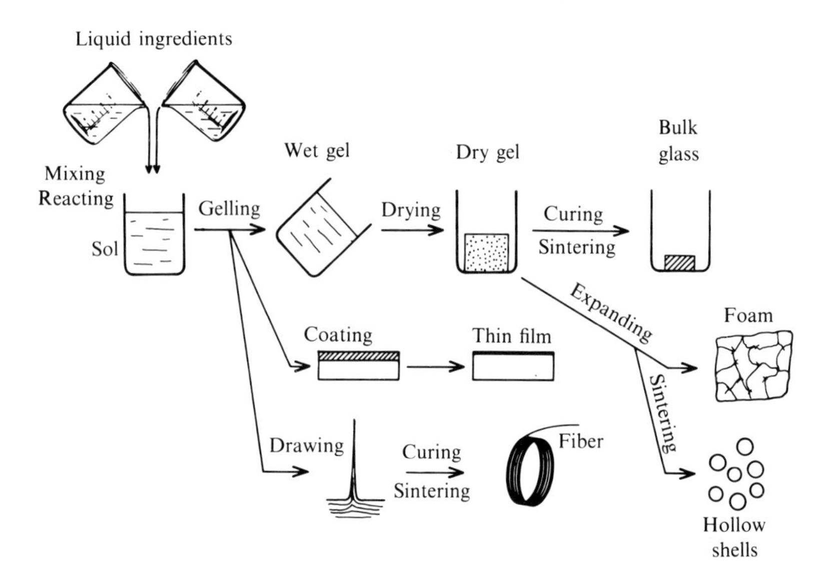

The excellent treatise of Iler[512] should be consulted for details on the SiO_2 sols and gels.

18.3.1 *Gel formation from sols*

Colloidal solutions of SiO_2 (SiO_2 *sols*) are mostly prepared either by chemical condensation methods, acidifying solutions of sodium silicates, potassium silicates, ammonium silicates or from hydrolysable products such as $SiCl_4$ or $Si(OR)_4$ where R is an alkyl group. The formation of silicic acid in aqueous solutions is followed by polymerization of monomers $Si(OH)_4$ when its concentration exceeds 100 ppm, the limiting solubility in water at 25 °C.

The polymerization reaction is based on the condensation of silanol groups with elimination of water:

$$-Si-OH + OH-Si- \rightarrow -Si-O-Si- + H_2O \quad (18.1)$$

Amorphous spheroidal groupings of about 1–2 nm are formed by a nucleation process similar to that which occurs in the formation of crystalline precipitates.

At low pH values, particle growth stops once the size of 2–4 nm is reached. Above pH = 7, particle growth continues at room temperature until particles of about 5–10 nm in diameter are formed, then it slows down. At higher temperatures particles are negatively charged and they repel each other, growth continues without aggregation, resulting in the formation of stable *sols*.

Commercial SiO_2 hydrosols (e.g. Ludox®, Nalcoag®, Nyalcol® and Snowtex®) are stable sols with 20–50 wt.% SiO_2. They are made up of dense SiO_2 particles with an average diameter of between 7 and 21 nm. The pH is between 9 and 11.

To obtain a *gel* from a stable sol the latter must be destabilized either by temperature increase or by the addition of an electrolyte. Increase of temperature reduces the amount of intermicellar liquid by evaporation and increases thermal agitation which induces collisions between particles and chain formation by condensation of surface hydroxyls (Fig. 18.3).

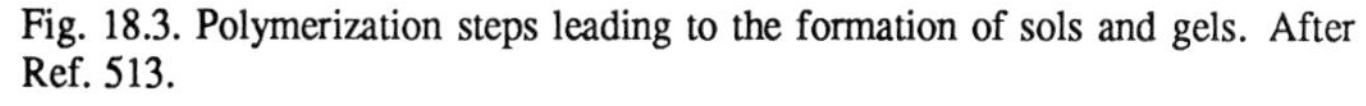
Fig. 18.3. Polymerization steps leading to the formation of sols and gels. After Ref. 513.

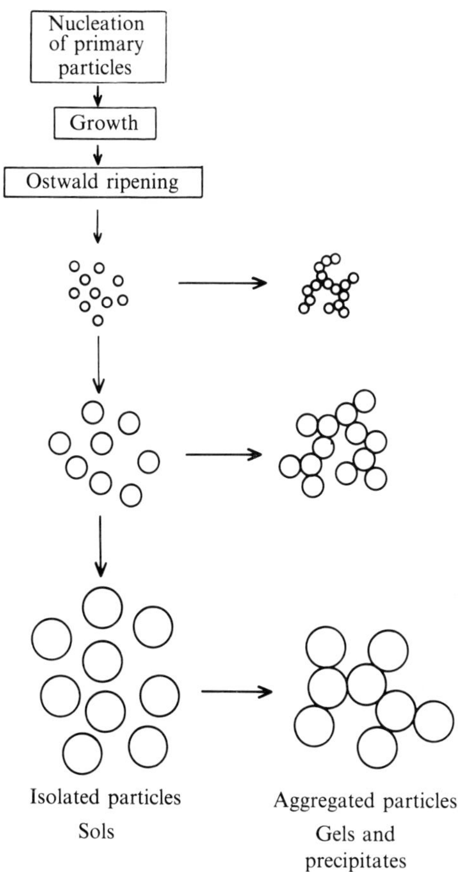

The sol–gel transition should be distinguished from a precipitation (or flocculation) mechanism in which separate aggregates are formed in contrast to the gelling where a continuous three-dimensional particle network invades the volume of the sol (Fig. 18.4).

Fig. 18.4. Difference between gel formation and precipitation from a sol. After Ref. 513.

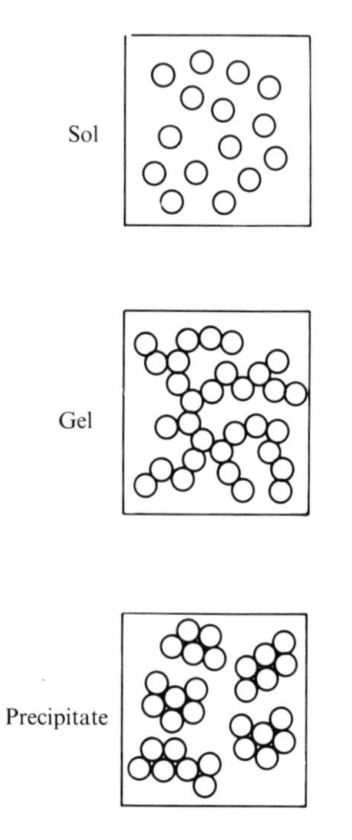

The pH of the sol may be modified by electrolyte addition in order to reduce the electric repulsion between the particles, (depending on the zeta potential). This is accomplished by adding an acid to lower pH to 5–6 to induce gel formation by aggregation. This conversion of sol into gel is progressive, the growing aggregates (microgel) gradually invading the whole volume originally occupied by the sol. When about half of the SiO_2 has entered the gel phase, a rapid increase in viscosity is noted (Fig. 18.5).

The mechanism of interparticle bonding leading to microgels and gels involves the attachment of two neighboring SiO_2 particles via the formation of Si–O–Si bonds (Equation (18.1)).

Fig. 18.5. Viscosity increase for SiO_2 sols close to gelation point, shown as a function of reduced gelation time T/T_{gel} for the different [TMOS]/[ethanol] volume ratios indicated (TMOS = tetramethoxysilane). After Ref. 514.

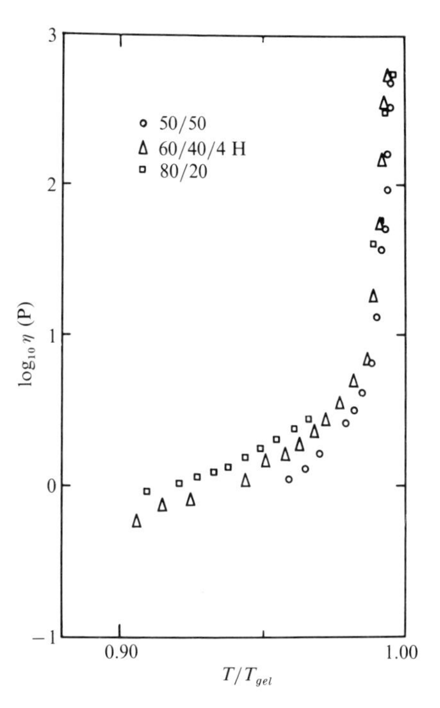

Colloidal particles will form gels only if there are no active forces which would promote coagulation into aggregates with a higher SiO_2 concentration than the original sol. Metal cations, especially the polyvalent ones, may cause precipitation rather than gelling: this is encountered with some multicomponent gels. In that case at least one gellifying constituent (generally SiO_2 sol) is required. Other constituents may be added in the form of soluble salts (nitrates, sulfates, etc.) or organometallic compounds.

By adjusting the temperature, concentration and especially the pH of the resulting sol, a homogeneous solution is obtained which may then be gelled in a controlled way in order to avoid precipitation. According to the composition, pH and temperature, the gelling time may vary from minutes to months.

18.3.2 *Gel formation from alkoxides*

Metal alcoholates, also called metal alkoxides $M(OR)_n$, where M is a metal (e.g. Si) and R an alkyl group (e.g. CH_3 or C_2H_5), react with water and undergo hydrolysis and polycondensation reactions which lead to the progressive formation of metal oxide. The overall reaction scheme consists globally of at least two steps:

$$M(OR)_n + H_2O \rightarrow M(OH)_n + nR(OH) \tag{18.2}$$

$$pM(OH)_n \rightarrow pMO_{n/2} + \frac{pn}{2} H_2O \tag{18.3}$$

The resulting oxide is produced in the form of extremely small particles ($\sim 2\,nm$) which may link to form a gel.

In reality the situation is more complex; reactions (18.2) and (18.3) proceed simultaneously and are generally incomplete. Hydrolysis may be achieved using a smaller quantity of water than that required by stoichiometry and a number of radicals R remain unreacted. Polycondensation is incomplete and the final product corresponds rather to the formula:

$$(MO)_x\,(OH)_y\,(OR)_z \tag{18.4}$$

In the case where several different compounds, e.g. $M(OR)_n$, $M'(OR)_n$, are reacted, a complexing step may precede reactions (18.2) and (18.3). In this way complex networks involving several different cations M, M′, may be produced e.g.

$$\text{–M–O–M}'\text{–O–M–} \tag{18.5}$$

The use of alkoxides of Si, B, Ti, Zr, etc. leads to the formation of complex gels which are composed of small particles and which produce the network of corresponding oxide glasses.

Table 18.1 gives a list of the organometallic compounds most frequently used in the synthesis of glasses by this method. As alcoholates and water are immiscible, the reagents are dissolved in alcohols, generally CH_3OH or C_2H_5OH. The use of a common solvent can, however, be avoided by subjecting the mixture alcoholate–water to the action of ultrasound. The "sonogels" obtained in this way are more dense due to the absence of a solvent and their structural properties are different from those of "classic" gels.[515,516]

The water necessary for hydrolysis can be taken from the atmosphere (as in the case of thin coatings) or added to the solution in a controlled amount. Other cations may also be introduced in the form of alcoholic or aqueous solutions of salts (nitrates, acetates, etc.). A carefully controlled amount of a catalyst, either an acid (HCl, HNO_3, CH_3CO_2OH) or a base

Table 18.1. *Alkoxides used in gel synthesis.*

M	$M(OR)_n$
Si	$Si(OCH_3)_4$ $Si(OC_2H_5)_4$
Al	$Al(O\text{–iso}\,C_3H_7)_3$ $Al(O\text{–sec}\,C_4H_9)_3$
Ti	$Ti(O\text{–}C_2H_5)_4$ $Ti(O\text{–iso}\,C_3H_7)_4$ $Ti(O\text{–}C_4H_9)_4$ $Ti(O\text{–}C_5H_7)_4$
B	$B(OCH_3)_3$
Ge	$Ge(O\text{–}C_2H_5)_4$
Zr	$Zr(O\text{–iso}\,C_3H_7)_4$ $Zr(O\text{–}C_4H_9)_4$
Y	$Y(O\text{–}C_2H_5)_3$
Ca	$Ca(O\text{–}C_2H_5)_2$

(NH_3, amines, etc.), is added. The gelling time depends on the pH, temperature, the amount of H_2O and the nature of the catalyst.

The structure of the gel depends very much on the nature of the catalyst used: acid catalysis leads to filamentous structures with a low degree of reticulation of the –M–O–M– chains, while base catalysis produces more compact spheroidal particles with a higher internal degree of reticulation.

18.3.3 *Redispersion methods*

Fine dry SiO_2 particles such as the commercial "fumed SiO_2" CAB-O-SIL®, or AEROSIL® obtained by flame oxidation of $SiCl_4$, can be mechanically redispersed in water using a shear blender. A sol is formed at pH = 2.7 which will gell in a few hours. SiO_2 particles form agglomerates which are linked by hydrogen bonds. SiO_2 dispersions in organic liquids such as chloroform or n-decanol were also prepared; they can be readily gelled by the action of amines or NH_3 vapor. Not only pure SiO_2 but also SiO_2–TiO_2 or SiO_2–GeO_2 particles, obtained by flame oxidation of mixtures of gaseous $SiCl_4$, $TiCl_4$, $GeCl_4$, were successfully gelled. Alternatively, $B(OH)_3$ may be added in aqueous solution to introduce the B_2O_3 component.

18.3.4 *Aging effects*

The freshly prepared "wet" gel consists of a network of particles holding an interstitial liquid – the solvent trapped during the gelling step: water in the case of *hydrogels*, mixtures of alcohols and water for the *alcogels*.

The interstitial liquid still contains unreacted sol particles which progressively attach themselves to the network. Furthermore, there is a transport of SiO_2 from convex to concave parts due to solubility effects based on Thomson's relation (cf. Section 6.4) which smooths the inequalities between linked particles and converts the chains into filaments (Fig. 18.6). Additional deposition of SiO_2 from solution may further "nourish" and thus stiffen the chains (Fig. 18.7). The *syneresis* effect then sets in whereby the network slowly contracts and tends to progressively expel the interstitial liquid in order to reduce the internal interface. All these effects bring about a progressive increase in the Young's modulus of the wet gel; with time (Fig. 18.8) the gel stiffens progressively and this occurs even if evaporation of the interstitial liquid is prevented.

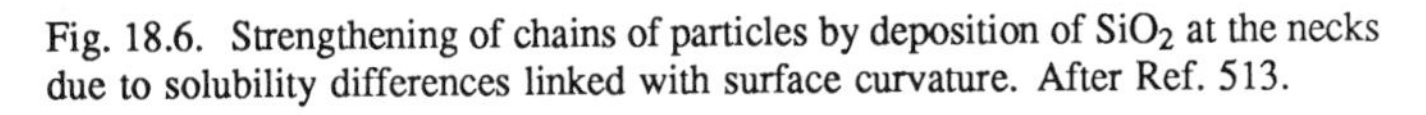

Fig. 18.6. Strengthening of chains of particles by deposition of SiO_2 at the necks due to solubility differences linked with surface curvature. After Ref. 513.

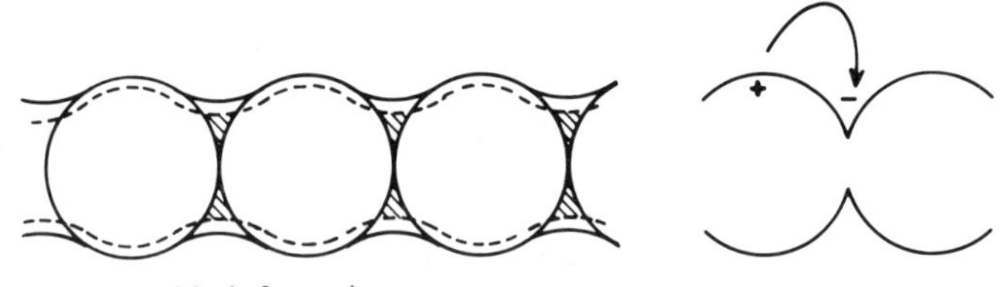

Freshly gelled wet SiO_2 gel is an extremely fragile material which behaves as a brittle solid with conchoidal fracture. Gels prepared from silicon alkoxides are purely elastic while those from colloidal solutions (Ludox) show partly viscoelastic or viscoplastic properties. Figure 18.9 shows the critical stress concentration factor K_{IC} of some SiO_2 gels and Fig. 18.10 the corresponding fracture surface energy.

Fig. 18.7. Transformation of particulate chains into fibrillar structures by secondary SiO_2 deposition. After Ref. 513.

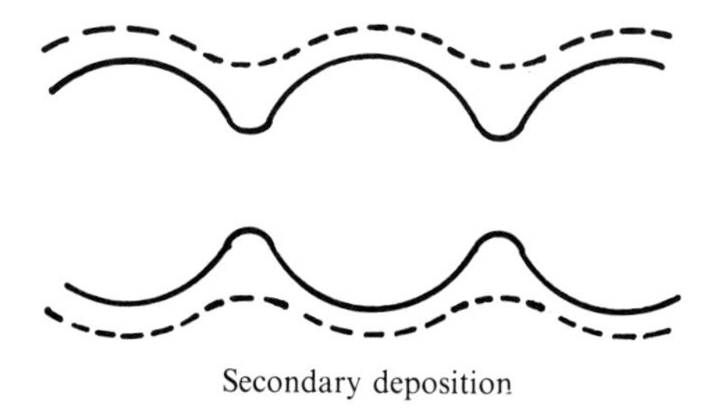

Fig. 18.8. Evolution of Young's modulus E corrected for Poisson's ratio ν for wet LUDOX gels as a function of time. After Ref. 517.

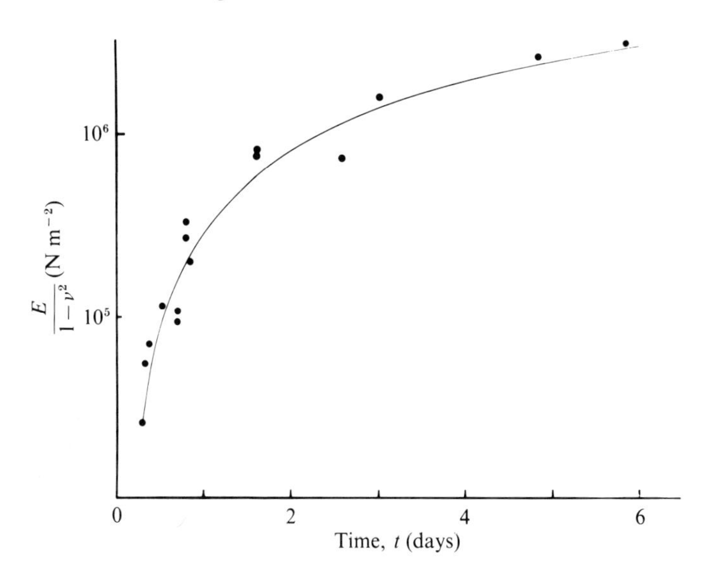

18.3.5 *Structural aspects*

The gelling process and the structure of the resulting wet gels have been the subject of numerous studies on various systems, the detailed description of which is outside the scope of the present chapter. More recent material can be found in Refs. 504, 506, 510. The use of SAXS combined with spectroscopic methods (IR, Raman, NMR, EXAFS) permitted the various aggregation theories to be tested and this led to structural models.

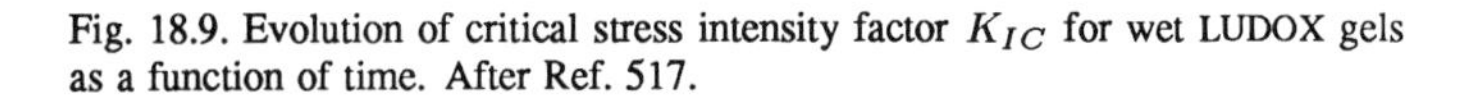

Fig. 18.9. Evolution of critical stress intensity factor K_{IC} for wet LUDOX gels as a function of time. After Ref. 517.

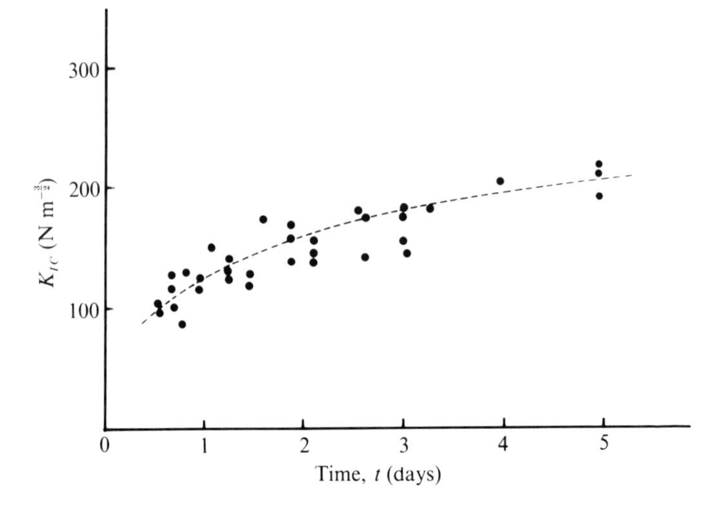

Fig. 18.10. Evolution of fracture surface energy for wet LUDOX gels as a function of time. After Ref. 517.

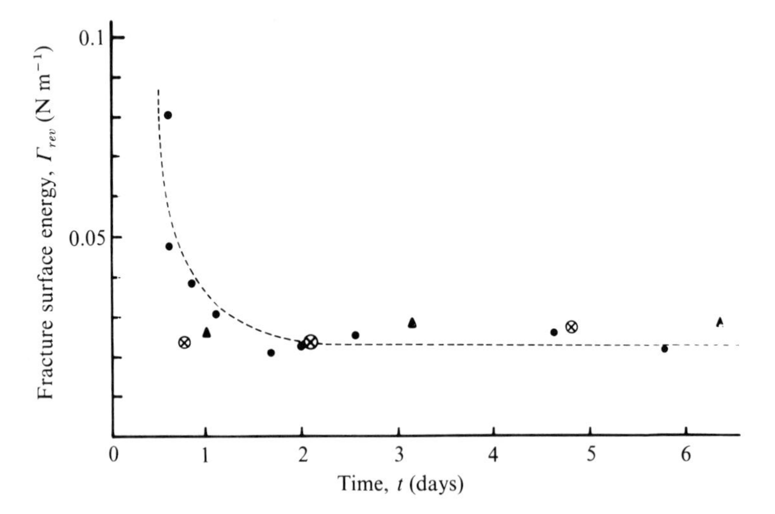

Gelling may be explained either by applying Flory's theory (familiar in polymer chemistry) or by percolation theories (more favored by physicists). The structure of growing aggregates and of the resulting gels may be described using *fractal* concepts.[518] A fractal cluster has a structure which becomes increasingly wispy as its dimensions increase. In particular its mass scales as r^D, where r is the radius of the cluster and D the fractal dimension which is smaller than 3. The fractal dimension D may be obtained from SAXS results using the scaling relations for the high-Q region. See Ref. 519 for a general review.

18.4 Drying

The drying stage is necessary to isolate the solid network of the gel from the accompanying interstitial liquid phase.

18.4.1 *Rôle of capillary forces*

Elimination of the liquid phase leads to dry gels, namely *xerogels*. When a "wet" gel is dried the following sequence of events is generally observed on a macroscopic scale:

- progressive shrinkage and hardening;
- stress development;
- fragmentation.

The chief difficulty is encountered when specimens of gel without any cracks are required to prepare bulk pieces of glass. The problems of producing *monolithic gels* have been the subject of intensive research.[513]

Cracking during the drying stage is the result of non-uniform shrinkage of the drying body as is well known in ceramic technology. The stresses arise not only from the local differences in expansion coefficient due to variable water content but, in the first place, from the action of *capillary forces* which become operative when the pores start to empty and a liquid–air interface is present in the form of menisci distributed in the pores of the drying gel.

The magnitude of these forces is given by Laplace's formula:

$$\Delta p = (2\gamma \cos\theta)/r$$

The pressure Δp is proportional to the specific surface energy γ at the liquid–air interface and inversely proportional to the pore radius r; θ is the contact angle at the liquid–solid–air boundary line. Considerable stresses may be generated in this way: $\Delta p = 7.3 \times 10^7\ \mathrm{N\ m^{-2}}$ for a pore radius $r = 2\,\mathrm{nm}$ filled with water, assuming perfect wetting (Fig. 18.11).

Differential stresses due to adjacent pores then induce breaking if the tensile strength is exceeded (Fig. 18.12). A detailed treatment of drying comprising the analysis of stress redistribution due to interstitial liquid transfers was proposed by Scherer.[520]

Fig. 18.11. Capillary pressure, Δp, as a function of pore radius, r. After Ref. 513.

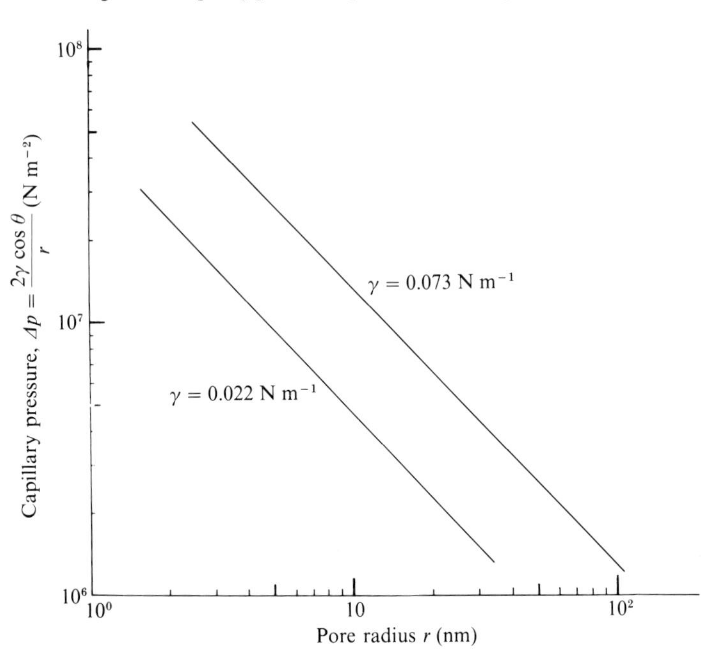

18.4.2 *Monolithic gels*

In practice, a very long drying time is necessary to preserve monolithic character. Drying times of hundreds of hours may be necessary to produce even small specimens (i.e. with a surface area of a few square centimeters). Much effort has been spent, therefore, to find more economical ways of drying gels which preserve their integrity.

All actions which tend to minimize the capillary stress and increase the mechanical resistance of the network should enhance the probability of keeping the gel monolithic. The following are possible:

- strengthening the gel by reinforcement (aging);
- enlarging the pores;
- tending towards monodispersity of the pores;
- reducing the surface tension of the liquid;
- making the surface hydrophobic;
- evacuating the solvent by freeze-drying;
- operating in hypercritical conditions where the liquid–vapor interface vanishes.

The last two methods represent the most efficient ways of eliminating the

Fig. 18.12. Formation of differential strains at the pores during drying: (*a*) before, and (*b*) after the onset of capillary forces. After Ref. 513.

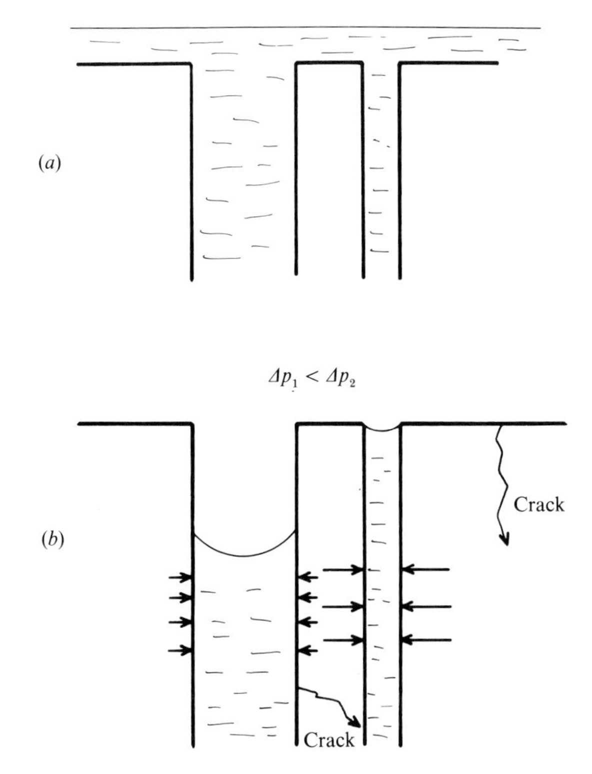

destructive action of the surface tension of the liquid by suppressing the liquid–vapor interface.

Hypercritical solvent evacuation consists of treating the gel in an autoclave in conditions which are *hypercritical* for the solvent. It is directly applicable to alcogels produced from alkoxides. Hydrogels cannot be directly treated in this way because SiO_2 becomes soluble under conditions hypercritical for water. This has led to the importance of the alkoxide method. On the other hand, redispersion methods (using a double redispersion treatment) give wet gels with large pores which facilitate solvent evacuation without cracking.

Figure 18.13 is a schematic representation of the process showing the equilibrium curve between the liquid and the gas phase of the solvent. In order to ensure the continuity of the liquid–gas transition, the path of the thermal treatment must not cross the equilibrium curve. To circumvent the critical point C a path such as *abde* may theoretically be used.

Fig. 18.13. Principle of drying by hypercritical solvent evacuation (see text).

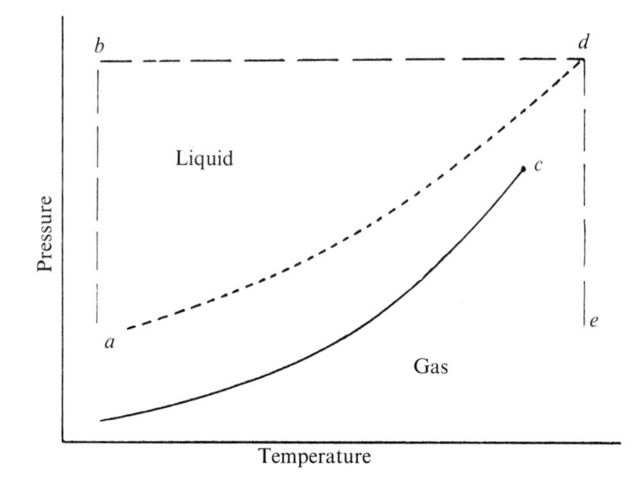

In practice this path is modified in the following way: the open container which contains the gel is placed inside an autoclave and, to obtain hypercritical conditions, a given quantity of solvent (e.g. CH_3OH) is added to the autoclave. This is closed and electrically heated. When the critical temperature of CH_3OH is exceeded, successive flushing with dry Ar eliminates the last trace of alcohol. The autoclave is then cooled down and the gel removed at ambient temperature (path *ade*). The gel, the pores of which are filled with air, is termed *aerogel.*

A number of experiments has shown that monolithism depends on many variables, namely:

- speed of heating;
- proportion of the additional solvent;
- concentrations of organometallic compounds and water of hydrolysis;
- geometry of the sample and its size;
- previous aging of the gel.

Optimizing these variables, monolithic samples can be obtained with 100 % certainty.[513]

The aerogels are hydrophobic due to partial surface esterification and contain an appreciable percentage of organic adsorbed radicals. The mechanical resistance of the aerogels is, however, sufficient to permit the elimination of the residual impurities by thermal treatment without loss of monolithic character. The purified aerogels can then be converted into a clear glass of excellent optical quality.

Aerogels can be made with an exceedingly high porosity close to 99 % with pore size of a few nanometers which makes them excellent thermal insulators. For SiO_2 aerogels thermal conductivity of the order of $0.01\ W\ m^{-1}\ K^{-1}$ has been demonstrated. Their refractive index may be close to 1 and they can be made translucent – hence their application in Cerenkov radiation detectors. (For applications see Ref. 521.)

Another method consists in adding a suitable additive to the sol before gelling which influences its subsequent behavior during drying. The most successful of these Drying Control Chemical Additives (DCCA) is formamide, the effect of which seems to be the tendency towards formation of gels with fine but much more uniformly distributed pores[522] which favor drying in monolithic form due to a lesser differential evaporation and more homogeneous stress distribution.

18.5 Curing–sintering

18.5.1 *Chemical effects*

The final structure of the dry gel will depend on the structure of the wet gel originally formed in solution; it is a contracted or distorted version of the latter. The constituent particles are coated with residual OH groups which are partly eliminated during the transition from a particulate texture towards a continuous solid; they may be detected and analyzed by conventional IR spectroscopic techniques.

To transform the particulate structure of a dried gel into continuous glass, the elementary particles must weld together resulting in progressive pore elimination. This is achieved by heating the gel in order to promote diffusion phenomena and viscous flow. During this heat-treatment the residual OH and OR groups will tend to be eliminated in the form of H_2O and ROH accompanied by an additional polymerization of the system:

$$-\mathrm{Si{-}OR} + \mathrm{OH{-}Si} \rightarrow -\mathrm{Si{-}O{-}Si} + \mathrm{ROH} \qquad (18.8)$$

The escape of residual products from *closed* pores may be a problem; the organic residues are finally carbonized at a higher temperature causing a coloration of the gel and leaving carbonaceous particles in the glass. It is therefore important to favor the escape of residues before closure of the pores, and oxidation treatments are often necessary to eliminate certain organic groups. For pure SiO_2 gels this oxidation treatment is carried out at 300–400 °C.

It is important to define the heating schedule in each particular case in order to eliminate the unwanted residues without impairing the monolithic character before the onset of the viscous flow phenomena.

Occluded OH and H_2O may provoke *bloating* on heating at high temperatures – even if the specimen remained monolithic up to this stage. Residual OH groups may be eliminated by chlorination treatments if very low OH levels are required in the final glass (e.g. for optical fiber applications). On the other hand, occluded water may be used in foaming processes, e.g. blowing gel particles into microballoons.

18.5.2 *Viscous flow sintering*

Densification is essentially a *sintering* process by which the pores of a dry gel are eliminated and the material progressively converted into clear bulk glass. After the elimination of residues, the driving force in this process is supplied by the *surface energy* of the porous gel. It tends to reduce the interface, thus eliminating the pores, the collapse being governed, in the case of glasses, by *Newtonian viscous flow*. Extra pressure, as in hot-pressing may be applied externally to speed up the process.

To study this transformation a simple model must be adopted to represent the texture of the porous solid. Two models are currently used in the case of gels:

1. The closed-pore model proposed by Mackenzie and Shuttleworth (MS) in their theory of sintering.[523]
2. The open-pore model devised by Scherer to study "lattice-like" less dense textures encountered in the sintering of "soots" in optical fiber technology.[524]

These two models which idealize the situation of Figs. 18.14(*a*) (*b*) are shown in Figs. 18.15(*a*) and (*b*), respectively.

(a) *Closed-pore model*

In this model it is assumed that the pores are identical spheres of initial radius r_i; their number n per unit volume of solid phase is supposed to remain constant during densification.

If the relative density $D = \rho/\rho_s$ is defined as the ratio of the apparent density ρ of the porous solid (gel) to the density ρ_s of the solid phase (glass), the relation between n and r_i is:

$$n\frac{4\pi}{3}r_i^3 = \frac{1-D}{D}$$

or:

$$r_i = \left(\frac{3}{4\pi}\right)^{1/3}\left(\frac{1-D}{D}\right)^{1/3}\frac{1}{n^{1/3}}$$

Fig. 18.14. Texture of gels (schematic): (*a*) dense agglomerate of particles with closed pores; (*b*) lattices of particles with open pores. After Ref. 525.

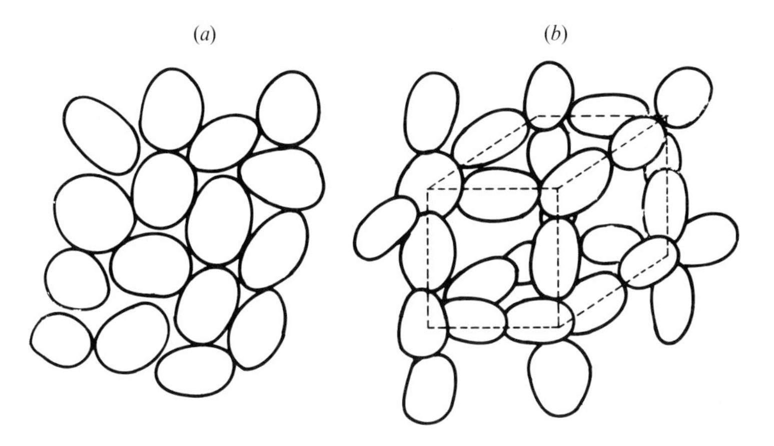

Fig. 18.15. Models of gel textures (schematic): (*a*) closed pores; (*b*) open pores. After Ref. 525.

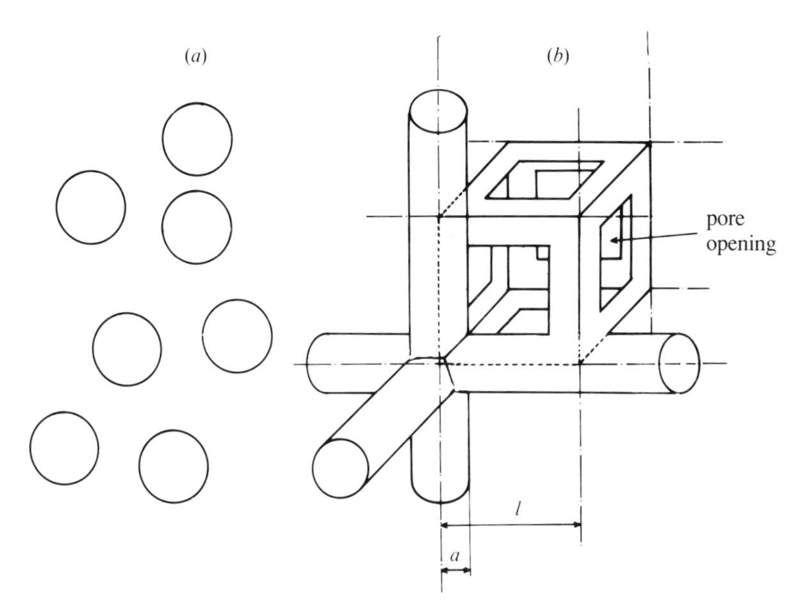

To evaluate $n^{1/3}$ it is most convenient to consider the total surface $S = 4\pi r_i^2 n$ of the pores per unit volume of solid phase:

$$S = 4\pi \left(\frac{3}{4\pi}\right)^{2/3} n^{1/3} \left(\frac{1-D}{D}\right)^{2/3}$$

Experimentally, using various techniques such as BET, SAXS, etc. a specific surface $S_{sp} = S/\rho_s$ is measured (expressed generally in $m^2\ g^{-1}$). $n^{1/3}$ may be considerable, as e.g. in the case of a SiO_2 gel for:

$$S_{sp} = 220\,m^2\ g^{-1}$$

$$D = 0.5$$

$$\rho_s = 2.20\,g\ cm^{-3}$$

$n^{1/3} \sim 10^6$ is obtained which corresponds to $r_i \sim 60$ Å.

In the case of the sintering of gels the solid phase is a glass which has a Newtonian viscosity η independent of the rate of strain and a surface energy γ. The kinetic equation of sintering is then:[523]

$$\frac{dD}{dt} = \frac{3}{2}\left(\frac{4\pi}{3}\right)^{1/3} \frac{\gamma n^{1/3}}{\eta}(1-D)^{2/3} D^{1/3}$$

which may be integrated to

$$\frac{\gamma n^{1/3} \Delta t}{\eta} = \frac{2}{3}\left(\frac{3}{4\pi}\right)^{1/3} \int_0^D \frac{dD}{(1-D)^{2/3} D^{1/3}}$$

where Δt is the time necessary to reach the reduced density D.

A *reduced time* t_r is defined by the relation:

$$t_r = \frac{\Delta t \gamma n^{1/3}}{\eta}$$

If an external pressure P is applied during sintering, γ has to be replaced by the expression:

$$\gamma + P r_i/2$$

or

$$\gamma\left[1 + b\left(\frac{1-D}{D}\right)^{1/3}\right]$$

Fig. 18.16. Kinetics of sintering for different values of parameter b (see text). The full curves represent results for the MS closed-pore model. The broken curve represents the Scherer's open-pore model for $b = 0$. After Ref. 525.

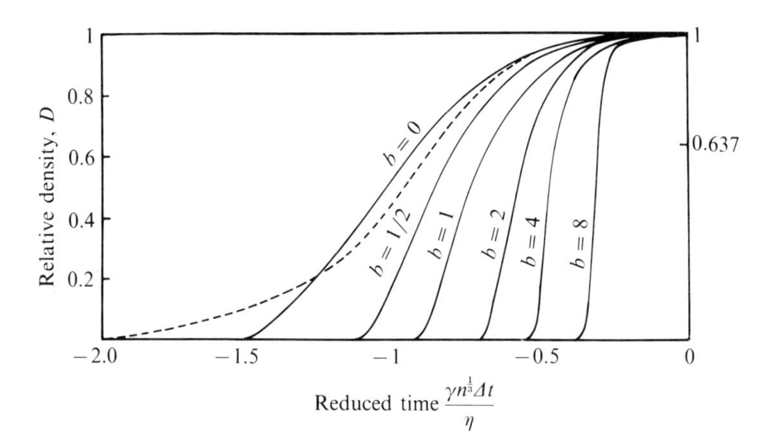

where:

$$b = \left(\frac{3}{4\pi}\right)^{1/3} \frac{P}{2\gamma n^{1/3}}$$

This slightly complicates the evaluation of the integral but in both cases it can be shown that sintering to $D = 1$ occurs in a finite time. Figure 18.16 shows the results of these calculations from Ref. 501. Sintering without applied pressure corresponds to $b = 0$. It can be seen in particular that the reduced sintering time from $D = 0.5$ to $D = 1$ is nearly equal to unity.

The effect of viscosity on the sintering time t may be evaluated for example for a reduced time $t_r = 1$ by the formula

$$\Delta t = \eta / \gamma n^{1/3}$$

Assuming that in the sintering range the viscosity η can be represented by a formula:

$$\log_{10} \eta = A + \frac{B}{T}$$

where A and B are constants and T is the absolute temperature; for specimens with specific surfaces S_1, and S_2, the same sintering time would correspond to temperatures T_1 and T_2 related by:

$$\log_{10}\left(\frac{S_1}{S_2}\right) = B\left(\frac{1}{T_1} - \frac{1}{T_2}\right)$$

The increase of the specific surface of the gel may thus substantially decrease the sintering temperature.

(b) *Open-pore model*

For gels with an open "lattice-like" texture which implies an open porosity, the MS model is no longer applicable. Scherer[524] proposed a model which consists of a regular cubic lattice of intersecting cylinders of radius a, the edge of the lattice being l.

This model idealizes the situation of Fig. 18.14(*b*): the two parameters a and l being related respectively to the radius of elementary particles and to a "spacing" of the pseudo-lattice. It is valid for the values of the ratio $x = a/l < 1/2$. For $x = 1/2$ the pores are closed and the MS model is again applicable.

It has been shown that the densification kinetics derived from this model follow rather closely the results of the MS model for values of $D > 0.3$ (Fig. 18.16). The model presents an advantage for small values of D; the (arbitrary) choice of a cubic lattice only slightly influences the final results. In this two-parameter model the relative density D is:[524]

$$D = 3\pi x^2 - 8\sqrt{2}x^3$$

and the specific surface:

$$S_{sp} = \frac{1}{\rho_s}\frac{6\pi - 24\sqrt{2}x}{lx\left(3\pi - 8\sqrt{2}x\right)}$$

Scherer's model thus constitutes a convenient extension towards the low D region where the significance of the MS model becomes doubtful.

In principle a knowledge of D and S_{sp} is sufficient to determine the parameters $x = a/l$ and l. In practice, additional determinations of the equivalent diameter of the interconnecting pores (situated on the sides of the cell) are made using Hg penetration porosimetry as well as electron microscope observations to determine a. The model may be further developed assuming pore size distributions. It has been successfully used for the evaluation of densification of SiO_2 "soots" and gels.

(c) *Murray–Rodgers–Williams approximation*

Murray *et al.*[526] have simplified the MS model in the case where $p \gg 2\gamma/r_i$ i.e. for $b \gg 1$ equation (18.1) becoming in this case:

$$\frac{\mathrm{d}D}{\mathrm{d}t} = \frac{3P}{4\eta}(1 - D)$$

which may be integrated into:

$$\ln(1-D) = -\frac{3P}{4\eta}t + \ln(1-D_i)$$

The consequence, however, is that the time t for complete densification is no longer finite, it is necessary to specify the final density D_f.

18.5.3 *Devitrification kinetics – use of TTT diagrams*

During densification the gel will, at the same time, tend to crystallize (devitrify). The successful conversion of gel into glass therefore depends on a competition between phenomena which lead to densification and those which promote crystallization. The appropriate thermal treatment may be calculated using the preceding equations.

TTT (time–temperature–transformation) diagrams are a convenient way of studying the devitrification versus compaction problem.[527] The TTT diagrams show the time t_y to reach a given crystallized fraction y as a function of the temperature T. Treating y as a parameter, a set of C_y curves is obtained which represents the kinetic behavior of the system. In particular if y_0 corresponds to the smallest crystallized fraction detectable by analytical techniques, the curve C_{y_0} represents a frontier which must not be crossed during a thermal treatment schedule if crystallization is to be avoided (generally $y_0 = 10^{-6}$ is adopted) (cf. Section 3.2.7(b)).

The relative positions of the thermal treatment path during densification and of the C_{y_0} curve of the gel determine the possibility of obtaining glassy or crystallized materials at the end of the compaction program. For example (Fig. 18.17), if for a gel corresponding to C_1 there is no danger of devitrification using path (*a*), this is no longer true for the curve C_2 and the same path would lead to crystallized material. The solution would then be either to shorten the sintering time, e.g. by applying a suitable pressure P (path (*b*)) or to increase the temperature for a short time using the technique of "flash-pressing" (path (*c*)).

In the case of gels the positions of the curves C strongly depend on the purity of the material and, most of all, on the water content which influences the viscosity η as well as the surface tension γ of the material.

It has been shown[527] that the nucleation–growth phenomena which lead to devitrification depend essentially on the ratio T/η. A method has been proposed to determine the new positions of C_y curves when the viscosity η of the material is modified. This "viscosity–time equivalence method" has been applied in particular to the SiO_2 gels 1 and 2 prepared respectively from LUDOX and alkoxides, the viscosity graphs of which are given in Fig. 18.18.

Fig. 18.17. TTT diagrams and thermal paths for compaction. After Ref. 527.

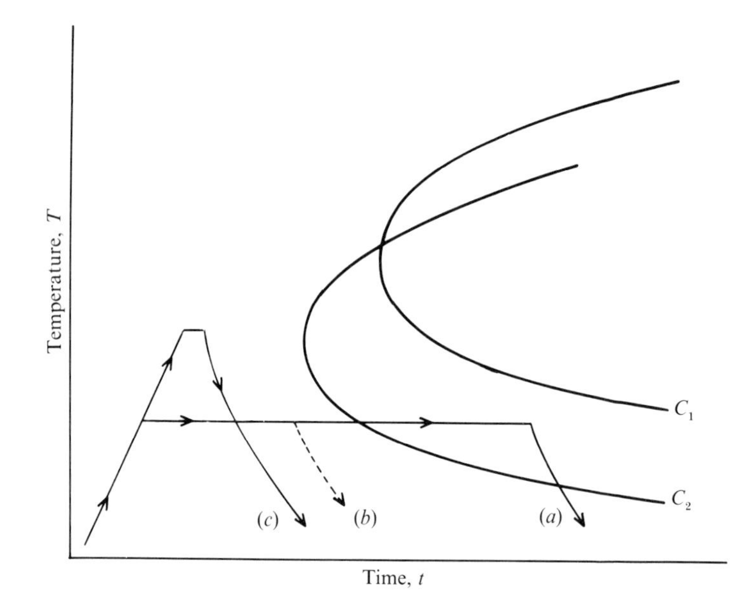

Figure 18.19 shows the TTT diagrams of vitreous SiO_2 and of SiO_2 gels 1 and 2. The application of the method leads to curves C_1' and C_2' which should be correct if the devitrification behavior were influenced by variations of viscosity only.

It can be seen that these curves are very nearly on the same temperature level as the curves C_1 and C_2 derived from experimental results but the latter are considerably shifted to the left.

As the modification of the nucleation barrier would be totally inadequate to explain this considerable decrease in the time for devitrification (by about 6 orders of magnitude) it seems that a heterogeneous nucleation mechanism linked with the specific surface of the gels may be responsible for the effect.

18.6 Potential of the sol–gel approach

In the laboratory, pure SiO_2 glasses, as well as those combining SiO_2 with other oxides, e.g. B_2O_3, TiO_2, GeO_2, P_2O_5, ZrO_2, etc. have been successfully prepared. Glasses containing alkali metals and alkaline earth oxides sometimes prove more difficult to obtain because of the tendency of gels to devitrify during the sintering stages.

The combination of organic–inorganic systems is another important ap-

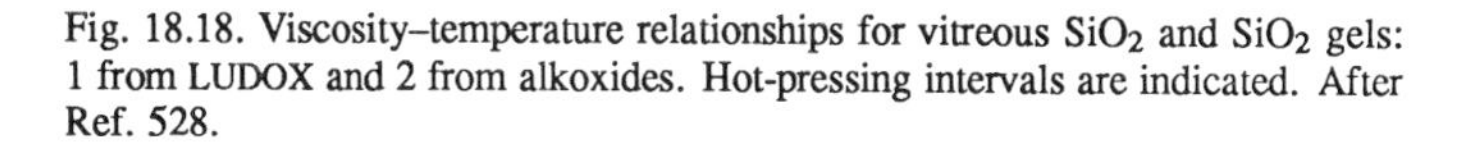
Fig. 18.18. Viscosity–temperature relationships for vitreous SiO_2 and SiO_2 gels: 1 from LUDOX and 2 from alkoxides. Hot-pressing intervals are indicated. After Ref. 528.

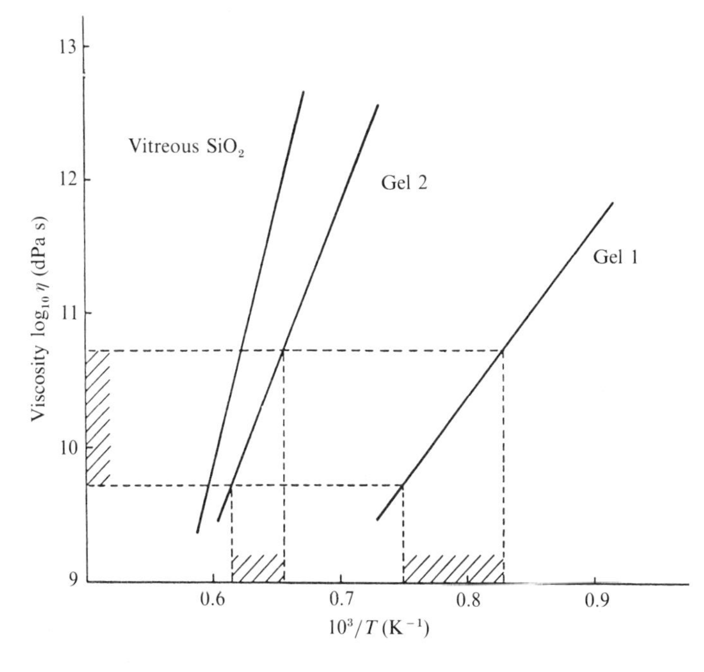

proach which led to the formation of intermediate materials, e.g. Organically Modified Silicates (Ormosils).[529]

Attempts to apply the sol–gel technique to chalcogenide glasses are indicated where the inherent heterogeneity of melt-obtained glasses could be improved or new systems produced. Investigations have also begun on halide gels and thus new gel systems are likely to be produced in the near future. In industrial practice, attempts at producing bulk gel-made SiO_2 optical glass have been reported. Advanced glasses for optical fiber preforms (SiO_2 doped by GeO_2, P_2O_5 or B_2O_5) have been successfully made but industrial applications have not yet followed. At the present time, the only recognized industrial application is the production of thin glass coatings using the alkoxide method to modify the spectral transmission of flat glass for architectural applications. Spreading of dilute solutions by dip-coating or spinning techniques followed by hydrolysis of alkoxides by atmospheric moisture and baking to consolidate the thin glass film thus formed, permits successive layers of $\sim$ 100 nm thickness to be applied without cracks. Attempts to produce glass fibers by direct spinning of

Fig. 18.19. TTT diagrams for vitreous SiO_2 and SiO_2 gels: 1 from LUDOX and 2 from alkoxides. The curves correspond to the crystallized fraction $y = 10^{-6}$. The broken curves represent the theoretical curves for gels 1 and 2 deduced from the curve of vitreous SiO_2 to be expected if the viscosity variations also were responsible for the differences observed in the devitrification behavior. After Ref. 527.

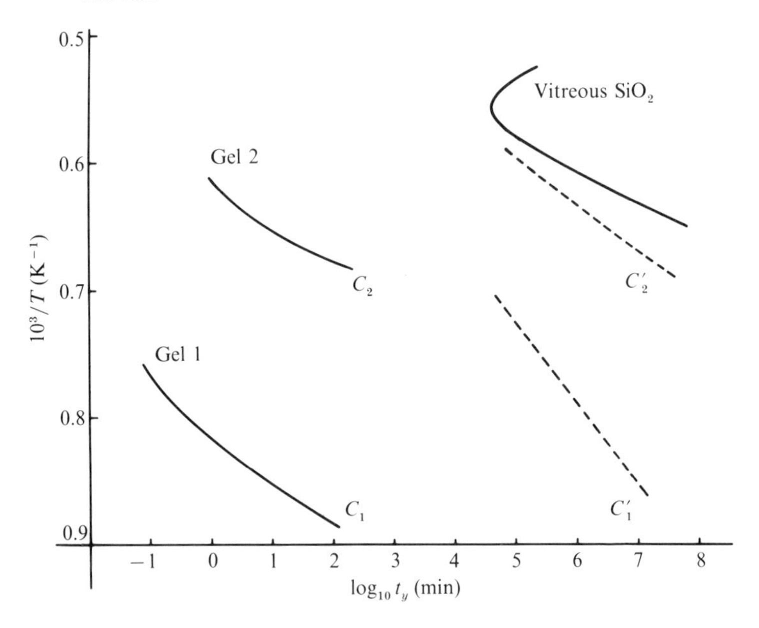

SiO_2–TiO_2 and SiO_2–ZrO_2 gel solutions have also been reported.

Hollow glass spheres used either as fillers for paints or in more sophisticated applications as targets for controlled fusion atomic experiments have been produced by expansion and flash vitrification of gel particles falling freely in a heated zone. In trying to ascertain the future trends, the original character of the sol–gel approach should be kept in mind.

What really differentiates the sol–gel route from the classic igneous route is essentially the fact that an inorganic gel is a *two-phase* system where each of the two phases can be influenced separately by the preparation methods.

In the classical way of obtaining glasses, the igneous melt is produced by progressive digestion of different crystalline components and the resulting melt is a collection of polymerized anions and fractions of chains with interdispersed accompanying cations. For a given melt the distribution of these entities depends essentially on the equilibrium conditions in the melt, i.e. on the temperature and atmosphere of the furnace which controls the oxidation–reduction changes. This high-temperature situation is preserved during quench and the polyanionic distribution reflects the equilibrium conditions in the melt. Apart from systems which undergo phase separation (either in the liquid or sub-liquidus), the quenched melt is a *one-phase* system with frozen-in local compositional fluctuations. The equilibrium conditions cannot be influenced to a great extent except by initial compositional changes.

On the other hand, if we consider the processes leading to the formation of a gel, the various steps by which the disordered network is built-up are essentially chemical polymerization, polycondensation, cross-linking, etc. which are familiar in polymer chemistry practice. In this way elementary particles, filaments, etc. are produced which then progressively link-up to produce first a colloidal solution – a sol – and then, when these unite into a reticulated network spanning the volume available – a gel, which is essentially a system consisting of *two* phases: a solid *backbone* immersed in an *interstitial liquid*, the two phases being intimately mixed on a very fine scale. This opens up extremely interesting possibilities as it has been shown that the two phases are largely independent and it is thus possible to subject the system to various manipulations.

18.6.1 *Influencing the backbone*

The fact that the backbone is not merely the result of the coupling of fragments present in the melt in thermal equilibrium but is formed by chemical polymerization opens up additional possibilities of sequencing, block copolymerization, etc. The resulting final *inorganic* backbone may benefit from the full power of methods of *organic* chemistry. Starting with several precursors, it is not only possible to manipulate the composition but the structure of the elementary particles may also be influenced. Taking e.g. SiO_2-based gels, the elementary particles of the SiO_2 sol may be produced *in situ* using silicon-alkoxides as sources of SiO_2 which, according to the catalyst used, will produce SiO_2 particles in the 1–10 nm range. Alternatively, larger SiO_2 particles (30–100 nm) can be introduced using SiO_2 sols (e.g. LUDOX). Still larger particles (0.1–1 μm) may be grown separately or produced by other methods (e.g. flame deposition) and subsequently gelled.

All these methods may be used concurrently using the *sol-mixing* approach. Intermediate organic–inorganic backbones which lead to products benefitting from both classes of materials (e.g. Ormosils) are also possible.[529]

18.6.2 *Modifying the interstitial liquid*

The nature and distribution of the second phase in the gel, i.e.

of the interstitial liquid, can also be influenced (Fig. 18.20). In aquagels this will be mainly a water solution of salts and in alcogels a complex mixture of water, alcohol and residual organic and inorganic compounds. The space occupied by the liquid will range from that essentially between the elementary filaments (meso- and macro-pores) to more individualized interparticle voids.

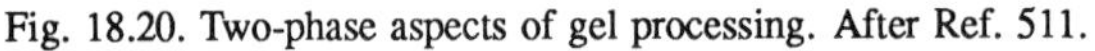

Fig. 18.20. Two-phase aspects of gel processing. After Ref. 511.

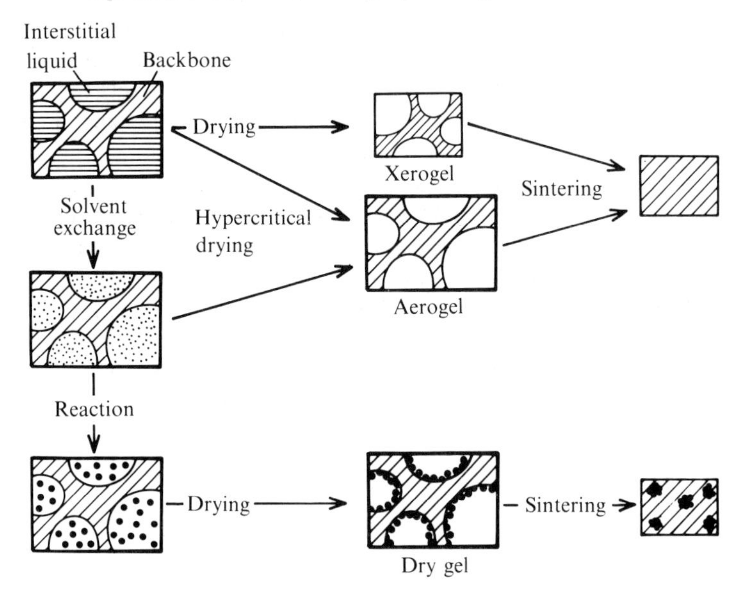

It is possible to change the nature of this second phase by various operations. In the drying process the liquid is simply replaced by a gas–solvent vapor and air leading to aerogels.

Using *exchange* processes, one liquid may be substituted for another: e.g. water may be replaced by alcohol to enable hypercritical treatment of aquagels. Foreign reactants may be diffused into gels *via* the interstitial liquid whereby chemical reactions, precipitation, etc. may be produced within the second phase or also within the backbone if this is done during the gel formation. Infiltration of foreign substances into dry gels permits the air voids (pores) to be filled, e.g. by colored or index matching substances. In this way a whole range of glass composites on a very fine scale, depending on the backbone configuration, may be obtained; the scale range may be as low as a few nanometers hence the name of *nanocomposites* which was proposed for this class of materials.[530] Glass may not be the ultimate goal of the process but only a secondary precursor for a *glass-ceramic* with improved grain size or locally modified properties.

Even if in some cases glass formation cannot be achieved by the gel route, the process leads to a micro-crystalline material, the grain size of which can be as low as 10–20 Å, i.e. even smaller than that of the best glass-ceramics. These "precursor" materials can be converted by sintering into ceramics of extra-fine grain size with many properties improved, especially mechanical resistance and toughness. The powder synthesis of ceramics from metal–organic precursors has recently attracted much attention.[502–6] The extremely reactive powders obtained in this way readily sinter into ceramic monolithic bodies if the primary agglomeration of particles and polydispersity effects can be minimized.

The method is applicable not only to oxides but also to carbides, nitrides, oxynitrides, etc. The grain size of classic ceramic being of the order of 10 μm or more and that of glass-ceramic obtained by controlled crystallization of glasses of the order 0.1–1 μm, the gel process leads to a new class of materials with a grain size 100 times smaller: "ultraceramics", or *ceramics of the third generation.*

This shows the great potential of the methods described. "Soft minerallurgy" is still a relatively new science and many unexpected developments are likely to occur in this field.

References

GENERAL REFERENCES

1. Tammann, G., *Der Glaszustand*, Voss, Leipzig, (1933).
2. Stanworth, J. E., *Physical Properties of Glass*, Clarendon Press, Oxford, (1950).
3. Morey, G. W., *The Properties of Glass*, (2nd ed.), Reinhold NY, (1954).
4. Mackenzie, J. D., *Modern Aspects of the Vitreous State*, Butterworth, London, Vol. I (1960), Vol. II (1962), Vol. III (1964).
5. Weyl, M. A. and Marboe, E. C., *The Constitution of Glasses: A Dynamic Interpretation*, Wiley, NY, Vol. I (1962), Vol. II (1965).
6. Eitel, W., *Silicate Science*, Academic Press, NY, Vol. 2, (1965).
7. Rawson, H., *Inorganic Glass-Forming Systems*, Acad. Press, NY, (1967).
8. Scholze, H., *Glas*, Vieweg, Braunsweig (1965), trad. fr. Inst. du Verre (1969) (2nd ed.) Springer, Berlin, (1977).
9. Jones, G. O., *Glass*, (2nd ed.) Chapman and Hall, London, (1971).
10. Mott, N. F. and Davies, E. A., *Electronic Processses in Non-Crystalline Materials*, Clarendon, Oxford, (1971).
11. Vogel, W., *Structure and Crystallization of Glasses*, Pergamon, Oxford, (1971).
12. Pye, L. D., Stevens, H. J. and Lacourse, W. D. *Introduction to Glass Science*, Plenum, NY, (1972).
13. Doremus, R. H., *Glass Science*, Wiley, NY, (1973).
14. Holloway, D. G., *Physical Properties of Glass*, Wykeham, London, (1973).
15. Wong, J. and Angell, C. A., *Glass Structure by Spectroscopy*, Dekker, NY, (1976).
16. Baltă, P. and Baltă, E., *Introduction to the Physical Chemistry of the Vitreous State*, Abacus Press, Turnbridge Wells, Kent, (1976).
17. Babcock, C. L., *Silicate Glass Technology Methods*, Wiley, NY, (1977).
18. Tooley, F. V., editor, *Handbook of Glass Manufacture*, Vol. I and II, Ogden, NY, 2nd ed., (1961).
19. Bartenev, G. M., *The Structure and Mechanical Properties of Inorganic Glasses*, Wolters-Nordhoff, Groningen, (1970).
20. Shand, E. B., *Glass Engineering Handbook*, McGraw, NY, (1958).
21. Douglas, R. W. and Frank, S., *A History of Glassmaking*, Foulis Co., London, (1972).
22. Zallen, R., *The Physics of Amorphous Solids*, Wiley, NY, (1983).
23. Uhlmann, D. R. and Keidl, N. J., editors, *Glass Science and Technology*, Vol. I, *Glass-forming Systems* Academic Press, NY, (1983).
24. Uhlmann, D. R. and Kreidl, N. J. editors, *Glass Science and Technology*, Vol. 5, *Elasticity and Strength of Glasses*, Academic Press, NY, (1980).
25. Uhlmann, D. R. and Kreidl, N. J., editors, *Glass Science and Technology*, Vol. 2, *Processing*, I, Academic Press, NY, (1984).
26. Tomozawa, M. and Doremus, R. H., editors, *Treatise on Materials Science and Technology*, Vol. 12 (1977), Vol. 22 (1982), Academic Press, NY.
27. Adler, D., Fritsche, H., Ovshinsky, S. R., editors, *Physics of Disordered Materials*, Plenum, NY, (1985).

COLLOQUIA, ETC.

28. Frechette, V. D., editor, *Non Crystalline Solids*, Wiley, NY, (1960).

29. Prins, J. A., editor, *Physics of Non-Crystalline-Solids*, North Holland, Amsterdam, (1965).
30. Douglas, R. W. and Ellis, B., editors, *Amorphous Materials*, Wiley, NY, (1972).
31. Frischat, G. H., editor, *Non-Crystalline Solids*, Trans. Tech. Publ., Aedermannsdorf, (1977).
32. Porai-Koshits, E. A., editor, *The Structure of Glass*, Vol. 5 (1965), Vol. 6 (1966), Vol. 7 (1966). Transl. by Consultants Bureau, Academic Press, NY.
33. Gaskell, P. D., editor, *The Structure of Non-Crystalline Materials*, Taylor and Francis, London, (1977).
34. Pantelides, S. T., editor, *The Physics of SiO_2 and Its Interfaces*, Pergamon, NY, (1978).
35. Mackenzie, J. D. and Varner, J. R., editors, *Frontiers of Glass Science*, Proc. Int. Conf. on Frontiers of Glass Science, Los Angeles, 16-18 July (1980), *J. Non-Cryst. Sol.*, **42**, 1-666, (1980).
36. Tomozawa, M., Levy, R. A., MacCrone, R. K.,and Doremus, R. H., editors, *Electrical, Magnetic and Optical Properties of Glasses*, Proc. 5th Intern. Conf. on Glass Science, *J. Non-Cryst. Sol.*, **40**, 1-640, (1980).
37. Fuxi, G. and Mackenzie, J. D., editors, *Beijing International Symposium on Glass*, Proc. 1st Int. Symp. on Glass, Beijing, 25-28 Aug. 1981, *J. Non-Cryst. Sol.*, **52**, 1-630, (1982).
38. Zarzycki, J., editor, *Physics of Non-Crystalline Solids*, Proc. 5th Int. Conf. Montpellier, 5-9 July 1982, J. de Phys. C-9, **43**, 699 pp., (1982).
39. Adler, D. and Bicerano, J., editors, *The Theory of the Structures of Non-Crystalline Solids*, Proc. Intern. Conf. on the Theory of the Structures of Non-Crystalline Solids, Bloomfield Hills, (Mich) USA, 3-6 June 1985, *J. Non-Cryst. Sol.*, **75**, 1-516, (1985).
40. Bayoumi, O. El and Uhlmann, D. R., editors, *Glass Science and Technology, Problems and Prospects for 2004*, Conf., Kreidl, Vienna, 1-5 July 1984, *J. Non-Cryst. Sol.*, **73**, 1-706, (1985).
41. Wright, A. F. and Dupuy, J., editors, *Glass, Current Issues*, (NATO Advanced Study Institute 2-13/4/1984 Tenerife, Spain), Martinus Nijholf, Publ. Dordrecht, 718 pp., (1985).
42. Adler, D., Fritzsche, H. and Ovshinsky, S. R., editors, *Physics of Disordered Materials*, Plenum, NY, 850 pp., (1985).
43. Fuxi, G. and Bray, P. J., editors, *International Symposium on Glass*, Beijing 3-7 Sept., (1984), *J. Non-Cryst. Sol.*, **80**, 712 pp., (1986).
44. Baró, M. D. and Clavaguera, N., editors, *Current Topics on Non-Crystallline Solids*, (Proc. 1st Int. Workshop on Non-Crystalline Solids, May 23-30, 1986, St. Feliú de Guixols, Spain), World Scientific, Singapore, 457 pp., (1986).
45. LaCourse, W. C., Pye, L. D., Varshneya, A. K., Shelby, J. E. and Stevens, H. J., editors, *The Physics and Chemistry of Glass and Glassmaking*, *J. Non-Cryst. Sol.*, **84**, 490 pp., (1986).

CURRENT REFERENCES

46. Topol, L. E. and Happe R. A., *J. Non-Cryst. Solids*, **15**, 116, (1974).
47. Sarjeant, P. T. and Roy R., in *Reactivity of Solids*, Mitchell, J. W., editor), pp. 725-33, Wiley, NY, (1969).
48. Duwez, Pol., Willems, R. H. and Klement, W., *J. Appl. Phys.*, **31**, 1136, (1960).
49. Pietrokowsky, P., *Rev. Sci. Instr.*, **34**, 445, (1963).
50. Sarjeant, P. T. and Roy, R., *J. Am. Cer. Soc.*, **50**, 500, (1967).
51. Suzuki, T. and Anthony, A., *Mat. Res. Bull.*, **9**, 745, (1974).
52. Coutures, J., Sibieude, F., Rouanet, A., Foex, M., Revcolevschi, A. and Collongues, R., *Rev. Int. Hautes Temp. and Refr.*, **11**, 263, (1974).
53. Mushakanana, S. C. and Vedam, K., *Surface Sci.*, **96**, 319, (1980).

54. Dislich, H., *Glastechn. Ber.*, **44**, 1, (1971).
55. McCarthy, G. J. and Roy, R., *J. Am. Cer. Soc.*, **54**, 639, (1971).
56. Mukherjee, S. P., Zarzycki, J. and Traverse, J. P., *J. Mater. Sci.*, **11**, 341, (1976).
57. Decottignies, M., Phalippou, J. and Zarzycki, J., *C. R. Acad. Sci.*, **285 C**, 265, (1977).
58. Weeks, R. A., Kinser, D. L., Kordas, G., editors, *1st Intern. Conf. on Effects of Modes of Formation on the Structure of Glass*, Nashville, July 9-12, 1984, *J. Non-Cryst. Sol.*, **71**, 1-456, (1985).
59. Weeks, R. A. and Kinser, D. L., editors, *2nd Intern. Conf. on Effects of Modes of Formation on the Structure of Glass*, Nashville, June 8-11, 1987, *Diffusion and Defect Data*, **53-4**, 9-20 (1989).
60. Holland, L., *Vacuum Deposition of Thin Films*, Wiley, NY, (1956).
61. Platakis, N. S. and Gatos, H. C., *J. Electrochem. Soc.*, **123**, 1409, (1976).
62. Anderson, G. S., Mayer, W. N. and Wehner, G. K., *J. Appl. Phys.*, **33**, 2991, (1962).
63. Davidse, P. D. and Maissel, L. I., *J. Appl. Phys.*, **37**, 574, (1966).
64. Wada, Y. and Ashikawa, M., *Japan J. Appl. Phys.*, **15**, 1725, (1976).
65. Kern, W., *Solid State Technol.*, **18**, 25, (1975).
66. Kern, W., Schnable, G. L. and Fischer, A. W., *RCA Review*, **37**, 3, (1976).
67. Kern, W. and Rosler, R. S., *J. Vac. Sci. Technol.*, **14**, 1082, (1977).
68. Secrist, D. R., and Mackenzie, J. D., *Bull. Am. Cer. Soc.*, **45**, 784, (1966).
69. Scherer, G. W., *J. Am. Cer. Soc.*, **60**, 236, (1977).
70. Dalton, R. H. and Nordberg, M. E., *US Patent*, 2,239,551, (1941).
71. Primak W., *J. Phys. Chem. Soc.* **13**, 279, (1960).
72. DeCarli, P. S. and Jamieson, J. C., *J. Chem. Phys.*, **31**, 1675, (1959).
73. Stoffler, D., *Fortschr. Mineral*, **51**, 256, (1974).
74. Masumoto, T., p. 73-86 in Ref. 44.
75. Scherer G. W. and Schultz P. C., p. 49-103 in Ref. 23.
76. Makino, Y., *The Applications of Non-crystalline Materials in Japan*, p. 87-100 in Ref. 44.
77. Ubbelohde, A. R., *Melting and Crystal Structure*, Clarendon, Oxford, (1965).
78. Scholze, H., in *Reactivity of Solids*, Anderson J. S., Roberts, M. W., and Stone, F. S., editors, p. 160, Chapman and Hall, London, (1972).
79. Winter, A., *J. Am. Cer. Soc.*, **21**, 259, (1938).
80. Simon, F. E., and Lange, F., *Z. Physik*, **38**, 227, (1926).
81. Kauzmann, W., *Chem. Rev.*, **43**, 219, (1948).
82. Turnbull, D. and Cohen, M. H., *J. Chem. Phys.*, **34**, 120, (1961).
83. Fox, T. G. and Flory, P. J., *J. Appl. Phys.*, **21**, 581, (1950), *J. Phys. Chem.*, **55**, 221, (1951), *J. Polymer Sci.*, **14**, 315, (1954).
84. Williams, M. F., Landel, R. F. and Ferry, J. D., *J. Am. Chem. Soc.*, **77**, 3701, (1955).
85. Doolittle, A. K., *J. Appl. Phys.*, **22**, 1471, (1951).
86. Cohen, M. H. and Turnbull, D., *J. Chem. Phys.*, **31**, 1164, (1959).
87. Fowler, R. and Guggenheim, E. A., *Statistical Thermodynamics*, p. 340, Cambridge Univ. Press, NY, (1952).
88. Gibbs, J. H. and Di Marzio, E. A., *J. Chem. Phys.*, **28**, 373, (1958).
89. Di Marzio, E. A. and Gibbs, J. H., *J. Chem. Phys.*, **28**, 807, (1958).
90. Adam, G. and Gibbs, J. H., *J. Chem. Phys.*, **43**, 139, (1965).
91. Bestul, A. B. and Chang, S. S., *J. Chem. Phys.*, **40**, 731, (1964).
92. Angell, C. A., *J. Phys. Chem.*, **70**, 2793, (1966).
93. Cohen, M. G. and Grest, G. S., *Phys. Rev.*, **B 20**, 1077, (1979).
94. Tool, A., *J. Am. Cer. Soc.*, **29**, 240, (1946).
95. Scherer, G. W., *Relaxation in Glass and Composites*, Wiley, NY, 331 pp., (1986).

96. Davies, R. O. and Jones, G. O., *Proc. Roy. Soc.*, **A 217**, 26 (1953), *Acta Physica*, **2**, 370, (1953).
97. Prigogine, I. and Defay R., *Chemical Thermodynamics*, Longman, London, (1954).
98. Goldstein, M., in Ref. 4, Vol. III, p. 90.
99. Goldstein, M. and Simha, R., editors, *The Glass Transition and the Nature of the Glassy State*, Annals, NY, Acad. Sci., **279**, (1976).
100. Robredo, J., *L'analyse thermique différentielle en verrerie* Publ. Commission Intl. du Verre, (1967).
101. Goldschmidt, V. M., Skrifter Norske Videnskaps Akad. (Oslo), *I. Math. naturwiss.* Kl, Nr 8, 7, 156, (1926).
102. Zachariasen, W. H., *J. Am. Chem. Soc.*, **54**, 3841, (1932).
103. Smekal, A., *J. Soc. Glass Technol.*, **35**, 411T, (1951).
104. Hägg, G., *J. Chem. Phys.*, **3**, 42, (1935).
105. Stanworth, J. E., *J. Soc. Glass Technol.*, **30**, 54T, (1946); **32**, 154T, 366T, (1948); **36**, 217T, (1952).
106. Pauling, L., *The Nature of the Chemical Bond*, (3rd ed.) Cornell, Ithaca, (1960).
107. Sun, K. H., *J. Am. Cer. Soc.*, **30**, 277, (1947).
108. Rawson, H. in *IV° Congrès International du Verre*, Paris, Imp. Chaix, Paris, (1956), p. 62-9.
109. Volmer, M. and Weber A., *Z. Phys. Chem.*, **119**, 277, (1925).
110. Becker, R. and Döring, W., *Ann. Phys.*, **24**, 719, (1935).
111. Walton, A. G., in *"Nucleation"*, Zettlermoyer, A. C., editor, p. 225, Dekker, NY, (1969).
112. Turnbull, D., p. 41-56 in Ref. 29.
113. Turnbull, D. and Fisher, J. C., *J. Chem. Phys.*, **17**, 71, (1949).
114. Turnbull, D., *Contemp. Phys.*, **10**, 473, (1969).
115. Hillig, W. B., in *Symposium on Nucleation and Crystallization in Glasses and Melts*, Reser, M. K., Smith, G., Insley H. editors, Am. Cer. Soc. (1962) p. 77.
116. Hammel, J. J., p. 489 in Ref. 111.
117. Collins, F. C., *Z. Elektrochem.*, **59**, 404, (1955).
118. Turnbull, D., in *"Solid State Physics"*, Seitz F. and Turnbull, D. editors., Vol. 3, p. 226-306, Academic Press, NY, (1956).
119. Turnbull, D. and Cohen, M. H., *J. Chem. Phys.*, **29**, 1049, (1958).
120. Jackson, K. A., in *Progress in Solid State Chemistry*, Vol. 4, Pergamon, NY, (1969).
121. Uhlmann, D. R., *J. Non-Cryst. Solids*, **7**, 337, (1972).
122. Burke, J., *The Kinetics of Phase Transformation in Metals*, Pergamon, Oxford (1965).
123. Grange, R. A. and Kiefer, J. M., *Trans. ASM*, **29**, 85, (1941).
124. Hopper, R. W., Scherer, G. and Uhlmann, D. R., *J. Non-Cryst. Solids*, **25**, 45, (1974).
125. Uhlmann, D. R., *J. Non-Cryst. Solids*, **25**, 43, (1977).
126. Sakka, S. and Mackenzie, J. D., *J. Non-Cryst. Solids*, **6**, 145, (1971).
127. Cohen, M. H. and Turnbull, J. D., *J. Chem. Phys.*, **34**, 120, (1960), *Nature* **203**, 964, (1964).
128. Turnbull, J. D. and Cohen, M. H., p. 38 in Ref. 4, Vol. I.
129. Strnad, Z. and Douglas, R. W., *Phys. Chem. Glasses*, **14**, 33, (1973).

130a. Mazeau, J. P. and Zarzycki, J., *Faraday Discussion, N. 61, Precipitation*, 110, (1976).

130b. James, P. F., *Phys. Chem. Glasses*, **15**, 95, (1974).

131. Swift, H. R., *J. Am. Cer. Soc.*, **30**, 165, (1947).
132. Levin, E. M., Robbins, C. R. and McMurdie, H. F., *Phase Diagrams for Ceramists*, Am. Cer. Soc., Columbus, (1964) (2nd ed.) (1969).
133. Hosemann, R. and Bagchi, S. N., *Direct Analysis of Diffraction by Matter*, North Holland, Amsterdam, (1962).

134. Van Hove, L., *Phys. Rev.*, **95**, 249, (1954).
135. Debye, P. *Ann. Phys.*, **46**, 809, (1915).
136. Leadbetter, A. J. and Wright, A. C., *J. Non-Cryst. Sol.*, **7**, 23, (1972).
137. Wright, A. D.,"The structure of amorphous solids by X-ray and neutron diffraction" in *Advances in Structure Research and Diffraction Methods*, Vol. 5, Pergamon, Oxford, (1974).
138. Zernike, F. and Prins, J. A., *Z. Phys.*, **41**, 184, (1927).
139. Norman, N., *Acta Cryst.*, **10**, 370, (1957).
140. Warren, B. E., Krutter, H. and Morningstar, O., *J. Am. Cer. Soc.*, **19**, 202, (1936).
141. Zarzycki, J. *Trav. IV° Congrès Intern. de Verre*, Paris, p. 323, (1956).
142. Finbak, Ch., *Acta Chem. Scand.*, **3**, 1279, 1293, (1949).
143. Zarzycki, J., *Verres et Réfractaires*, **11**, 3, (1957).
144. Zarzycki, J., *J. Mat. Sci.*, **6**, 130, (1971).
145. Zarzycki, J., p. 117-43 in Ref. 28 and *J. Phys. Rad. (Suppl. Phys. Appl*, **18**, 65 **A**, (1957); **19**, 13 **A**, (1958).
146. Domenici, M. and Pozza, F., *J. Mat. Sci.*, **5**, 746, (1970).
147. Loshmanov, A. A., Sigaev, V. N. and Khodakovskaya, R., *Fiz. Khim. Stekla*, **1**, 35, (1975).
148. Henniger, E. H. and Buschert, R. C., *J. Chem. Phys.*, **44**, 1758, (1966).
149. Porai Koshits, E. A., in *Proceedings Conf. Structure of Glass*, Leningrad, 1953, Transl. Consultants Bur., NY, (1958).
150. Zarzycki, J., *J. Phys. Chem. Glasses*, **12**, 97, (1971).
151. Block, S. and Piermarini, G. J., *Phys. Chem. Glasses*, **5**, 138, (1964).
152. Sadoc, J. F. and Dixmier, J., in Ref. 33, p. 85.
153. Waser, J. and Schomaker, V., *Rev. Modern Phys.*, **25**, 671, (1953).
154. Warren, B. E., *X-ray Diffraction*, Addison-Wesley, Reading, Mass. (1969).
155. Mozzi, R. L. and Warren, B. E., *J. Appl. Cryst.*, **2**, 164, (1969).
156. Mozzi, R. L. and Warren, B. E., *J. Appl. Cryst.*, **3**, 251, (1970).
157. Warren, B. E. and Mavel, G., *Rev. Sci. Inst.*, **36**, 196, (1965).
158. Wright, A. C. and Sinclair, R. N., in Ref. 34, p. 133.
159. Janot C. and Wright, A. F., editors, "Third Intern. Conf. on the Structure of Non-Cryst. Materials", (Proc. Conf. Grenoble (France) 8-12 July, 1985) *J. de Phys* **46**, C-8, (1985).
160. Rindone, G. E., Pantano, C. G. and White, G. E. editors, *Glass Microstructure: Surface and Bulk*, (Proc. 6th Inter. Conf. on Glass Science, Penn State Univ. (USA), 29-31 July, 1981). *J. Non-Cryst. Sol.*, **49**, 1-548, (1982).
161. Price, D. L., editor, *Research Opportunities in Amorphous Solids with Pulsed Neutron Sources*, (Proc. Workshop on Research Opportunity). *J. Non-Cryst. Sol.*, **76**, 1-214, (1985).
162. Zarzycki, J. and Naudin, F., *Verres et Réfractaires*, **14**, 113, (1969).
163. Bell, R. J., Bird, N. F. and Dean, P., *J. Phys. Chem.* **1**, 299, (1968).
164. Lyon, R. J. P., *Nature*, **196**, 266, (1962).
165. Zarzycki, J., p. 525 in Ref. 29.
166. Tarte, P., p. 549 in Ref. 29.
167. Zarzycki, J. and Naudin, F., *J. Chimie Phys.*, **58**, 830, (1961).
168. Hilton, A. R. and Jones, C. E., *Phys. Chem. Glasses*, **7**, 112, (1966).
169. Adams, R. W., *Phys. Chem. Glasses.*, **2**, 39, (1961).
170. Neuroth, N., *Glastechn. Ber.*, **28**, 411, (1955).
171. Konijnendijk, W. L., *Philips, Res. Repts.* Suppl. **I**, (1975).
172. Wong, J. and Angell, C. A., *Appl. Spectrosc. Rev.*, **4**, 155, (1971).
173. Brodsky, M. H. and Cardona, M., *J. Non-Cryst. Solids*, **31**, 81, (1978).

174. Galeener, F. A., *Solid State Communic.* **44**, 1037 (1982).
175. Pake, G. E., *Solid State Physics*, Academic Press, NY, Vol. II, (1956).
176. Ellis, B. and McDonald, M. P., *J. Non Cryst. Sol.* **1**, 186, (1969).
177. Herzog-Cance, H., Potier, J., Potier, A., Beny, J. M., Sombret, B. and Wallart, F., *Advances in Molecular Relaxation and Interaction Processes*, **15**, 1-23, Elsevier, Amsterdam, (1979).
178. Mosel, B. D., Müller-Warmuth, W. and Dutz, H., *Phys. Chem. Glass.*, **15**, 154, (1974).
179. Bray, P. J., p. 65 in Ref. 31.
180. Baugher, J. F. and Bray, P. J., *Phys. Chem. Glasses*, **10**, 77, (1969).
181. Kim, K. S., Bray, P. J. and Merrin, S., *J. Chem. Phys.*, **64**, 4459, (1976).
182. Bray, P. J. and O'Keefe, J. G., *Phys. Chem. Glasses*, **4**, 37, (1963).
183. Bray P. J. and Dell, J. W., *J. de Phys.*, C-9, **43**, 131, (1982).
184. Bray P. J. and Mulkern, R. V., *J. Non-Cryst. Sol.*, **80**, 181, (1986).
185. Bray, P. J. and Silver, A. H., in Ref. 4, Vol. I, chap. 5 p. 92.
186. Müller-Warmuth, W., *Glastechn. Ber.*, **38**, 121, 405, (1965).
187. Bray, P. J., in *Magnetic Resonance*, Plenum, NY, (1970).
188. Bray, P. J., in *Proc. X. Intern. Glass Conf.*, Kyoto (Japan) sect. 13, p. 1 (1973).
189. Bates, T., p. 195 in Ref. 4, Vol. II.
190. Low, W., *Solid State Phys.*, Suppl. 2, (1960).
191. Stevels, J. M. and Kats, A., *Philips Res. Rep.*, **11**, 103, (1956).
192. Weeks, R. A., *J. Appl. Phys.*, **27**, 1376, (1956).
193. Stevels, J. M., p. 422 in Ref. 28.
194. Griscom, D. L., *J. Non-Cryst. Solids*, **13**, 241, (1973-74).
195. Griscom, D. L., *Defects and their Structure in Nonmetallic Solids*", p. 323, Henderson, B. and Hughes, A. E., editors, Plenum, NY, (1976).
196. Griscom, D. L., in Ref. 34, p. 232.
197. Frauenfelder, H., *The Mössbauer Effect.*, Benjamin, NY, (1962).
198. Wertheim, G. K., *Mössbauer Effect: principles and applications*, Academic Press, NY, (1964).
199. Kurkjan, C. R., *J. Non-Cryst. Solids*, **3**, 157, (1970).
200. Bartenev, G. M., Suzdalev, I. P. and Tsyganov, A. D., *Phys. Status Solidi*, **37**, 73, (1970).
201. Pelah, I. and Ruby, S. L., *J. Chem. Phys.*, **51**, 383, (1969).
202. Sugisaki, M., Suga, H., and Seki S., *Bull. Chem. Soc. Japan*, **41**, 2586, (1968).
203. Sakka, S., 25° *Annual Meeting of Japan Chem. Soc.*, Tokyo, (1971).
204. Dodd, C. G. and Glen, G. L., *J. Am. Cer. Soc.*, **53**, 322, (1970).
205. Sayers, D. E., Stern, E. A. and Lytle, F. W., *Phys. Rev. Letters*, **27**, 1204, (1971).
206. Sayers, D. E., Lytle, F. W., and Stern, E. A., *J. Non. Cryst. Solids*, **8**, 401, (1972).
207. Theo, B. K., *Inorganic Chemistry Concepts*, **9**, Springer, Berlin, (1986).
208. Lee, P. A., Citrin, P. H., Eisenberger, P, and Kincaid, B. M., *Rev. Mod. Phys.*, **53**, 769-806, (1981).
209. Biancon, A., Ino Ccia, L., and Stipchich, editors, EXAFS *and Near-Edge Structure* **II**, Springer, Berlin, (1983).
210. Hodson, K. O., Hedman, B. and Penner-Hahn, J. E., editors, EXAFS *and Near-Edge Structure*, **III**, Springer, Berlin, (1984).
211. Greig, J. W., *Am. J. Sci.*, **13**, 133, (1927).
212. Zarzycki, J., in *Discussions of the Faraday Soc.*, No. 50, p. 122, (1970).
213. Levin, E. M. and Block, S., *J. Am. Cer. Soc.*, **95**, (1957).
214. Zarzycki, J. and Naudin, F., *Phys. Chem. Glasses*, **8**, 11, (1967).
215. Porai Koshits, E. A. and Averjanov, V. I., *J. Non Cryst. Solids*, **1**, 29, (1968).
216. Charles, R. J., *J. Am. Cer. Soc.*, **49**, 55, (1966).

217. Zarzycki, J. and Naudin, F., *J. Non-Cryst. Solids* **5**, 415, (1971).
218. Tomozawa, M., *J. Am. Cer. Soc.*, **56**, 378, (1973).
219. Warren, B. E. and Pincus, A. G., *J. Am. Cer. Soc.*, **23**, 301, (1940).
220. Levin, E. M. and Block, S. *J. Am. Cer. Soc.*, **40**, 95, 113, (1957); **41**, 49, (1958).
221. Becker, R. *Am. Phys.*, **32**, 128, (1938); *Proc. Phys. Soc.*, **52**, 71, (1940).
222. Van der Toorn, L. J. and Tiedema, T. Y., *Acta Met.*, **8**, 711, (1960).
223. Borelius, G., *Ann. Phys.*, **28**, 507, (1937). *Trans. AIME*, **191**, 477, (1951).
224. Hobstetter, J. N., *Trans. AIME*, **180**, 121, (1949).
225. Scheil, E., *Z. Metallk.*, **43**, 40, (1952).
226. Cahn, J. W. and Hilliard, J. E., *J. Chem. Phys.*, **28**, 258, (1958); **31**, 688, (1959). *Acta Met.*, **9**, 795, (1961); **10**, 179, (1962).
227. Cahn, J. W., *Chem. Phys.*, **42**, 93, (1965).
228. Cahn, J. W. and Charles, R. J., *Phys. Chem. Glasses*, **6**, 181, (1965).
229. Cahn, J. W., *Trans. AIME*, **242**, 166, (1968).
230. Haller, W., *J. Chem. Phys.*, **42**, 686, (1965).
231. Porter, D. A. and Easterling, K. E., Phase Transformations in Metals and Alloys, Van Nostrand Reinhold Australia, Victoria (1981).
232. Lifschitz, I. M. and Slyozow, V. V., *J. Phys. Chem. Solids*, **19**, 35, (1961).
233. Wagner, C., *Z. Elektrochem.*, **65**, 581, (1961).
234. Mazurin, O. V., Streltsina, A. S. and Totesh, A. S., *Phys. Chem. Glasses*, **10**, 63, (1969).
235. De Hoff, R. T. and Rhines, F. N., *Microscopie Quantitative*, Masson, Paris, (1972).
236. James, P. F. and McMillan, P. W., *Phys. Chem. Glasses*, **8**, 132, (1967).
237. Guinier, A. and Fournet, G., *Small Angle Scattering of X-rays*, Wiley, NY, (1955).
238. Hammel, J. J., *J. Chem. Phys.*, **46**, 2234, (1967).
239. Lumsden, J., *Thermodynamics of Alloys*, Inst. Metals, London, (1952).
240. Zarzycki, J. and Naudin, F., *C. R. Ac. Sc.*, **265 B**, 1456, (1967).
241. Tomozawa, M, MacCrone, R. K. and Herman, H., *Phys. Chem. Glasses*, **11**, 136, (1970).
242. Naudin, F. and Zarzycki, J., *C. R. Acad. Sci.*, **266 C**, 729, (1968).
243. Zarzycki, J. and Naudin, F., *C. R. Acad. Sci.*, **266 B**, 145, (1968).
244. Roth, M. and Zarzycki, J., *J. Non Cryst. Sol.*, **16**, 93, (1974).
245. Larché, F., Roth, M. and Zarzycki, J., in Ref. 33, p. 13.
246. Zarzycki, J., *Proc. X Intern. Congress on Glass*, Kyoto, Japan, No. 12, p.28, (1974).
247. Crewe, A. V., *Science*, **168**, 1338, (1970).
248. Cowley, Y. M., *Diffraction Physics*, North Holland, Amsterdam, (1975).
249. Howie, A., *J. Non-Cryst. Solids*, **31**, 41, (1978).
250. Krivanek, O. L., Gaskell, P. H. and Howie, A., *Nature*, **262**, 454, (1976).
251. Gaskell, P. H. and Mistry, A. B., *Phil. Mag.*, **A, 39**, 245, (1979).
252. Gaskell, P. H., Private communication.
253. Chaudhari, P., Graczyk, J. F. and Charbnau, H. P., *Phys. Rev. Letters*, **29**, 425, (1972).
254. Krivanek, O. L. and Howie, A., *J. Appl. Cryst.*, **8**, 213, (1975).
255. Howie, A., Krivanek, O. L. and Rudee, M. L., *Phil. Mag.*, **27**, 235, (1973).
256. Bando, Y. and Ishizuka, K., *J. Non-Cryst. Solids*, **33**, 375, (1979).
257. Landau, L. and Liftschitz, E., *Physique Statistique*, **5**, 414, Ed. MIR, Moscow, (1967).
258. Laberge, N. L., Vasilescu, V. V., Montrose, C. J. and Macedo, P. B., *J. Am. Cer. Soc.*, **56**, 506, (1973).
259. Schroeder, J. in Ref. 26, Vol. 12.
260. Schroeder, J., Results cited in Ref. 259.
261. Zarzycki, J., p. 201 in Ref. 311.

262. Buccaro, J. A. and Dardy, H. D., *J. Appl. Phys.*, **45**, 2121, (1974).
263. Weinberg, D. L., *J. Appl. Phys.*, **33**, 1012, (1962); *Phys. Lett.*, **7**, 324, (1963).
264. Levelut, A. M. and Guinier, A., *Bull. Soc. Fr. Mineral.*, **90**, 445, (1967).
265. Pierre, A. and Uhlmann, D. R., *J. Appl. Cryst.*, **5**, 216, (1972).
266. Porai-Koshits, E. A., in Ref. 311, p. 183.
267. Bockris, J. O. M. and Kojonen, E., *J. Am. Chem. Soc.*, 4493, (1960).
268. Porai-Koshits, E. A., *J. Non-Cryst. Sol.*, **25**, 86, (1977).
269. Zarzycki, J. and Naudin, F., *C. R. Acad. Sc.*, **C 266**, 1005, (1968).
270. Zarzycki, J., *Rev. Pure Appl. Chem.*, **18**, 227, (1968).
271 Zarzycki, J. and Naudin, F., *J. Non-Cryst. Sol.*, **5**, 415, (1971).
272. Debye, P., in Ref. 28, p.1.
273. Zarzycki, J. and Mezard, R., *Phys. Chem. Glasses*, **3**, 163, (1962).
274. Zarzycki, J., *J. Chimie Phys.*, **66**, 153, (1969).
275. Bernal, J. D., *Nature*, **185**, 68, (1960); *Proc. Roy. Soc.*, **280A**, 299, (1964).
276. Scott, G. D., *Nature*, **188**, 633. (1960); **194**, 956. (1962).
277. Bernal, J. D. and Masson, Y., *Nature*, **188**, 910, (1960).
278. Finney, J. L., *Proc. Roy. Soc.*, **319A**, 479, (1970).
279. Bennett, C. H., *J. Appl. Phys.*, **43**, 2727, (1972).
280. Polk, D. E., *Scripta Met.*, **4**, 117, (1970); *Acta Met.*, **20**, 485, (1972).
281. Whittaker, E. J. W., *J. Non-Cryst. Sol.*, **28**, 293, (1978).
282. Sadoc, J. F., Dixmier, J. and Guinier, A., *J. Non-Cryst. Sol.*, **12**, 46, (1973).
283. Boudreaux, D. S. and Gregor, J. M., *J. Appl. Phys.*, **48**, 152, (1977).
284. Bell, R. F. and Dean, P., *Phil. Mag.*, **25**, 1381, (1972).
285. Bell, R. F. and Dean, P., Private communication.
286. Evans, D. L. and King, S. V., *Nature*, **212**, 1353, (1966).
287. Evans, D. L. and Teter, M., p. 53 in Ref. 33.
288. Polk, D. E., *J. Non-Cryst. Sol.*, **5**, 365, (1971).
289. Polk, D. E. and Boudreaux, D. S., *Phys. Rev. Lett.*, **31**, 92, (1973).
290. Steinhardt, P., Alben, R. and Weaire, D., *J. Non-Cryst. Sol.*, **15**, 199, (1974).
291. Shevchik, N. J. and Paul, Y., *J. Non-Cryst. Sol.*, **13**, 1, (1973/74).
292. Henderson, D. and Herman, F., *J. Non-Cryst. Sol.*, **8-10**, 359, (1972).
293. Renniger, A. L., Rechtin, M. D. and Averbach, B. L., *J. Non-Cryst. Sol.*, **16**, 1, (1974).
294. Leadbetter, A. J. and Wright, A. C., *J. Non-Cryst. Sol.*, **7**, 23, 156, (1972).
295. Frank, F. C., *Proc. Roy. Soc.*, **A 215**, 43, (1952).
296. Tilton, L. W., *J. Res. Nat. Bur. Std.*, **59**, 139, (1957).
297. Robinson, H. A., *J. Phys. Chem. Sol.*, **26**, 209, (1965).
298. Hoare, M. R. and Barker, J. A., p. 175 in Ref. 33.
299. Hoare, M. R., *Am. NY Acad. Sci.*, **279**, 186, (1976).
300. Farges, J., Thesis, Univ. Paris, (1977).
301. Chen, M. S., *Chaotic Order in Quasicrystals and Metal Glasses*, in Ref. 44, p. 21
302. Penrose, W., *Bull. Inst. Math*, **10**, 266, (1974).
303. Gaskell, P. H., *Phil. Mag.*, **32**, 211, (1975).
304. Woodcock, L. V., Angell, C. A. and Cheeseman, P. A., *J. Chem. Phys.*, **65**, 1565, (1976).
305. Amini, M. and Hockney, R. W., *J. Non-Cryst. Sol.*, **31**, 447, (1979).
306. O'Keefe, J. and Weiss-Kirchner, W., *Glastechn. Ber.*, **43**, 199, (1970).
307. Levi, C., Private communication. Also Levi, C., Barton, J. L., Guillemet, C., Le Bras, E. and Lehuede, P., *J. Mat. Sc. Letters*, **8**, 337 (1989).

308. Pye, L. D., O'Keefe, J. A. and Frechette, V. D., editors *Natural Glasses*, Proc. Int. Conf. on Glass in Planetary and Geological Phenomena, August 14-18, 1983, Alfred, NY, *J. Non-Cryst. Sol.*, **67**, 1-662 (1984).
309. Sosman, R. B., *Properties of Silica*, Reinhold, NY, (1927); *Phases of Silica*, Rutgers Univ. Press, New Brunswick (1965).
310. Brückner, R. J., *J. Non-Cryst. Sol.*, **5**, 123, 177, (1970).
311. Pye, L. D., Frechette, V. D. and Kreidl, N. J., editors, *Borate Glasses*, Plenum, NY, (1970).
312. Schultz, P. C. Vapor phase materials and processes for glass optical waveguides, in *Recent Advances in Fiber Optics*, Mitra, S. S. and Bendow, B., editors, Plenum, NY, (1979).
313. Baldwin, C. M., Almeida, R. M. and Mackenzie, J. D., *J. Non-Cryst. Sol.*, **43**, 309, (1981).
314. Vogel, W. and Gerth, K., *Glastechn. Ber.*, **31**, 15, (1958).
315. Deganello, S., *J. Am. Cer. Soc.*, **55**, 584, (1972).
316. Weber, M. J., Cline, C. F., Smith, W. L. , Milam, D., Heiman, D. and Hellwarth, R. W., *Appl. Phys. Lett.*, **32**, 403, (1978).
317. Poch, W., *Glastech. Ber.*, **7**, 261, (1967).
318. Poulain, M., Lucas, J., *J. Mat. Res. Bull.*, **10**, 243, (1975). *Verres et Refract.*, **32**, 505, (1978).
319. Lucas, J., *Halide Glasses for* IR *Fiber Optics: The First Ten Years*, Proc. NATO Adv. Res. Workshop, Vilamoura, Portugal, Mar. 31-Apr. 4, 1986, Nijhoff, Dordrecht, (1986).
320. Lucas, J., *Fluoride Glasses: Glass Formation Concept. Structure and Optical Properties*, pp.141–160 in Ref. 44.
321. Lucas, J. and Moynihan, G. T., editors *Materials Science Forum, Vol. 6*, 3rd Intl. Symp. on Halide Glasses, Rennes, France, June, 1985.
322. Lucas, J., Proc. 4th Rare Earth Conf., Zurich, March 1985; *J. Less Common Met.*, **112**, 27, (1985).
323. Lucas, J., *Halide Glasses*, pp. 307-316 in Ref. 41.
324. Lucas, J., *Zirconium-free Fluoride Glasses: Halide Glasses for* IR *Optics* in Ref. 319.
325. Jacoboni, A., Le Bail, A. and De Pape, R., *Glass Tech.*, **24**, 167, (1983).
326. Sun, K. H., *J. Am. Cer. Soc.*, **30**, 277, (1947).
327. Schröder, J., *Angew. Chem.*, **76**, 344, (1964).
328. Drexhage, M. G., *Heavy Metal Fluoride Glasses*, Treatise on Materials Science and Technology, Vol. 26, p. 155, Tomozawa, M. and Doremus, R. H., editors, Academic Press, NY, (1985).
329. Tran, D. C., Sigel, G. H. and Bendow, B., *J. Lightwave Tech.*, **2**, 566, (1984).
330. Miyashita, T. and Manabe, J., *J. Quant. Elect.*, **18**, 1432, (1982).
331. Tcheichivili, L., summary in *Phys. Chem. Glass*, **10**, 38A, 312, (1969).
332. Weber, M. J., Cline, C. F., Smith, W. L., Milam, D., Heiman, D. and Hellwasrth, R. W., *Appl. Phys. Lett.*, **32**, 403, (1978).
333. Krause, J. T., Kurkjan, C. R., Pinnow, D. A. and Sigety, E. A., *Appl. Phys. Lett.*, **17**, 367, (1970).
334. Kurkjan, C. R., Krause, J. T. and Sigety, E. A., *IX° Int. Congr. on Glass*, Versailles, **1**, 503, (1971).
335. Anthonis, H. and Kreidl, N. J., *J. Non-Cryst. Sol.*, **11**, 257, (1972).
336. Betts, F., Bienenstock, A. and Ovshinsky, S. R., *J. Non-Cryst. Sol.*, **4**, 554, (1970).
337. Luborsky, F. E., editor, *Amorphous Metallic Alloys*, Butterworth, London, (1983).
338. Hasegawa, R., editor, *Glassy Metals, Magnetic, Chemical and Structural Properties*, CRC Press, Boca Raton, USA, (1983).

339. Amantharaman, T. R., editor, *Metallic Glasses: Production, Properties and Applications*, Trans. Tech. Publ., Switzerland, (1983).
340. Kaneyoshi, T., *Amorphous Magnetism*, CRC Press, Boca Raton, USA, (1983).
341. Guntherodt, H. J. and Beck, H., editors, *Glassy Metals*, Springer, NY, I, (1981), **II**, (1983).
342. Wagner, C. N. J. and Johnson, W. L., editors, *Liquid and Amorphous Metals*, **V**, Proc. 5th Intl. Conf. on Liquid and Amorphous Metals, Los Angeles, USA, 15-19 Aug., 1983. *J. Non-Cryst. Sol.*, **61, 62**, pp. 1450 (1984).
343. Chen, H. S., *Rev. Mod. Phys.*, **43**, 353, (1980).
344. Cahn, R. W., *Physical Metallurgy* (3rd ed.), Cahn, R. W., editor, p. 1780, North Holland, Amsterdam, (1981).
345. Herman, H., editor, *Ultrarapid Quenching of Liquid Alloys*, Vol. **20**, Treatise on Materials Science and Technology, Academic Press, NY, (1981).
346. Raskin, D. and Smith, C. H., *Amorphous Metallic Alloys*, Ch. 20, Luborsky, F. E., editor, Butterworth, London, (1983).
347. Ferry, J. D., *Viscoelastic Properties of Polymers*, Wiley, NY, (1970).
348. Glasstone, S., Laidler, K. J. and Eyring, H., *The Theory of Rate Processes*, McGraw Hill, NY, (1941).
349. Dietzel, A. and Brückner, R. *Glastechn. Ber.*, **34**, 49, (1961).
350. Littleton, J. T., *J. Am. Cer. Soc.*, **10**, 259, (1927).
351. After various authors collected in Ref. 8, p. 131.
352. Cabarat, R., *Rev. Metall.*, **146**, 617, (1949).
353. Loehman, R. E., *J. Non-Cryst. Sol.*, **42**, 433, (1980).
354. De Bast, J. and Gilard, P., *C. Rend. Rech. IRSIA*, No. 32, (1965).
355. Kurkjian, C. R., *Phys. Chem. Glasses*, **4**, 128, (1963).
356. Larsen, D. C., Mills, J. J. and Sievert, J. L., *J. Non-Cryst. Sol.*, **14**, 269, (1974).
357. Mills, J. J., *J. Non-Cryst. Sol.*, **14**, 255, (1974).
358. Read, B. E. and Williams, G,. *Trans. Faraday Soc.*, **576**, 1979, (1961).
359. Nowick, A. S. and Berry, B. S., *I.B.M. J. Res. Dev.*, **5**, 297, 312, (1961).
360. Day, D. E., p. 39 in Ref. 30.
361. Day, D. E. and Stevels, J. M., *J. Non-Cryst. Sol.* **14**, 178, (1974).
362. Phalippou, J., Masson, S., Boyer, A. and Zarzycki, J., *J. Non-Cryst. Sol.*, **14**, 178, (1974).
363. Phalippou, M., Boyer, A., Groubert, E. and Zarzycki, J., *Rev. Phys. Appl.* **10**, 437, (1975).
364. Doremus, R. H., p. 1 in Ref. 4, Vol. II.
365. Jost, W. and Haudde, K., *Diffusion*, D. Steinkopf, Darmstadt, (1972).
366. Jost, W., *Diffusion in Solids, Liquids, Gases*, Academic Press, Oxford, (1952).
367. Crank, J., *The Mathematics of Diffusion*, Clarendon Press, Oxford, (1957).
368a. Haul, R. and Dümbgen, G., *Z. Elektrochem.*, **66**, 636, (1962).
368b. Sucov, E. W., *J. Am. Cer. Soc.*, **46**, 14, (1963).
368c. Williams, E. L., *J. Am. Cer. Soc.* **48**, 191, (1965).
368d. Kingery, W. D. and Lecron, J. A., *Phys. Chem. Glass.*, **1**, 87, (1960).
368e. Hagel, W. C. and Mackenzie, J. D., *Phys. Chem. Glass.*, **5**, 113, (1964).
368f. Schaeffer, H. A. and Oel, H. J., *Glastechn. Ber.*, **42**, 493, (1969).
369. Choudhury, A. Oakler, D. W., Ansel, G., Curien, H. and Baruch, P., *Solid State Commun.*, **3**, 119, (1965).
370. Frischat, G. H., *Glastechn. Ber.*, **43**, 174, (1970).
370a. Frischat, G. H., *J. Am. Cer. Soc.*, **52**, 625, (1969).
371. Johnson, J. R., Bristow, R. H. and Blau, H. H., *J. Am. Cer. Soc.*, **34**, 165, (1951).
372. Varshneya, A. K. and Cooper, A. R., *J. Am. Cer. Soc.*, **55**, 220, (1972).

373. Frischat, G. H., *Ionic Diffusion in Oxide Glasses*, Trans. Tech. Publ. Aedermannsdorf, (1975).
374. Cantor, B. and Cahn, R. W., in *Amorphous Metallic Alloys*, p. 487, Luborsky, F. E., editor, Butterworth, London (1983).
375. Cahn, R. W., p. 3-20 in Ref. 44.
376. Cantor, B., in *Proc. Electric Power Research Inst. Acta Metallurgica Workshop*, Jaffee, R. I. and Haasen, P., Editors, *Acta Met.*, **34**, (1986).
377. Tallan, N. M., editor, *Electrical Conductivity in Ceramics and Glass*, Vol. B, Dekker, NY, (1974).
378. Littleton, J. T. and Morey, G. W., *The Electrical Properties of Glass*, Wiley, NY, (1933).
379. Mazurin, O. V., *Electrical Properties and Structure of Glass*, Consultants Bur. NY, (1965).
380. Tomozawa, M., p. 283 in Ref. 26, Vol. 12.
381. Fulda, M., *Sprechsaal*, **60**, 769, 789, 810, (1927)
382. Ravaine, D. and Souquet, J. L., *Phys. Chem. Glass.*, **18**, 27, (1977).
383. Lengyel, B and Boksay, Z. Z., *Phys. Chem.*, **204**, 157, (1955).
384. Hendrickson, J. R. and Bray, P. J., *Phys. Chem. Glass.*, **13**, 43, 107, (1972).
385. Mott, N. F. and Davis, E. A., *Electronic Processes in Non-Crystalline Materials*, OUP, Oxford, (1971).
386. Gaskell, P. H. and Mackenzie, J. D., Eds. "Electronic Properties and Structure of Amorphous Solids," in *J. Non-Cryst. Sol.*, **32**, [1-3], 1–444, (1979).
387. Mott N., editor, "Amorphous and Liquid Semiconductors," *J. Non-Cryst. Sol.*, **4**, (1970).
388. Cohen, M. H. and Lucovsky, G., editors, "Amorphous and Liquid Semiconductors," *J. Non-Cryst. Sol.*, **8-10**, 1-1050 (1972).
389. Cargill, G. S., III and Chaudari, P., editors, "Atomic Scale Structure of Amorphous Solids," *J. Non-Cryst. Sol.*, **31**, 1-286, (1978).
390. Pantelides, S. T., editor, *The Physics of* SiO_2 *and its Interfaces*, Pergamon, NY, (1978).
391. Paul, W. and Kastner, M., editors, "Amorphous and Liquid Semiconductors," Proc. 8th Intl. Conf. on Amorphous and Liquid Semiconductors, Cambridge, Mass. 27-31 Aug. 1979. *J. Non-Cryst. Sol.*, **35**, **36**, 1-1328, (1980)
392. Tanaka, K. and Shimizu, T., editors, Proc. 10th Intl. Conf. on Amorphous and Liquid Semiconductors, Tokyo, Japan, 22-26 Aug. 1983, *J. Non-Cryst. Sol.*, **59**, **60**, 1–1326, (1983).
393. Fritzsche, H. and Kastner, M. A., editors, Proc. Intl. Conf. on Transport and Defects in Amorphous Semiconductors, Bloomfield Hills, Mich., 22-24 March 1984, *J. Non-Cryst. Sol.*, **66**, 1–392, (1984).
394. Evangelish, F. and Stuke, J., editors, Proc. 11th Intl. Conf. on Amorphous and Liquid Semiconductors, Rome, 2-6 Sept. 1985, *J. Non-Cryst. Sol.*, **77**, **78**, 1540 pp. (1985).
395. Somogyi, I. Kòsa, editor, Proc. 8th Intl. Conf. on Non-Crystalline Semiconductors, *J. Non-Cryst. Sol.*, **90**, 1–688, (1987).
396. Proc. 12th Intl. Conf. on Amorphous and Liquid Semiconductors, Prague, Czechoslovakia, 4-28 Aug. 1987. *J. Non-Cryst. Sol.*, **97/98**, 1-1524, (1987).
397. Cohen A. E. and Spear, W. E., *Phys. Chem. Glass.*, **17**, 174, (1976).
398. Cohen, M. H., p. 391 in Ref. 387.
399. Cohen, M. H., Fritzsche, H. and Ovshinsky, S. R., *Phys. Ref. Letters* **22**, 1065, (1969).
400. Pearson, A. D., Dewald, J. F., Northover and Pecd, W. F. Jr. in *Advances in Glass Technology*, Tech. Papers of the IV Intl. Cong. on Glass, Washington, DC, 1962. p. 357, Plenum, NY, (1962).
401. Kolomiets, B. T. and Lebedev, E. A., *Radio Eng. Electron. USSR*, **8**, 1941, (1963).
402. Ovshinsky, S. R., Phys. Rev. Letters, **21**, 1450, (1968).

403. Dekker, A. J., *Solid State Physics*, McMillan, London, (1969).
404. Stevels, J. M., p. 412 in Ref. 28.
405. Guyer, F. M., *J. Am. Cer. Soc.*, **16**, 607, (1933).
406. Kim, C. and Tomozawa, M., *J. Am. Cer. Soc.*, **59**, 127, (1976).
407. Taylor, H. E., *J. Soc. Glass Technol.*, **43**, 124, (1959).
408. Isard, J. O., *Proc. Inst. Elec. Eng.* Suppl. 22, **109**, part B, 440, (1952).
409. Namikawa, M. and Kumata, K., *J. Ceram. Assoc. Japan* **76**, 10, (1968).
410. Stevels, J. M., in *Handbuch der Physik*, Flügge, S., editor, Vol 20, p. 350, Springer, Berlin.
411. Taylor, H. E., *J. Soc. Glass Technol.*, **43**, 124 T, (1969).
412. Charles, R. J., *J. Suppl. Phys.*, **32**, 1115, (1961).
413. Sillars, R. W., *J. Inst. Elec. Eng.*, **80**, 378, (1937).
414. Charles, R. J., *J. Am. Cer. Soc.*, **46**, 235, (1963).
415. Stirling, J. F. *J. Soc. Glass Technol.*, **39**, 134T, (1955).
416. Takamori, T., *J. Am. Cer. Soc.*, **46**, 366, (1963).
417. Takamori, T. and Tomozawa, M., p. 123 in Ref. 26, Vol. 12.
418. Bartholomew, R. F., pp. 75-127 in Ref. 26, Vol. 22.
419. Klein, R. M., p. 285 in Ref. 25.
420. Fanderlik, I., "Optical Properties of Glass," *Glass Science and Technology*, **5**, Elsevier, Amsterdam, (1983).
421. Bamford, C. R., Nicoletti, F., Gliemeroth, G., Russo, V. and Marechal, A., editors, Proc. Conf. on Optical Properties of Glass and Optical Materials, Florence, Italy, 6-8 April, 1982, parts I and II *J. Non-Cryst. Solids*, **47**, 1-296, (1982).
422. Weyl, W. A., *Coloured Glasses*, Soc. Glass Technol., (1951).
423. Sigel, G. H., p. 5, in Ref. 26, Vol. 12.
424. Mie, G., *Ann. Phys.*, **25**, 377, (1908).
425. Doremus, R. H., *J. Chem. Phys.*, **40**, 2389, (1964).
426. Morain, M and Barton, J. L., *Symp. sur la surface du verre et ses traitements modern*, (Symp. on the suface of glass and its modern treatment.), (USCV), Luxembourg, p. 1-25, (1967).
427. Araujo, R. F. p. 91 in Ref. 26, Vol. 12.
428. Snitzer, E., *Bull. Am. Cer. Soc.*, **52**, 516, (1973).
429. Stokowski, S. E. and Weber, M. J., "Nd-doped Laser Glass, Spectroscopic and Physical Properties, *Laser Glass Handbook*, LLNL, M-95, Lawrence Livermore Laboratory, (1978), (1979).
430. Guy, R., *Verres et Refractaires*, **19**, 185, (1965).
431. Böttger, H., *Phys. Stat. Sol.* (B), **62**, 9, (1974).
432. Phillips, W. A., *J. Non-Cryst. Sol.* **31**, 267, (1978).
433. Weaire, D. and Alben, R., *Phys. Rev. Lett.* **29**, 1505, (1972).
434. King, C. N., Phillips, W. A. and Neufville, J. P., *Phys. Rev. Lett.*, **32**, 538, (1974).
435. Stephens, R. B., *Phys. Rev.*, **B8**, 2896, (1973).
436. Kingery, W. D., *J. Am. Cer. Soc.*, **42**, 617, (1959). Ratcliffe, E. H., *Glass Technol.*, **4**, 113, (1963).
437. Klemens, P. G., p. 508 in Ref. 28.
438. Anderson, P. W., Halperin, B. I. and Varma, C. M., *Phil. Mag.*, **25**, 1, (1972).
439. Phillips, W. A., *J. Low Temp. Phys.* **7**, 351, (1972).
440. Lasjaunias, J. C., Ravex, A., Vandorpe, M. and Hunklinger, S., *Solid State Comm.* **17**, 1045, (1975).
441. Vukcevich, M. R., *J. Non-Cryst. Sol.*, **11**, 25, (1972).
442. Wiederhorn, S. M., in *Mechanical and Thermal Properties of Ceramics*, Wachtman, J. B., editor , Nat. Bur. Stds. Special Pub. 303, p. 217, (1969).

443. Griffith, A. A., *Phil. Trans. Roy. Soc.* (London), **221 A**, 163, (1920): *Proc. First Intl. Congr. Appl. Mech.*, Delft, p. 55, (1924).
444. Inglis, C. E., *Trans. Inst. Naval Architects* (London), **53**, 219, (1913).
445. Doremus, R. H., *J. Appl. Phys.*, **47**, 1833, (1976).
446. Irwin, G. R., *J. Appl. Mech.*, **24**, 361, (1957), Orowan, E., *Weld, Res. Supp.*, **34**, 157s, (1955).
447. Williams, M. L., *J. Appl. Mech.* **24**, 109, (1957).
448. Evans, A. G., in *Fracture Mechanics of Ceramics*, Vol.I, Bradt, R. C. Hasselmann, D. P. H. and Lange, F. F., editors, Plenum, NY, p. 17, (1974).
449. Bui, H. D. *Mécanique de la rupture fragile*, (Mechanics of brittle failure), Masson, Paris, (1978).
450. Dugdale, D. S., *J. Mech. Phys. Solids*, **8**, 100, (1960).
451. Barenblatt, G. I., *Adv. Appl. Mech.*, **7**, 55, (1962).
452. Mott, N. F., *Engineering*, **165**, 16, (1948).
453. Berry, J. P., *J. Mech. Phys. Solids*, **8**, 194, (1960).
454. Vernaz, E. and Zarzycki, J. C., *R. Acad. Sci.*, **285 B**, 1, (1977).
455. Vernaz, E., *Influence de la composition sur la ténacité des verres* (Influence of composition on the toughness of glass), Thesis, Dr. Eng., Montpellier, (1978).
456. Kennedy, C., *Strength and Fracture Toughness of Binary Alkali Silicate Glasses*, Thesis, Pennsylvania State Univ., (1974).
457. Marsh, D. M., *Proc. Roy. Soc.*, **279 A**, 420, (1964), **282 A**, 33, (1964).
458. Mackenzie, J. D., *J. Am. Cer. Soc.*, **46**, 461, 470, (1963).
459. Mould, R. E., *Fundamental Phenomena in Materials Science*, Vol. 4, p. 119, Plenum, NY, (1967).
460. Charles, R. J. and Hillig, W. B., *Symp. Mechanical Strength and Ways of Improving It.* Florence, USCV, p. 25, (1962).
461. Wiederhorn, S. M., *J. Am. Cer. Soc.*, **50**, 407, (1967).
462. Wiederhorn, S. M., *Reliability, Life Prediciton and Proof Testing of Ceramics*, NBS Rep. 74-486, (1974).
463. Weibull, W. A., *A Statistical Theory of the Strength of Materials*, Ingeniör Vetenskaps Adademinen Handlingar Nr. 151, (1939).
464. Wiederhorn, S. M., Evans, A. G. and Roberts, D. E., p. 829 in Ref. 448, Vol. II.
465. Schardin, H. and Struth, W. *Z. techn. Phys.*, **18**, 474, (1937).
466. Kerkhof, F., *Bruchvorgänge in Gläsern*, DGG Frankfurt, (1970).
467. Frechette, V. D., p. 433 in Ref. 12.
468. Le Boiteux, H. and Boussard, R., *Elasticité et Photoélasticimétrie*, (Elasticity and Photoelasticimetry), Hermann, Paris, (1940).
469. Garfinkel, H. M., *Glass Ind.* **50**, 28, 74, (1969).
470. Guillemet, C., *Interférométrie á ondes multiples, appliquée á la détermination de la répartition de l'indice de réfraction dans un milieu stratifié.*, Thesis, Dr. Ing., Paris, (1968).
471. Adams, L. H. and Williamson E. D., *J. Franklin Inst.*, **190**, 597, 835, (1920).
472. Kurkjian, C.R., editor, *Strength of Inorganic Glasses*, Nato Adv. Workshop, 21-25 March, 1988, Algarve, Portugal, Plenum, NY, (1983) pp. 643.
473. Schröder, H., *Verres et Réfractaires*, **18**, 19, (1964).
474. Ernsberger, F. M., *Ann. Rev. Mater. Sci.*, **2**, 529, (1972).
475. Day, D. E., editor, "Proc. IV Rolla Mater. Sci. Conf.," *J. Non Cryst. Sol.*, **19**, (1975).
476. Frechette, V. D., Lacourse, W. and Burdick, V. L., editors, *Surfaces and Interfaces of Glass and Ceramics*, Mat. Sci. Res., Vol. 7, Plenum Press, NY, (1974).
477. Colombin, L. *Analyse Physico-chimique des zones superficielles du verre plat*, Thesis, Namur, (1979).

478. Pantano, C. G., Dove, D. B. and Onoda, G. Y., p. 41 in Ref. 475.
479. Thomas, S., *J. Appl. Phys.*, **45**, 161, (1974).
480. Powell, C. J., *Surf. Sci.*, **44**, 29, (1974).
481. Penn, D. R., *Phys. Rev.* B, **13**, 5248, (1976).
482. Flitsch, R. and Raider, S., *J. Vac. Sci. Technol.*, **12**, 305, (1975).
483. Jenkins, R., *An Introduction to X-ray Spectrometry*, Heyden, London, (1974).
484. Herglotz, M. K. and Birks, L. S., editors, *X-ray Spectrometry*, Dekker, NY, (1978).
485. Bertin, E. P., *Principles and Practice of X-ray Spectrometric Analysis*, Plenum, NY, 2nd edition, (1975).
486. Deconninck, G., *Introduction to Radioanalytical Physics*, Elsevier, Amsterdam, (1978).
487. Bach, H. Z. *Angew. Physik*, **28**, 239, (1970); *Schott. Inform.*, **4**, 6, (1972).
488. Nebraska Symposium, *Third Intern. Conf. on Ellipsometry*, in *Surf. Sci.*, **56**, (1976).
489. Zarzycki, J. and Mezard, R., *Phys. Chem. Glass*, **3**, 163, (1962).
490. Hench, L. L.,*J. Non Cryst. Sol.*, **25**, 343, (1977).
491. Hench, L. L. and Paschall, H. A., *J. Biomed. Materials Res.*, No. 4, 25, (1973).
492. Clark, A. E., Pantano, C. G., Hench, L. L., *J. Biomed. Materials Res.*, Symposium on Materials for Reconstructive Surgery, Clemson Univ., (1975).
493. Stookey, S. D., *Glastechn. Ber.*, (5th Intl. Glass Congress), **32 K**, V, (1959).
494. Cahn J. W., *J. Amer. Cer. Soc.*, **52**, 113, (1969).
495. Myiata, N. and Jinno H., *J. Mater. Sci.*, **7**, 973, (1972).
496. McMillan, P. W. *Glass Ceramics*, Acad. Press, London, (1964) 2nd ed., pp. 285 ff, (1979).
497. Chyung, C. K., Beall, G. H. and Grossman, D. G. in *Structure of Glass Lectures*, Rennselaer Poly. Inst., Troy, USA, p. 140, (1972).
498. Pernot, F., Zarzycki, J., Bonnel, F. I, Rabischong, P. and Baldet, P., *J. Mater. Sci.*, **14**, 1694, (1979).
499. Piganiol, editor, *Les Industries Verrières* (Glass Industries), Dunod, Paris, 1966).
500. Zarzycki, J., "Processing of Gel Glasses", p. 209-49 in Ref. 25.
501. Sakka, S., p. 129-67 in Ref. 26, Vol. 22.
502. Hench, L. and Ulrich, D. G. editors, *Ultrastructure Processing of Ceramics, Glasses and Composites* (1st Conference, Gainesville, Fla, Feb. 13-17), 1983, Wiley, NY, (1984).
503. Ibid, (2nd Conference, Palm Beach, Fla, Feb 25-Mar. 1, 1985), Wiley, NY, (1986).
504. Mackenzie, J. D. and Ulrich, D. G., editors, *3rd Conference on Ultrastructure Processing of Ceramics, Glasses and Composites*, San Diego, Cal., Feb. 23-27, 1987, Wiley, NY, (1988).
505. Brinker, C. J., Clark, D. E. and Ulrich, D. R., editors, *Better Ceramics Through Chemistry*, **32**, Mat. Res. Soc. (Albuquerque Symposium 1984) North Holland, Amsterdam, (1984).
506. Brinker, C. J., Clark, D. E. and Ulrich, D. R., editors, *Better Ceramics Through Chemistry II*, Vol. 73, Materials Res. Soc. (1986).
507. Gottardi, V. editor, 1st International Workshop "Glasses and Glass Ceramics from Gels", Padova (Italy) Oct. 8-9, 1981, *J. Non Cryst. Sol.*, **48**, 1-230, (1982).
508. Scholze, H. editor, 2nd International Workshop "Glasses and Glass Ceramics from Gels", Wurzburg (Germany) July 1-2, 1983, *J. Non Cryst. Sol.* **63**, 1-300, (1984).
509. Zarzycki, J. editor, 3rd International Workshop "Glasses and Glass Ceramics from Gels", Monpellier (France) Sept. 12-14, 1985, *J. Non Cryst. Sol.*, **82**, 1-436, (1986).
510. Sakka, S. editor, 4th International Workshop "Glasses and Glass Ceramics from Gels", Kyoto (Japan) July 13-15, 1987, *J. Non Cryst. Sol.*, **100**, 1–554, (1988).

511. Zarzycki, J., "Advanced glass by sol-gel process", in Proc. 1st International Symposium on New Glass, Dec. 1-2 (1987), Tokyo, Japan, The Assoc. of New Glass Industries, p. 35-42, (1988).
512. Iler, R. K. *The Chemistry of Silica*, Wiley, NY, (1979)
513. Zarzycki, J., Prassas, M. and Phalippou, J., *J. Mater. Sci.* **17**, 3371, (1982).
514. Mizuno, T., Phalippou, J. and Zarzycki, J., *Glass Technol.*, **26**, 39, (1985),
515. Esquivias L. and Zarzycki, J., "Sonogels, an alternative method in sol-gel processing", p. 255-70 in Ref. 504.
516. de la Rosa Fox, N., Esquivias L. and Zarzycki, J., "Glasses from Sonogels", p. 363-73 in Ref. 59.
517. Zarzycki, J., *J. Non. Cryst. Sol.*, **100**, 359, (1988).
518. Mandelbrot, B. B., *The Fractal Geometry of Nature*, Freeman NY (1983).
519. Zarzycki, J., "Fractal properties of gels", in Proc. 6th International Conf. on Physics of Non-Crystalline Solids, Kyoto 6-10 July 1987, *J. Non Cryst. Sol.*, **95-96**, 173, (1987).
520. Scherer, G. W., *J. Non Cryst. Sol.*, **87**, 199,(1986); **89**, 217, (1987); **91**, 83, 101, (1987); **92**, 375 (1987).
521. Fricke, J., *Aerogels*, Springer, Berlin, (1986).
522. Hench, L. L., "Use of drying control chemical additives (DCCAs) in controlling sol-gel processing", p. 52 in Ref. 503.
523. Mackenzie, J. K. and Shuttleworth, R., *Proc. Phys. Soc. (London)*, **62**, 833-852, (1949).
524. Scherer, G., *J. Am. Ceram. Soc.*, **60**, 236, (1977).
525. Zarzycki, J., *J. Non Cryst. Sol.*, **48**, 105, (1982).
526. Murray, P., Rodgers E. P. and Williams, A. E., *Trans. Br. Ceram. Soc.*, **53**, 474, (1954).
527. Zarzycki, J. "Nucleation in glasses from gels" in Symp. Amer. Cer. Soc., Washington, April 1981, *Advances in Ceramics*, **4**, 204, (1982).
528. Decottignies, M., Phalippou, J. and Zarzycki, J.,*J. Mater. Sci.* **13**, 2605, (1978).
529. Schmidt, H., *J. Non Cryst. Sol.* **73**, 681, (1985).
530. Roy, R., Komarneni, S. and Roy, D. M., "Multi-phasic ceramic composites made by sol-gel technique", p. 347 in Ref. 505.

Index